ELEMENTS OF CARTOGRAPHY

ELEMENTS OF CARTOGRAPHY

SIXTH EDITION

ARTHUR H. ROBINSON
University of Wisconsin—Madison
Lawrence Martin Professor Emeritus of Cartography

JOEL L. MORRISON
U.S. Geological Survey
(formerly, Professor , University of Wisconsin—Madison)

PHILLIP C. MUEHRCKE
University of Wisconsin—Madison
Professor of Geography

A. JON KIMERLING
Oregon State University
Professor of Geography/Geosciences

STEPHEN C. GUPTILL
U.S. Geological Survey

JOHN WILEY & SONS, INC.
New York • Chichester • Brisbane • Toronto • Singapore

Acquisitions Editor	Frank Lyman
Marketing Manager	Catherine Faduska
Senior Production Editors	Sandra Russell/Bonnie Cabot
Designer	Laura Nicholls
Manufacturing Manager	Susan Stetzer
Photo Researcher	Lisa Passmore
Illustration Coordinator	Ed Starr

This book was set in 10/12 pt ITC Cheltenham Light by Digital Production Group

Recognizing the importance of preserving what has been written, it is a policy of John Wiley & Sons, Inc. to have books of enduring value published in the United States printed on acid-free paper, and we exert our best efforts to that end.

The paper in this book was manufactured by a mill whose forest management programs include sustained yield harvesting of its timberlands. Sustained yield harvesting principles ensure that the number of trees cut each year does not exceed the amount of new growth.

Library of Congress Cataloging in Publication Data:
Main entry under title:
Elements of cartography.
Revision of: Robinson, A.H. Elements of cartography.

6th ed. © 1995.
Includes index.
1. Cartography. 1. Robinson, Arthur Howard, 1915–

GA105.2.E43 1995 526—dc20 94-33102
ISBN 0-471-55579-7

Printed in the United States of America

10 9 8 7 6 5 4 3 2

PREFACE

We introduced the preface to the previous edition by stating:

"Cartography is in transition. Where the changes will lead is uncertain, but change in the discipline is pervasive, and the rate of change seems to be accelerating." Now, nearly a decade later, the process continues. Many changes predicted in the fifth edition have occurred. Others have not. The pace of change has been particularly hard to predict. Forecasting easily falls victim to the complex interplay of forces for and against change.

The impetus for change is twofold. First, our modern society faces environmental problems of growing complexity. We need vast amounts of spatial data to resolve or manage these environmental issues. Second, the electronic revolution provides the only means to handle the immense amounts of spatial data in the time frames demanded. Telecommunications, satellites, computers, and related technologies already pervade both personal and professional aspects of our lives. Even greater future change is certain. It is the marriage of this demand for data and the proliferation of electronic devices that we now commonly refer to as the information age.

Since electronic technology greatly enhances spatial data-handling opportunities, it is understandable that cartography is greatly affected. New map products and mapping applications serve our need to know and manage the environment. More than in previous technological revolutions, cartography is being swept along with society as a whole. The effects are dramatic with respect to such things as what gets mapped, who produces maps, who uses maps, and how maps are created and used.

We can compare the current technological revolution to the invention of the printing press in terms of its impact on cartography. When all maps were manuscripts, there were few copies and a limited user group. The printing press revolutionized cartography by making many copies of a map possible, which greatly expanded the audience of map users. The current technological revolution goes a step further, permitting everyone to be a mapmaker. This means that diverse map users are no longer forced to make do with identical copies of a printed map. They can construct or tailor maps to fit individual interests.

In writing this edition, we have addressed the changes that have taken place in the past 10 years. We have also pointed the reader in directions the discipline appears to be taking. The challenge in writing each subsequent edition to a classic textbook has several dimensions. We have had to let go of a significant portion of the past in which we have had a vested interest. Old material loses utility in periods of rapid change. But we also realize that loyal supporters of previous editions are often among the most conservative when it comes to embracing change. Therefore, we have tried to strike a balance between futurist musings and the reality of most people's current needs.

What we have retained from previous editions is the emphasis on the theory and practice of cartography. This book does not focus on one or another aspect of the mapping process. It is not designed to teach you how to use a particular computer platform or mapping software package. Instead, it stresses cartographic principles that span the breadth of the discipline and also serves as a reference manual on cartography.

To accomplish our goals, we have shifted emphasis between the fifth and sixth editions in several ways. In response to information-age demands, mapping increasingly is conducted within the context of geographical information systems (GIS) technology. Therefore, we have explicitly linked GIS and cartography throughout the book. Since integration and flexibility lie at the heart of GIS technology, we have had to expand the scope of the sixth edition. Whereas the previous edition stressed the design and production of small-scale thematic maps, we have given new emphasis to reference mapping and considered mapping throughout the possible range of scales. Since the information-systems emphasis has raised database issues such as data formats, structuring, accuracy, and exchange standards, we have addressed these topics in this edition. The impact of electronic technology has also made it appropriate to expand material on image-based mapping, and to broaden coverage to include topics such as surveying and positioning methods, dynamic mapping, and cartographic modeling.

The transition to electronic mapping has progressed dramatically but is still not complete. Even when electronic technology is fully implemented, other technologies continue to be used. For example, the photographic and printing processes are as relevant today as they were several decades ago. Therefore, we have not entirely eliminated material on manual and photochemical procedures found in the previous edition, but we have cut and shifted some of this material to appendixes.

These changes in emphasis have required a restructuring of material in this edition. You will find more (but shorter and better focused) chapters. Material is graded and segmented by using boxes to provide the more technical information. In response to many requests, we gave top priority to making the text more consistent and readable throughout.

Illustrations for the sixth edition provide an example of the degree of change that has occurred since the previous edition. All artwork for the fifth edition was drawn manually using photomechanical procedures, yet for this edition, artwork for all revised and new illustrations was created on Apple MacIntosh computers. In most cases compilation artwork was scanned into the system, although some material was retrieved from existing digital databases. The digital files were output to a 600 dpi laser printer for proofing and to a 2400 dpi film recorder for separation negatives.

Cartographic theory and practice are not the only things that have changed since the previous edition. Randall D. Sale died in the interim, and his involvement is sadly missed. Since Arthur H. Robinson is now retired, his contribution to this edition was primarily of an advisory and editorial nature. In recognizing the changing nature of cartography and the need to bring in fresh perspectives, A. Jon Kimerling and Stephen C. Guptill were added as coauthors.

As with previous editions, the authors are deeply indebted to many people without whom production of the sixth edition would have been more difficult, if not impossible. Research findings published by many scholars and the expertise of colleagues were immensely useful, as were comments and suggestions made directly to the authors. The many students at the University of Wisconsin and elsewhere who were exposed to the organization and reorganization of the material in this book over the years deserve much credit for their feedback to us.

Several individuals provided special assistance. Among these people were Cynthia Brewer, G. O. Tapper, and Patricia Gilmartin, who commented on draft manuscript supplied by the publisher. Kenneth Parsons, working in the University of Wisconsin Cartography Laboratory under the supervision of Onno Brouwer, coordinated the production of the hundreds of illustrations that appear in the sixth edition. Jill Muehrcke served as coordinating editor and helped immensely in bringing the work of the various coauthors to a common style and reading level.

Finally, as always in the production of such works, the families of the authors must sacrifice. Collectively to our spouses and children, the authors express their appreciation and thanks for their support and help.

A.H.R
J.L.M
P.C.M
A.J.K
S.C.G

CONTENTS

PART I

INTRODUCTION

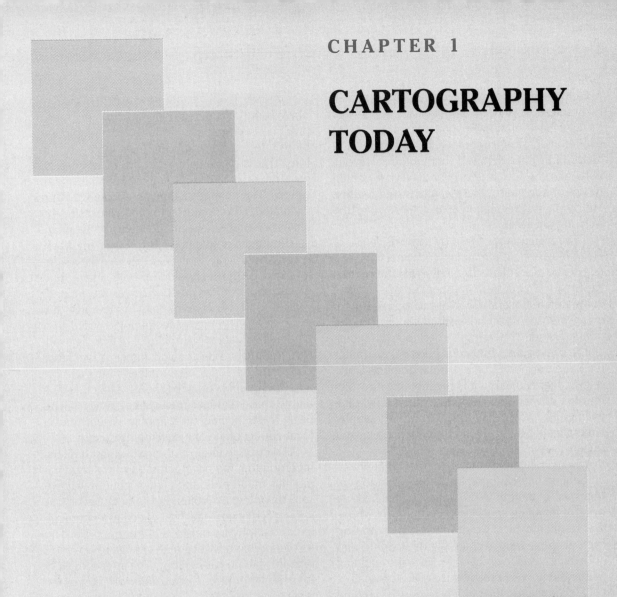

CARTOGRAPHY TODAY

Cartography is in the midst of a revolution in technology. This book appears about midway through the revolution. Revolutions, as opposed to evolutions, are more traumatic for a discipline and for the individuals involved. Technological revolutions are not new in the over 2000-year written history of cartography. But something truly is different about this revolution that goes beyond the normal upheavals.

This revolution is caused by a pervasive use of electronics and, in particular, computers. We will refer to the use of this technology as **digital cartography**. Society and cartographers have been using computers for several decades, but today computers are individualized, not institutionalized. It is the popularization of inexpensive desktop computers, epitomized by the ubiquitous personal computer (PC) and the increasingly popular graphics workstation, that is making this technological revolution so different. This revolution is not only influencing the highly technical field of cartography but is affecting the relation between mapping and society in general.

When photography and then plastic films were introduced into cartography, these technological "marvels" of their time dramatically changed the work of mapping specialists. But the procedures that resulted, which we now refer to as **analog cartography**, were technically demanding and not available to society at large. Since the mapping public did not have ready access to analog mapping technologies, their ability to create mapping products was rather limited. That is not true of digital cartography today. Electronic mapping systems are available to anyone with a desktop computer, making it easy for even nonspecialists to create cartographic products. Thus, the impact of electronic technology on the practice of cartography (especially by the novice) is more general and immediate than was true with the introduction of previous technologies.

Increased demand for cartographic products is another factor driving the current cartographic revolution. At least within the United States, which is the world's technological superpower at this time, society is reawakening to the field of geography. Not only is it increasingly easy for everyone to produce cartographic products, but people are more interested in such products than ever before. This is particularly true of professionals in governmental, industrial, and commercial organizations. This change in attitude, encouraged and facilitated by computer technology, is causing the biggest revolution ever in the realm of mapping.

The intent of this book is to cover the basic tenets of cartography and to introduce the reader to a mix of the technologies, both analog and digital, that will be practiced over the next 10 years. (Ten years is an eternity in the midst of a revolution). The mix of technologies used in a particular situation is a matter of balancing available resources. Time, capital, material, and labor costs must be considered to determine the most appropriate mix. Thus we should expect, for example, that a national intelligence agency will be using a much different mapping system than a local county office or a developing country.

As will be evident in the following chapters, incorporating the new technology into cartography has made profound changes in the way we "do" mapping. As is often the case, however, our knowledge of potential uses of digital technology in cartography far outstrips our current use. The revolution has only begun to change how we are organized to do cartography. The impact of electronics on the current organization of cartography is yet to occur.

CARTOGRAPHY NOW

Over many years, cartographers have defined basic elements of their science. Increasingly, they have been able to perform cartography more accurately and to make the resulting information more widely available. Throughout this long period, one product, the map, has been the central focus of the science of cartography. The map has provided two important functions:

1. It serves as a storage medium for information which humanity needs.
2. It provides a picture of the world to help us understand the spatial patterns, relationships, and complexity of the environment in which we live.

The computer revolution in cartography is preserving the basic elements of the science. But digital cartography is providing two distinct products, one to satisfy each of the former functions that maps alone once served:

1. The **digital database** is replacing the printed map as the storage medium for geographic information.
2. Cartographic **visualizations** on many different media now satisfy the second function served previously by printed maps.

Today the computer hardware available to cartographic scientists is capable of replicating all analog methods used previously in cartography. Software algorithms can nearly replicate all standard analog methods. Some of the more subjective analog techniques, such as feature generalization and geographic name placement, are still rather crudely replicated by computer software, but the use of digital technology has greatly increased the possibilities of new methods. Some of these new methods were ideals in the analog cartographic world, goals which time and human skills did not permit the cartographer to reach.

A serious shortcoming in the use of digital technology in current cartography is the lack of data in readily usable forms. At this time in our history, cartographers have more data than ever before, with the anticipation of even greater quantities in the future. But at the same time, much available information lacks portability because it is coded in incompatible hardware or software specific formats. Other data are too specialized in content to be of general interest. The costs to convert these data to usable form can be enormous.

Another problem at this stage in the transition to digital cartography is a lack of skilled professionals. The cadre of cartographers trained in digital technology is still too small to take full advantage of new opportunities. Thus, cartographic development lags significantly behind technological potential. To address this education gap, the cartographic curriculum needs revision. We need to train many more people to use and maintain the two principal products of digital cartography.

The potential now exists for every individual to create a cartographic visualization from a spatial database using specialized mapping software on desktop computers. What this means is that mapping chores will increasingly be shifted from the professional cartographer to the map user. This shift in responsibility creates a need for training in the public schools and continuing education for adults. A need to focus on the nonspecialist is new for cartographic educators and will require much more attention than it has received in the past.

CARTOGRAPHY TOMORROW

When this revolution has run its course, the science of cartography will have been transformed. We will discuss some of these transformations under the heading *Implications* later in this chapter. But more important, society will have been transformed as well, and society's demands for cartography will be quite different.

Maps will not disappear, but they will take on new forms and encourage new uses (and users). We envision many new cartographic visualizations in the future. Likewise, individual access and use of very large spatial databases will become commonplace. The revolution has already run through punched cards and magnetic tapes as media of choice, and we are now using floppy disks, CD-ROMs, cassette tapes, and optical disks to store and convey digital spatial data. Undoubtedly, newer media will soon be discovered and existing media will be perfected. Currently within the United States we are installing fibre-optic cable at a fast pace, and we are interlacing the use of telephones, televisions, video cassette recorders, fax machines, and personal digital assistants. Dynamic, photorealistic presentations of spatial data that have a three-dimensional character soon will be in widespread demand.

People will call for larger, faster, and less expensive computers. They will demand efficient software algorithms using data sets of different lineages, and they will use data in quantities never before thought of.

But it will not be necessary for each individual to take direct physical possession of all mapping resources. Inexpensive desktop machines will be linked over sophisticated electronic networks, which will tie map makers to map users. Distributed resources (people, computers, software, and data) will characterize these networks. A person in New York City may have data from San Francisco sent to a computer in Denver for processing, and the final cartographic output telecommunicated to a printer in Boston. Although this example is fictitious, it does suggest how electronic communication will enhance an individual's access to the mapping power now available only to experts.

These developments in mapping will force changes to our society, to businesses that produce cartographic products, and to cartography itself. Let's look specifically at the implications of these changes for cartographers.

IMPLICATIONS

One of the major changes is that cartographers are losing the control they exercised in the past. When the printed map was the sole product, cartographers controlled every bit of information. Users could extract from a map only what cartographers chose to include. With digital technology, the situation is entirely different. Users can select the information they want to include in a visualization.

Thus, the mapping process is being decentralized. To a large extent the expertise of cartographers will be codified in the mapping software and data structures available to nonprofessionals. In other words, the professional cartographer's role is shifting to the earlier preparatory stages of mapping, whereas the map user is taking on more of the actual map production chores.

For example, traditional cartography placed a high importance on scale. An axiom of traditional cartography was: "Always compile a map from larger to smaller scales." Cartographic scientists can still follow that advice in digital cartography, but individual users may or may not adhere to it. In fact, many uses of digital spatial data can be accommodated by a file collected at one scale and used for visualization at a range of scales. This range may include larger scales than would formerly have been justified by the resolution of the data. We can speculate, therefore, that this change in technology will generate new requirements for data. It will also require greater knowledge on the part of novice map users and, thus, an expanded emphasis on cartographic education.

One of the most important concerns for future cartographers will be **standards**. Increasingly, local units will feed data up to central (state, national, global) coordinating organizations. These organizations will be responsible for setting standards and facilitating data distribution. Standards are needed for data quality, data exchange, hardware and software interoperability, and data collection procedures. Knowledge of data models, features, attributes, and data set lineage are examples of concepts that cartographers must now learn. An increased emphasis on error propagation during analysis and synthesis will assume a greater importance in cartography in the future.

To efficiently provide the two principal cartographic products, the database and the visualization, cartographers will have to reorganize current cartographic institutions. In an effort to survive during a revolution, institutions tend to overreact to events. New and differently organized institutions are born, and many perish. It is only near the end of the revolution that it becomes clear what institutional structure will survive. We are not yet at the stage where we can identify with confidence the institutions that will flourish in the fully digital cartographic world of the 21st century.

OUTLINE OF BOOK

Chapters 2 and 3 of this text will introduce you to the nature and history of the exciting field of cartography. Material in the remainder of the book is organized under broad section headings. Chapters 4, 5, and 6 present the basic geometric and measurement concepts necessary to transform and use information about the earth in a two- or three-dimensional context. Chapters 7, 8, and 9 deal with collecting data to create cartographic products. Chapters 10 through 17 introduce concepts basic to an understanding of cartography in a digital technological environment. Chapters 18 through 23 discuss the principles underlying the design and subsequent perception of cartographic visualizations. Chapters 24 through 29 detail methods for abstracting information contained in databases so that it can be used effectively in visualizations. Finally, Chapters 30 and 31 explore the processes of producing and reproducing printed maps. The appendixes provide valuable supplemental information.

Despite our emphasis on digital cartography in this edition, we have not abandoned information on analog cartography. We have included material on analog mapping procedures in the last two chapters and the appendixes. It is our contention that printed maps will continue to be in demand by society for the foreseeable future.

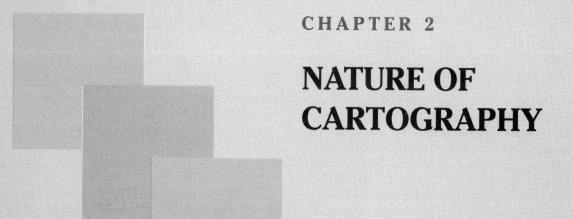

CHAPTER 2

NATURE OF CARTOGRAPHY

What is a map, and what is cartography? To answer these questions, we will first look at how cartography fits into our endeavor to know and communicate.

Next, we will consider the characteristics maps share. We will then explore the purposes maps serve and see how special needs have led to a number of map types and forms.

Finally, we will discuss the benefits of viewing cartography as a series of information transformations, starting with environmental data and ending with an image of the environment in the mind of the map user.

FORMS OF REPRESENTATION

The beginnings of our struggle to know and communicate are lost in prehistory. But we can assume that the earliest ways of knowing and communicating used utterances and drawings to create the mental images involved in understanding objects and their relationships. From these sounds developed the spoken and written natural languages and mathematics of today, and the sketches evolved into present-day graphics. We can think of these vehicles of thought and communication as forms of **knowledge representation**. We use these representations as powerful mental tools.

Written and verbal languages allow us to develop ideas and express them in a variety of ways, ranging from tightly structured scholarly treatises to literary creations and dramatics which evoke emotional responses. The use of the written language is called **literacy**, while the use of spoken language is called **articulacy**.

Mathematics, or **numeracy**, is a way of symbolizing and dealing with relationships among abstractions, sets, numbers, and magnitudes. Just as literacy ranges from dealing with highly emotive novels to staid scientific matters, mathematics can range from abstract relationships to precise calculations.

Graphicacy is a fourth way of communicating. Graphic methods extend from drawing and painting to the construction of plans and diagrams. The use of graphics to capture the spatial structure of the environment is the subject of this book.

Our desire for spatial imagery of things in our environment is as normal as breathing. This imagery can be simple, as when we are concerned with such basic relations as inside or outside, near or far. Or it can be quite sophisticated, as when it involves such abstract concepts as distribution of air pollution.

When we communicate with someone by describing a spatial relationship, we want our description to evoke a similar image in that person's mind. The best way to be sure that will happen is to provide a visual representation of the image. This graphic representation of the geographical setting is what we call a **map**. **Cartography** is the making and study of maps in all their aspects.

Cartography is an important branch of graphics, since it is an extremely efficient way of manipulating, analyzing, and expressing ideas, forms, and relationships that occur in two- and three-dimensional space. In the broad sense, cartography includes any activity in which the presentation and use of maps is a matter of basic concern. This includes teaching the skills of map use; studying the history of cartography; and maintaining map collections with the associated cataloging and bibliographic activities. It also includes the collection, collation, and manipulation of data and the design and preparation of maps, charts, plans, and atlases.

Although each of these activities involves highly specialized procedures and requires particular training, they all deal with maps. It is the unique character of the map as the central intellectual object that unites all cartographers.

NEED FOR MAPS

Cartography is concerned with reducing the spatial characteristics of a large area—a portion or all of the earth, or another celestial body—and putting it in map form to make it observable. The same techniques can be used to enlarge microscopic things to make them visible. Although it is uncommon to refer to these enlarging activities as cartography, the resulting images are sometimes called maps.

Just as spoken and written languages allow us to express ourselves without having to point to everything, a map extends our normal range of vision, so to speak. A map lets us see the broader spatial relations that exist over large areas or the details of microscopic particles.

Even an ordinary map is much more than a mere reduction. It is a carefully designed instrument for recording, calculating, displaying, analyzing, and understanding the interrelation of things. Nevertheless, its most fundamental function is to bring things into view.

Maps range from tiny portrayals on postage stamps to enormous mural-like wall maps used by civilian and military security groups to keep track of events and forces. They all have one thing in common: to add to the geographical understanding of the viewer.

Since classical Greek times, curiosity about the geographical environment has steadily grown, and ways to represent it have become more and more specialized. Today there are many different kinds of mapmaking, and the objectives and techniques seem very different. It is important to realize, however, that all maps have the same goal of communicating spatial relationships and forms. However dissimilar the maps may seem, the cartographic methods are fundamentally alike.

A detailed map of a small region, depicting its landforms, drainage, vegetation, settlement patterns, roads, geology, or a host of other detailed distributions, communicates the relationships necessary to plan and carry on many types of work. Building a road, a house, a flood-control system, or almost any other construction requires prior mapping.

Less detailed maps of larger areas showing floodplain hazards, soil erosion, land use, population character, climates, income, and so on, are indispensable to understanding the problems and potentialities of an area. Highly ab-

stract maps of the whole earth indicate generalizations and relationships of broad earth patterns with which we may study the course of past, present, and future events.

BASIC CHARACTERISTICS OF MAPS

The radar map on the television weather report seems very unlike the map in the travel brochure proclaiming the "glories of ancient Greece"; yet, they have much in common. All maps are concerned with two elements of reality: locations and attributes. **Locations** are positions in two-dimensional space, such as places with the coordinates *x,y*. **Attributes** are qualities or magnitudes, such as languages or temperatures.

From these two basic elements, many relationships can be formed. Some examples are:

Relationships among locations when no attributes are involved, such as the distances or bearings between origins and destinations needed for navigation.

Relationships among various attributes at one location, such as temperature, precipitation, and soil type.

Relationships among the locations of the attributes of a given distribution, such as the variation of precipitation amounts from place to place.

Relationships among the locations of derived or combined attributes of given distributions, such as the relation of per capita income to educational attainment, as they vary from place to place.

All sorts of topological and metrical properties of relationships can be identified and derived, such as distances, directions, adjacency, insidedness, patterns, networks, and interactions. A map is therefore a very powerful tool of spatial analysis.

All geographical maps are **reductions**. Thus, the map is smaller than the region it portrays. Each map has a defined dimensional relationship between reality and the map; this relationship is called **scale**. Because of the relative "poverty" of map space, the scale sets a limit on the information that can be included.

All maps involve geometrical **transformations**. For instance, it is common to transform a spherical surface (essentially the shape of the earth) to a surface that is easier to work with, such as a computer screen or a flat map sheet. Such a systematic transformation is called a **map projection**. The choice of a map projection affects how a map should be used. It is often convenient to use map referencing systems called plane coordinate grids. These coordinate systems help us calculate distances and directions from the map. Coordinate systems depend on map projections for their accuracy.

All maps are **abstractions** of reality. The real world is so complex that merely reducing it or putting a small part of it in image form would make it even more confusing. Consequently, maps portray only the information that has been chosen to fit the use of the map. This information is subjected to a variety of operations, such as classification and simplification, to make it easier to understand.

All maps use **signs** to stand for elements of reality. The meanings of these signs make up the **symbolism** of cartography. Few symbols used on maps have universal meanings, just as few words mean the same thing in all languages.

All map symbols used to portray data consist of various kinds of **marks**, such as lines, dots, colors, tones, patterns, and so on. When using a map, then, we must constantly compare the symbols with those in a legend. Whether the marks are on a luminous cathode ray tube or on a piece of paper, their selection and the way they are assembled (how the map is designed graphically) greatly affect the communicability of the map.

PURPOSES MAPS SERVE

Although maps share basic characteristics, they can vary dramatically in appearance. The look of a map depends largely on its intended use. Since map design principles are flexible enough to accommodate a wide range of needs, we have specialized maps for many purposes.

The primary objective of some maps is to store geographic information in spatial format. Other maps serve mobility and navigation needs. Another type of map is designed for analytical purposes involving measuring and computing. Still other maps are used to summarize voluminous statistical data and, thereby, assist in spatial forecasting and spotting trends. Sometimes maps are used to visualize what otherwise would be invisible. And, finally, maps may be intended to help stimulate spatial thinking.

Most maps are designed to serve at least several of these objectives simultaneously, of course. There are few single-objective maps. This blurring of purpose makes it difficult to develop one neat taxonomy of map types. Thus, rather than focus on a single classification scheme, let's consider the implications of categorizing maps in different ways.

CATEGORIES OF MAPS

The number of possible combinations of map scales, subject matter, and objectives is astronomical. Consequently, there is an almost unlimited variety of maps.

Nevertheless, there are recognizable groupings of objectives and uses for maps, which permit us to catalog them to some degree. One of the problems in doing so is distinguishing between the objectives of the mapmaker and the responses of the viewer. As observed by R.A. Skelton (1972), "Maps have many functions and many faces, and each of us sees them with different eyes."

The chart that provides bearings, depths, and coastal positions for a ship's navigator will, for another person, conjure up visions of coconut palms and an idyllic beach life. We cannot, therefore, predict responses. We can be sure only of the cartographer's objectives.

To help us appreciate the similarities and differences among maps and cartographers, we will look at maps from three points of view: (1) their scale, (2) their function, and (3) their subject matter.

CLASSED BY SCALE

Each cartographer must change the dimensions of reality to serve the function of the map. The ratio between the dimensions of the map and those of reality is called the **map scale**.

When a small sheet is used to show a large area (such as a map of the United States, or even the world on a sheet the size of this page), that map is described as being a **small-scale map**. If a map the size of this page showed only a small part of reality (for example, less than 1 km²), it would be described as a **large-scale map**.

The terms large and small when combined with scale refer to the relative sizes at which objects are represented, not to the amount of reduction involved.* Accordingly, when little reduction is involved and features such as roads are large, the map is termed a large-scale map (Figure 2.1). When great reduction has been employed, as for a small-scale map, most of the smaller features on the earth cannot be shown at a size proportional to the amount of reduction, but must be greatly magnified and symbolized to be seen at all. Consequently, reality must be portrayed selectively and with considerable simplification on small-scale maps (Figure 2.2). On the other hand, although selection is also characteristic of large-scale maps, such maps can portray many aspects of reality in the actual proportion of the amount of reduction used.

There is no consensus on the quantitative limits of the terms small, medium, and large scale, and there is no reason why there should be, since the terms are relative. Most cartographers would agree,

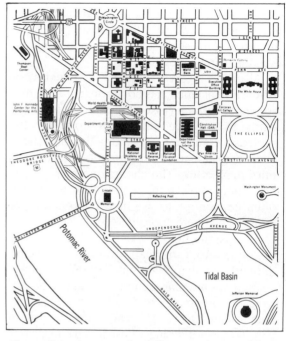

Figure 2.1 An example of a large-scale map. (From *George Washington University Bulletin*, courtesy George Washington University.)

however, that a map with a reduction ratio of 1 to 50,000 or less (for example, 1 to 25,000) would be a large-scale map, and maps with ratios of 1 to 500,000 or more (for example, 1 to 1,000,000) would be considered small-scale maps. (For more on map scale, see Chapter 6).

CLASSED BY FUNCTION

As we have just seen, the range from large-scale to small-scale is a continuum with no clear divisions separating the classes of map scale. Similarly, if we try to divide maps into classes based on their function, we find a great difference between extremes, but the transition from one class to another is gradual. We can recognize three main classes of maps: general reference maps, thematic maps, and charts. Let's take a look at each.

---

*We can also think of map scale as the ratio between map and ground distance. Thus, one inch on a map to 100 inches on the ground is a large ratio, whereas one inch on a map to 1 million inches on the ground is a small ratio.

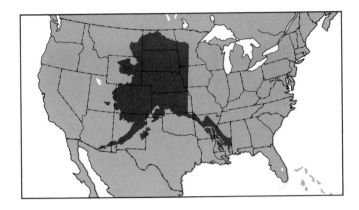

Figure 2.2 An example of a small-scale map. (After F. Van Zandt, "Boundaries of the United States and the Several States," U.S. Geological Survey Professional Paper 909, Washington, D.C.: U.S. Government Printing Office, 1976.)

General Reference Maps

With some maps, the objective is to show the locations of a variety of different features, such as water bodies, coastlines, and roads. These are called **general reference maps**.

Large-scale general reference maps of land areas are called topographic maps (Figure 2.3). They are usually made by public agencies, using photogrammetric methods, and are issued in series of individual sheets.

Maps of much larger scale are required for site location and other engineering purposes. Great attention is paid to their accuracy in terms of positional relationships among the features mapped. In many cases they have the validity of legal documents and are the basis for boundary determination, tax assessments, transfers of ownership, and other such functions that require great precision. In the United States and other countries, official map accuracy standards have been established for such general, large-scale maps. The map accuracy standards apply only to the metrical horizontal and vertical qualities of the maps, not to nonmetrical factors such as labeling blunders, incompleteness, or being out of date.

Small-scale general reference maps are typified by the maps of states, countries, and continents in atlases. Such maps show similar phenomena to those on large-scale general reference maps. But, because they must be greatly reduced and generalized, they cannot attain the detail and positional accuracy of large-scale maps.

Thematic Maps

A second class of maps, called **thematic** or **special purpose maps**, concentrates on the distribution of a single attribute or the relationship among several. Thematic maps range from satellite cloud cover images to shaded maps of election results. They are typified by maps of precipitation, temperature, population, atmospheric pressure, and average annual income (Figure 2.4).

Just because a map deals mainly with a single class of phenomenon does not mean that it is a thematic map. Maps showing soils, rocks, or population density can be classified as general reference maps if the objective is to show the locations of these features. On the other hand, maps of the same features may be called thematic maps if they

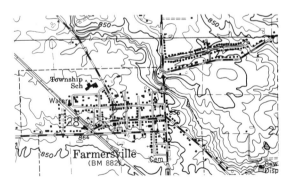

Figure 2.3 A section of a modern topographic map. (From USGS 1:24,000 Farmersville, Ohio, quadrangle, 1974.)

focus attention on the structure of the distribution.

In the past, thematic maps tended to be of small scale. One reason was that available data were relatively coarse in resolution. Also, great reduction is necessary to show geographical distributions occurring over large areas. At small scales, it is more important to capture the basic structure of the distribution than to show individual map positions.

As better data have become available and our need for accurate spatial information has grown, thematic maps have grown larger in scale. When the area of interest is a city, for example, there is demand for maps to show the structure of individual phenomena at a level of detail suitable for making site-specific decisions. These maps may need to be of relatively large scale (Figure 2.5).

At these larger scales, however, the spatial fidelity of thematic maps can be a matter of great concern. A concept such as "soil" represents a composite of several variables. Furthermore, soils maps are based on relatively sparse data. For these reasons, a map of this type of phenomena is more meaningful for making regional comparisons than for site-level decisions. Thus, for a critical building project, an engineer is well advised to forego an existing soils map in favor of on-site soil testing, regardless of the scale of map available.

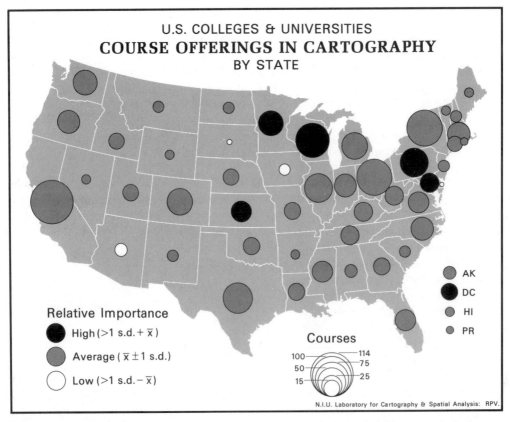

Figure 2.4 Distribution of course offerings in cartography by state. Data from the Mapping Sciences Education Data Base. AK = Alaska, DC = District of Columbia, HI = Hawaii, PR = Puerto Rico. "Relative importance" is the number of cartography courses as a percentage of all mapping sciences courses in that state. (From Dahlberg, 1981, courtesy Richard E. Dahlberg.)

Charts

Maps especially designed to serve the needs of navigators, nautical and aeronautical, are called **charts**. Although it is an oversimplification, one distinction is that maps are to be *looked at*, while charts are to be *worked on*. On charts, navigators plot their courses, determine positions, mark bearings, and so on. It should be noted that navigators also use general reference maps. The marine equivalent of the topographic map is the bathymetric map.

Nautical charts include sailing charts for navigation in open waters; general charts for visual and radar navigation offshore using landmarks; coastal charts for near-shore navigation; harbor charts for use in harbors and for anchorage; and small-craft charts. All these charts show such features as soundings, coasts, shoal waters, lights, buoys, and radio aids (Figure 2.6). Their scales vary depending upon the detail necessary; unlike topographic maps, chart series are not made at a uniform scale. Charts are designed to show accurate locations and to be easy to read and to mark on.

There are two types of aeronautical charts, those for visual flying and those for instrument navigation. Aeronautical charts for visual flying are similar to general reference maps showing such features as cities, roads, railroads, airports, and beacons (Figure 2.7). Charts for instrument navigation include radio facility en route charts, high-altitude en route charts, terminal arrival charts, and taxi charts.

Although it is not called a chart, the familiar road map is really a chart for land navigation. It supplies information about such factors as routes, distances, road qualities, stopping places, and hazards, as well as incidental information such as regional names and places of interest.

There are few "pure" general reference maps, thematic maps, or charts. Most combine functions to some extent. For example, the green printing often seen on topographic maps shows the distribution of forested areas, and the representation of terrain shows the landform. Thus, while we usually classify topographic maps as general reference maps, they may have thematic components.

Similarly, most thematic maps include boundaries, cities, rivers, and other basic reference information, so that the user can more easily fix the location of the subject distribution. Therefore, they have general-purpose as well as thematic functions. Charts, on the other hand, are more likely to have one specific function.

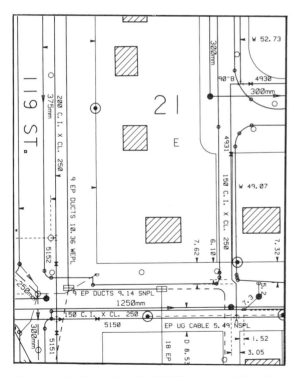

Figure 2.5 A portion of a 1:1000 (large-scale) utility cadastre. (Reproduced by permission from E.A. Kennedy and R.G. Ritchie, "Mapping the Urban Infrastructure." *Proceedings of Auto-Carto V*, copyright 1983 by the American Congress on Surveying and Mapping and the American Society of Photogrammetry.)

CLASSED BY SUBJECT MATTER

The variety of geographical phenomena and the myriad uses to which maps may be put combine to cause an enormous variety. Although they may all be classed by scale and can usually be placed

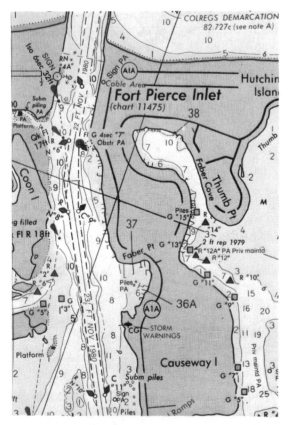

Figure 2.6 A section of the Fort Pierce–Fort Pierce Inlet areas on Nautical Chart 11472, Inset 1, Side A, by the National Ocean Survey. Soundings in feet at mean low water.

somewhere in the continuum from general reference to thematic, it is sometimes useful to group maps on the basis of their subject matter. Several important categories may be recognized.

Probably among the earliest "permanent" maps were drawings to accompany the **cadastre**—the official list of property owners and their land holdings. These drawings, called **cadastral maps**, showed the geographic relationships among land parcels. They are common today, and they record property boundaries much as they did several thousand years ago (Figure 2.8). The fact that cadastres are used to assess taxes helps explain why cadastral maps have always been with us.

Closely allied to cadastral maps, but more general in nature, is a category of large-scale maps called **plans**. These are detailed maps showing buildings, roadways, boundary lines visible on the ground, and administrative boundaries. Plans of urban areas are likely to be very large scale. In countries that have a well-integrated mapping program, such as the Ordnance Survey of the United Kingdom, the very large-scale plans form the basis for the topographic map series.

There is no limit to the number of classes of maps that can be created by grouping them according to their dominant subject matter. Thus, there are soil maps, geological maps, climatic maps, population maps, transportation maps, economic maps, statistical maps, and so on without end.

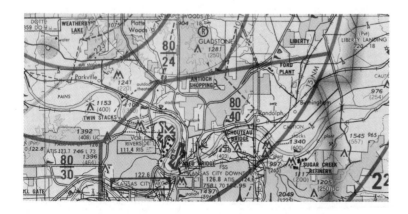

Figure 2.7 A portion of a VFR (Visual Flying Rules) Terminal Area Chart—Kansas City, by the National Ocean Survey.

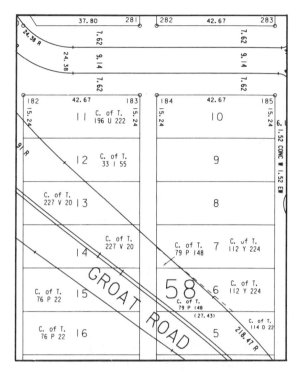

Figure 2.8 A portion of a typical 1:1,000 (large-scale) cadastral map. (Reproduced by permission from E.A. Kennedy and R.G. Ritchie, "Mapping the Urban Infrastructure." *Proceedings of Auto-Carto V*, copyright 1983 by the American Congress on Surveying and Mapping and the American Society of Photogrammetry.)

It is a mistake, however, to think that all such maps are alike. There may be more difference between a large-scale map of surface bedrock and a small-scale map of plate tectonics—both geological maps—than there would be between a soil map and a vegetation map of a given area. Cartography is independent of subject matter.

EMPHASIS ON CARTOGRAPHIC REPRESENTATION

The principal task of cartography is to communicate environmental information. As we explore the field, we will consider such topics as mapping costs, accuracy, and aesthetics. But the primary theme that ties the material together is map effectiveness in thought and communication (Figure 2.9).

Mapping effectiveness is best achieved by treating the making and using of maps equally. The task of the map designer is to enhance the map user's ability to retrieve information. The burden of the map user is to understand the mapping process. The challenge to representational effectiveness comes in thematic mapping, where the goal is to create a general impression of a phenomenon's spatial distribution rather than to provide information about individual places.

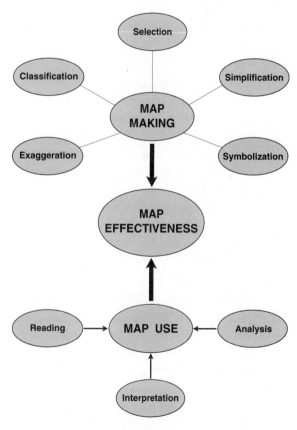

Figure 2.9 The theme of map effectiveness is used to organize material in this book.

The mapping process is a series of information transformations, each of which has the power to alter the appearance of the final product (Tobler, 1979) as shown in Figure 2.10. In data collection, environmental information is distorted through the filters of ground survey, census, remote sensing, or compilation procedures. Mapping further modifies this information by the abstraction processes of selection, classification, simplification, exaggeration, and symbolization.

Finally, the use of the map leads to the distorting effects of map reading, analysis, and interpretation.* The effect of a map is in large part a function of the user's skill, experience, and perceived needs.

The great power of the mapping process lies in its ability to provide fresh, insightful perspectives on our environment. Clearly, there are many possible maps of the same information, each of which will possess certain communication advantages as well as limitations. The cartographer's task is to explore the ramifications of each mapping possibility and to select the most appropriate for the intended task.

THE SCOPE OF CARTOGRAPHY

We can liken cartography to a drama played by two actors, the map maker and map user, with two stage properties, the map and the **data domain** (all potential information that might be put on a map). The map maker selects information from the data domain and puts it into map format. The user then observes and responds to this information (Figure 2.11).

Thus, there are four processes in cartography:

1. Collecting and selecting the data for mapping
2. Manipulating and generalizing the data, designing and constructing the map
3. Reading or viewing the map
4. Responding to or interpreting the information

In order to master these processes, a cartographer must be familiar with all mapping activities, including those associated with the other mapping sciences (geodesy, surveying, photogrammetry, remote sensing, and geographic information systems). A skilled cartographer must also know a great deal about human thought and communication (cognitive science) and the disciplines associated with the environmental features being mapped (Figure 2.12). It takes knowledge, skill, and experience to express in map form the essential characteristics of environmental data.

Geographers are the primary users of maps, but they aren't the only ones. People in the sciences, engineering, and humanities also see the map as

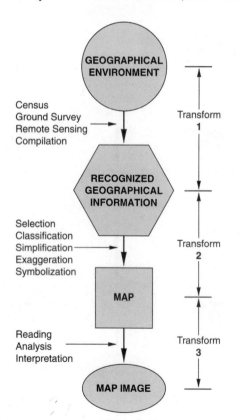

Figure 2.10 Fundamental information transformations in cartography.

*These activities associated with using maps are not covered in this book. For more information see: P.C. Muehrcke and J.O. Muehrcke, *Map Use: Reading, Analysis and Interpretation*, 3rd ed. Madison, WI: JP Publications, 1992.

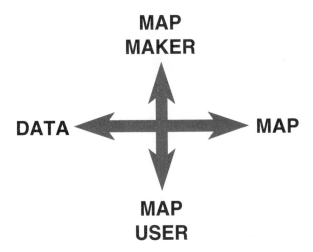

Figure 2.11 We can liken cartography to a drama played by two actors (map maker and user) with two stage properties (data and map).

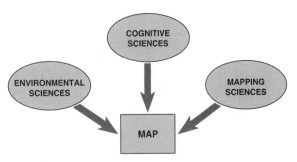

Figure 2.12 Skilled cartographers not only will have mastered the principles of mapmaking but also will have a firm grasp of environmental sciences, understand the cognitive processes of thought and communication, and be familiar with the other mapping sciences, such as geodesy, surveying, photogrammetry, remote sensing, and geographic information systems.

a valuable way to organize and express ideas. Cartographers must be sensitive to the mapping needs of these diverse fields.

SELECTED REFERENCES

Balchin, W.G.V., "Graphicacy," *The American Cartographer*, 3 (1976), 33–8.

Board, C., "Cartographic Communication," *Cartographica*, 18 (1981), 42–78.

Bos, E.S., "Another Approach to the Identity of Cartography," *ITC Journal* (1982) 104–8.

Guelke, L. (ed.), "The Nature of Cartographic Communication," *Cartographica*, Monograph No. 19. Toronto: University of Toronto Press, 1977.

Kadmon, N., "Cartograms and Topology," *Cartographica*, 19 (1982), 1–17.

Keates, J.S., *Understanding Maps*, London: Longman Group Limited, and New York: Halstead Press, John Wiley & Sons, 1982.

Morrison, J.L., "Towards a Functional Definition of the Science of Cartography with Emphasis on Map Reading," *The American Cartographer*, 5 (1978), 97–110.

Muehrcke, P.C., "Maps in Geography," *Cartographica,* 18 (1981) 1–41.

Muehrcke, P.C., and J.O. Muehrcke, *Map Use: Reading, Analysis and Interpretation*, 3rd ed. Madison, WI: JP Publications, 1992.

Petchenik, B. B., "From Place to Space: The Psychological Achievements of Thematic Mapping," *The American Cartographer*, 6 (1979), 5–12.

Robinson, A.H., "The Uniqueness of the Map," *The American Cartographer* (1978), 5–7.

Robinson, A. H., and B. B. Petchenik, *The Nature of Maps: Essays Toward Understanding Maps and Mapping*, Chicago: University of Chicago Press, 1976.

Salichtchev, K.A., "Cartographic Communication: Its Place in the Theory of Science," *The Canadian Cartographer*, 15 (1978), 93–9.

Skelton, R.A., *Maps: A Historical Survey of Their Study and Collecting,* Chicago: University of Chicago Press (1972), 3.

Tobler, W.R., "A Transformational View of Cartography," *The American Cartographer*, 6 (1979), 101–6.

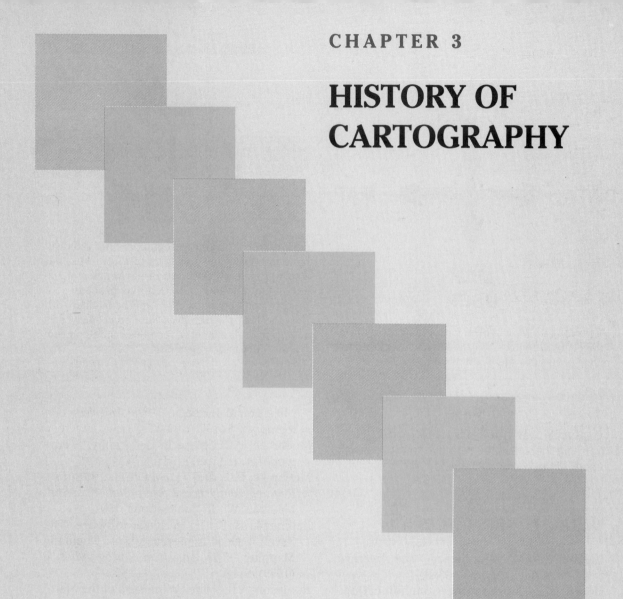

CHAPTER 3

HISTORY OF CARTOGRAPHY

SEQUENCE OF DEVELOPMENT

IMPACT OF CHANGING IDEAS
  CONCEPT OF REPRESENTATION
  GEOMETRY
  RECONCILING CONFLICTING
    INFORMATION
  GLOBALISM

SCIENCE AND MEASUREMENT
CONCEPT OF DISTRIBUTION
SYSTEMS/ECOLOGICAL THINKING

IMPACT OF CHANGING TECHNOLOGY
  TECHNICAL ADVANCES
  INTEGRATING TECHNOLOGIES

INFORMATION AGE MAPPING

*I*n this chapter, we will consider the impact of changing ideas and technological advances on the history of cartography. As with most histories, the story of mapping is one of repeated cycles of revolution and evolution. Ideas provide the primary driving force. Throughout the history of cartography, new ideas have expressed as well as changed our relationship with the environment.

The influence of changing ideas was sometimes direct. The concept of a spherical earth, for example, had a direct influence on global exploration. At other times, new ideas had an indirect effect, first expressed in the form of a technological advance and then through the application of this new technology. The change in navigational practices following the invention of the magnetic compass is an example of this indirect influence of new ideas.

Prevailing ideas and associated technologies define the character of society in a given historical period. Furthermore, the way we think, live, work, play, and communicate are strongly reflected in the mapping of the time. Thus, developments in mapping through the centuries closely parallel developments in society in general.

In this chapter, we will focus on major turning points in the history of mapping. First we will consider the sequence of map development. Next we will examine shifts in thinking that had a direct impact on the field. Finally, we will look at the mapping ramifications of emerging technologies. We understand, of course, that ideas and technology go hand in hand; a change in one usually is quickly followed by a change in the other.

SEQUENCE OF DEVELOPMENT

Development and progress are synonymous terms for many people. Those who hold this view tend to see the history of mapping as an **evolution** in linear fashion from primitive to more sophisticated forms. We might call this the **ladder concept** of cartographic development (Figure 3.1A). When discontinuities or gaps occur in the se-

quence, they are commonly referred to as **missing links**. Proponents of this concept hope that existing gaps in knowledge will someday be bridged by new evidence. This belief has given impetus to a global search for "undiscovered" old maps.

The problem with the ladder concept of mapping development is that the different map types we know today do not fit neatly into a linear sequence from primitive to sophisticated. If we plot the emergence of map types, the result looks more like the branches of a tree (Figure 3.1B). With this **tree concept** of development, a **revolution** in mapping creates a branching. Each revolution leads to a new map type, represented in the diagram by the branches themselves. Map types develop in evolutionary fashion, at least until the next revolutionary change occurs.

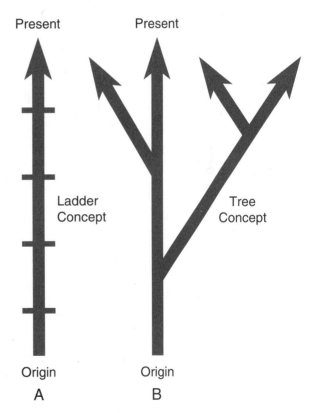

Figure 3.1 Cartographic development has followed a branching course rather than a linear sequence.

A tree conception of mapping history is outlined schematically in **Figure 3.2**. The branches are labeled by map type, and branching represents a significant turning point in cartographic development. You can place revolutionary events in temporal perspective by using the timeline found at the bottom of the diagram. As you proceed through this chapter, refer to this tree diagram to help put things into proper historical context. We will begin by considering some ideas that significantly changed cartographic development.

IMPACT OF CHANGING IDEAS

In this section, we will demonstrate the close association between conceptual developments in so-

ciety and the history of cartography. No attempt is made to be exhaustive. The brief discussions that follow are only intended to provide a general sense of the importance of changing attitudes on mapping development.

CONCEPT OF REPRESENTATION

The cartographic importance of the idea of representation was outlined in the previous chapter. Suffice it to say here that this is the concept upon which the entire field of mapping is built. Once people embrace the idea of symbols, the elaboration of mapping techniques was limited only by our imagination.

As Figure 3.2 illustrates, starting with the basic idea of representation, we have devised a wide variety of mapping methods. People created some

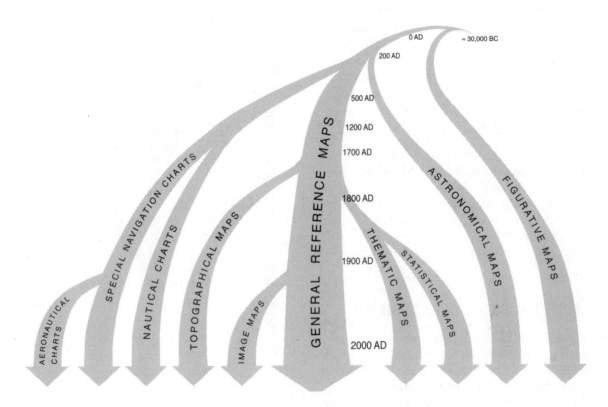

Figure 3.2 The diverse map types we know today emerged through a long process of cartographic revolution and evolution.

of these mapping techniques in response to needs for a new type of map. Other methods of mapping followed directly from new ideas or technical advances.

Early maps are often more figurative than literal. The environment was chaotic and mysterious for these pre-literate mappers. It is not surprising, then, that they commonly chose something other than physical space to structure their environmental representations.

Most of these figurative maps were probably used for ceremonial and ritual purposes. Hence, their lack of spatial fidelity may have been more an asset than a liability. Furthermore, some figurative maps did have practical applications. South Sea Islanders, for instance, were able to navigate their canoes between distant islands with the aid of **stick charts** that apparently captured the patterns of prevailing winds and currents (**Figure 3.3**).

GEOMETRY

A big change in mapping occurred following the development of geometrical concepts by Greek

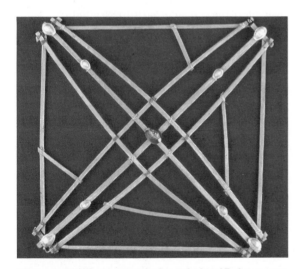

Figure 3.3 Abstract stick charts helped Polynesians navigate between remote South Sea Islands by somehow representing essential characteristics of prevailing winds and currents.

scholars. Geometry provided a means for determining the shape and size of the earth, and for determining the relative position of environmental features (see Chapter 7).

Geometrical concepts also provided a foundation for the development of locational reference systems, such as latitude/longitude and various rectangular grids (**Figure 3.4**). These concepts are discussed in considerable detail in Chapters 4 and 6.

In general, the idea of geometry led to a dramatic increase in mapping accuracy, and greatly facilitated mapping and map use processes. Maps became much more literal in their representation of geographical relations than they had been before. This emphasis on spatial relations quickly became synonymous with mapping, and with few exceptions has dominated the field to the present day.

RECONCILING CONFLICTING INFORMATION

The most dramatic challenge to the dominance of geometry in mapping occurred in medieval times. This period in Western civilization is called the **Dark Ages**, because it is characterized by a general regression in thought and culture. During this period, church dogma prevailed over math and science. Many earlier discoveries were lost or ignored.

In cartography, medieval thinking called for a melding of geographical experience with religious teachings.* The price of failure at this chore was the cartographer's life. Since the differences between geographical knowledge and church beliefs could not be reconciled, figurative world maps were the result (**Figure 3.5**).

Although these so-called **church maps** have been harshly judged for their lack of geographical fidelity, this criticism is not altogether fair. Faced with the same impossible task today, you might

*It is unlikely that church dogma dominated the design of large-scale maps used in land transactions, trade, and commerce. But no maps serving these practical purposes are known to exist.

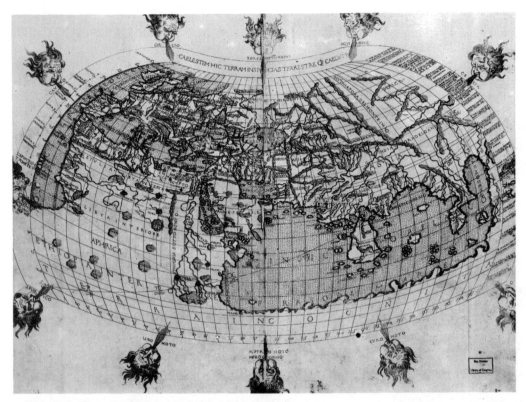

Figure 3.4 This is one of the maps constructed in the fifteenth century from Ptolemy's written directions and descriptions, and reflects geographical knowledge of the known world in the second century A.D. (From Berlinghieri's edition of Ptolemy's *Geography*, Florence, 1492. From the Library of Congress collection.)

come up with something similar. It is more meaningful to evaluate these maps on the basis of their success at representing conflicting ideas about the environment. Indeed, the practice of figurative, nongeographical mapping continues today.

GLOBALISM

The introspective character of medieval world maps stands in marked contrast to the outward perspective the Renaissance brought to mapping. This was a time of global exploration and conquest by European fortune hunters, colonists, and missionaries. Commerce and trade quickly engulfed the planet.

Mapping grew dramatically during this period to keep abreast of the rapid expansion of geographic information. Never before nor since has knowledge of the planet grown so quickly nor the demand for new maps been greater. The need to represent data in the unfamiliar global context even led to the creation of novel projection geometries (see Chapter 5).

SCIENCE AND MEASUREMENT

The next major conceptual shift in mapping came with the idea of Western science. Initially, the scientific method was based on the concept of **order** (the opposite of randomness) and the no-

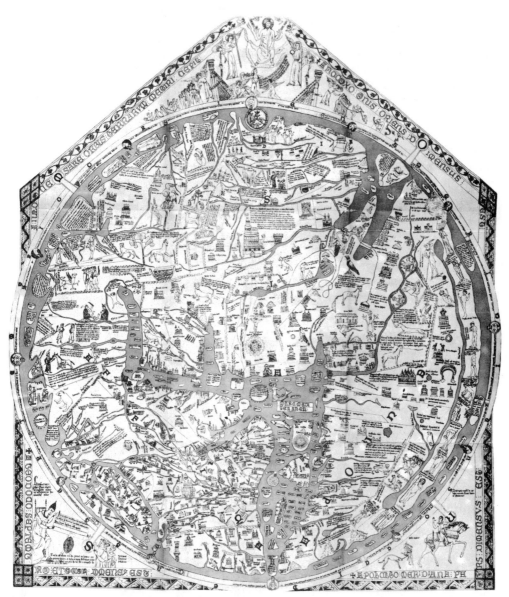

Figure 3.5 This map was made in the thirteenth century to serve as an altarpiece in the cathedral at Hereford, England. It illustrates the degree to which much cartography had become figurative and allegorical compared to that in the time of Ptolemy, 1000 years earlier. The map is oriented with east at the top and Jerusalem at the center. (Photo from the Library of Congress collection.)

tion that order can be explained by **cause-effect** relations. Some time later, scientists added the idea of **chance** or **probability** to their basic tools.

Careful observation and measurement achieved high status in many fields during this period in the West, called the **Enlightenment**. The foundations of modern mapping were laid at this time.

A concern with positional accuracy sent mapping professionals into the field to take rigorous measurements of the shape and size of the earth and the location of features such as coastlines, rivers, and cities (Figure 3.6). At the same time, scientists went into the field to gather and classify specimens of animals, plants, and rocks. Other scientists searched the earth taking measurements of various physical and cultural phenomena.

CONCEPT OF DISTRIBUTION

Until the Enlightenment, maps depicted features such as rivers, mountains, and towns. The aim was to locate these individual environmental features on a map of a given scale as accurately as possible. The focus is on **place**, and the perspective is analytical. Representations of this type are called **general reference maps**.

As data built up from environmental observations and measurement during the Enlightenment,

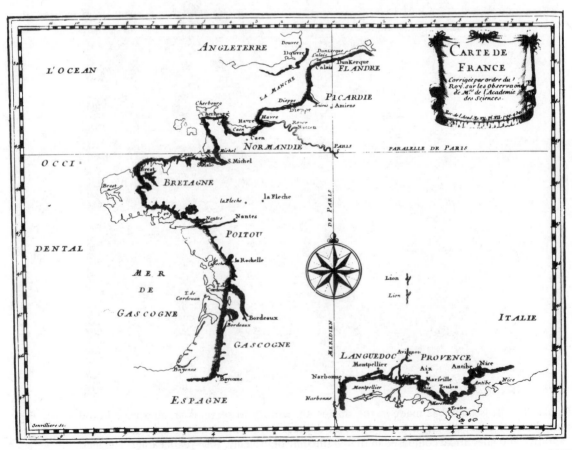

Figure 3.6   This map, published by the French Academy of Sciences, shows the more accurate 1693 coastlines of France (shaded) compared with their delineation on a 1679 map.

attention shifted from place to **space**. Focus shifted from analytical concern with the position of features to holistic concern with the spatial extent and variation of features. Thus, the idea of **distribution** was born.

The conceptual leap from place to space led to distributional representations called **thematic maps**. Vegetation, rock types, climate, ocean currents, population density, and a multitude of other distributions were soon mapped (**Figure 3.7**). In each case, positional accuracy was subservient to spatial variation. Many distributions, such as soils or climate, had to be derived from other phenomena, since they could not be observed or measured directly.

SYSTEMS/ECOLOGICAL THINKING

All manner of environmental distributions have been identified and mapped since the inception of thematic mapping in the late seventeenth century. Each scientific discipline had its own selection of cartographic documents. These special maps served the needs of various environmental professions. Mineral exploration, for example, relied heavily on geological maps.

By the mid-twentieth century, however, it was apparent that identifying and mapping still another distribution component of the environment was not sufficient. Complex problems such as species survival, environmental contamination, and quality of life could not be addressed adequately by breaking the environment into parts watched over by specialists.

People realized that the environment could best be understood and managed by viewing it as a system of interrelated processes. In this **ecological model**, a change in one component or process is seen to lead to changes in others as well.

Since the **systems approach** to the environment called for reintegrating what had previously been separated, this way of thinking had a profound effect on cartography. The concept of **cartographic modeling**, in which environmental

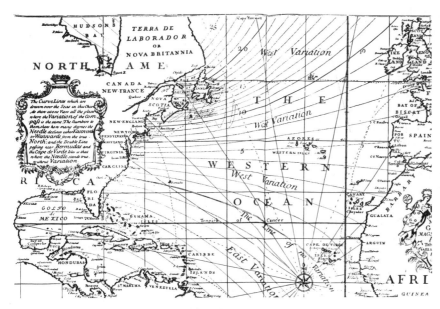

Figure 3.7 This is a portion of Halley's isogonic chart of 1701 showing "curve lines" (isarithms or isolines), which are intended to connect points with the same variation of the compass. (From the Library of Congress collection.)

phenomena are selected, weighted by importance, and linked together to form a numerical index, ushered in a new type of mapping (see Chapter 28). Thus, we now have maps of susceptibility, trafficability, sustainability, suitability, and similar abstract concepts, each of which is created by forming a weighted composite of individual environmental measures.

IMPACT OF CHANGING TECHNOLOGY

Now that we have explored the close link between new ideas concerning the environment and cartographic change, it is time to shift attention to the role changing technology plays in mapping. As with ideas, there has been a close relationship between mapping and the prevailing state of technological development. Cartographers have usually been quick to borrow and adopt technological innovations. How data for mapping are selected, how map design is conceived, how maps are produced and reproduced, and how maps are used have always strongly reflected the technological achievements of the period.

One reason for this close correlation between technological advances and cartographic achievement is that each succeeding generation of cartographers has had to face the same two goals. Society has been unrelenting in its demand for maps that are more timely, accurate, and complete, and at the same time there has been a continual demand for greater accessibility to lower-cost maps. These dual forces have been at play regardless of cartographic developments in the previous generation. Indeed, to a large extent it has been the constant struggle to meet these goals that has led to the kinds of maps we know today.

TECHNICAL ADVANCES

A wide array of technical developments has been important in cartography. Advanced knowledge of the principles of mechanics, optics, chemistry, metallurgy, electromagnetism, and electronics has found application in the mapping process.

To a large extent, this knowledge has been made possible by continual refinement in engineering and manufacturing processes. Indeed, the two—understanding and engineering—have gone hand in hand. An advance in one usually soon leads to an advance in the other.

This seesaw growth in knowledge and technology has meant that cartographers in each new generation have had the benefit of better tools, machines, and materials than the last. Today the products of technology that find cartographic application are of greater sensitivity, accuracy, speed, and durability; are easier to handle and use; and are generally of much higher quality than ever before. As a result, contemporary cartographers enjoy an unprecedented flexibility in their mapping endeavors.

Although constant adjustment to technological change has long been a fact of life for the cartographer, two fundamental trends are evident. First, the magnitude of change brought about by each new technology seems to be greater than the impact of the one that came before.

Second, the rate of change seems to be accelerating (**Figure** 3.8). Manual techniques based on simple hand-held tools dominated mapmaking for thousands of years. Magnetic Technology came next. Then, beginning with the development of the printing press and movable type, in the twelfth century in China and the fifteenth century in the Western world, mechanical technology came to dominate mapping activities.

The next change came with the improvement of optics. For the past 175 years, cartographers have been integrating photo-chemical technology into their work. Currently the revolution in electronic technology is rapidly affecting the way cartographers think and work, what gets mapped, and how maps are used. The profession is already fully immersed in high technology, and there seems to be no end to cartographic applications of this latest technology.

On many occasions in the history of cartography, new machines have been introduced to per-

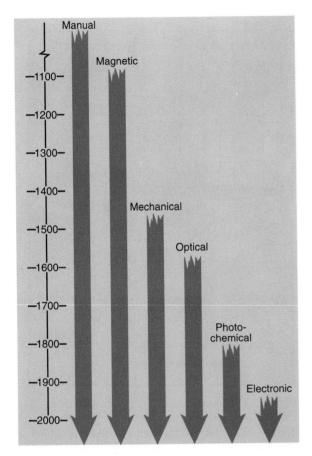

Figure 3.8 Mapping in the Western world has undergone six major technological revolutions since the time that cognitive images were first transcribed into tangible cartographic products. Each technology has had a dramatic impact, both in terms of the scope and depth of the discipline, and with respect to the ease, speed, and cost of mapping. The timing is somewhat different in the Eastern world.

form some of the necessary time-consuming, exacting, and repetitive operations. In each case, basic cartographic principles have not been substantially changed. Rather, the way these principles are applied has in many respects been modified to take advantage of the new technology.

However, rarely has one technology completely replaced another. Over the years the cartographer has simply been given more options from which to select. Thus, the modern cartographer is privileged to pick and choose procedures from the full array of the technologies so far devised. Cartographers also can take advantage of nearly every conceivable combination or hybrid technology.

The breadth of technologies currently available means that there are usually several ways to achieve the same end in mapping. There are also many alternative map designs and forms from which to choose. More than ever before, then, the cartographer enjoys the opportunity to tailor the mapping process to the situation at hand. The particular approach adopted for a given mapping project should represent a balance between available time, acceptable costs (labor, equipment, materials), and type and quality of the product (map) desired. For this reason, great differences in mapping practices currently exist among various mapping establishments.

Clearly, by introducing more things to consider, technological innovations serve to make the cartographer's decision-making responsibilities more rather than less difficult. The real value of technological advance is that it opens to the cartographer options or alternatives that can lead to more effective and efficient mapping. But to take advantage of technology's potential benefits requires a thorough understanding of the central role that technology plays in the cartographic process.

The chapters that follow treat this topic in detail. The aim at this point is to provide an overview to put subsequent discussion into perspective. To this end, the next sections of this chapter are organized around the broad headings taken from Figure 3.8. In each section, the impact of technology is assessed with respect to all phases of the mapping process. These include the capture of raw data for mapping, the arrangement or compilation of these data into a draft map format, the design or specification of graphic factors that will give the map its visual character, the physical production or execution of the map design in an appropriate medium, and the duplication or reproduction of the original map.

Manual Technology

Manual mapping procedures were dominant during the longest period in the recorded history of cartography. Craftspeople became incredibly skillful, using only hand-held tools such as brushes, quills, and the stylus, and working on papyrus, silk, parchment, and even in clay and metal. The miniature **mappae mundi** in medieval illuminated manuscripts and the later **portolan charts** used by navigators of the Mediterranean region were intricate and colorful. As a matter of fact, it is suggested that mapmaking became a recognized profession when the demand for skillfully made charts and other navigational instruments burgeoned in the fourteenth and fifteenth centuries.

Manual techniques did not go out of fashion with the development of printing. People skilled in making relief woodcuts (nonprinting areas carved away) and intaglio copper plates (printing areas incised) carried on the manual tradition (**Figure 3.9**). To these groups were added the lithographic and wax engraver in the nineteenth century.

Some of the great names in the history of cartography are associated with the "handmade map." Mercator was a skilled engraver; Abraham Ortelius, who produced the first modern atlas, started out as a map colorist; and August Petermann, who became a leading German cartographer–geographer and who founded a leading periodical, *Petermann's Geographishche Mitteilungen*, began his career as a skilled engraver–lithographer.

Today, thousands of years after the first extant cartographic map, manual procedures are still responsible for a sizable portion of mapping activity. Smaller mapping firms, and those lacking capital for expensive equipment purchases, still have a strong manual orientation.

Continual improvements in materials, tools, and techniques have helped to ensure a place for cartographers who are skilled in manual procedures, even in modern mapping activities. Indeed, for many purposes, it is difficult to match the speed, flexibility, and skill of the human craftsperson and decision maker in the mapping

Figure 3.9   A *formschneider* (one who carves woodcuts) at work in front of a window. In those days there was no satisfactory substitute for daylight. From Amman and Hans Sachs, *Eygentliche Beschreibung aller Stände auf Erden*, Frankfurt, 1568. (Courtesy of the John M. Wing Foundation on the History of Printing of the Newberry Library, Chicago.)

process. Thus, rather than disappearing with the introduction of new technologies, manual methods have been incorporated into the new ways. Computer-aided cartographic design, for example, is now widely practiced (see *Electronic Technology* later in this chapter).

Magnetic Technology

The second generation of technological innovation in cartography began with the invention of the **magnetic compass**. This useful device was brought to the Western world from China sometime in the twelfth century. It contains a free-floating magnetized needle that aligns itself with the earth's magnetic field (see Chapter 7), providing a baseline from which angles can be measured. Navigators—and, later, surveyors—found the magnetic compass to be a most useful tool.

More recently, magnetic technology has proven invaluable as a way to store data in our electronic age (see *Electronic Technology* later in this chapter). Magnetic tape (audio and video) and disk storage media are used in a variety of consumer products. These same media have also made it possible to store vast quantities of environmental information in a form that is readily accessible by digital computer

Mechanical Technology

The third generation of major technological innovation in cartography involved applying principles of mechanics. Machine power augmented and magnified human muscle power (Figure 3.10). The result was a major increase in the speed and efficiency of the mapping process, with a commensurate reduction in mapping cost. Since this was just what cartographers wanted, mechanical technology was widely adopted, ranging from engraving machines for producing closely spaced, parallel lines to engine-driven presses. Maps became far more accessible to a much wider audience than was possible with manual technology, in which every map, original and copy, had to be drawn by hand.

The adoption of mechanical technology also started a trend toward greater need for more expensive capital equipment compared to what was required for manual methods. The impact of mechanical technology on the different phases of the mapping process has varied considerably, however. The reproduction phase has been by far the most drastically affected. Map compilation and production procedures in large part have become slave to the choice of subsequent map reproduction method. In fact, the introduction of mechanical technology has changed the way the cartographer must think about the mapping process.

Optical Technology

The accuracy of mapping and map use benefited greatly from advances in lens grinding technology. Telescopic sighting instruments and magnifying lenses effectively extended human sight and enhanced visual acuity. Thus, optical technology vastly improved environmental data collection.

More recent advances in light projection have substantially reduced the labor and improved the accuracy of image transfer operations associated with photo-chemical processes (see next section). Optical technology is also being used in the form of laser beams to make highly accurate distance and direction measurements (see Chapter 7).

Finally, laser technology has made massive data storage and retrieval possible in the form of compact disks (CD-ROMs) and optical disks. Modern mapping systems commonly make use of cartographic data stored on these optical media.

Photo-Chemical Technology

The development of lithography and photography and the application of etching techniques to mapmaking in the early nineteenth century stimulated the fifth major technological revolution in cartography. Lithography, which was first called chemical printing, produced duplicates from a flat surface by employing the mutual repulsion of oil and water to form a printing image (see Chapter 30). It was much cheaper and more manageable than copperplate engraving and led to color printing by mid-century. Today lithography is the most widely used method for printing maps.

Photography, first applied to mapmaking in the mid-nineteenth century, rapidly had a major impact. Field photography, begun in earnest in the late 1800s and currently referred to as environmental remote sensing, gave cartographers a dramatic new map form, the photomap (see Chapter 8). The variety and uses of this image-based map form now seem unlimited.

The photographic process also provided cartographers with a powerful new technique for carrying out map compilation, production, and

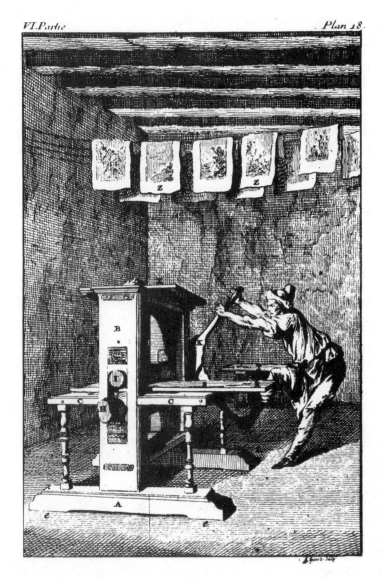

VI. Partie *Plan 18.*

Figure 3.10 Printing from a copper-plate engraving with the rolling press was hard work. From Abraham Bosse, *De la manière de graver...*, Paris, 1745. (Courtesy of The John M. Wing Foundation on the History of Printing of The Newberry Library, Chicago.)

reproduction tasks in the laboratory (see Chapter 31 and Appendix D). Some chores characteristic of manual and mechanical mapping practice, such as image enlargement or reduction and the transfer of a guide image from the surface of one material to that of another, have been greatly modified. Other tasks, such as hand engraving of printing plates, have been altogether eliminated.

Still other chores, such as those associated with the processing and preparation of photographic images (for example, aerial photos) for reproduction, are new additions to the cartographer's list of responsibilities.

The basic components of the photographic process are similar in both remote sensing (field) and laboratory applications. There must be an object to be photographed, a light source, a light-sensitive medium, a device for controlling exposure, and a means for chemically developing the photographic image.

Actual equipment and procedures used in the laboratory differ in significant ways from those used in the field, however. In laboratory work, called graphic arts photography, artwork replaces the ground scene as the object to be copied. Illumination is achieved with a light source that can be precisely regulated rather than by the sun. And devices for making controlled exposures can use light-reflection as well as light-transmission principles. These factors combine to provide the cartographer working with graphic arts photography with a wide array of materials and procedures.

The fact that photo-chemical technology could be so effectively applied in the field as well as in the laboratory has had diverse consequences in the mapping process. Unlike mechanical technology, which has a strongly unbalanced influence on the different phases of mapping, the impact of photo-chemical technology is quite evenly distributed. Indeed, it has become an integral part of all phases of mapping. Even today new applications are constantly being discovered.

Even more than mechanical technology, photo-chemical technology has forced the cartographer to think backward through the stages of the mapping process. To a large degree, reproduction method alters production procedures. For example, map artwork would be correctly scribed right-reading for letterpress printing, whereas it should be scribed in reverse (wrong-reading) for lithographic printing.

Similarly, production method alters compilation procedure. Thus, on a compilation worksheet designed to be traced in subsequent ink drafting, the use of colored pencils on paper might be the best means of distinguishing one category of feature from another. In contrast, a worksheet to be used as a guide image in scribing might better be constructed in black ink on drafting film in order to facilitate the photographic transfer of the image to the scribecoat surface.

What these examples suggest is that in order to take full advantage of the alternative mapping strategies made possible by photographic technology, the cartographer must plan ahead carefully. Failure to do so generally leads to inefficiency and added costs in time, labor, and materials.

Electronic Technology

Cartographers began to explore the power of electronics in the early 1950s and now find themselves in the midst of a high-technology revolution. The change in mapping has been phenomenal. In less than 50 years an entirely new practice of computer-assisted mapping has been invented and implemented. It has concentrated on doing by machine many operations that were formerly painstakingly done by hand.

The increased flexibility and capabilities of the new technology will enable even further extensions of the field which, at present, are only partially conceived. The jargon of the electronic age has become part of the cartographer's language.

This sixth major technological revolution to affect cartography promises to alter the mapping process as nothing before. The **analog image**—that is, the visible two-dimensional map—that characterized previous technologies is being replaced by a **digital record** or **file** in which all locations and characteristics are coded in a binary system. In the process, the way that cartographers think about maps and go about data collection, compilation, production, and reproduction is being changed dramatically. The conventional graphic map is no longer the only—or ultimate—product to be generated in the cartographic process.

The changes brought about by electronic technology are to date quite uneven. In some aspects the developments have been profound, but in others relatively little has occurred. Initial attention was directed at integrating electronics with older technologies. This was highly successful and resulted in a period of imitation and replication rather than true innovation. More recently, significant steps have been taken toward fully electronic mapping. Already this has led to the creation of photo image maps based on digital data sensed from nonvisible parts of the electromagnetic spectrum and displayed as dynamic video images.

Although other innovations will undoubtedly soon follow, computer-assisted cartography is still a better term than automated cartography to describe the present state of affairs. Computer-assisted mapping involves the development and integration of the three components of any computer system: (1) the machines that perform the operations, or hardware; (2) the instructions that tell the machine what to do (what, when, where, how), or software; and (3) the data to be manipulated by the machines under software control (**Figure 3.11**). The success of computer-assisted mapping rests on the skill of the cartographer and the development and application of each of these computer system components within a cartographic environment.

Hardware The hardware aspects of a computer-assisted mapping system are outlined in Figure

3.12. At the heart of the system is a high-speed digital computer or **central processing unit (CPU)**. Its function is to receive data in digital form and process them into a form that the cartographer needs as output. Control units range widely in speed, power, and function.

Relatively small capacity microcomputers (or microprocessors) form the core of **personal computers**, which have brought the potential for computer-assisted mapping into the office and home. For serious desktop mapping, however, more powerful **workstations** dedicated to high-quality computer graphic displays are presently the platform of choice (**Figure. 3.13**). Still higher capacity **minicomputers** are used in larger mapping establishments to store large databases and serve networks of desktop machines. Very large **mainframe** and **supercomputers** are now used to support the most demanding mapping chores and the immense data manipulation needs of cartographic research establishments.

In support of the CPU are a host of peripheral devices for data entry, storage, and output (see Figure 3.12). Most of these devices serve a wide range of users, but some are of special interest in computer-assisted cartography. These include data entry devices, such as digitizers and scanners. They also include special graphic output devices

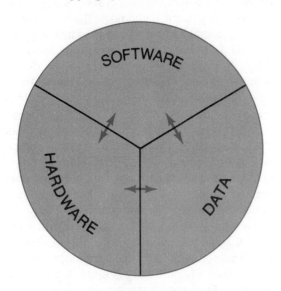

Figure 3.11 Progress in computer-assisted mapping depends upon the development and integration of the hardware, software, and data components of the system. Of course, the central character is the cartographer.

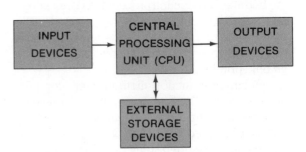

Figure 3.12 The hardware configuration of a standard computer system consists of data input devices, a central processing unit (CPU), external storage devices, data output devices, and appropriate linkages for communicating data from one unit to another.

Figure 3.13 Multipurpose computer graphics workstations, such as this dual screen system, make it possible to perform several mapping chores on-screen at the same time. Fully configured systems permit dramatic special effects, such as animation, zooms, fades, and a host of other image enhancements and manipulative possibilities. (Courtesy Intergraph Corp.)

such as printers, plotters, and display screens suitable for map presentation. Since these data input and graphic output devices have become important mapping tools, they will be discussed in detail in Chapter 11.

Software The software component of a computer mapping system is somewhat less tangible but no less important than the hardware component. Computers require explicit instructions. Thus, sets of machine instructions, called **programs**, are needed to carry out the processing and reformatting (restructuring) of data that cartographers heretofore have had to do by nonelectronic means.

Before such programs can be written, each cartographic task must be analyzed and then structured as a logical step-by-step procedure, known as an **algorithm**. Most common procedures used by cartographers have been translated into software programs written in special computer **languages**, such as FORTRAN or C. Indeed, many mapping procedures have been programmed more than once. Since this usually has been done by individuals working independently at different lo-

cations, the programs are commonly written using different algorithms, computer languages, operating systems, and machine specifications.*

Today's well-rounded cartographer is routinely involved with these "canned" (prewritten) mapping programs and should possess sufficient computer literacy to be able to understand the cartographic implications of each algorithm being used. Not all possible programs have been created, of course. Thus, modern-day cartographers may also be directly involved in computer programming tasks in the course of performing day-to-day duties. The professional cartographer should, therefore, have a working knowledge of at least one computer language.

Data Computer hardware and software are of little value to cartographers unless the data to be mapped are available in computer-compatible form. This necessity points out the importance of the data component of a computer mapping system and helps explain why data entry devices such as digitizers and scanners have assumed such a prominent role in modern mapping.

But there is more to making data computer-compatible than merely transforming them into digital records. The raw numbers must be edited, labeled, tagged with the necessary reference information, and formatted and structured in a fashion that preserves their spatial character and facilitates ease of future use. Once created, the resulting files or databases must be kept up to date and otherwise maintained to serve the needs of cartographers both now and in the future.

To date, the data component of computer mapping systems has been the weakest of the three major parts. Database construction has lagged far behind hardware advances and software development. Progress is being made at a more rapid rate than in the past, however, and a number of

---

*One of the more frustrating aspects of computer-assisted mapping is that programs written for use on one manufacturer's equipment may have to be extensively rewritten before they will function properly on another company's computer.

government agencies, including the U.S. Geological Survey, have committed themselves to building and maintaining databases that will be of great value to cartographers for some years to come.

INTEGRATING TECHNOLOGIES

The power, operating speed, and functional capabilities of modern computers can handle a vast array of cartographic chores, but at this time several things are still lacking. One is the availability of adequate files of computer-compatible data to suit the broad range of mapping scales and topics. The other is sufficiently versatile software to enable the cartographer to perform the host of data and graphic manipulations involved in the mapping process. Although the supply of digital data and cartographic software is improving steadily, it will take some time before data and programs are available to cover the full range of mapping tasks and be workable on each of the computer systems currently in use.

Already, however, electronic technology is having a melding influence on cartography. For one thing, applications of electronics are blurring past distinctions among various mapping technologies. Cartographers have worked with three broad generations of electronic devices during the last quarter century (**Figure 3.14**). Beginning in the mid-1950s there was a mating of digital and mechanical technologies. A mating of digital and photographic technologies followed in the early 1960s. And, finally, fully digital mapping had its origins in the mid-1980s.

Currently, machines representing every possible hybrid of the various technologies are being used. The integration with the most sophisticated modern devices is so complete that it has become meaningless to try to identify or separate the contribution of the individual technologies.

Mapping procedures have become just as difficult to categorize by technology as the machines that are used in mapping. Old boundaries are fast breaking down. Since certain strengths and weaknesses are inherent to each technology,

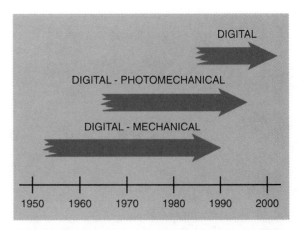

Figure 3.14 Electronic map production devices have gone through three broad developmental stages during the past quarter century. Although all three generations of equipment are currently in use, the trend is toward more fully digital units.

modern cartography commonly involves procedures based on several approaches. Manual, optical-mechanical, photo-chemical, and electronic activities may all be integrated in an attempt to attain a mapping goal. The idea is to blend procedures based on the different technologies in such a way that each will contribute its own special advantages with respect to speed, cost, accuracy, and other desirable qualities.

In similar fashion, electronic technology is creating fuzzy boundaries between the stages in the mapping process. Previously, it made sense to think about mapping in terms of the distinct phases of data gathering, compilation, production, and reproduction. But if data are collected directly in digital form, the stages of compilation, production, and reproduction in some cases tend to merge into one operation. A mere touch of a button (switch) moves the cartographer from one phase to the next. This breakdown of old boundaries is particularly true when only a single copy of a map is made for the benefit of a single user, or when a map image is telecommunicated to several display terminals for direct viewing and is then erased.

This blurring of mapping technologies, procedures, and stages is changing the way maps are made and in the process is altering the relationship between mapmaker and map user. When maps were manually created and only one or a few copies were ever made, the map user was normally in some way directly involved with their construction. But during the period when sophisticated mechanical and photo-chemical technologies dominated mapping activities, the map user tended to be left out. Professional cartographers were responsible for designing, constructing, and reproducing maps and map series, which were then stored in depositories and libraries to be made available later to potential users. Until recently, for the most part, the map user had to make do with whatever printed maps were available.

Today, electronic technology is fast making it possible for the map user to be directly involved, especially if mapping is being done on a display screen or with microcomputer-driven plotting devices. The result, of course, is that the knowledgeable user can better tailor mapping decisions to requirements at hand and need not rely on the judgment of a professional cartographer, who may know neither what a map is to be used for nor how sophisticated the map user might be.

Thus, electronic technology is also changing the role of the professional cartographer. Rather than pursuing the goal of building up the stock of printed maps for potential future use, cartographers will increasingly need to concern themselves with making environmental data and mapping programs more accessible to non-professional mapmakers. There will also need to be a major new emphasis on raising the level of cartographic sophistication among these professional map users, as well as the populace at large, if the promise of modern mapping is to be realized.

INFORMATION AGE MAPPING

We commonly refer to the current state of ideas and technology as the **information age**. Global systems thinking prevails in the form of international trade and law, bio-diversity, environmental contamination, global warming, earth-systems science, sustainable economies, and so forth. The technologies that support these concepts include computers, earth-orbiting satellites, and telecommunications.

One of the most powerful tools in our quest to inventory and manage our planet is **information**. Unfortunately, old technologies for handling information are not adequate for today's needs. We need access to vast volumes of data from diverse sources. We also need to be able to sort through and extract key items from this material at a moment's notice. These requirements are addressed by modern **information systems**, which employ electronic technologies to provide for the integrated storage, retrieval, manipulation, analysis, and presentation of critical data.

An information system specially designed to serve environmental needs is called a **geographic information system (GIS).** In this case, a spatial reference is associated with the data. Data manipulation, analysis, and representation have spatial dimensions as well (**Figure 3.15**). Although the term GIS is commonly used in reference to computer software designed to serve spatial information system functions, the term more appropriately covers the system as a whole, including hardware, software, and data.

Maps play a key role in GIS, and geographic information systems are crucial in modern mapping. The bulk of the data used in a GIS are digitized or scanned from existing map sources. The output from a GIS is also largely cartographic.

Conversely, the need for integration of hardware, software, and data in modern mapping in order to avoid work-stopping incompatibilities means that a GIS is becoming the cartographic tool of choice in a growing number of situations. A trend toward greater analytical use of cartographic data has also increased the popularity of GIS technology. For these reasons, we will be making frequent reference to GIS technology throughout this book.

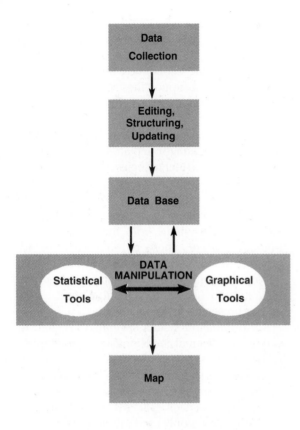

Figure 3.15 The mapping process is greatly facilitated with GIS technology, which integrates data collection, editing, storage, processing, and display.

SELECTED REFERENCES

Bagrow, L., *History of Cartography*, revised and enlarged by R.A. Skelton, London: C.A. Watts & Co., Ltd., 1964.

Boorstin, D. J., *The Discoverers: A History of Man's Search to Know His World and Himself*, New York: Vintage Books, A Division of Random House, 1985.

Clarke, K. C., *Analytical and Computer Cartography*, Englewood Cliffs, N.J.: Prentice-Hall, Inc., 1990.

Cromley, R. G., *Digital Cartography*, Englewood Cliff: Prentice-Hall, 1992.

Hall, S. S., *Mapping the Next Millennium: The Discovery of New Geographies*, New York: Random House, 1991.

Maguire, D. J., M. F. Goodchild, and D.W. Rhind, *Geographical Information Systems: Principles and Applications*, New York: John Wiley & Sons, 1991.

Monmonier, M., *Technological Transition in Cartography*, Madison: University of Wisconsin Press, 1985.

Peuquet, D. J., and D. F. Marble, *Introductory Readings in Geographic Information Systems*, New York: Taylor & Francis, 1990.

Robinson, A.H., *Early Thematic Mapping in the History of Cartography*, Chicago: University of Chicago Press, 1982.

Robinson, A. H., J. L. Morrison, and P.C. Muehrcke, "Cartography 1950-2000," Transactions of the Institute of British Geographers, New Series 2, 1977, pp. 3–18.

Taylor, D. R. F. (ed.), *Geographical Information Systems: The Computer and Modern Cartography*, New York: Pergamon Press, 1991.

Thrower, N. J. W., *Maps and Man*, Englewood Cliffs, N.J.: Prentice-Hall, Inc., 1972.

Wilford, J. N., *The Map Makers: The Story of Great Pioneers in Cartography from Antiquity to the Space Age*, New York: Vantage Books, A Division of Random House, 1982.

Wurman, R. S., *Information Anxiety*, New York: Doubleday, 1989.

PART II

EARTH-MAP RELATIONS

BASIC GEODESY

*M*apping involves determining the geographic locations of features on the earth, transforming these locations into positions on a flat map through use of a map projection, and graphically symbolizing these features. Geographic locations are specified by geographic coordinates called latitude and longitude. To establish a system of geographic coordinates for the earth, we first must know its shape and size.

The earth is a very smooth geometrical figure. Much of the earth's surface appears rugged and rough to us, but even the highest peaks and deepest ocean trenches are barely noticeable irregularities on the smoothly curving surface. To see this, imagine the earth reduced to a "sea level" ball 10 in. (25.4 cm.) in diameter. Mt. Everest would be but a 0.007-in. (0.176 mm.) bump and the Mariana Trench but a 0.0085-in. (0.218 mm.) scratch in the ball. Since the earth's average land elevation and ocean depth is much less than these extremes, we are safe in saying that if the earth were the size of a bowling ball, it would be smoother than any bowling ball yet made! We will begin this chapter by examining three ever more accurate approximations to the earth's shape: the sphere, the ellipsoid, and the geoid.

SPHERICAL EARTH

More than 2,000 years ago most educated people knew that, if we disregard such features as hills and valleys, the earth is spherical in shape. This understanding was due in part to the teachings of Pythagoras (6th century B.C.) that humans must live on a body of the "perfect shape"—a perfect sphere. More compelling, however, were Aristotle's (4th century B.C.) arguments for a spherical earth. He noted that sailing ships always disappear from view hull first, mast last, rather than becoming ever smaller dots on the horizon of a flat earth. Astronomical observations such as the immutable circular appearance of the sun and moon and the circular shadow cast upon the sun or moon during all eclipses were powerful physical evidence. These

arguments prevailed, and the earth's spherical shape became widely accepted in ancient Greece and later civilizations with access to Greek writings.

Determining the spherical earth's size was another matter. Again, the first calculation was made by a Greek scholar. About 250 B.C., Eratosthenes, head of a great Egyptian library in Alexandria, came close to the figures for the earth's circumference we now accept. The story is that Eratosthenes read of a deep well in Syene (modern Aswan). Sunlight lit the bottom of the well only on June 21, the summer solstice.*

Eratosthenes reasoned that here the sun must be directly overhead on the solstice, meaning that Syene must be located on the Tropic of Cancer (Figure 4.1). On the next solstice he measured the angle above the southern horizon of the noon sun at Alexandria, thought to be due north of Syene along a meridian. By measuring the shadow length of a vertical column of known height, represented by the vertical line at Alexandria, the vertical angle was found to be 82°48'. If the vertical lines at both locations were extended to the earth's center, they would form an angle of 7°12', meaning that the arc distance between the two cities relative to the earth's circumference must be 7°12'/360°, or 1/50th of the circumference.

Eratosthenes next estimated from travelers' reports that the distance between Alexandria and Syene was 5,000 stadia, about 925 km. (575 mi.), since that Greek unit of measure is thought to equal about 185 m. Since this corresponded to 1/50th of the circumference, he multiplied by 50, thus estimating the earth's circumference as

---

*The summer solstice is the day on which the sun appears highest in the sky at noon. The orbits of all planets except Pluto are aligned in a plane extending outward from the sun's equator. The plane bisecting the earth through the equator is tilted about 23½° from the solar plane, creating the cyclical upward and downward shifting of the noon sun during each year. The Tropic of Cancer (23½°N) is the parallel on which the sun appears directly overhead on the northern hemisphere summer solstice.

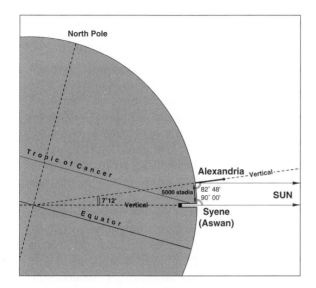

Figure 4.1 The geometrical relationships that Eratosthenes used to calculate the circumference of the earth. See text for explanation.

250,000 stadia (46,250 km. or 28,750 mi.), which is only about 15 percent too large. His method was sound, but his measurements and assumptions were somewhat in error. Fortunately, the errors tended to compensate one another.

As we shall see in the next section, the earth is not a perfect sphere, but rather slightly ellipsoidal in shape. However, cartographers still use a sphere of the same surface area as the ellipsoid, called an **authalic sphere**, as the basic figure for mapping. The dimensions of the most recently determined authalic sphere are given in Table 4.1. The 6,371-km radius is of considerable importance, since this is a fundamental constant in map projection and other often used equations.

Table 4.1
The Authalic Sphere (Based on the WGS 84 Ellipsoid)

	Kilometers	Statute Miles (U.S.)
Radius	6,371	3,959
Circumference	40,030.2	24,875.1

ELLIPSOIDAL EARTH

Until the late 1600s, the earth was thought to be perfectly spherical in shape. The change came around 1670, when Isaac Newton proposed that a consequence of his theory of gravity would be a slight bulging of the earth at the equator due to the greater centrifugal force generated by the earth's rotation. This equatorial bulging would produce a slight flattening at the poles, predicted by Newton to be about 1/300th of the equatorial radius. Newton's prediction was confirmed by measurements taken from 1735 to 1743 by expeditions sent to Ecuador and Finland to measure the ground distance for one degree of angular change (one degree of latitude) in equatorial and polar regions. As Newton predicted, the polar distance was found to be slightly greater due to flattening.

In geometrical terms (Figure 4.2), if we sliced the earth from pole to pole through its center, we would see a slightly elliptical cross section. Rotating this ellipse about the polar axis would outline

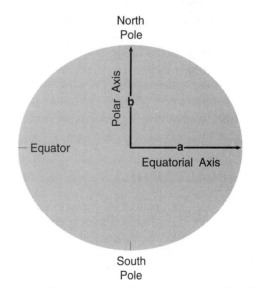

Figure 4.2 The shape of an oblate ellipsoid is determined by the relative length of the equatorial (semi-major) axis (a) and the polar (semi-minor) axis (b). The actual amount of flattening is greatly exaggerated in this diagram.

the three-dimensional figure of the earth called an **oblate ellipsoid** (or sometimes oblate spheroid).

The amount of polar flattening (oblateness) is given by the ratio $f = (a-b)/a$ where a is the equatorial radius and b is the polar radius.* Flattening is usually expressed as $1/f$, so that Newton's predicted value, for example, would be 1/300. Modern values based upon satellite orbital data are around 1/298.

From 1800 to the present, at least 20 determinations of the earth's radii and flattening have been made from measurements taken at widely different locations. Values for 11 different ellipsoids used as the basis for mapping in various parts of the world are listed in Table 4.2. Slight differences in the lengths of radii and flattening values are the result of varying accuracy in the measurements made, and of slight variations in curvature from continent to continent due to irregularities in the earth's gravity field.

The **WGS (World Geodetic System) 72 and 84 ellipsoids**, determined from satellite orbital data, are considered more accurate than the earlier ground measurement determinations, but may not give the best fit for a particular part of the earth. The **Clarke 1866 ellipsoid**, based on measurements taken in Europe, India, Peru, Russia, and South Africa, is of special interest in the United States, since it has been used for mapping in North America until recently. North American cartographers are now rapidly switching to the WGS 84 ellipsoid, which is intended to be a global standard.

GEOIDAL EARTH

An even more faithful figure of the earth, called the **geoid** (meaning earthlike), deviates ever so slightly from the ellipsoid in an irregular manner. The geoid is the three-dimensional shape that would be approximated by mean sea level in the oceans and the surface of a series of hypothetical sea-level canals crisscrossing the continents. In more technical terms, it is a sea level **equipotential** surface—the surface on which gravity is everywhere equal to its strength at mean sea level. If the earth were of uniform geological composition and devoid of mountain ranges, ocean basins, and other vertical irregularities, the geoid surface would match the ellipsoid exactly. However, due primarily to variations in rock density and topographic relief, the geoid surface deviates from the ellipsoid by up to 300 ft. (100 m.) in certain locations.

The minute undulations on the geoid can be seen on a "contour map" of variations between its surface and that of the WGS 84 ellipsoid (Figure 4.3). The lines of equal deviation (in meters) were modeled from millions of gravity observations taken throughout the world. Note that the "hills and valleys" on the geoid do not correspond with continents and oceans. Indeed, the highest point on the geoid is 75 meters above the ellipsoid in New Guinea, and the lowest point is 104 meters below at the southern tip of India.

The most detailed geoidal model for the conterminous United States and immediate surroundings (see Color Figure 4.1) shows deviations relative to the GRS 80 or WGS 84 ellipsoids. This map was created by combining more than one million gravity observations with raster digital elevation data (see Chapter 11). Notice that the geoid dips below the ellipsoid everywhere in the conterminous United States, in contrast to the positive deviations found in Alaska and Hawaii.

CARTOGRAPHIC USE OF THE SPHERE, ELLIPSOID, AND GEOID

Cartographers use these three approximations to the earth's true shape in different ways. The authalic sphere is the reference surface for small-scale maps of countries, continents, and larger areas. This is because the difference between sphere and ellipsoid is negligible when mapping large areas in a

---

*Another important measure is the eccentricity of the ellipsoid. This is defined by the equation $e = [(a^2-b^2)/a^2]^{1/2}$. Most eccentricities are around 0.08.

Table 4.2
Official Ellipsoids (from J. Snyder, *Map Projections—A Working Manual*)

Name	Date	Equatorial Radius *a* (meters)	Radius *b* (meters)	Polar Flattening
WGS 84	1984	6,378,137	6,356,752.3	1/298.257
GRS 80*	1980	" "	" "	" "
WGS 72	1972	6,378,135	6,356,750.5	1/298.26
Australian	1965	6,378,160	6,356,774.7	1/298.25
Krasovsky	1940	6,378,245	6,356,863	1/298.3
Internat'l	1924	6,378,388	6,356,911.9	1/297
Clarke	1880	6,378,249.1	6,356,514.9	1/293.46
Clarke	1866	6,378,206.4	6,356,583.8	1/294.98
Bessel	1841	6,377,397.2	6,356,079.0	1/299.15
Airy	1830	6,377,563.4	6,356,256.9	1/299.32
Everest	1830	6,377,276.3	6,356,075.4	1/300.8

*Geodetic Reference System 1980, adopted by the International Association of Geodesy.

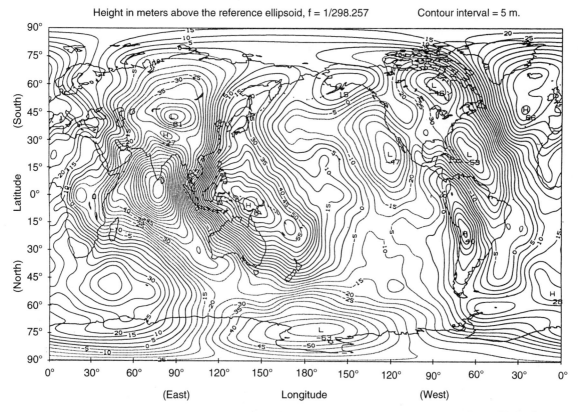

Figure 4.3 Geoid surface computed from the GEM-T3 gravity model by the NASA/Goddard Space Flight Center.

general manner on page-size maps. There is also a significant increase in the complexity of map projection equations for the ellipsoid. In addition, the spherical and spheroidal equations for a particular map projection give essentially the same results for small-scale maps. Thus, it makes sense to use the sphere for these small-scale maps.

With large-scale maps, however, it's a different story. With detailed, large-scale maps of small areas, such as topographic maps and nautical charts, the differences between locations on the spherical and ellipsoidal approximations can be significant, and we need to take the earth's oblateness into account. Distances, directions, and areas measured on these detailed maps would be incorrect at individual locations if the authalic sphere were used. Therefore, cartographers use the ellipsoid as the reference surface for these large-scale maps. Using the ellipsoid also ties in well with modern data collection methods for large-scale mapping. Global positioning satellite receivers (see Chapter 7), for example, compute latitude, longitude, and elevation using the WGS 84 ellipsoid as the reference surface.

The geoid is the reference surface for ground surveyed horizontal and vertical positions. Horizontal positions are *adjusted* to the ellipsoid surface, however, since the irregularities on the geoid would make map projection and other mathematical computations extremely complex. On the other hand, elevations are determined relative to the mean sea level geoid, as we shall see later in this chapter.

GEOGRAPHICAL COORDINATES

The geographical coordinate system, employing latitude and longitude, can be traced to the 2nd century B.C. astronomer-geographer Hipparchus of Rhodes. The geographical coordinate system is the primary locational reference system for the earth. It has always been used in cartography and for all basic locational reckoning, such as navigation and surveying. A second system, called plane

coordinates, or plane rectangular coordinates, is the subject of Chapter 6.

The geographical coordinate system was devised to make possible a unique statement of location for each earth feature. The north and south poles, where the axis of rotation intersects the earth's surface, are the starting points on which to base the system. Specifying a location on the earth requires determining **latitude**, the north-south angular distance from the equator, and **longitude**, the east-west angular distance from a prime meridian. All points on the earth having the same latitude form a line called a parallel; all points of the same longitude form a meridian line.

LATITUDE

The latitude system for locating our north-south position depends on the regular curvature of the earth's surface. The equator, the line on the earth formed by points halfway between the two poles, is the natural starting place for latitude.

Authalic Latitude

If you are making a small-scale map based on the spherical earth, you should use authalic latitude. Authalic latitude may be defined as the angle formed by a pair of lines extending from the equator to the center of the earth, and then from the center to our position (Figure 4.4). It ranges pole to pole from 90°N to 90°S, or from +90° to –90° when using digital databases and map projection equations. It is normally given in degrees, minutes, and seconds, using the sexagesimal (base 60) number system developed by ancient Babylonian mathematicians. Computer calculations often require the decimal degree system, with which, for example, 54° 30´N is expressed as 54.5°.

The north-south distance on the sphere between each degree of authalic latitude is identical, and only depends on the circumference of the sphere. For the WGS 84 authalic sphere of circumference 40,030.2 km. (24,875 mi.), the distance between each degree of latitude is 111.20 km. (69.11 mi.). On the Clarke 1866 authalic sphere, the circumference

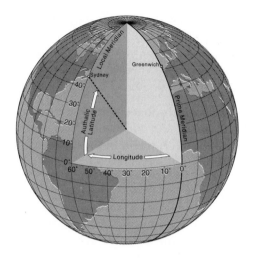

Figure 4.4   Authalic latitude and longitude.

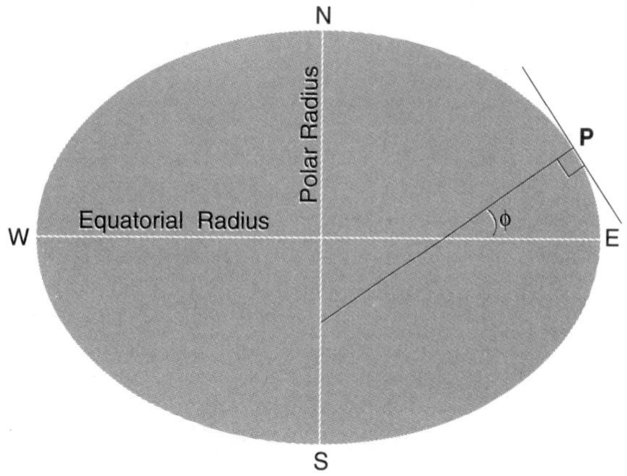

Figure 4.5   Geodetic latitude (ϕ) on the ellipsoid, with flattening greatly exaggerated.

is also 40,030.2 km., meaning that these two, and most other, authalic spheres can be used interchangeably for small-scale mapping.

Geodetic Latitude

Latitude on the ellipsoid is called **geodetic latitude**. We define geodetic latitude as the angle formed by a line from the equator toward the center of the earth, and a second line perpendicular to the ellipsoid surface at one's location (Figure 4.5). Notice that the perpendicular line intersects the first line at the center of the ellipsoid only at geodetic latitudes of 0 and 90 degrees.

The north-south distance between degrees of geodetic latitude is nearly the same, but not quite. We saw earlier that the distance is greater in polar areas and less near the equator. Furthermore, the distance steadily increases in a predictable manner from equator to pole. Table 4.3 is a list of distances for the WGS 84 ellipsoid, showing the steady increase from 110.57 km. (68.71 mi.) at the equator to 111.69 km. (69.40 mi.) at the poles. Values for the Clarke 1866 ellipsoid are essentially identical, varying at most by 10 meters. Also notice that the 111.20 km. distance for a degree of latitude on the authalic

sphere matches the geodetic distance only at around 50° latitude.

This difference of about 1 km in 110 km. is of little significance when making small-scale maps, but cannot be neglected when producing large-scale maps and charts. Geodetic latitude should always be used for large-scale mapping, whereas geodetic latitudes of ground features can be thought of as equivalent to authalic latitudes for small-scale mapping.

Table 4.3
Length of a Degree of Geodetic Latitude (on the WGS 84 Ellipsoid)*

Latitude	Kilometers	Miles
0°	110.57	68.71
10°	110.61	68.73
20°	110.70	68.79
30°	110.85	68.88
40°	111.04	68.99
50°	111.23	69.12
60°	111.41	69.23
70°	111.56	69.32
80°	111.66	69.38
90°	111.69	69.40

*Length of a degree of arc centered on the named latitude.

LONGITUDE

Longitude, our east-west position on the earth, is associated with an infinite set of meridians, arranged perpendicularly to the parallels. Unlike the equator in the latitude system, no meridian has a natural basis for being the starting line from which to reckon east-west positions. The choice of the starting line, called the **prime meridian**, has always been a matter of international importance and national pride.* During the last century, many nations began to accept the meridian of the Royal Observatory at Greenwich near London, England, as 0°. In 1884, it was universally agreed upon at the International Meridian Conference in Washington, D.C.

The universal choice of Greenwich as the prime meridian established the (0°, 0°) "point of origin" of the geographical coordinate system as being a point in the Gulf of Guinea. It also became the starting point for the international system of time zones. The position of the 180° meridian in the Pacific Ocean opposite the prime meridian provides a convenient place for the International Date Line. Days on earth must begin and end somewhere, and only a few deviations from 180° are needed to keep from separating sparsely populated areas into time zones with different days.

Longitude can be thought of as the angle formed by a line going from the intersection of the prime meridian and the equator to the center of the earth, and then back to the intersection of the equator and the "local" meridian passing through the position (see Figure 4.4).

Longitude ranges from 180°W to 180°E of the prime meridian (from −180° to +180° for digital databases and map projection computations).

Length of a Degree of Longitude

We are all aware that meridians converge toward the poles. As a consequence, the east-west distance along a parallel between two meridians one degree apart becomes progressively less as the pole is approached. On the authalic sphere, this distance equals the ground distance between one degree of longitude at the equator multiplied by the cosine of the parallel's latitude. For example, on the GRS authalic sphere, one degree of longitude at the equator is 111.20 km. (69.11 mi.), the same distance as one degree of latitude. At a latitude of 60°, however, the length of one degree of longitude is 55.6 km. (34.55 mi.), since the cosine of 60° is 0.5. The length of one degree of longitude on the ellipsoid almost follows the above formula, but not precisely.* Table 4.4 lists lengths for the WGS 84 ellipsoid.

PROPERTIES OF THE GRATICULE

The imaginary network of parallels and meridians on the earth is called the **graticule,** as is their projection onto a flat map. The graticule has certain geometrical properties, some of which cartographers may try to preserve when making a map projection for part or all of the earth. These properties deal with distance, direction, and area. To simplify our discussion, we will describe each property assuming the earth to be spherical. In the next chapter, we will examine these properties in relation to specific map projections.

---

*In addition to estimating the earth's circumference, Eratosthenes is known for placing one of the earliest prime meridians through his home city of Alexandria. Later, the Canary Islands, the then westernmost point of the known world, emerged as the prime meridian and continued to be used widely until the 17th century. Beginning in the mid-1600s, numerous countries, each with patriotic ambition, published their own maps and charts with the prime meridian running through their capitals. The result obviously was much confusion.

---

*The ellipsoidal equation is distance = $a/(1 - e^2\sin^2(\phi))$ ½ · cos (ϕ) · $\delta \lambda$, where a and e are the semi-major axis and eccentricity of the ellipsoid, ϕ is the geodetic latitude of the parallel, and $\delta \lambda$ is the difference in longitude (one degree, converted to radians, in this example).

Table 4.4
Length of One Degree of Longitude (on WGS 84 Ellipsoid)

Latitude	Kilometers	Statute Miles
0°	111.32	69.17
10°	109.64	68.13
20°	104.65	65.03
30°	96.49	59.95
40°	85.39	53.06
50°	71.70	44.55
60°	55.80	34.67
70°	38.19	23.73
80°	19.39	12.05
90°	0.00	0.00

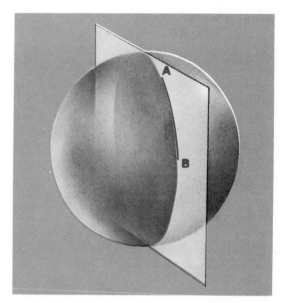

Figure 4.6   The trace of the intersection of a plane and a sphere is a great circle whenever the plane includes the center of the sphere. Any two points on a sphere, such as A and B, can be connected by a great circle arc, and that arc is the shortest route over the surface between the two points.

DISTANCE ON THE SPHERE AND GREAT CIRCLES

The shortest distance between two points is a straight line. On the curving three-dimensional surface of the spherical earth, however, it is obviously impossible to follow such a straight line. The shortest "straight line" course over the surface between any two points on a sphere is the arc on the surface directly above the true straight line. This arc is formed by the intersection of the spherical surface with the plane passing through the two points and the center of the earth (Figure 4.6). The circle established by the intersection of such a plane with the surface divides the earth into hemispheres and is called a **great circle**.

The equator is the only complete great circle in the graticule. Since all meridians are one half a great circle in length, pairs of meridians also make up great circles. All parallels other than the equator are called **small circles**. Their circumference C is computed from:

$$C = 6371 \text{ km} \cdot 2\,\pi \cdot \cos\theta \text{ or } C = 3959 \text{ mi} \cdot 2\,\pi \cdot \cos\theta$$

where θ is the parallel's geographic latitude. Calculating the great circle distance between any two points on the sphere is explained in Box 4.A.

It is possible to make a map projection so that all meridians are the same length and are one half the length of the equator. We can also project the grati-cule so that all the equator and all parallels become straight lines proportional in length to their great or small circle circumference. However, doing either requires distorting other properties of the graticule.

DIRECTION

Directions on the earth are entirely arbitrary, since a spherical surface has no edges, beginning, or end. By convention, north-south is defined as direction along any meridian and east-west is defined as direction along any parallel. Because of the arrangement of the graticule, these two directions are everywhere perpendicular except, of course, at the poles. The directions determined by the orientation of the graticule are called **geographic** or **true directions** as distinguished from two other kinds of direction, magnetic and grid (see Chapter 7).

Box 4.A
Great Circle Distance Calculation

The great circle arc distance (D) on the sphere between any two points A and B can be calculated using the standard formula in spherical trigonometry:

$$\cos D = (\sin a \sin b) + (\cos a \cos b \cos |\delta\lambda|)$$

where a and b are the geographic latitudes of A and B, and $|\delta\lambda|$ is the absolute value of the difference in longitude between A and B. Note that if A and B are on opposite sides of the equator, the product of the sines will be negative.

Example: great circle distance between Washington, D.C. (38°50′N, 77°00′W) and Moscow (55°45′N, 37°37′E).

$$\cos D = \sin(38.833) \cdot \sin(55.75) + \cos (38.833) \cdot \cos(55.75) \cdot \cos (|-77.0 - 37.62|)$$

$$\cos D = 0.627 \cdot 0.827 + 0.779 \cdot 0.563 \cdot -0.417$$

$$\cos D = 0.518 - 0.183 = 0.335$$

$$D = \cos^{-1}(0.335) = 70.43°$$

$$D_{miles} = 70.43° \cdot 69.11 \text{ mi./}° = 4867 \text{ miles.}$$

As discussed in Chapter 7, the needle of a magnetic compass aligns itself with the earth's field of magnetic force, which in most places is not aligned parallel with the local meridian. The reason is that the magnetic field poles do not coincide with the poles of the earth's rotation (90°N and S).

In fact, the magnetic poles slowly change position over time so that the north magnetic pole currently is located at approximately (78°N, 103°W), about 800 miles (1300 km.) south of the geographical pole. Consequently, there is usually a difference between true and magnetic north. This difference is called **compass variation** on nautical charts and **magnetic declination** on topographic maps.

Cartographers usually show the angular difference between true and magnetic north, along with grid north, by drawing a declination diagram (Figure 4.7). Furthermore, the slow but predictable changes in the earth's magnetic field make the declination value correct only for the date the map was issued. Often a statement of the amount of annual change in declination is included with the diagram.

The direction of a line on the earth is called many things: bearing, course, heading, flightline, or azimuth. Their meanings are essentially the same, differing largely in the context in which they are used. Two direction specifications of special importance in cartography are true azimuth and constant azimuth.

True Azimuth

As is apparent from observing the globe, directions on the earth, established by the graticule,

UTM GRID AND 1980 MAGNETIC NORTH
DECLINATION AT CENTER OF SHEET

Figure 4.7 Arrows showing the differences among true north, magnetic north, and grid north at the center of the USGS Paradise East, California, quadrangle, 1:24,000 (1980). A mil is an angular division equal to 1/6,400 part of the circumference of a circle.

are likely to change constantly as we move along the arc of a great circle (Figure 4.8). Only along a meridian or the equator does direction remain constant. We can designate a true azimuth by measuring the clockwise angle the arc of the great circle makes with the meridian at the starting point. The computation of the true azimuth (Z) between any two points A and B on the sphere is explained in Box 4.B.

Since arcs of great circles are the shortest courses between points, movement along them is of major commercial importance. Cartographers construct special map projections which maintain these directional relations as much as possible (Figure 4.9A).

Constant Azimuth

A **constant azimuth** (also called a **rhumb line** or **loxodrome**) is a line that intersects each meridian at the same angle. All meridians have a constant azimuth of 0° (north) or 180° (south), depending on the direction of travel. The equator and all other parallels

have a constant azimuth of 90° (east) or 270° (west). All other constant azimuths cross meridians at the same angle, tracing out a spiral known as a loxodromic curve (Figure 4.10). If we were to continue along this curve, we would spiral toward the pole along a spherical helix with the pole the limit.

A great circle is the most economical route to follow when traveling on the earth. However, it is not practical for pilots to change course continuously. In order for ships and aircraft to follow the great circle route between two points as closely as possible, movement is directed along a series of constant azimuths that approximate the great circle route. Cartographers making charts for air and sea navigation often employ a special map projection, the Mercator, on which all plotted straight lines will be constant azimuths (Figure 4.9B).

AREA

The surface area of quadrilaterals, areas bounded by pairs of parallels and meridians on the sphere,

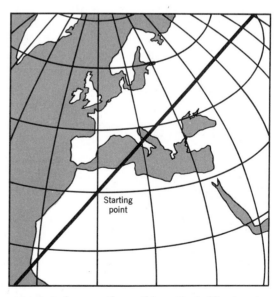

Starting point

Figure 4.8 How a true azimuth is given. The drawings show a great circle arc on the earth's graticule. The drawing on the right is an enlarged view of the center section of the drawing on the left. The azimuth, from the starting point, of any place along the great circle arc to the northeast is stated as the angle between the meridian and the great circle arc, reckoned clockwise from the meridian. Notice that the great circle arc intersects each meridian at a different angle.

Box 4.B
Computing the True Azimuth Between Two Points

The true azimuth (Z) between points A and B on the earth may be computed from the spherical trigonometry equation:

$$\cot Z = (\cos a \tan b \csc |\delta\lambda|) - (\sin a \cot |\delta\lambda|)$$

where a and b are the latitudes of points A and B, and $|\delta\lambda|$ is the absolute value of the difference in longitude between A and B.

Example: true azimuth between Washington, D.C. (38°50´N, 77°00´W) and Moscow (55°45´N, 37°37´E).

$$\cot Z = \cos (38.833) \cdot \tan (55.75) \cdot \csc$$
$$(|-77.0 - 37.62|) - \sin (38.833) \cdot \cot$$
$$(|-77.0 - 37.62|)$$

$$\cot Z = 0.779 \cdot 1.469 \cdot 1.100 - 0.627 \cdot -0.458$$

$$\cot Z = 1.259 + 0.287 = 1.546$$

$$\tan Z = 1.0/1.546 = 0.647$$

$$Z = \tan^{-1} (0.647) = 32.9°$$

varies in a predictable manner. Looking at the globe, we see that all quadrilaterals in an east-west band bounded by two parallels, such as 30° and 40°N, and equally spaced meridians have the same surface area. We also see that the areas of quadrilaterals in bands of equal width, such as 0°–10°N, 10°–20°N, and so on, decrease in area from equator to pole. This systematic decrease in surface area (S) may be computed using the equations:

$$S_{\text{sq. km.}} = 6371^2 \cdot (\sin a \cdot \sin b) \cdot \delta\lambda \text{ or}$$

$$S_{\text{sq. mi.}} = 3959^2 \cdot (\sin a \cdot \sin b) \cdot \delta\lambda$$

where a and b are the latitudes of the upper and lower bounding parallels and $\delta\lambda$ is the difference in longitude between the bounding meridians, expressed in radians. Table 4.5 illustrates this systematic decrease for 10° × 10° quadrilaterals.

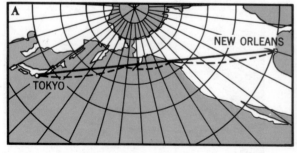 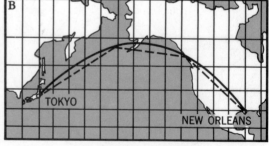

Figure 4.9 Two maps showing the same great circle arcs (solid line) and rhumbs (dashed lines). Map A is a gnomonic map projection in which the great circle arc appears as the shortest distance between Tokyo and New Orleans, that is, as a straight line, while the rhumbs appear as somewhat longer "loops." This is their correct relationship. In map B, a Mercator map projection, the representation has been reversed so that the rhumbs appear as straight lines, with the great circle "deformed" into a longer curve on the map.

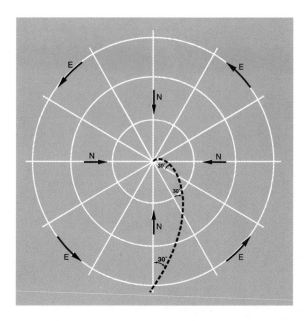

Figure 4.10 A constant heading of 30° will trace out a loxodromic curve.

Cartographers often prepare map projections that either preserve the equality of surface area within latitude bands, or that go further and preserve the areas of all quadrilaterals on the sphere. However, both require the distortion of the shapes of features within the area, as will be explained in the next chapter.

Table 4.5
Surface Area of 10° × 10° Quadrilaterals

Lower Latitude	Area (km²)	Area (mi²)
0°	1,224,480	472,771
10°	1,188,528	458,890
20°	1,117,359	431,412
30°	1,011,480	390,532
40°	875,136	337,889
50°	711,510	274,713
60°	525,312	202,823
70°	322,195	124,399
80°	108,584	41,924

GEODETIC POSITION DETERMINATION

Maps showing individual earth features cannot be made without knowing the latitude, longitude, and often the elevation or depth of each. Before mapping can begin, a network of accurately measured geodetic control points must be defined on the ground. **Geodetic control points** influence the accuracy of all further mapping in their vicinity. Consequently, they are the fixed starting points which surveyors (see Chapter 7), photogrammetrists (see Chapter 12), and others use to determine the two or three dimensional positions of the natural and cultural features that appear on maps.

In this section, we will see how latitude and longitude are determined and then examine the separate horizontal and vertical control networks used in North America. The instruments and data collection methods used to determine the control points are discussed in Chapter 7.

GEODETIC LATITUDE AND LONGITUDE DETERMINATION

Surveyors can use one of several methods to determine **geodetic latitude**. The oldest method (and one still widely used) is to observe the altitude (angle above the horizon) of Polaris, the North Star. The popularity of this ancient method lies in the one-to-one relationship between altitude angle and latitude—that is, the altitude angle equals the geodetic latitude[*] (Figure 4.11A).

This method can be extended to the sun by measuring the sun's altitude at noon, when it is highest in the sky. If the sun were always directly above the equator, geodetic latitude would simply

---

[*]This astronomical measurement method can be extended to any star or planet by using the information found in an ephemeris. Here the altitude of the celestial body is recorded, and appropriate corrections can be made, based on the time of observation, to obtain the geodetic latitude.

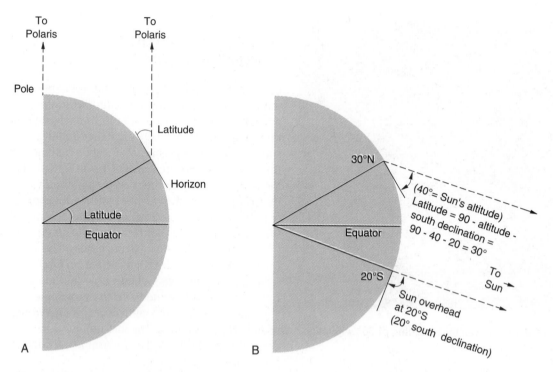

Figure 4.11   Latitude determination through observation of Polaris (A) and the sun (B).

equal 90 degrees minus the noon altitude. However, this only occurs on the fall and spring equinoxes (September 22 or 23 and March 21 or 22). On all other days, the sun's **declination** must be taken into account (Figure 4.11B) by using the equation: latitude = 90° − altitude + declination.

Solar declination is the latitude at which the sun is directly overhead on a particular day. Declinations are listed in nautical almanacs or star almanacs for surveyors, or they may be read from a figure-eight shaped graph called an analemma that is found on many globes.

Longitude determination is basically a time difference problem, since the earth revolves 360° of longitude every 24 hours, or 15° per hour. If, for

example, we know the time at the prime meridian when the sun is at its noon high point at our location, we can find our longitude simply by multiplying the time difference in hours by 15. Hence, Greenwich is important as the international standard starting point for time, called Greenwich Mean Time (GMT). Very accurate clocks called **chronometers** set to GMT made possible the time difference measurements needed to accurately determine longitude.* Today radio transmitters continuously broadcast GMT worldwide, allowing even more accurate longitude calculation.

HORIZONTAL CONTROL NETWORKS

The United States is covered by a network of over 200,000 points of precisely known geodetic latitude and longitude. Until the development of GPS methods, each point was measured on the geoid,

---

*The first chronometer for both land and sea was developed by John Harrison in England during the middle of the 18th century, solving the longitude determination problem.

and then adjusted for its elevation so as to lie on the surface of the ellipsoid. Points generally are spaced three to eight kilometers (two to five miles) apart in urban areas, and three to five times this far apart in rural and mountainous regions. The physical location of each control point is marked by a small bronze **survey monument** sunk into the ground.

Networks have been constructed at various **orders of accuracy**, which are described in Box 4.C. First-order horizontal control data are required for large-scale engineering plans showing exact feature locations. Large-scale topographic and cadastral maps are normally compiled using second-order, class II, or higher accuracy data. Medium to small-scale topographic map compilation requires at least third-order accuracy.

Historically, control networks have been created using the method of triangulation. Triangulation is based on the geometrical principle that if we know one side (base line) and the two adjacent angles in a triangle, we can compute the remaining angle and two sides (see Chapter 7 for more detailed descriptions of triangulation and more modern control point determination methods).

Box 4.C
Orders of Accuracy for Geodetic Control Networks

Horizontal Control

	First-Order	Second-Order		Third-Order	
		Class I	Class II	Class I	Class II
Minimum* Accuracy	1 part in 100,000	1 part in 50,000	1 part in 20,000	1 part in 10,000	1 part in 5,000
General Title	Primary Horizontal Control	Secondary Horizontal Control	Supplemental Horizontal Control	Local Horizontal Control	
Role in National Network	Comprises Network	Strengthen Primary Network	Contribute to Primary Network	Local Surveys Referenced to Network	

Vertical Control

	First-Order		Second-Order		Third-Order
	Class I	Class II	Class I	Class II	
Minimum** Accuracy	$0.5_m mK^{1/2}$	$0.7_m mK^{1/2}$	$1.0_m mK^{1/2}$	$1.3_m mK^{1/2}$	$2.0_m mK^{1/2}$
Role in Network	Comprises Network	Strengthen Primary Network	Secondary Framework For Network	Local Densification	Local Control Tied to Network

*Minimum relative accuracy between directly connected adjacent points.

**Minimum relative accuracy between directly connected points or benchmarks, where K is the distance in kilometers between points.

Furthermore, we can calculate the latitude and longitude of the triangle's third vertex if we know the first two vertexes. A triangulation network is built up by having one or both computed sides in the initial triangle serve as base lines for new triangles, with each new triangle adding a control point to the network .

The United States Coast Survey, later renamed the U.S. Coast and Geodetic Survey, established the first base line in 1833 near the Connecticut– New York boundary. From this base line grew a network that eventually ran from Maine to Louisiana. In the following decades, independent networks were created from New York to the Great Lakes, along the Gulf Coast, from the Atlantic to the Pacific Coast, and along the Pacific Coast. Additional north-south and east-west running networks were soon completed in the United States, Canada, and Mexico, so that by the early 1900s a number of North American networks existed.

In 1927, these separate networks were "adjusted" so as to be integrated into a single network originating at a highly accurate base line (Figure 4.12) beginning at Meades Ranch, Kansas (39°13'26.686"N, 98°32'30.506'W). This new geodetic control network based on the Clarke 1866 ellipsoid became known as the **North American Datum of 1927**. Here the term datum means a set of numerical values that serves as a reference or base for mapping.

The **North American Datum of 1983 (NAD 83)** is an adjustment of the 1927 datum that reflects the higher accuracy of geodetic surveying in the intervening decades, as well as the destruction of survey monuments and horizontal shifts in the earth's crust. Consequently, published coordinates for neighboring first-order control points have been found in error by up to one part in 15,000. This is an unacceptably large error for many large-scale mapping projects.

The new datum was created through a least squares adjustment* of nearly two million geodetic control points in the United States and throughout North America. Not only are geodetic latitude and longitude determined more accu-

rately, the new North American Datum is also tied into a global network through a change to the WGS 84 ellipsoid. Latitude and longitude coordinates for all mapped features have changed slightly. Meades Ranch has lost its importance, since all control points are now determined relative to a single point, the center of the GRS 80 (or WGS 84) ellipsoid. This means that GPS measurements made on this same ellipsoid inherently fit into the new datum (see Chapter 7).

VERTICAL CONTROL

Topographic maps, nautical charts, and many other cartographic products must accurately show the three-dimensional form of the earth's surface. A network of vertical control points with latitude, longitude, and elevation known to high accuracy serves as a reference point for contour line (see Chapter 12) determination. Each vertical control

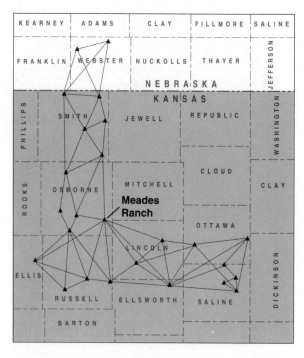

Figure 4.12 Horizontal control network near Meades Ranch, Kansas.

point is marked by a small bronze **bench mark** set into the ground (Figure 4.13).

For centuries, elevations have been measured relative to the surface of the geoid by starting vertical control networks at mean sea level.** Hence, mean sea level serves as the mean datum (starting level) for vertical measurements.

Elevations measured by leveling (see Chapter 7) are not made in isolation, but in a series that form a "level line" between two endpoints, one often a datum point. In the United States, the first level line was run in 1856 from a tidal gauge in New York harbor up the Hudson River to Albany. Over the next 70 years, level lines were run from the east coast to the Great Lakes, down the Atlantic Coast, across the continent to Seattle, and to many other places. In general, these elevation points did not coincide with the horizontal control network points being determined by triangulation. Thus, we currently have two control networks—horizontal and vertical.

With additional mean sea level beginning or ending points, such as Seattle, the vertical control network formed by the interconnected level lines was adjusted (corrected so as to be internally consistent) every few years. This continued until 1929, when 75,000 km. (47,000 mi.) of level lines in the United States and 32,000 km. (20,000 mi.) in Canada, totaling over 500,000 vertical control points, were adjusted based on mean sea level at 21 gauge stations. The result was called the National Geodetic Vertical Datum of 1929.

Since 1929, over 625,000 km. (388,000 mi.) of level lines in the United States alone have been added to the network. The result is a dense set of vertical control points at different orders of accuracy (see Box 4.C). These control points are suit-

Figure 4.13   A bench mark showing the ground location of a vertical control point.

able for map compilation from aerial photography and other data sources.

First-order vertical control is required for engineering plans and other maps showing precise elevation data. Second-order vertical control is sufficient for large-scale topographic and cadastral mapping. Third-order control can be used in medium to small-scale topographic map compilation.

As with horizontal control, advances in measurement techniques, coupled with vertical crustal movements, led to a readjustment of the network in 1983. All previously mapped elevations changed slightly as a result, creating the new National Vertical Datum of 1983.

Elevations are now accurately determined from GPS measurements made relative to the GRS 80 or WGS 84 ellipsoid surface. These ellipsoidal elevations must be adjusted so as to be relative to the mean sea level datum (Figure 4.14). This adjustment involves using the simple equation $H = h - N$,

---

*See the Mikhail and Gracie (1981) reference for a clearly written and profusely illustrated introduction to least squares adjustment of horizontal and vertical survey measurements.

**Mean sea level is the average of high and low tides at a location over a 19-year period. This corresponds with one metonic cycle, the pattern in the moon's movement around the earth that repeats every 235 lunar months.

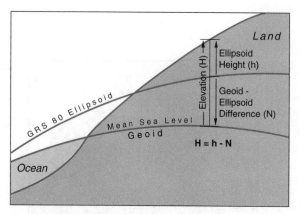

Figure 4.14 The relationship between ellipsoid height, geoid-ellipsoid height difference, and elevation.

where *H* is the elevation, *h* is the ellipsoidal height from the GPS system, and *N* is the vertical difference between the geoid and ellipsoid surfaces at the control point.

SELECTED REFERENCES

Berry, R. M., "History of Geodetic Leveling in the United States," *Surveying and Mapping*, 36 (1976): 137–53.

Burkhard, R. K., *Geodesy for the Layman*, Aeronautical Chart and Information Center (now Defense Mapping Agency, Aerospace Center), St. Louis, Mo., 1968.

Langley, R. B., "Basic Geodesy for GPS," *GPS World*, 3 (1992): 38–43.

Maling, D. H., *Coordinate Systems and Map Projections*, London: George Philip and Son Ltd., 1973.

Meade, B. K., "Latitude, Longitude, and Ellipsoid Height Changes NAD–27 to Predicted NAD–83," *Surveying and Mapping*, 43 (1983): 65–71.

Mikhail, E. M. and G. Gracie, *Analysis and Adjustment of Survey Measurements*, New York: Van Nostrand Reinhold Co., 1981.

Milbert, D. G., "GPS and Geoid 90—The New Level Rod," *GPS World*, 3 (1992): 44–49.

Snyder, J. P., *Map Projections—A Working Manual*, U.S. Geological Survey Professional Paper 1395, Washington, D.C., 1987.

U.S. Naval Observatory, *The American Ephemeris and Nautical Almanac*, Washington, D.C.: U.S. Government Printing Office, issued annually.

Wallis, H. M. and A. H. Robinson eds., *Cartographical Innovations*, Map Collector Publications Ltd., 1987.

CHAPTER 5

MAP PROJECTIONS

As we saw in Chapter 4, the earth is essentially spherical, especially when it is reduced in scale so that we can use it for reference. One simple way of mapping the earth without distortion is to map it on a globe or (if a much larger scale is desired) on a spherical segment of a globe. When we do so, all we change is the size (scale). Relative distances, angles, and areas, as well as azimuths, rhumbs, and great circles, are all retained without any additional distortion. Globes are indispensable for helping us appreciate global strategic and geopolitical relationships.

On the other hand, globes have many practical disadvantages. They are expensive to make, difficult to reproduce, cumbersome to handle, awkward to store, and difficult to measure and draw on. Less than half the globe is visible at any one time (Figure 5.1).

All these drawbacks are eliminated when a map is prepared on a flat surface. Nevertheless, constructing a map on a flat surface does require an important operation in addition to altering scale: The spherical surface must be transformed to a plane (flat) surface. This combination of a scale alteration and a system of transformation results in a map projection.

If we assume that the term "map" encompasses all sorts of photographs, perspective representations, satellite views, side-looking radar images, and so on, then the subject of map projections can involve some complex geometry. In this chapter, we will explore the types of transformations necessary to create a flat map from the earth's curved surface. Such a map, whether it is drawn by hand or by computer, or is created by transforming the perspective geometry of an air photograph, involves a basic assumption—namely, that the viewer has an **orthogonal** (looking straight down) relationship with all parts of the earth's surface and to the map portraying it.

The geometry of the earth's surface discussed in Chapter 4 is important in many ways. We often need to know distances between places; areas of counties, states, and parcels of land; and directions of electronic signals, winds, and headings for

Figure 5.1 Comparison of globe and a flat map of the earth.

navigation. For these and many other kinds of data, we use flat maps instead of globes, and that necessitates a map projection.

A spherical surface cannot be transformed to a plane without modifying the surface geometry. Luckily, there are a great many transformations that retain one, or even several, of the geometric qualities of the globe. Sometimes, however, for a given map no simple geometric property needs to be preserved. Increasingly, statistical properties of the distortions resulting from the transformations are of greatest interest to map users. An extremely useful transformation may be made by minimizing one or more statistical properties of the distortion. For instance, you might minimize the areal distortion for just the land area on a map.

The significance of the geometric qualities that

can be retained depends on the extent of the region being mapped. Some transformation systems are suitable for representations of the whole earth. In that case, our interest is likely to be in global topological relationships rather than azimuths or precise distances. At the other end of the scale range, with maps of small areas, we may be very much concerned with geometric qualities so as to minimize scale variation on the map.

Some transformations can even make the geometry of the map more useful than the globe for certain purposes. A good example is the projection introduced in 1569 by Gerardus Mercator, a Flemish cartographer and mathematician. Mercator's projection helped solve a major problem of early navigators. These 16th-century mariners had a rough idea of where lands were and had the compass to help them. But they had no precise way to determine the bearing of courses that would take them to their destinations. They had to sail along a line of constant bearing because that was the only course they could readily maintain.

Mercator's solution was to transform the spherical surface to a plane so that a straight line on the resulting chart, anywhere in any direction, was a line of constant bearing. To use the chart, mariners needed only draw a straight line (or a series of straight lines) from the starting point to their destination. If they made allowances for currents, winds, and compass declinations, they had a good chance of arriving near their destination.

Mercator's projection demonstrates how a transformation can be useful for one purpose but totally unfit for another. Navigators don't care much about the relative sizes of land areas, but people concerned with geography and world affairs do. In order to show all rhumbs as straight lines, Mercator's projection severely strains the spherical surface, so that regions in middle and higher latitudes appear grossly enlarged. For example, on a Mercator chart Alaska and Brazil appear to be about the same size, but in fact, Brazil is more than five times larger (Figure 5.2). For non-navigational purposes, Mercator's projection has quite unsuitable geometric characteristics.

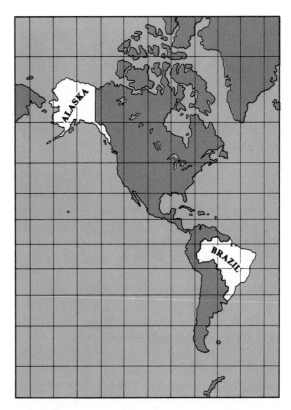

Figure 5.2 How Mercator's projection enlarges areas in higher latitudes. Note that Alaska in North America appears about the same size as Brazil in South America, whereas in fact Brazil is more than five times larger than Alaska.

SCALE FACTOR AND TRANSFOR-MATIONS

The best way to understand how a map projection is created is to see it as a two-stage process. First, assume that the earth has been mapped on a globe reduced to the size (scale) chosen for the flat map. We will call such a hypothetical globe the **reference globe**. Second, suppose that the globe's surface is mathematically transformed, point by point, onto a flat surface. The three-dimensional information on the globe's surface is now displayed on a two-dimensional flat surface.

The reference globe will have a given representative fraction (RF), called the **principal scale**. (See Chapter 6 for a detailed explanation of the RF). To derive the principal scale, we divide the earth's radius by the radius of the globe.

On the reference globe, the actual scale anywhere will be the same as the principal scale. The **scale factor (SF)** is the actual scale divided by the principal scale. Thus, by definition, the scale factor will be 1.0 everywhere on the globe. When all or part of the globe surface is transformed to a flat map, however, the actual scale at various places on the map will be larger or smaller than the principal scale. This is because the sphere and the plane are not **applicable**. That is, one cannot be transformed to the other without stretching, shrinking, or tearing. Consequently, the SF will always vary from place to place on a flat map.

To visualize what happens, imagine a pattern of equally distant points on the reference globe. Then picture corresponding points established on a flat map. The mathematical scheme used to specify positions of the points on the flat map defines the method of transformation. Since the two surfaces are not applicable, distance relationships among the points on the flat map must be modified. Consequently, it is impossible to devise a transformation from the reference globe surface to a plane so that any figure drawn on one will appear exactly the same on the other. Nevertheless, by suitably varying the SF, we can (1) retain some angular relationships, or (2) retain relative sizes of figures. But what if those are not the qualities we want in a projection? What if we want some other attribute, such as straight-line azimuths from one point to all others? In that case, most angular relationships will usually be changed, and areas of regions on the two surfaces will not have a constant ratio to each other.

To understand how these facts apply to map projections, keep in mind two important details: (1) SF values may occur at a point. (2) SF values may be different in different directions at a point. Let's look more closely at each of these elements.

To visualize the first of these propositions, pic-

ture an arc of 90°, projected orthographically (at right angles) to a straight line tangent at a. (See Figure 5.3). Imagine that a, b, c,...j are the positions on the arc at 10° intervals. Their respective positions after transformation to the line tangent at a are indicated by points a, b', c',...j'. Line aj' therefore represents arc aj. As Figure 5.3 shows, the intervals on the straight line, starting at the point of tangency, a, become progressively smaller as j' is approached. If SF = 1.0 along the arc, then on the transformation of the arc, the SF is gradually reduced from 1.0 at a to 0.0 at j'. The rate of change is graphically indicated by decreasing the spaces between points. Since it is a continuous change, every point on aj' must have a different SF.

Now let's explore the second proposition—that scale at a point may be different in different directions. To visualize this fact, picture a rectangle $abcd$. Imagine that it is projected to $ab'c'd$ so that side ad coincides in $abcd$ and its projection $ab'c'd$. (See Figure 5.4A).

Next, take a look at Figure 5.4B. It is an orthogonal view of rectangle $abcd$ and its projection superimposed with line ad of each coincident. If the SF of $abcd$ is assumed to be 1.0 and the length ad is the

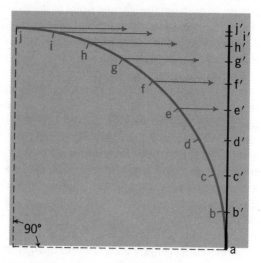

Figure 5.3 Orthographic projection of an arc to a tangent straight line.

same in each rectangle, there has been no change in scale in that direction. However, the length *ab´* is half the length *ab*, and it is evident from the method of transformation that the change has been made in a uniform fashion. Thus, the SF along *ab´* must be 0.5. Furthermore, by projection, line *ac* has become line *ac´*. The ratio of lengths *ac´* to *ac* constitutes the SF along *ac´*. It is clear that this ratio is neither the 1.0 ratio along *ad* nor the 0.5 ratio along *ab´*; it is somewhere in between. Any other diagonal from point *a* to a position on side *bc* would have its corresponding place of intersection on *b´c´*. The ratio of lengths of similar diagonals on the two rectangles would be different for each such line. Hence, the scale at point *a* in rectangle *ab´c´d* is different in every direction.

Those two propositions—that SF values may occur at a point and that SF values may be different in different directions at a point—are very important. They provide the basis for analyzing how a projection has transformed distances, directions, angles, and areas on the sphere.

On the reference globe, there is at each point an infinite number of paired orthogonal directions, such as N-S with E-W, NE-SW with NW-SE, and so on. When transformation to the plane is made, the paired orthogonal directions on the globe will be represented by paired directions on the map projection. However, these pairs will not necessarily remain orthogonal.

A theorem formulated by the 19th century French mathematician, A. Tissot, states that *whatever the system of transformation, there is at each point on the spherical surface at least one pair of orthogonal directions which will also be orthogonal on the projection.* The paired directions retained as orthogonal on the projection may be called the **principal directions**. What we are primarily interested in knowing is not precisely what those principal directions are. Rather, our prime concern is the SF in the principal directions at the point. The SF will usually be more or less than 1.0 in different directions at each point on a projection. Tissot proved that the maximum deviations of the SF will be in the two principal directions. The greater value of the SF deviation is denoted *a* and the lesser, *b*. For a globe, therefore, $a = b = 1.0$. With those values, you can calculate the angular and area distortions caused by the system of transformation at any point. (See Box 5.A).

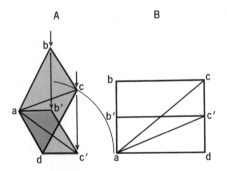

Figure 5.4 Projection of rectangle *abcd* to rectangle *ab´c´d* with side *ad* held constant. In drawing (A) the perspective view shows the geometric relation of the two rectangles. Drawing (B) shows the relation of the two rectangles when they are each viewed orthogonally.

DISTORTIONS RESULTING FROM MAP TRANSFORMATIONS

Whenever the spherical surface is transformed to a plane, one thing is certain. All of the geometrical relationships on the sphere, such as parallel parallels, converging meridians, and perpendicular intersection of parallels and meridians, can not be entirely duplicated. The major alterations have to do with angles, areas, distances, and directions. Therefore, let's take a closer look at these important distortions.

TRANSFORMATION OF ANGLES

The compass rose appears the same at each point on the globe's surface (except at the poles). That is, at each point, the cardinal directions are always 90° apart. It is possible to retain this property of

Box 5.A

Tissot's Indicatrix

Tissot used a graphic device, the **indicatrix,** to illustrate the angular and areal distortions that occur at points as a result of transformation. At any point on the globe, the SF is, by definition, the same in every direction. Therefore, SF = a = b = 1.0.

Tissot represented a point by an infinitely small circle with a radius of 1.0. He demonstrated that in any system of transformation, the magnitudes of a and b will ordinarily be different than 1.0—except for standard points or points lying on a standard line. When a is not equal to b, the indicatrix circle is transformed into an ellipse, with a as the semi-major axis and b as the semi-minor axis. By analyzing geometric changes in the indicatrix that occur when the circle is transformed into an ellipse, we can determine how much angular and/or areal distortion has occurred at any point.

As an example, an indicatrix is shown in Figure 5.A.1. The gray circle represents a point on the globe. The SF in every direction is 1.0, so the circle is constructed with $OA = 1 =$ the radius r. In the transformations, the directions OA and OB are the principal directions—that is, the pair of orthogonal directions on the globe that are retained as orthogonal by the transformation. In the illustration, the circle of the indicatrix is transformed into the tinted ellipse with a semi-major axis of $OA' = a = 1.25$, indicating that the SF has been increased in that direction, and with a semi-minor axis of $OB'' = b = 0.80$, showing a reduction of scale in that direction. Such a transformation would be termed **equal-area**, since the product $ab = (1.25) \times (0.8) = 1.0$. Because a does not equal b, conformality has been lost.

Alternatively, consider the gray circle representing the point O on the reference globe. Imagine that this gray circle is being transformed

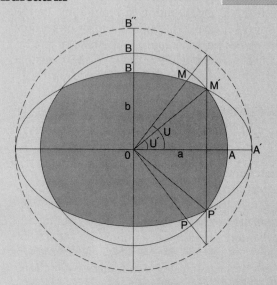

Figure 5.A.1 The indicatrix shows a transformation of the circle to which the following indexes apply: $OA = OB = 1.0$. The corresponding indexes of the ellipse are: $OA' = a = 1.25$, and $OB' = b = 0.80$. For the larger dashed circle, the indexes are $OA' = a = 1.25$, and $OB'' = b = 1.25$.

into the dashed outer circle in the illustration. The semi-major axis $OA' = a = 1.25$, and the semi-minor axis $OB' = b = 1.25$. The product $ab = (1.25) \times (1.25) = 1.5625$. Such a transformation would not be equal-area but would be **conformal,** since $a = b$. The area in a minute vicinity of point O would be enlarged on the projection by a factor of 1.5625.

Now let's use Tissot's indicatrix to analyze the angular distortion at point O. First, note that all points on the circumference of the circle in Figure 5.A.1 have their counterparts on the ellipse. Considering the tinted ellipse resulting from the equal-area transformation, point A has been shifted to point A', B to B', M to M', and P

Box 5.A (continued)

to $P´$ as a result of the transformation. By definition, no angular deformation has taken place in shifting points A and B, since angle BOA = angle $B´OA´$, the angle formed by the principal directions. All other points on the arc between B and A will, when projected to the ellipse, be shifted a greater or lesser amount in their direction from point O. The point subjected to the greatest distortion is identified in the circumference of the circle by M, and it has its counterpart on the ellipse in point $M´$. The angle MOA on the globe thus becomes $M´OA´$ on the projection. If we denote angle $MOA = U$ and angle $M´OA´ = U´$, then $U-U´$ denotes the maximum angular distortion within one quadrant. The value of $U - U´$ is designated as **omega (Ω)**.

Next, imagine that an angle such as MOP has its sides located in two quadrants and that they occupy the position of maximum change in both directions. The angle in question would then be changed to $M´OP´$. The angle would thus incur the maximum deflection for each quadrant. Consequently, the maximum possible angular distortion that may occur becomes 2Ω. Since the values of 2Ω will range from 0 in the principal directions to a maximum somewhere between them, it is not possible to state

an average angular distortion. The scale distortion along the principal directions can be used to calculate values for Ω. Note the example given in Figure 5.A.1. When $a = 1.25$ and $b = 0.80$, then $\Omega = 12° 42´ 05´´$. The maximum distortion, 2Ω, would equal $25° 24´ 10´´$.

Changes in the representation of areas likewise may occur as a consequence of transformation. Refer again to Figure 5.A.1. Consider the outer dashed circle resulting from the conformal transformation. You can readily establish the magnitude of area change by comparing the area of the gray circle from the reference globe with that of the outer dashed circle. Clearly, the original circle has been enlarged to the dashed circle by the transformation. The gray circle's radius is 1.0. The radius of the outer circle is 1.25. The area change is {(1.25) × (1.25)}/1.0 = 1.5625.

In general, the area of a circle is πr^2, whereas the area of an ellipse is $ab\pi$. Therefore, since the ellipse axes are based on the original circle, whose radius was unity, and since π is constant, the product ab compared to unity (one) expresses how much the representation of the area has been changed. The product of ab is usually designed as **S**.

angular relations on a map projection. When correct angles are retained, the projection is termed conformal or orthomorphic. Both words imply "correct form or shape." It is important to understand that these terms apply to the directions or angles that occur at points in infinitesimally small areas. This attribute of correct shape does not apply to regions of any significant size.

On the reference globe, by definition the SF is 1.0 in every direction at every point. In any transformation, distortion of some kind must occur. Thus, the SF

must either vary from point to point or have different values in different directions about a point. It is possible to arrange the stretching and compression so that at each point on a projection the SF is constant in every direction at every point, (that is, $a = b$ and the projection is conformal), but then the SF will not be equal to 1.0 at every point (that is, the projection won't be equal-area). When this condition occurs, all directions around each point will be represented correctly, and the parallels and meridians will intersect at 90°. It must be emphasized that this desirable quality is limited to directions at points and does not necessarily apply to directions between distant points on the projection. It is also important to realize that just because a projection shows perpendicular parallels and meridians, it does not necessarily have the property of conformality.

TRANSFORMATION OF AREAS

It is also possible on a map projection to retain representation of areas so that all regions will be shown in correct relative size. Such a projection is said to be equal-area or equivalent. This property is obtained by arranging the SF in the principal directions so that the product of the SFs equals 1.0 (that is, $ab = 1.0$) at every point. The SF varies in every other direction about a point (that is, the projection is not conformal).

It is evident that scale requirements for conformality ($a = b$) and for equivalence, ($ab = 1.0$ for all points) are contradictory. (Theoretically, this condition, a = b = 1.0, is met at standard points or at every point on a standard line as noted in the following section). Consequently, no map projection can be both conformal and equivalent. Thus, all conformal transformations will present similar earth regions with unequal sizes, and all equal-area transformations will deform most earth angles.

TRANSFORMATION OF DISTANCES

Representing distance correctly is a matter of maintaining consistency of scale. That is, for a distance between two points to be truly represented on a map projection, the scale must be uniform along the entire extent of the line joining the two points. The scale must also be the same as the principal scale on the reference globe from which the transformation was made. We have two options when representing distance on a map:

1. A scale factor of 1.0 may be maintained along one or more parallel lines, but only along the lines. When this is done, such lines are called **standard lines** or **standard parallels**.

2. A scale factor of 1.0 may be maintained in all directions from one or two points, but only from those points. The resulting map projections are called **equidistant** and the points are called **standard points**.

TRANSFORMATION OF DIRECTIONS

Just as it is impossible to represent all earth distances with a consistent map scale, it is also impossible to represent all earth directions correctly with straight lines. It is true that we can arrange the SF distribution so as to show rhumbs or arcs of great circles as straight lines. But no projection can show direction so that all great circles are straight lines with the same angular relations to the map graticule that they have with the globe graticule. For example, the oft-stated assertion that Mercator's projection "shows true direction" applies only to the fact that constant bearings are shown as straight lines. Such an assertion is erroneous, since true direction on a sphere is along a great circle, not along a rhumb (Figure 5.5).

We can think of correct direction on a map projection as being a great circle shown as a straight line. At its starting point on both reference globe and map projection, this line will set out from the meridian at the same azimuth (or angle). Given these conditions, several representations are possible:

1. Great circle arcs between all points may be shown by straight lines for a very limited area. However, angular intersections of great circles with meridians (azimuths) will not be shown correctly. To do this causes such a strain on the transformation process that it is not possible to extend it to even a hemisphere. (See the discussion of the gnomonic projection later in this chapter.)

2. Great circle arcs with correct azimuths may be shown as straight lines for all directions from *one* or, at the most, *two* points. Such projections are called *azimuthal*.

ANALYSIS AND VISUALIZATION OF DISTORTION

There are several ways to compare map projections in terms of amount and distribution of distortion. Some comparisons are entirely graphic and provide a visual display of the area and/or amounts of deformation. Others are mathematically involved and result in a quantitative measure of the distortion. Often a visual pattern of a quantitative measure of distortion is used.

As we learned in Chapter 4, a sphere has a uniform surface. To identify locations on this unvarying sphere, we use the graticule—the earth's coordinate system of latitude and longitude. The graticule is so useful that we often include it automatically when we think of the earth's surface. Consequently, we tend to identify projections by the way they represent the graticule. Furthermore, because the cardinal directions are so important to our lives, we ordinarily

arrange the graticule by making east-west parallels or north-south meridians appear as straight lines or arcs.

The graticule merely provides a handy series of reference points, however. No matter how we position the transformation with respect to the reference globe, the distortion characteristics of the projection are the same. The pattern and amounts of distortion do not change. Figure 5.6 shows how the graticule appears when the transformation for the sinusoidal projection is centered at various places. The resulting projections are all the result of the same transformation and have the same pattern of distortion.

Conventionally, we transform the reference globe so that standard points or lines on the projection coincide with prominent points or lines of the graticule. For instance, standard great circles may coincide with the equator or meridians. Or standard small circles may coincide with parallels. This results in the graticule being displayed in a simple pattern. Remember, however, that such an arrangement of the transformation is a convenience, not a necessity.

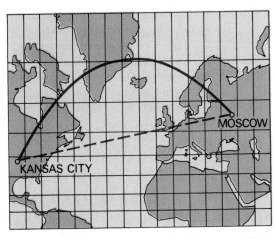

Figure 5.5 The great circle (solid line) and the rhumb (dashed line) between Kansas City in the United States and Moscow in Russia as they appear on Mercator's projection. The great circle arc shown is the "true direction," since it is the shortest route from Kansas City to Moscow.

VISUAL ANALYSES

You can evaluate many projections effectively just by "looking"—by comparing the graticule on the reference globe with the way it appears on the projection. You may find the following list helpful in making such an evaluation. It outlines important visual characteristics of the earth's coordinate system as portrayed on a globe:

1. Parallels are parallel.

2. Parallels when shown at a constant interval are spaced equally on meridians.

3. Meridians and great circles on a globe appear as straight lines when viewed orthogonally (looking straight down), which is the way we look at a flat map.

4. Meridians converge toward the poles and diverge toward the equator.

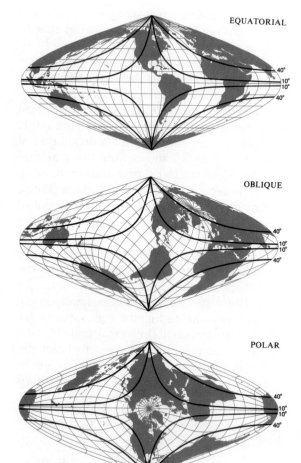

EQUATORIAL

OBLIQUE

POLAR

Figure 5.6 Different centerings of the sinusoidal projection produce different appearing graticules. Nevertheless, the arrangement or pattern of the deformation is the same on all, since the same system of transformation is employed. Isarithms of distortion are 2Ω.

5. Meridians when shown at a constant interval are equally spaced on the parallels, but their spacing decreases from the equator to the pole.

6. When both are shown with the same intervals, meridians and parallels are equally spaced at or near the equator.

7. When both are shown with the same intervals, meridians at 60° latitude are half as far apart as parallels.

8. Parallels and meridians always intersect at right angles.

9. The surface area bounded by any two parallels and two meridians (a given distance apart) is the same anywhere between the same parallels.

QUANTITATIVE MEASURES

Using values from Tissot (see Box 5.A) on all conformal projections, $a = b$ everywhere on the projection. When $a = b$, the value of 2Ω is 0°. Hence there is no angular deformation at points on conformal projections. Because the values of a and b vary from place to place, however, the product of ab will also vary from place to place. Consequently, all conformal transformations enlarge or reduce areas relative to one another. The product ab at various points provides an index of the degree of areal change.

On equal-area projections, the scale relationships at each point are such that the product of ab always equals 1.0. Any difference between the values of a and b produces a value of 2Ω greater than 0°. Consequently, all equal-area projections deform angles. The value of 2Ω at various points provides an index of the degree of angular distortion.

On projections that are neither conformal nor equal-area, a will not equal b. Nor will the product of $ab = 1.0$. Therefore, on such projections, both the values of S and 2Ω will vary from place to place and serve as quantitative measures of the distortion.

GRAPHIC PORTRAYAL OF DISTORTIONS

Equal values of Tissot's indicatrix for a transformation can be connected by lines to create a visual pattern of the distortion. (See Figure 5.6). Similarly, a small circle or ellipse scaled to the values of a and b can denote the values of distortion at selected points on a map. These techniques are useful in making geometric analyses of the properties of map projection grids. They are not used, however, when geographic information is being mapped. Since you won't see these distortion indicators on most maps, you will find it useful to learn to visualize them. This will help you (1) select a suitable map projection and (2) evaluate information already in a map format.

You can also derive a mean value for the distortion, either for the entire projection or for a portion of it, such as land areas. You can do so by statistically calculating values at a series of points. A comparison of mean values for several similar projections is helpful in evaluating their relative merits.

Tissot's method of analysis is limited to values at a point. It is not much help in evaluating the inevitable distortion of distance and angular relations among widely spaced points. Nor is it helpful in evaluating distortion of areas over large portions of the earth, such as continents. To identify this type of distortion, you can use various graphic devices. For instance, you can plot the outline of a human head on different projections. Since you're familiar with a head's shape, this outline helps you detect distortions of shapes in the projection (Figure 5.7). Another way to help visualize shape distortions is to cover the globe with nearly equilateral triangles or equally spaced dots and then reproduce the same shapes on the different projections. Likewise, you can show distortions of directions by plotting various great circles in different parts of a projection (Figure 5.8).

CHOOSING A MAP PROJECTION

Cartographers need to be thoroughly familiar with map projections. They must understand the effects different transformations have on the representation of angles, areas, distances, and directions. Only then can they make proper allowances when making measurements or analyses on maps. (For example, never measure areas on Mercator's projection!)

Furthermore, cartographers frequently transfer data from one projection to another. Knowing the distortion characteristics of each is necessary to maintain accuracy during the transfer. Computers can be of great aid in this data transfer process.

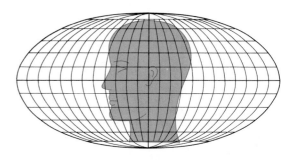

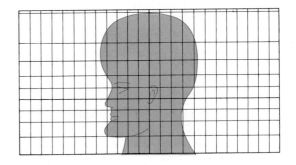

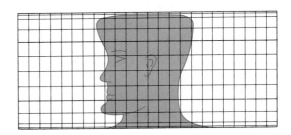

Figure 5.7 A head drawn on the Mollweide projection (top) has been transferred to Mercator's projection (center) and to the cylindrical equal-area projection with standard parallels at 30° (bottom). Just because the profile looks most natural on Mollweide's projection, that projection is not necessarily "better." The natural profile could have been drawn on any projection and then plotted on the others.

Cartographers must also choose the best projection for a particular map or map series. This task requires fitting the characteristics of a projection to the maps' objectives.

Until recently, cartographers were also responsible for calculating and drawing projections—both tedious tasks. Happily, computers and plotters can now complete these chores at relatively low cost. The ease with which such operations can be performed enhances the cartographer's primary task—selecting the proper projection. The greater the number of options, the more important is the choice.

Many diverse factors may influence the choice of map projection. Geographers, historians, and ecologists are likely to be concerned with relative sizes of regions. Navigators, meteorologists, astronauts, and engineers are generally concerned with angles and distances. The atlas mapmaker often wants a compromise, while the illustrator is usually limited by a prescribed format. The maker of a series of maps is interested in how the individual maps may be made to fit together well.

Because all map projections involve distortion, there is a tendency to think of them as necessary but poor substitutes for the globe surface. On the contrary, projections have many advantages over globes. Projections let us map distributions and convey concepts that would be undesirable or impossible on a globe. Furthermore, the notion that one projection is by nature better than another is unwarranted. There are systems of projection for which no useful purpose is now known, but there

is no such thing as a bad projection—there are only good and poor choices.

You should keep several guidelines in mind when selecting a transformation to create your map projection. The first thing to consider is the projection's major property, such as conformality, equivalence, azimuthality, reasonable appearance, and so on. Projection attributes such as parallel parallels, localized area distortion, and rectangular coordinates may also contribute to a map's success. For example, a small-scale map of temperature distributions over large areas will be more effective if the parallels are parallel. The map will be even more expressive if the parallels are straight lines that allow for easy north-south comparisons. This is because temperatures normally decrease with increasing latitude.

A second important element is the amount and arrangement of distortion. Mean distortion, either maximum angular (2Ω) or area (S), is the weighted arithmetic mean of the values that occur at points over the projection. When derived for similar areas on different projections, a comparison of the mean distortion values provides one index of the relative efficiency of the projections. Therefore, a good match between the shape of the region being mapped and the shape of the area of low distortion on the projection is desirable. Certain general classes of projections have specific arrangements of the distortion (see Box 5.B). Knowing these patterns helps considerably in choosing and using a particular system.

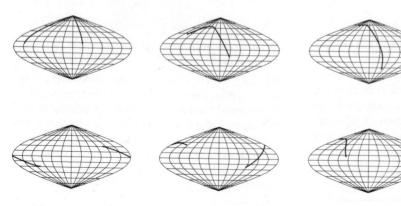

Figure 5.8 Selected great circle arcs on an equatorial case of the sinusoidal projection showing their departures from straight lines. Each uninterrupted and interrupted arc is 150° long. (Courtesy W.R. Tobler.)

Box 5.B
Common Distortion Patterns on Map Projections

The pattern of distortion is the arrangement of either S or 2Ω values on a projection. To visualize this pattern, think of the quantities as representing the z values of an S or 2Ω third dimension above the surface. Then isarithms (contours) (see Chapter 26) of this surface will show the arrangement of the relative values. By their spacing, the isarithms will indicate the gradients or rates of change. Certain classes of projections have similar patterns of distortion. These are diagrammed in Figures 5.B.1 to 5.B.4. Within the shaded areas, the heavy lines are the standard lines. The darker the shading, the greater the deformation.

An azimuthal pattern occurs if the transfor-mation takes place from the globe to a tangent or an intersecting (secant) plane. The trace of the intersecting plane and sphere will, of course, be circular. Lines of equal distortion are concentric around the point of tangency or the center of the circle of intersection (see Figure 5.B.1).

A conical pattern results if the initial transformation is made to the surface of a true cone tangent at a small circle or intersecting at two small circles on the globe. Lines of equal distortion parallel the standard small circles (see Figure 5.B.2).

A cylindrical pattern occurs on all map projections that, in principle, are developed by first transforming the spherical surface to a

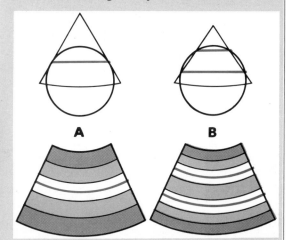

Figure 5.B.1 Azimuthal patterns of deformation. (A) The pattern when the plane is tangent to the sphere at a point, and (B) the pattern when the plane intersects the sphere. In B, the trace of the intersection will be a small circle on the globe, not necessarily a parallel of latitude. The standard point in A and the standard line in B, where the SF = 1.0 are shown in red.

Figure 5.B.2 Conical patterns of deformation. (A) The pattern when the cone is tangent to one small circle, and (B) the pattern when the cone intersects the sphere along two small circles. The small circles may or may not be parallels of latitude. Standard lines, where the SF = 1.0, are shown in red.

continued on next page

Box 5.B (continued)

tangent or secant cylinder. In all cases the lines of equal distortion are straight lines parallel to the standard lines (see Figure 5.B.3). Distortion will increase away from the standard lines, and the greatest gradient will be in the directions normal (at right angles) to the standard line.

A fourth group of projections is used mainly to portray the entire world. These projections have patterns of distortion less systematic than azimuthal, conical, or cylindrical patterns. In the conventional or equatorial case, these projections usually have straight-line parallels that vary in length, with curved meridians equally spaced along them. Dozens of such projections have been proposed. Some are widely known and often used. None are conformal or azimuthal, but many are equal-area.

The pattern of angular distortion (2Ω) varies among these projections. As you might expect, it increases away from the central areas. In the equatorial case, maximum distortion is in the polar regions. (See Figure 5.B.4). As with all projections, the graticule may be arranged independently of the pattern of distortion.

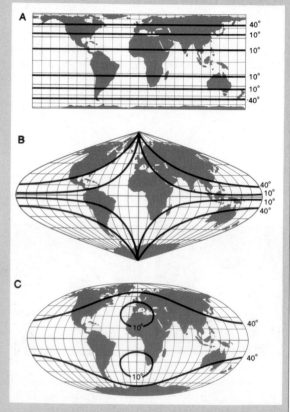

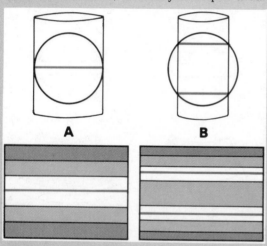

Figure 5.B.3 Cylindrical patterns of deformation. (A) The pattern when the cylinder is tangent to a great circle resulting in one standard line, and (B) the pattern when the cylinder is secant and the standard lines are two parallel small circles. The tangent great circle or the parallel small circles, shown in red, need not be a meridian or parallels of latitude.

Figure 5.B.4 A few of the many equivalent world map projections. (A) cylindrical equal-area with standard parallels at 30°N and S latitude; (B) sinusoidal projection; and (C) Mollweide's projection. The black lines are values of 2Ω.

Maps to be made in series, such as sets for atlases or topographic series, have special projection requirements. For such map series, you will want to choose a projection that shows the same pattern of distortion for large areas as for small areas. Thus, the larger-scale maps in the series will have the same geometric characteristics as the small-scale maps in the series. Most projections in which the meridians are straight lines that meet the parallels at right angles satisfy this requirement.

The overall shape of an area on a projection may be of great importance. Many times the format (shape and size) of the page or sheet on which the map is to be made is prescribed. By using a projection that fits a format most efficiently, you can often increase the scale considerably. This may be a real asset to a map with many details.

You can also use the computer to create map projections. To do so, you specify the mathemati-

cal and statistical criteria that you want the transformation to achieve. Based on indicatrix values, you can specify that distortion be minimized for the specific region of interest. For example, you might instruct the computer to create a transformation that results in minimum areal distortion for Alaska. (See Figure 5.9) Or you could request a transformation that minimizes error within a map of the 48 conterminous United States. (See Figure 5.10).

There are a great many projections that can only be classed as "miscellaneous," since each tends to be unique. Many were devised for special purposes and then put to other, sometimes inappropriate uses. Some projections with no great merit, such as Gall's cylindrical, Van der Grinten's, or the Peter's cylindrical, for some reason caught the fancy of mapmakers for awhile and were often used when some other projection

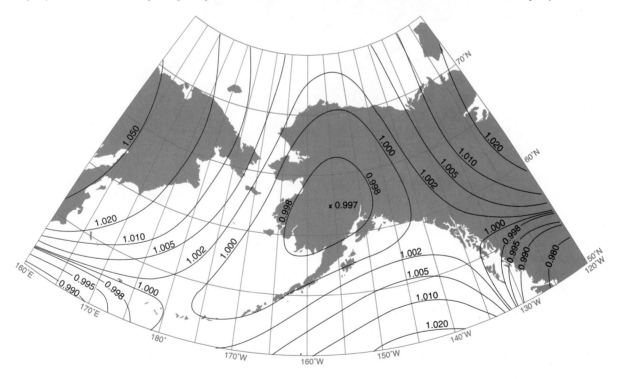

Figure 5.9 Modified-stereographic conformal projection of Alaska, with lines of constant scale superimposed. Scale factors for Alaska range from 0.997 to 1.003, one-fourth the range for a corresponding conic projection.

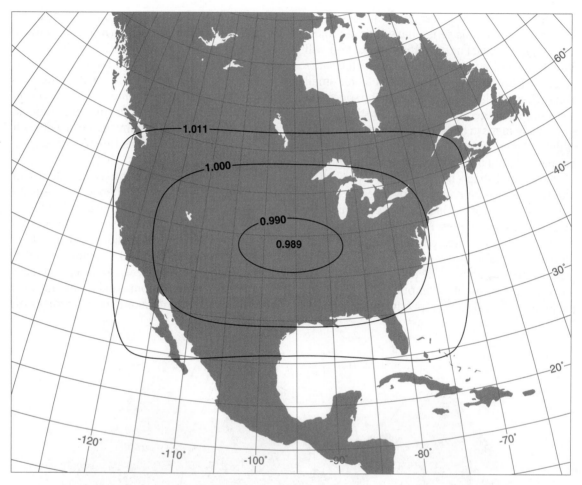

Figure 5.10 Modified-stereographic conformal projection of 48 United States, bounded by a near rectangular area of constant scale. Three lines of constant scale are superimposed. Region inside near rectangle has minimum error.

may have been a better choice. The ease with which cartographers can now create projections will undoubtedly continue these cartographic projection fads (See Box 5.C.)

COMMONLY USED MAP PROJECTIONS

As we have seen, all projections of the earth to a flat map involve spatial distortion. But there is a great deal of variation in the nature and pattern of this distortion between projections. Let's consider some of the important spatial properties you will find on different map projections.

CONFORMAL PROJECTIONS

Maps to be used for analyzing, guiding, or recording motion and angular relationships require the use of **conformal** projections. In these categories fall the navigational charts of the mariner or aviator, the plotting and analysis charts of the meteorologist, and the general class of topographic maps.

The concept of distortion on a conformal projection is a matter of definition. Since all that changes is the SF, one point is as "accurate" as another. Only the scales are different. Thus we can refer to the principal scale as "correct," while other parts of the projection will be relatively enlarged or reduced. On the other hand, variation in S is obviously distortion. The systematic change of S from place to place can result in noticeable distor-

---

Box 5.C
Software for Map Projections

There are many software packages for map projections available at reasonable cost. Improvements in versatility, speed, and ease of use are occurring almost daily. While we cannot name them all, we will describe one to illustrate the type of package that is available. Other packages are competing successfully with this one, and by the time this book is published new packages will undoubtedly be available.

Geocart, available through Terra Data Inc., Bramblebush, Croton-on-Hudson, NY 10520, provides the following capabilities:

- nine interruption schemes and over 100 map projections—user may add more
- 40,000 point composite world database with coastlines and international borders
- Based on sphere or custom or predefined ellipsoids with auxiliary latitudes
- Color, pattern, and weight specification for each line type
- Latitude/longitude coordinate calculation of mouse position for any projection
- Areal, angular, scalar, and distance distortion calculation at any point; Tissot indicatrix plotting
- RMS scale error computation of region within map boundaries
- Specifiable graticule resolution, including polar meridian pruning

- Automatic perfect detail selection and smooth plotting at any resolution, regardless of map properties
- Handles arbitrarily complex projections, including any interruption or aspect of any projection
- Discrete databases of coastlines, islands, hydrography, international borders, U.S. state and county boundaries
- Accepts your custom text databases, as well as MAPMAKER text and native latitude/longitude files
- Up to 1,000 databases per map, 20 maps per document, and any number of documents open at one time
- Export maps to Adobe Illustrator©, PICT, or EPSF files, also copy to clipboard in bitmap or PICT format
- Print to any printer or plotter compatible with Macintosh
- Loxodrome and great-circle plotting
- System 7 compatible
- Includes manual, *An Album of Map Projections* and *An Introduction to Map Projections*

It is easy to see that the job of creating a map projection has been greatly simplified. (See Figure 5.C.1). On the other hand, the job of selecting a map projection has been made more difficult.

continued on next page

Box 5.C (continued)

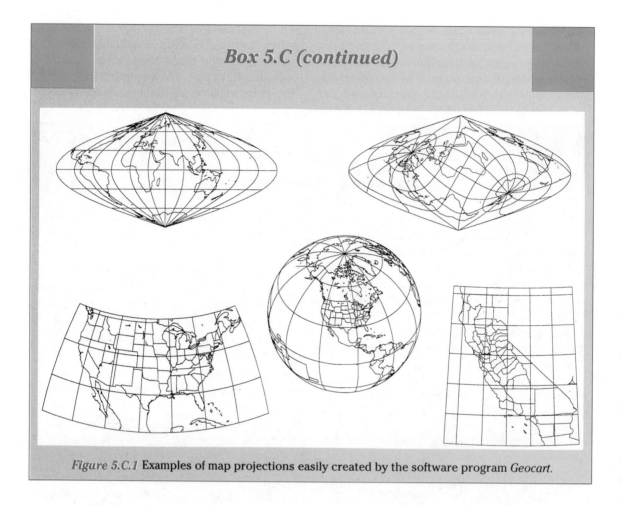

Figure 5.C.1 Examples of map projections easily created by the software program *Geocart*.

tion of area from one part of the map to another.

There are four conformal projections in common use: the Mercator, the transverse Mercator, Lambert's conformal conic with two standard parallels, and the stereographic azimuthal. Mercator's projection is the best known, because of its long history of usefulness to mariners. The stereographic projection, familiar to the classical Greeks, is also widely used. In recent times Lambert's conformal conic and the transverse form of Mercator's projection have become more popular among cartographers. (See Chapter 6.)

The **Mercator** projection (Figure 5.11) is probably the most famous map projection ever devised. As we discussed earlier, Mercator intro-

duced it specifically for nautical navigation, and it has served this purpose well. It has the property that all rhumbs appear as straight lines. Except for the meridians and equator, great circle arcs do not appear as straight lines. Thus, Mercator's projection does not show "true direction"; but such courses can be easily transferred from a projection (gnomonic) that does so. Figure 4.9 in Chapter 4 illustrates that the great-circle course can be approximated by a series of straight rhumbs.

In the normal form of Mercator's projection, the standard line is the equator along which the SF is 1.0. The rate of change of the SF values is relatively small for the lower latitudes, meaning that a zone

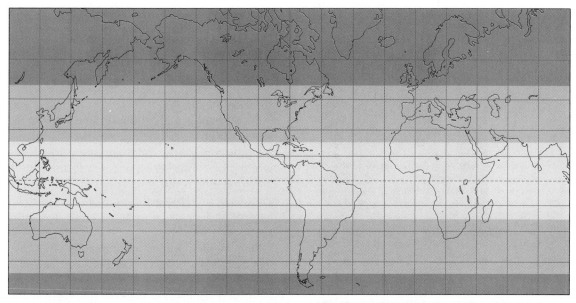

Figure 5.11 Mercator's projection. Values of lines of equal-scale exaggeration compared to the value $S = 1.0$ at the equator are 25 and 250 percent.

along the equator is well represented. For example, even at 10° latitude the SF would only be about 1.016. As is apparent in Figure 5.11, however, Mercator's projection enlarges areas at a rapidly increasing rate toward the higher latitudes, so it is of little use for purposes other than navigation.

Since the earth is spherical and the equator is like any other great circle, we can rotate Mercator's projection (or the sphere) 90° so that the standard line becomes a meridian (a great circle) that takes the place of the equator (Figure 5.12). When this is done, the projection is called the transverse Mercator.

The **transverse Mercator** projection is conformal, but it does not have the attribute that all rhumbs are straight lines. Consequently, since scale exaggeration increases away from the standard meridian, this projection is useful for only a small zone along that central meridian (Figure 5.12). In the secant case of the transverse Mercator projection, lines of equal-scale difference (parallels of the graticule in the normal aspect of Mercator's projection) are small circles parallel to the central meridian. (See Box 5.B.)

Thus, the earth is often represented by a series of narrow east-west but long north-south strips, each strip on a single secant case transverse Mercator map projection. In recent years the transverse Mercator projection has been widely used for topographic maps (for example in the United Kingdom and the United States) and as a base for the Universal Transverse Mercator (UTM) plane coordinate system. (See Chapter 6.)

Lambert's conformal conic projection with two standard parallels in its normal form has concentric parallels and equally spaced, straight meridians that meet the parallels at right angles (Figure 5.13). The SF is <1.0 between the standard parallels and >1.0 outside them. Area distortion between and near the standard parallels is relatively small. Thus the projection provides exceptionally good directional and shape relationships for an east-west mid-latitudinal zone (for example, the 48 conterminous states of the United States). Consequently, it is used for air navigation, topographic maps and for meteorological charts in these mid-latitudes.

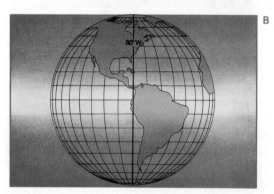

Figure 5.12 (A) The conceptual cylinder for the normal form of Mercator's projection is arranged parallel to the axis of the sphere, resulting in the equator (0°) being tangent and thus the standard line. (B) To develop the transverse Mercator projection, the cylinder is turned so that a meridional "great circle" is tangent and a meridian, in this case 80°W, becomes the standard line.

The **conformal stereographic** projection (Figure 5.20A) is in the azimuthal group. The distortion is arranged symmetrically around the center point. This is an advantage when the shape of the area to be represented is more or less compact.

EQUAL-AREA PROJECTIONS

For most maps used for instruction and for small-scale general reference maps, the property of equal area commands high priority. The comparative extent of geographical areas is of obvious significance on the maps.

Because the earth's entire surface area can be plotted within the bounding lines of a plane figure of almost any shape, there are many equal-area "world projections." For a world map, the pattern of deformation is the chief concern. The "better" portions of such projections can also be used for maps of continents or even for areas of smaller extent.

Many of our impressions of regions' sizes have been acquired subconsciously. Because non-equal-area projections are so frequently used for instructional and small-scale general reference maps, most people have developed erroneous conceptions of regions' comparative sizes. For instance, many people assume Greenland is much larger than Mexico when it is nearly the same size. They also believe that Africa is smaller than North America when, in reality, Africa is nearly 6 million sq. km. larger. These false impressions come from educators' use of projections not well suited to the task.

For some maps, the property of equivalence is more than a passive factor influenced by preconceptions. Some kinds of symbolization require equivalence in order to portray relative densities properly. For example, Figure 5.14 shows a map on which equivalence is mandatory.

As a rule, the smaller the region to be represented, the less significant is the choice of an equal-area projection. There are many ways to transform the reference globe to a plane while maintaining equivalence. Thus, there are more well-known equal-area projections than azimuthal and conformal projections. The choice of equal-area map projections depends on two important considerations:

1. The size of the region involved.
2. The distribution of angular deformation.

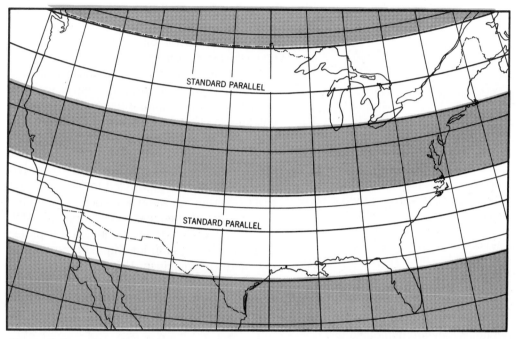

Figure 5.13 Lambert's conformal conic projection with two standard parallels. Values of lines of equal-scale exaggeration (*S*) are 2 percent.

The **Albers' equal-area** projection is widely used. When conventionally arranged, this projection has two standard parallels (Figure 5.15). Because it is conically derived, distortion zones are arranged parallel to the standard lines. Any two small circles reasonably close together may be chosen as standard. The closer together these circles are, the better the representation in their immediate vicinity. Albers' conic projection has a low distortion value. Also, when conventionally arranged, its appearance resembles the earth graticule, with straight converging meridians and concentric arc parallels that meet the meridians at right angles. Thus, this is a good projection for a middle-latitude area of greater east-west than north-south extent (the conterminous United States, for example). Outside the standard parallels, the SF along the meridians is progressively reduced— that is, the SF < 1.0. Between the standard parallels, the SF > 1.0. Parallel curvature ordinarily becomes undesirable if the projection is extended for much over 100° longitude. Clearly, this is an excellent projection for studying geographical distributions on maps of the United States. Thus, it has been selected for standard base maps by many U.S. government agencies, such as the Census Bureau.

Lambert's equal-area projection is also widely used. (See Figure 5.20B). This projection is azimuthal as well as equivalent. Distortion is symmetrical around the central point, which can be located anywhere. Therefore, the projection is useful for areas that have nearly equal east-west and north-south dimensions.

World Equal-Area Projections

The **cylindrical equal-area** projection (Figure 5.B.4A) usually uses two standard parallels. The two parallels may "coincide" and be a great circle (the equator). Or they may be any two standard

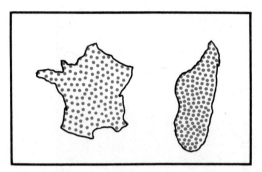

Figure 5.14 These two areas—France, left, and the island of Madagascar, right—are nearly the same size on the non-equal-area projection from which these outlines were traced. France is shown larger than it actually is in comparison to Madagascar. The same number of dots has been placed within each outline, but the apparent densities are clearly not the same.

parallels so long as they are homolatitudes (the same parallels in opposite hemispheres). Distortion is arranged parallel to the standard small circles. Although this projection looks "peculiar" to many people, it does in fact provide, when standard parallels just under 30° are chosen, the least overall mean angular distortion of any equal-area world projection.

The **sinusoidal** projection (Figure 5.B.4B), when conventionally oriented, has a straight central meridian and equator, along both of which there is no angular distortion. All parallels are standard. When a constant interval is portrayed, the parallels appear equally spaced on the projection. This illusion of proper spacing makes the projection useful when latitudinal relations are significant. The sinusoidal is particularly suitable when properly centered for maps of less-than-world areas, such as South America, for which the distribution of distortion is especially fortuitous.

Mollweide's projection (Figure 5.B.4C) does not have the excessively pointed polar areas of the sinusoidal, and thus it appears a bit more realistic. In order to attain equivalence within its oval shape, it is necessary to decrease the north-south scale in the high latitudes and increase it in the low lati-

tudes. The opposite is true in the east-west direction. Shapes are modified accordingly. The two areas of least distortion in the middle latitudes make the projection useful for world distributions when interest is concentrated in those areas.

Several projections for world maps have been prepared by combining the better parts of two. The best known of these is **Goode's homolosine** (Figure 5.16), an equal-area projection which combines the equatorial section of the sinusoidal and the poleward sections of Mollweide's. These two projections, when constructed to the same area scale (as from the same reference globe), have one parallel of identical length (approximately 40°), along which they may be joined. The projection is usually used in interrupted form (see below) and is popular in the United States. Its overall quality, as shown by a comparison of mean values of deformation in Figure 5.16, is not appreciably better than Mollweide's alone.

Interruption and Condensing

Sometimes you may be most interested in studying either the continents or the oceans. It is possible to display one or the other to better advantage on a projection by (1) interrupting the projection, and (2) condensing the map. Interruption allows you to repeat the better parts of the projection, as in Goode's homolosine projection (Figure 5.16). Condensing is accomplished by deleting the unwanted sections of the projection. You can thus attain greater scale for the areas of interest within a fixed page format (Figure 5.17).

AZIMUTHAL PROJECTIONS

As a group, azimuthal (sometimes called zenithal) projections have steadily increased in prominence. The advent of airplanes, development of radio electronics and satellites, mapping of other celestial bodies, and general increase in scientific activity all have contributed to this popularity. Azimuthal projections have a number of useful qualities not shared by other classes of projections.

All azimuthal projections are "projected" on a plane that may be centered anywhere with respect

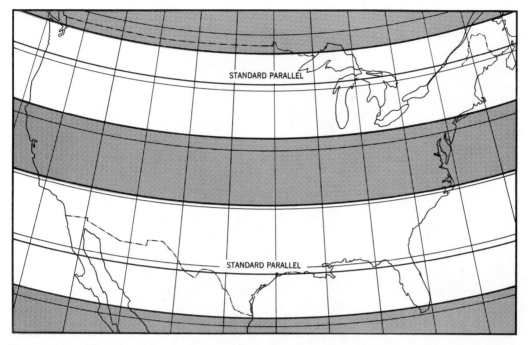

Figure 5.15 Albers' equal-area conic projection of the United States. Values of lines of equal maximum angular deformation (2Ω) are 1°.

to the reference globe. (See Figure 5.18.) A line perpendicular to the plane of projection will necessarily pass through the center of the globe. Consequently, the distortions characteristic of all azi-

muthal projections are symmetrical around the chosen center. (See Figure 5.B.1.) Furthermore, all great circles passing through the center of the projection will be straight lines and will show

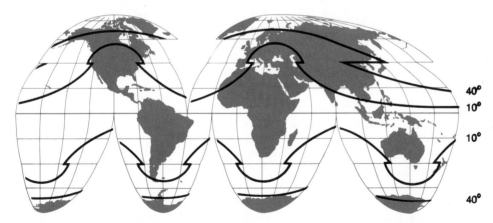

Figure 5.16 Goode's homolosine projection is an interrupted union of the sinusoidal projection equatorward of approximately 40° latitude and the poleward zones of Molllweide's projection. The black lines are values of 2Ω.

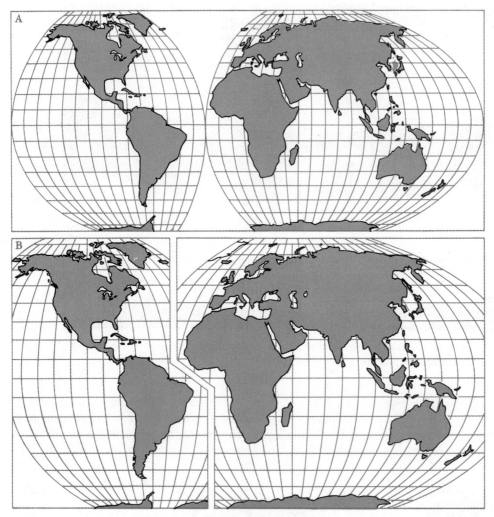

Figure 5.17 (A) An interrupted flat polar quartic equal-area projection of the entire earth. By deleting unwanted areas—that is, condensing as in (B)—additional scale is obtained within a limiting width.

correct azimuths from and to the center in relation to any point. It should be emphasized that only azimuths (directions) from and to the center are correct on an azimuthal projection.

At the center point, all azimuthal projections with the same principal scale are identical. Variation among these projections is merely a matter of how the scale changes along the straight great circles radiating from the center. Figure 5.19 illus-

trates this relationship. The fact that distortion is arranged symmetrically around a center point makes this class of projections useful for areas having more or less equal dimensions in each direction.

Although an infinite number of azimuthal projections is possible, only five are well known: the Lambert equal-area, the stereographic, the azimuthal equidistant, the orthographic, and the gno-

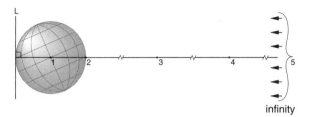

Figure 5.18 The hypothetical positions of the points of projection for the class of azimuthal projections: (1) gnomonic, (2) stereographic, (3) equidistant, (4) equivalent, and (5) orthographic.

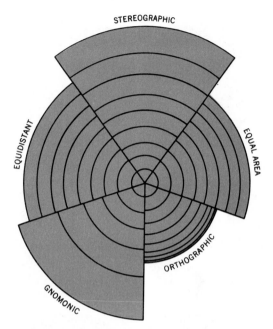

monic (Figure 5.20). We considered the Lambert equal-area (Figure 5.20B) in the section on equal-area projections, and we considered the conformal stereographic (Figure 5.20A) in the section on conformal projections. The stereographic exhibits a second property worth noting here. This attribute is unique among projections: Any circle or circular arc on the globe will plot as a circle or circular arc on the projection. With this projection, therefore, we can plot ranges of radio waves or aircraft using merely a compass.

The **azimuthal equidistant** projection (Figure 5.20C) also has a unique quality: The linear scale is uniform along the radiating straight lines through the center. Therefore, the position of every place is shown in simple relative position and distance from that center. Directions and distances between points whose great circle connection does not pass through the center are not shown correctly. Any movement directed toward or away from a center, such as radio impulses and seismic waves, can be shown well on this projection. The projection has an advantage over many of the other azimuthal projections in that it is possible to show the entire sphere (Figure 5.21).

Most azimuthals are limited to presenting a hemisphere or less. The bounding circle on this world projection is the point (antipode) on the sphere opposite to the center point. Thus, shape and area distortion in the peripheral areas are extreme.

Figure 5.19 Comparison of segments of five azimuthal projections, in this case all centered at the pole. Meridians (great circles), when shown at a constant interval, in all cases appear as equally spaced straight lines radiating from the center. The important thing to note is that the only variation among the projections is in the spacing of the parallels. That is, the only difference among them is the radial scale from the center.

The **orthographic** projection (Figure 5.20D) looks somewhat like a perspective view of a globe from a considerable distance. Distortion of areas and angles, although great around the edges, is not obvious to the viewer. Thus, this projection is useful for preparing illustrative maps on which the globe's sphericity is of major significance.

The **gnomonic** projection (Figure 5.20E) has the unique property that all great-circle arcs are represented as straight lines anywhere on the projection. Therefore, the projection is useful in marine navigation. Navigators need only join points of departure and destination with a

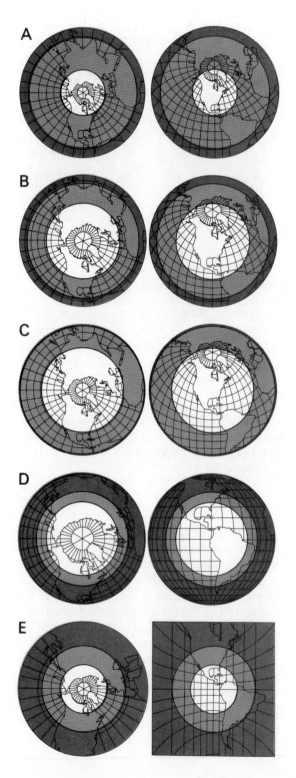

straight line on a gnomonic chart, and the great-circle course is displayed. Compass directions constantly change along most great circles, however. Thus, navigators transfer the course from the gnomonic graticule to a curved one on a Mercator projection. Finally, they approximate the curved course with a series of straight rhumbs.

OTHER SYSTEMS OF MAP PROJECTION

Some projections do not maintain a major earth property yet are indispensable for special purposes. These include two old projections known to the ancient Greeks—the plane chart and the simple conic with two standard parallels. In this section, we will briefly describe these projections. We will also look at three other projections—the polyconic, Robinson's and the space oblique Mercator—which have been devised more recently to solve specific problems.

Plane Chart

The **plane chart**, sometimes called the **equidistant cylindrical** projection, resembles a sheet of graph paper (see Figure 5.22). This projection is also called the **equirectangular** or **Plate Carrée** when the equator is made the standard parallel. It is one of the oldest and simplest map projections. Until replaced by Mercator's projection, it was widely used for navigational charts.

The plane chart is useful for city plans and base maps of small areas. It is easily constructed and, for a limited area, has small distortion. All meridians and any chosen central parallel are standard. The projection may be centered anywhere.

Figure 5.20 The five well-known azimuthal projections: (A) stereographic; (B) Lambert's equal-area; (C) azimuthal equidistant; (D) orthographic; and (E) gnomonic. In (A) the zones of areal exaggeration (S) are less than 30 percent, 30 to 200 percent, over 200 percent; in (B), (C), (D), and (E), zones of 2Ω are 0°, 10° to 25°, over 25°.

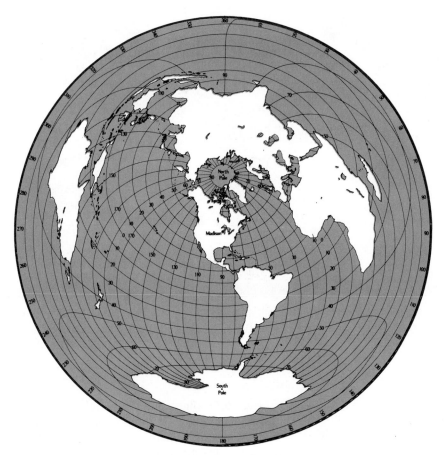

Figure 5.21 An azimuthal equidistant projection of the world centered on Madison, Wisconsin. Note the enormous deformation (both S and 2Ω at the outer edge. The bounding circle is the point (antipode) on the globe opposite the center point of the projection. (University of Wisconsin Cartographic Laboratory.)

Simple Conic

The **simple conic** is shown in Figure 5.23. Several arrangements of scale may be used to produce conic projections with two standard parallels. This projection is similar in appearance to Albers' and Lambert's conic projections, but it does not have their special properties. It does not greatly distort either areas or angles—if the standard parallels are placed close together and if the projection is not extended far north and south of the standard parallels. This projection has been widely used in atlases for areas in middle latitudes.

POLYCONIC PROJECTION

When choosing a projection for a set of large-scale topographic maps for a country, we need to consider several special factors. Conformality is desirable, overall distortion should be minimal, and adjacent maps should fit together.

When the first U.S. survey organization devised a system for large-scale mapping in the United States in 1819, conformality was judged less important than it is now. Thus, they chose a compromise projection, which was neither conformal nor equal-area.

The projection they chose was the **polyconic**

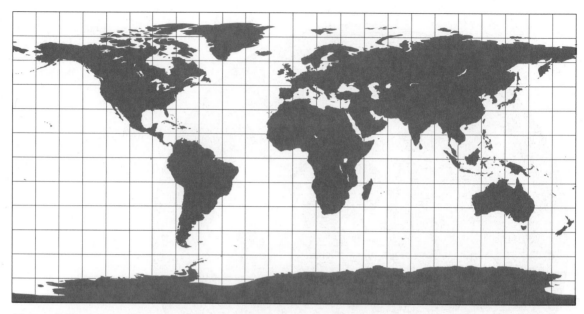

Figure 5.22 An equidistant cylindrical projection or plane chart. When the equator is chosen as the standard parallel, the equirectangular or Plate Carrée projection is one of the simplest to construct, as it is essentially a piece of graph paper.

projection (Figure 5.24). This projection permits an easy fit of adjacent sheets in one direction or another but not in all directions. Distortion within the area portrayed on one topographic sheet is insignificant for almost all uses of that quadrangle. (See Figure 5.25.) This scheme was used until the mid-1950s for standard topographic maps of the United States.

Robinson's Projection

In 1961, Rand McNally and Company commissioned a new projection for world maps at all scales. In their view, all the then-available projections for uninterrupted world maps had serious shortcomings, especially in the appearance of land masses. The objective was a world projection that minimized the appearance of angular and area distortion. The result is the **Robinson projection** shown in Figure 5.26.

The Robinson projection is neither conformal nor equal-area but a compromise between the two. Its pattern of angular distortion (2Ω) is similar to that on other world projections (see Figure 5.B.4). But Robinson's projection includes a much larger area of minimum values in the central regions than do the equivalent projections. Values of S vary according to latitude. Specifically, S increases with distance from parallels 38°N and 38°S (in the conventional case). These parallels may be considered the standard parallels when the projection is constructed for a globe of the same surface area. More than 3/4 of the earth's surface is shown on the projection with less than a 20 percent departure from its true size at scale.

Space Oblique Mercator Projection

So far we have viewed a map projection as a framework onto which we fit geographical data.

Figure 5.23 Simple conic projection with two standard parallels at 20° and 60° N. The parallels remain equidistant over the entire projection, as can be seen on this map of the North American continent.

The development of aerial photography turned this concept around. With aerial photography, data were recorded for the first time directly in a complex map projection mode. These static photographs are capable of being transformed by photogrammetric methods into simpler and more desirable projectional forms.

The development of continuously recording scanning devices in orbiting satellites introduced dynamics into the transformation process. This dynamism holds the prospect of continuous real-time mapping directly in a projection-based mode. For example, the electron scanner aboard Landsat (see Chapter 8) continuously records data in digital mode. These data are subject to more than a dozen geometric corrections before printout as imagery. The geometrically corrected imagery has a projection format with unique characteristics. The complex geometry of the image results from the relative motions of the scanning device, the

Figure 5.24   North America on a polyconic projection with the central meridian at 100° W. The parallels are arcs of circles which are not concentric. The meridians are complex curves formed by connecting points along the parallels at their true distances.

orbiting spacecraft, and the earth. Figure 5.27 illustrates the relationships involved.

The resulting projection, the **space oblique Mercator**, is essentially conformal, but it is significantly different from the transverse Mercator projection. The groundtrack of the satellite becomes the central line of the projection along which the SF is nearly 1.0. Instead of being a simple great circle on the reference globe, the groundtrack is slightly curved, and it crosses the equator at an angle (Figure 5.27).

The curved groundtrack results from the fact that as the satellite moves along its orbit, the earth is also rotating. The inclination of the orbit to the plane of the equator makes the limit of the groundtrack in the polar areas about 81°N and 81°S. The space oblique Mercator projection is not perfectly conformal, but within the scanning swath the departure is negligible. The graticule and the UTM plane coordinate grid can be fitted to Landsat imagery with no detectable anomalies.

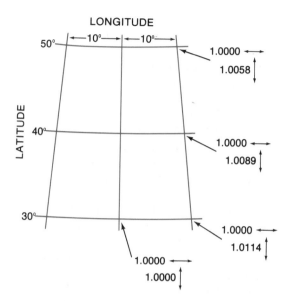

Figure 5.25 The distribution of scale factors on a polyconic projection in the vicinity of 40° latitude. North-south SF values away from the central meridian are approximate. Note that the section of the projection which is used for a standard 7.5-minute quadrangle map would be ⅛ degree east-west and north-south along the central meridian.

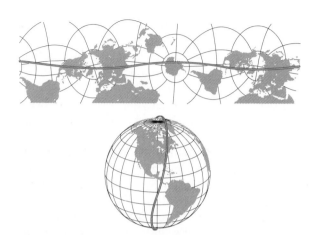

Figure 5.27 The central line of the space oblique Mercator projection is slightly curved and oblique to the equator. It crosses the polar areas at about 81°N and S latitude. Along this line, which represents the Landsat groundtrack, the SF is essentially 1.0. The conceptual basis for the projection is similar to that for the transverse Mercator projection, but the central line is not a great circle (compare Figure 5.12).

SELECTED REFERENCES

American Cartographic Association, *Matching the Map Projection to the Need*, Committee on Map Projections, Special Publication 3, ACA/ American Congress on Surveying and Mapping, Bethesda, MD, 1991.

American Cartographic Association, *Choosing a World Map: Attributes, Distortions, Classes, Aspects*, Committee on Map Projections, Special Publication 2, ACA/American Congress on Surveying and Mapping, Falls Church, VA, 1988.

American Cartographic Association, *Which Map is Best: Projections for World Maps*, Committee on Map Projections, Special Publication 1, ACA/American Congress on Surveying and Mapping, Falls Church, VA, 1986.

Canters, F., and H. Decleir, *The World in Perspective: A Directory of World Map Projections*,

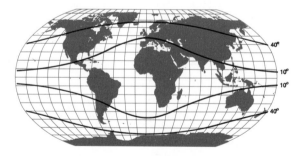

Figure 5.26 Robinson's projection. The lines denote values of 2Ω. The value of S would be 1.0 at approximately 38°N and S latitude. Isolines of values of S would parallel the parallels.

John Wiley & Sons Ltd., Chichester, U.K. (1989).

Dahlberg, R. E., "Evolution of Interrupted Map Projections," *International Yearbook of Cartography*, 2 (1962): 36–54.

Dyer, J. A., and J. P. Snyder, "Minimum-Error Equal-Area Map Projections," *The American Cartographer*, 16 (1989): 39–43.

Jankowski, P., and T. L. Nyerges, "Design Considerations for MaPKBS-Map Projection Knowledge–Based System," *The American Cartographer*, 16 (1989): 85–95.

Laskowski, P. H., "The Traditional and Modern Look at Tissot's Indicatrix," *The American Cartographer*, 16 (1989): 123–133.

Maling, D. H., *Coordinate Systems and Map Projections*, Oxford: Pergamon Press, 2nd ed., 1992.

Nyerges, T. L., and P. Jankowski, "A Knowledge Base for Map Projection Selection," *The American Cartographer*, 16 (1989): 29–38.

Pearson, F., *Map Projections: Theory and Applications*, Boca Raton, FL: CRC Press, 1990.

Robinson, A.H., "A New Map Projection: Its Development and Characteristics," *International Yearbook of Cartography*, 14 (1974): 145–155.

Snyder, J. P., and H. Stewart, "Bibliography of Map Projections," *U.S. Geological Survey Bulletin*, 1856 (1988).

Snyder, J. P., "Space Oblique Mercator Projection—Mathematical Development," *U.S. Geological Survey Bulletin*, 1518 (1981).

Snyder, J. P., "Map Projections Used by the U.S. Geological Survey," *U.S. Geological Survey Bulletin*, 1532 (1982).

Snyder, J. P., "Computer Assisted Map Projection Research," *U.S. Geological Survey Bulletin*, 1629 (1985).

Snyder, J. P., "Map Projections—A Working Manual," *U.S. Geological Survey Professional Paper*, 1935 (1987).

Snyder, J. P., "An Album of Map Projections," *U.S. Geological Survey Professional Paper*, 1453 (1989).

Tobler, W. R., "A Classification of Map Projections," *Association of American Geographers, Annals*, 52 (1962): 167–175.

CHAPTER 6

SCALE, REFERENCE, AND COORDINATE SYSTEMS

*I*n Chapter 4 we learned about the graticule—the system of latitude and longitude lines on the earth's surface that helps us reference locations. In Chapter 5 we learned the methodology that lets us transform a reference globe onto another surface, most often a flat surface which we call a **map**. In this chapter we will look at systems that have been devised to make reference to and between locations on a map easy to use.

From Chapter 5 we realize that transforming locations on a globe to a map results in scale variations. This transformation causes shapes, angles, distances, and directions on the map to be distorted. Yet we know that maps are useful, if not essential, for a wide variety of human activities. How can we make maps, with their inherent distortions, easy to use for our everyday activities? To do so, we must understand the concept of **map scale** and then learn about the reference and coordinate systems that cartographers have devised.

MAP SCALE

Maps, to be useful, are necessarily smaller than the areas mapped. Consequently, every map must state the ratio or proportion between measurements on the map to those on the earth. This ratio is called the **map scale** and should be the first thing the map user notices. Map scale is an elusive thing because, as we saw in Chapter 5, transformation from globe to map means that the map's scale will vary from place to place. It can even vary in different directions at one place.

The scale of a map may be shown in many ways. It can be specifically indicated by some statement or graphic device. Or it may be shown indirectly by the spacing of parallels and meridians and even subtly by the size and character of marks on the map. But some indication of the scale *must* be included in the information presented on any map.

STATEMENTS OF SCALE

Map scale is the ratio between a distance on the map and the corresponding distance on the earth, with the distance on the map always expressed as one. The map scale may be expressed in the following ways.

Representative Fraction (RF)

The **representative fraction** (or **RF** for short) is a simple fraction or ratio. It may be shown either as 1:1,000,000 or 1/1,000,000; the former is preferred. This means that (along particular lines) one millimeter or one centimeter or one inch on the map (the numerator) represents 1,000,000 millimeters, centimeters, or inches on the earth's surface (the denominator). The unit of distance in both the numerator and denominator of the fraction must be the same. We referred to the RF when we discussed reducing the earth to a reference globe for the purposes of transformation in Chapter 5.

Verbal Statement

Another way to show scale is with a statement of meaningful map distance in relation to meaningful earth distance. For example, the RF 1:1,000,000 indicates "one millimeter represents one kilometer" or "about one inch to 16 miles." Many older map series had this type of scale—for example, the "one-inch" or "six-inch" maps of the British Ordnance Survey (one inch to one mile, six inches to one mile).

Graphic or Bar Scale

The **graphic or bar scale** is a line symbol, subdivided to show map lengths of earth distance units. One end of the bar scale is usually subdivided further, allowing you to measure distances more precisely (Figure 6.1).

Area Scale

The **area scale** refers to the ratio of areas on the map to those on the earth. As we saw in Chapter 5, when

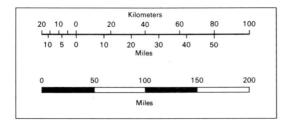

Figure 6.1 Examples of graphic or bar scales. Often, as illustrated in the upper bar scale, the left end of the bar is subdivided into smaller units to provide more precise estimation of distances.

the sphere is transformed to a plane so that all area proportions on the earth are correctly represented, the result is an equal area projection. On such a projection, one unit of area (square centimeters, square inches) is proportional to a particular number of the same square units on the earth. If the area scale were expressed as an RF, it would be shown, for example, as $1:1,000,000^2$. Usually, however, the fact that the denominator of the map scale is squared is assumed and not shown. The area scale can also be shown graphically by a square representing a stated number of square kilometers or square miles. (See Figure 6.2).

SCALE FACTOR (SF)

As we saw in Chapter 5, the **scale factor (SF)** at a point is computed by the following formula:

$$SF = \frac{\text{actual RF}}{\text{principal RF}}$$

This expresses the SF as a ratio. A SF of 2.0 would mean that the actual scale (on the map)

was twice as large as the principal scale (on the reference globe). That would be the case, for example, if the actual scale were 1:15,000,000 and the principal scale were 1:30,000,000. (Remember that the RF is a fraction, and the larger the denominator, the smaller the scale). Similarly, an SF of 0.50 would show that the actual scale was half that of the principal scale. That would be the case, for example, if the actual scale were 1:60,000,000 and the principal scale were 1:30,000,000.

Scale factors of the magnitudes used for the above statements occur only on small-scale maps. On large-scale maps, the SF at various places will vary only slightly from one (unity). As we will see later in this chapter, on large-scale maps using the transverse Mercator projection, SF magnitudes within a 6° longitude zone may vary only from 0.99960 to 1.00158. (Also see Figure 5.25 for scale factor values for a polyconic projection). When you make a map, you must decide the tolerances you can accept, depending on your objective.

DETERMINING THE SCALE OF A MAP

Sometimes you may want to determine the scale of a map on which no scale is given. Or, you may want to discover the scale of part of a map (since, as noted earlier, the scale is never the same over an entire flat map).

You can estimate map scale along a line by measuring map distance between two points that are a known earth distance apart and then computing the scale. Certain known distances are easy to use, such as the lengths of degrees of latitude or longitude (see Tables 4.3 and 4.4). Be sure to take the measurement in the direction the scale is to be used; frequently the distance scale of the map will not be the same in all directions from a point.

You can also determine map scale using area measurements. To do so, you first must find a region of known size. Then compute the area of that region on the map. You can then determine the proportional relation between map area and earth area. Remember that scales based on the relation between areas are conventionally

  Represents 10 hectares

Figure 6.2 Example of a graphic area scale.

expressed as the square root of the number of units on the right of the ratio. Suppose, for instance, that your measurement shows that one square unit on the map represents 25,000,000,000,000 of the same square units of the earth. You would record the RF as the square root, 1:5,000,000. This figure would approximate the linear scale.

TRANSFORMING THE MAP SCALE

You will often need to change a map's size by reducing or enlarging it. To do so, you can use a method similar to the technique for transforming one type of scale to another. This method is explained in Box 6.A. Once you are skilled at this sort of scale transformation, you

---

Box 6.A
Conversion of Map Scales

Assume the RF on the map used in examples 1 and 2 below is 1:75,000:

Example 1   To determine a verbal or written scale:

Metric units

The cm/km scale will be:

1. 1 cm (map) represents 75,000 cm (earth), and
2. 75,000 cm = 0.75 km
3. 1 km/0.75 km = 1.333
4. Therefore 1.333 cm represents 1.0 km or 1 cm represents 0.75 km

U.S. Customary units

The inch/mile scale will be:

1. 1 inch (map) represents 75,000 inches (earth), and
2. 1/75,000 inches = x/63,360 inches, therefore
3. x = 0.845
4. Therefore, 0.845 inch represents 1.0 mile or 1 inch represents 1.184 mile

Example 2   To construct a graphic scale, a proportion is established as:

Metric units

1. 1.333 cm:1.O km::x cm:10 km
2. x= 13.33 cm
3. 13.33 cm represents 10 km, which may be easily plotted and subdivided

U.S. Customary units

1. 0.845 inch:1.O mile::x inches:5 mile
2. x = 4.225 inches
3. 4.225 inches represents 5 miles, which may be plotted and subdivided

Example 3   To determine the RF:

Metric units

If the graphic scale shows by measurement that 1 cm represents 50 km, then:

1. 1 cm represents 50 x 100,000 cm, or
2. 1 cm to 5,000,000 cm and, therefore,
3. The RF is 1:5,000,000

U.S. Customary units

If the graphic scale shows by measurement that 1 inch represents 75 miles:

1. 1 inch represents 75 x 63,360, or
2. 1 inch to 4,752,000 inches and, therefore,
3. The RF is 1:4,752,000

TABLE 6.1
Common Map Scales and Their Equivalents

Map Scale	One Centimeter Represents		One Kilometer Is Represented by		One Inch Represents		One Mile is Represented by	
1:2,000	20 m		50	cm	56	yd.	31.68	in.
1:5,000	50 m		20	cm	139	yd.	12.67	in.
1:10,000	0.1	km	10	cm	0.158	mi.	6.34	in.
1:20,000	0.2	km	5	cm	0.316	mi.	3.17	in.
1:24,000	0.24	km	4.17	cm	0.379	mi.	2.64	in.
1:25,000	0.25	km	4.0	cm	0.395	mi.	2.53	in.
1:31,680	0.317	km	3.16	cm	0.500	mi.	2.00	in.
1:50,000	0.5	km	2.0	cm	0.789	mi.	1.27	in.
1:62,500	0.625	km	1.6	cm	0.986	mi.	1.014	in.
1:63,360	0.634	km	1.58	cm	1.00	mi.	1.00	in.
1:75,000	0.75	km	1.33	cm	1.18	mi.	0.845	in.
1:80,000	0.80	km	1.25	cm	1.26	mi.	0.792	in.
1:100,000	1.0	km	1.0	cm	1.58	mi.	0.634	in.
1:125,000	1.25	km	8.0	mm	1.97	mi.	0.507	in.
1:250,000	2.5	km	4.0	mm	3.95	mi.	0.253	in.
1:500,000	5.0	km	2.0	mm	7.89	mi.	0.127	in.
1:1,000,000	10.0	km	1.0	mm	15.78	mi.	0.063	in.

will have no difficulty enlarging or reducing maps. You will find such a skill especially useful when working with geographic information systems, which allow you to zoom "in" or "out" on a map displayed on a screen. A sense of scale, which comes with practice, will keep you from making costly mistakes.

As Box 6.A shows, transforming decimal metric scales is fairly simple. But U.S. Customary units are more bothersome. When working with U.S. Customary units, keep in mind that there are 63,360 inches in one mile (statute). Using this information, you can change each of the linear scales (RF, graphic, verbal) into one of the others. Examples are given in Box 6.A.

To change the scale of a map that has an area scale, convert the known area scale and the desired area scale to a linear proportion. Table 6.1 lists some common map scales and their metric and U.S. Customary equivalents.

REFERENCE SYSTEMS

We live in an increasingly complex world. It is human nature to try to simplify complexity in order to increase our understanding. One way we do so is to create reference systems. Reference systems introduce order into chaos.

Imagine a library without a reference system. The difficulty of finding a book would be overwhelming, rendering a library almost useless. Similarly, we order the world's complexity in its spatial component. Such spatial referencing can be of several types. Telephone area codes and postal zip codes represent two spatial referencing systems. (See Figure 6.3 and Figure 6.4). These two systems exhibit only rudimentary metric properties and do not give information about direction or size. Yet they make our complex society work better.

A more refined spatial referencing system is house numbering. An address tells you something

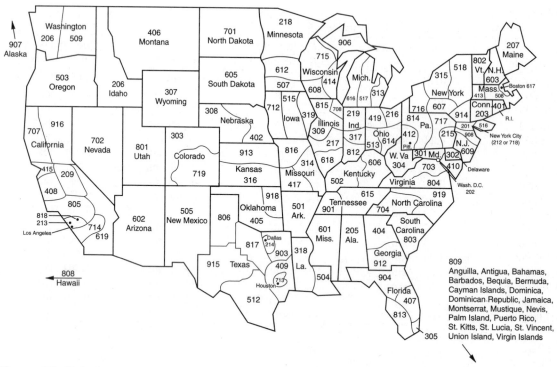

Figure 6.3 Telephone area code map. An example of a spatial referencing system without metric or topological properties.

about spatial location. You can speculate, for example, that 221 East Main Street is farther from some arbitrary center point than is 118 East Main Street. (See Figure 6.5). With some systems, you can tell that these two addresses are on opposite sides of Main Street, with one cross street intervening. Other city systems are not so refined and therefore connote less spatial information.

The Encyclopedia Britannica uses a spatial referencing system for information on a given page. This system divides each page into two columns of four equal spaces. It labels the four spaces on the left column *a*, *b*, *c*, and *d*. The four spaces on the right column are labeled *e*, *f*, *g*, and *h*. Each index entry can then refer to a volume number, a page number, and a spatial position on the page. For instance, 8:47*h* refers to the lower right hand 1/8 of page 47 in volume 8. This system speeds up your search for information.

Atlases and road maps use an areal spatial referencing system to help locate positions on maps. The most common form is to divide the perimeter of the map into equal sections. Then, one axis is labeled with letters of the alphabet and the other axis with numbers. Thus, the reference C-9 could connote the intersection of the third column with the ninth row. (See Figure 6.6).

To a limited extent, you can calculate distances and directions with the spatial referencing system used in these atlases and road maps. For instance, assume that numbering is from upper left corner to lower left corner, and lettering is from upper left corner to upper right. You can then tell that a point with coordinates C-9 is downward and left of a point with coordinates J-2. Going further, assume that each division of the map's perimeter is equivalent to one kilometer on the ground and that the map is oriented with north at the top. Then area

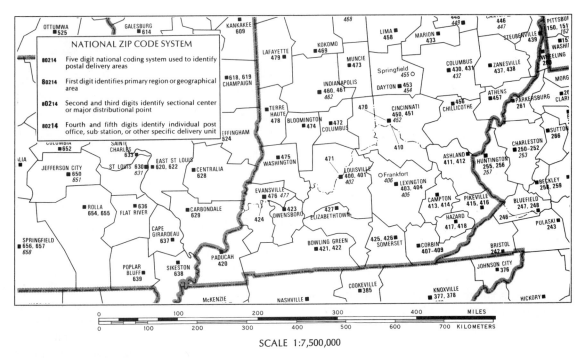

Figure 6.4 Three-digit zip code map. An example of a spatial referencing system with a minimum of topological and no metric properties.

J-2 is approximately 10 kilometers from area C-9, and J-2 is northeast of C-9. (See Figure 6.6). While the topological relationships in the plane are complete in this example, the metric properties are relative and approximate.

COORDINATE SYSTEMS

Primitive people reckoned distance and direction in relative terms, with respect to their own location. To express direction, they used such aids as the sun's position and such terms as forward and backward, left and right, and so on. They expressed distance in terms of the length of some body part or travel time.

To locate points precisely, however, you need to use a coordinate system which includes both topology and a strict metric form of measurement.

Two types of coordinate systems are now in general use. The older of the two is the **geographical coordinate system**, which uses latitude and longitude. It was first practiced by Greek philosopher-geographers before the Christian era. It was explained in Chapter 4. It is the primary system, since it is used for all basic locational reckoning, such as navigation and fundamental surveying. This system is useful for locating positions on the uniformly curved surface of the earth.

The second type of system, called **rectangular coordinates** or **plane coordinates**, is also old, at least in its basic form. This type of system is used to locate positions on a flat map representation of the earth's curving surface. Such systems were standard features of Chinese cartography after being included in the six principles of mapmaking by Pei Hsiu in the 3rd century A.D. In modern form the plane coordinate system evolved from Cartesian

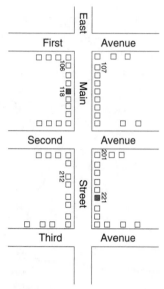

Figure 6.5 A spatial reference system with some metric properties and some topological properties.

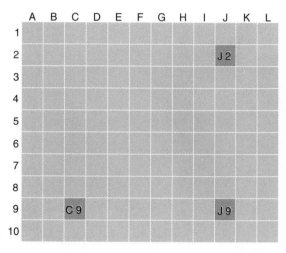

Figure 6.6 A spatial reference system with topology and some metric properties.

coordinates applied to maps for military needs. Such systems have since become extremely useful for a variety of other uses. Before discussing rectangular coordinates in detail, let's look at the Cartesian coordinate system on which they are based.

CARTESIAN COORDINATES

Cartesian coordinates are a mathematical construct defined by an **origin**—that is, an initial point—and a unit of distance. In a plane, two axes are established running through the origin and spaced so that they are perpendicular to each other. These axes, normally called X and Y, are subdivided into units of the specified distance from the origin. Points can then be located with respect to the origin and the axes. (See Figure 6.7).

The same type of spatial reference system is used in making graphs. The statistical graphs in your daily newspaper depend on your knowledge of Euclidean geometry. The axes are labeled with numbers. After plotting two points on a set of axes, you can calculate the distance and direction (slope

of the line) between them. (See Figure 6.8). Maps likewise depend on Euclidean geometry and Cartesian coordinates. Digitizing tables or tablets are simply a set of Cartesian coordinates in a plane. You can locate and record any point to the resolution of the digitizing table (see Chapter 11).

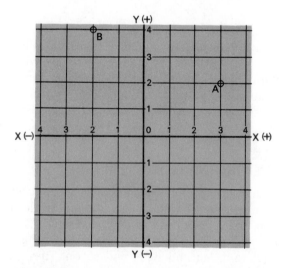

Figure 6.7 Cartesian coordinates. The origin is 0. The abscissa values are $-X$, 0, $+X$, and the ordinate values are $-Y$, 0, $+Y$. The position of point A is 3,2; the position of point B is $-2,4$.

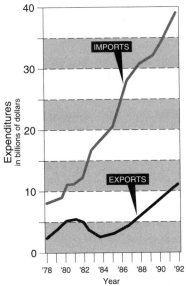

Figure 6.8 A typical business graphic plotted on a Cartesian coordinate grid.

Cartographers use Cartesian coordinates by specifying coordinate systems for use on maps in order to take advantage of the simplified calculations which are then possible. (See Box 6.B). For

limited areas on maps with the property of conformality, variation between planar and spherical calculations is not significant. You will want to use different coordinate systems depending on how accurate your calculations need to be. As your accuracy requirements increase, the size of the area of the earth that can be portrayed and analyzed using plane coordinates on any one map decreases. It is important therefore to know the accuracy of each coordinate system and the distortions resulting from the map projection used.

RECTANGULAR COORDINATES

On a limitless plane surface, there is no natural reference point. That is, every point is like every other point. We can create a reference point by using an arbitrary Cartesian coordinate system. To do so, we establish a point of origin at the intersection of two perpendicular axes and tie that origin to a known point on the plane. We then divide the plane into a grid with an infinite number of equally spaced lines parallel to each axis.

We may now state the position of a point, *A*, with reference to the point of origin. To do so, we

---

Box 6.B
Calculation of Distance and Direction in Two Dimensions

Since ancient Greek civilization, humans have known how to calculate distances and directions on a plane. The familiar Pythagorean theorem is taught to every school child. Most high school children are introduced to basic trigonometric functions. Thus, distances and directions on a plane are routine calculations for most educated adults. The same is not true for calculating distances and directions on a sphere or on the earth's surface. (See Boxes 4.A and 4.B). Therefore, working with the assumption of a planar surface simplifies and promotes the use of maps.

Calculation of Distance

The distance, d_{12}, between two points in a plane is given by the following formula where (x_1, y_1) and (x_2, y_2) are the locations of the two points in a Cartesian plane:

$$d_{12} = \{(x_1 - x_2)^2 + (y_1 - y_2)^2\}^{1/2}$$

Calculation of Direction

The direction, α_{12}, between the two points (x_1, y_1) and (x_2, y_2) is given by the following formula:

$$\tan \alpha_{12} = \{(y_1 - y_2)/(x_1 - x_2)\}$$

indicate the distance from each axis to point *A*, measured in each case parallel to the other axis, and expressed to any desired precision. In the familiar rectangular coordinate system (for example, cross-section paper or graph paper), the horizontal distance is called the *X* value or **abscissa**. The distance perpendicular to it is called the *Y* value or the **ordinate** (see Figure 6.7).

There is a good reason why coordinate systems came to appear on maps. With the increasing range of artillery in World War I, it became more and more difficult for an army to arrive at accurate azimuth (bearing or direction) and range (distance) calculations to a target. Until World War I, battles were hand to hand, or enemies were within sight of one another. The increased range of munitions meant that visual sighting of targets was no longer necessary, but trajectories had to be calculated in real time in the field. Thus, as the range of munitions increased, armies had to calculate where to shoot without being able to see the target. Calculations involving latitude and longitude were too involved for quick field calculation. To simplify the problem, the French were the first to construct a series of local plane, rectangular coordinate grids on their maps. This proved so useful that other nations quickly followed suit. Between World Wars I and II, a great many systems of plane rectangular coordinates were devised and put to use by cartographers. Today the use of rectangular grid systems is almost universal.

To establish a plane rectangular coordinate system, the procedure is as follows: First, a map is made by transforming the spherical surface to a plane (by a system of map projection). Next a rectangular plane coordinate grid is placed over the map. The coordinate grid is most often tied to the map by placing the origin of the coordinate grid somewhere near the center of interest on the map. The perpendicular axes of the coordinate grid are usually made to coincide with straight line meridians and/or parallels, if they exist. To locate a position, you need to specify the *X* and *Y* coordinates of that point. You can do so to whatever degree of precision you desire in decimal divisions of whatever earth distance units are used for the coordinate system.

Remember that in this process, the transformation from reference globe to map has introduced distortion. Thus, it is standard practice for large-scale maps to be on conformal projections. On such projections, coordinate reference grids maintain the accuracy you need to calculate directions and distances. Under such a system, every point on the earth surface has an unique pair of coordinates. The pair of coordinates (ordinate and abscissa) of a point are called a **grid reference**.

To assure accuracy of calculations, you must be sure that each map or limited set of maps has its own plane coordinate system. If you work on maps with different grid zones, you can have serious problems. It is difficult to calculate distances between points falling in different coordinate system zones.

There is a conventional way to read a grid reference in rectangular map coordinates. You always give the *X* value first. This value is called an **easting** (presuming the map is oriented with north at the top). You give the *Y* value (called a **northing**) second. A rule of thumb is that when using grid references, you always read "right, up." For example, see Figure 6.9. It shows that point P can be given an easting value of 14.5 and a northing value of 20.1 by decimal subdivision of the squares.

When used on a map, a grid reference is given as an even set of numbers run together. The first half of the group is the easting coordinate value, and the second half is the northing. Decimals are not shown, but are deduced from the numbering of the lines. Thus, the grid reference mentioned above for point *P* would be 145201. With lesser precision, it would be 1420; with greater precision it would be 14562011. If each square in Figure 6.9 represented 1 km² (1,000 meters on a side), the reference 145201 would be a statement that point *P* lies within a 100-meter square, the point read being the location of the southwest corner of the 100-meter square. Adding an additional digit to the *x* and *y* coordinates would narrow the position to within a 10-meter square.

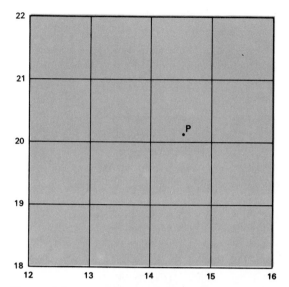

Figure 6.9 A portion of a rectangular grid. If the squares are one km. on a side, then point P may be located to within a $10m^2$ with the grid reference 14562011.

To further simplify calculation on a coordinate grid, only the upper right-hand quadrant of a plane coordinate system is normally used (see Figure 6.7). This is done so that both numbers in a grid reference are positive. Therefore, there will be no negative numbers nor repetition of numbers east and west or north and south of the axes. To accomplish this, an arbitrary large value, called a **false easting,** is given to the *Y*-axis origin. A second arbitrary large value, called a **false northing**, is given to the *X*-axis origin. This in effect moves the new origin (grid reference, 0,0) off the map to the southwest, resulting in all positive values for *X* and *Y* within the map area.

The accuracy with which you can use a plane coordinate system depends on the way the scale factors (SFs) are arranged on the map. This, of course, depends upon the map projection. In the early part of this century, a variety of projections were used for plane coordinate systems. Today, however, most plane coordinate systems are based

on only three map projections, all conformal: the transverse Mercator, the polar stereographic, and Lambert's conformal conic. If the United States straddled the equator, we might have used a system based on the Mercator projection.

Plane coordinates are not usually used on small-scale maps. The distortions that result from transforming the spherical surface to the plane are so great on small-scale maps that detailed calculations and positioning are difficult. Thus, plane coordinates are used only on large-scale maps. Most large-scale topographic maps show one or more systems of plane coordinates. In the United States most 1:24,000-scale topographic maps have four reference or coordinate grid systems indicated.

COORDINATE SYSTEMS USED ON MAPS IN THE UNITED STATES

In the rest of this chapter, we will discuss the common Cartesian coordinate systems used on maps in the United States. Two of these systems can be used to reference points. One system is used to reference areas. The fourth system is the standard geographical reference system, latitude and longitude.

UNIVERSAL TRANSVERSE MERCATOR (UTM) GRID SYSTEM

The **Universal Transverse Mercator (UTM)** grid system has been widely adopted for topographic maps, satellite imagery, natural resources databases, and other applications that require precise positioning. It is a metric system—that is, the meter is the basic unit of measurement.

In the UTM grid system, the area of the earth between 84°N and 80°S latitude is divided into north-south columns 6° of longitude wide. These columns are called **zones**. They are numbered from 1 to 60 eastward, beginning at the 180th meridian. Each column is divided into quadrilaterals of 8°

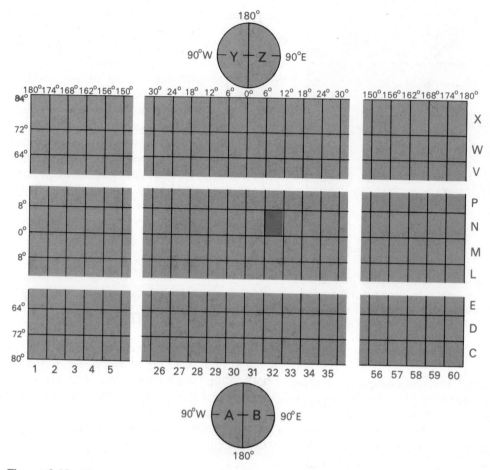

Figure 6.10 The system of UTM and UPS grid zone designations. Each quadrilateral is identified by its column number and row letter (*I* and *O* omitted). The tinted quadrilateral is designated 32N. The UPS grid zones in the polar areas are sectors on each side of the 0°–180° meridians.

of latitude. The rows of quadrilaterals are assigned letters *C* to *X* consecutively (with *I* and *O* omitted) beginning at 80°S latitude (Figure 6.10). (Row *X* is 12° latitude extending from 72°N to 84°N to cover all land areas in the northern hemisphere).

Each quadrilateral is assigned a number-letter combination. For example, 18S denotes a quadrilateral 6° in longitude, between 78°W and 72°W, and 8° in latitude, between 32°N and 40°N. As always, in giving a grid reference, you read right,

up. Each 6° by 8° quadrilateral is divided into 100,000-meter zones. These zones are designated by a system of letter combinations arranged so that the same two-letter combination is not repeated within 18°. Finally, within each 100,000-meter square, you can specify an easting of up to five digits and a northing of up to five digits. (See Box 6.C). Such a reference is theoretically correct to one-meter resolution.

For the UTM grid system, the transverse Mercator

Box 6.C
Use of the UTM Grid System

Depending on what part of the world is mapped, either the UTM or UPS grid system is used on all U.S. military maps. Figure 6.C.1 explains the UTM grid system as it appears at the bottom of the Alexandria 1:50,000-scale map produced by the Defense Mapping Agency. Beginning at the bottom of the explanation, the Grid Zone Designation is given by 18S. This number indicates a

quadrilateral of 8° of latitude (32°N to 40°N) by 6° longitude (78°W to 72°W). This quadrilateral appears on a secant transverse Mercator projection centered on 75°W longitude. The Cartesian coordinate grid is aligned with the 75°W line of longitude, and the 100,000 meter squares are measured from that line.

The middle part of the explanation uses a two-letter system to designate a 100,000-meter square within zone 18S. It uses the letters *UT* for the 100,000 meter square south of a line located 4,300,000 meters north of the equator. The letters *UU* are used to designate the 100,000-meter square north of that line. Next, the 100,000-meter square is broken into smaller units. A total of 10 digits, five representing the x coordinate and five representing the y coordinate, would locate a point within a one-meter square. An eight-digit set of coordinates locates a point within a 10-meter square; six digits, a 100-meter square; and four digits, a 1,000-meter square.

Now take a look at Figure 6.C.2. Suppose you wish to designate the location of the Washington monument in Washington, D.C. First, note that the monument is in the 1,000-meter square bounded by 06 and 07 to the north and 23 and 24 to the east. Interpolating, you would arrive at a coordinate location of 234062. This figure would designate the southwest corner of a 100-meter square on the earth's surface. The entire UTM grid reference for the Washington monument would then read 18S *UU* 234062.

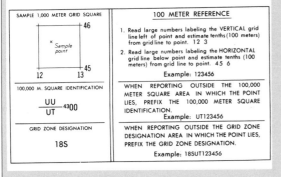

CONTOUR INTERVAL 10 METERS

SPHEROID . . . . CLARKE 1866
GRID . . . . 1000-METER UTM ZONE 18 (BLACK NUMBERED LINES)
10,000-FOOT STATE GRID TICKS, VIRGINIA (NORTH ZONE)
MARYLAND
PROJECTION . . . . UNIVERSAL TRANSVERSE MERCATOR·
VERTICAL DATUM . . . . NATIONAL GEODETIC VERTICAL DATUM OF 1929
HORIZONTAL DATUM . . . . 1927 NORTH AMERICAN DATUM
CONTROL BY . . . . USGS, NOS/NOAA, AND USCE
PREPARED BY . . . . U. S. GEOLOGICAL SURVEY
PRINTED BY . . . . DMAHTC, 10-79

FOR SALE BY U. S. GEOLOGICAL SURVEY, RESTON, VIRGINIA 22092

SAMPLE 1,000 METER GRID SQUARE

× Sample point

12   13

46

45

100 METER REFERENCE

1. Read large numbers labeling the VERTICAL grid line left of point and estimate tenths (100 meters) from grid line to point. 12 3
2. Read large numbers labeling the HORIZONTAL grid line below point and estimate tenths (100 meters) from grid line to point. 45 6

Example: 123456

100,000 M. SQUARE IDENTIFICATION

UU
──── 4300
UT

WHEN REPORTING OUTSIDE THE 100,000 METER SQUARE AREA IN WHICH THE POINT LIES, PREFIX THE 100,000 METER SQUARE IDENTIFICATION.
Example: UT123456

GRID ZONE DESIGNATION

18S

WHEN REPORTING OUTSIDE THE GRID ZONE DESIGNATION AREA IN WHICH THE POINT LIES, PREFIX THE GRID ZONE DESIGNATION.
Example: 18SUT123456

Figure 6.C.1 The legend explanation of the UTM grid system as it appears on a 1:50,000-scale Defense Mapping Agency quadrangle.

continued on next page

projection is used. Each of the 60 zones is mapped onto one transverse Mercator projection. The Y coordinate axis coincides with the central meridian of the 6° longitudinal zone. The SF is constant along each north-south coordinate grid line, but it

varies in the east-west direction. A secant case of the transverse Mercator projection is constructed for each zone to minimize variations in the SF over the entire projection. Thus, along the center grid line of each UTM grid zone, the SF is 0.99960. At the

Box 6.C (continued)

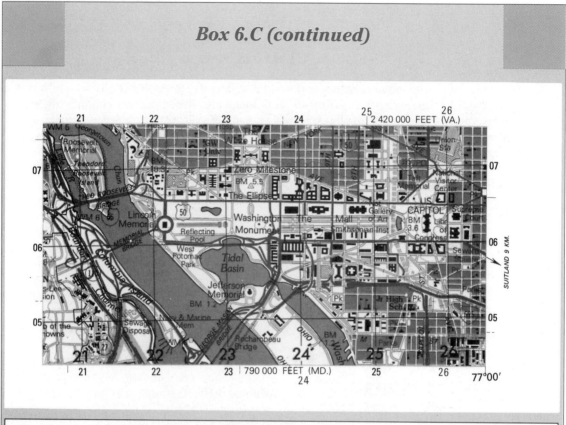

Figure 6.C.2 The portion of the Defense Mapping Agency, Alexandria, 1:50,000-scale quadrangle. Compare to Figure 6.12.

widest part (along the equator), about 363 kilometers from the center grid line, the SF is 1.00158. (See Figure 6.11). This positioning of the coordinate grid relative to the map results in an overall accuracy for the UTM system of one part in 2,500. Therefore, you can calculate distances and directions between two points in a UTM zone to an accuracy of one meter in 2,500 meters.

UNIVERSAL POLAR STEREOGRAPHIC (UPS) GRID SYSTEM

In order to cover the polar areas (south of 80°S latitude and north of 84°N latitude) with a coordinate grid as accurate as the UTM grid system, the

Universal Polar Stereographic (UPS) grid system is used. Each circular polar zone is divided in half by the line representing the 0° and 180° meridian. In the north polar zone, the west half (west longitude) is designated grid zone *Y*, the east half as grid zone *Z*. In the south polar zone. the west longitude half is designated *A*, the east half *B* (see Figure 6.10).

In the polar areas, the false northings and false eastings of the poles are given the value 2,000,000 meters in both zones. The 2,000,000-meter easting coincides with the 0°–180° meridian line. The 2,000,000-meter northing coincides with the 90°E-90°W meridian line. Grid north is parallel to true north along the 0° meridian and, therefore, also to

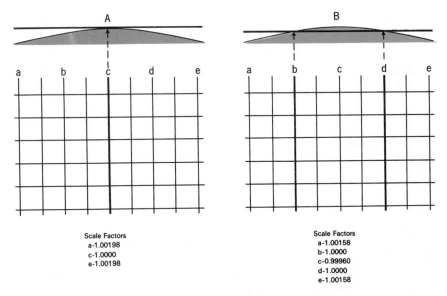

Scale Factors
a-1.00198
c-1.0000
e-1.00198

Scale Factors
a-1.00158
b-1.0000
c-0.99960
d-1.0000
e-1.00158

Figure 6.11   On the secant case transverse Mercator projection, used as a base for the UTM grid system, two lines are made standard. The standard lines, *b* and *d* (small circles parallel to the central meridian) are placed 180 km. east and west of the central meridian. The east-west zone shown here, that includes an overlap area, would be approximately 725 km. wide or about 6.5° of longitude near the equator.

true south along the 180° meridian. The UPS grid zones are divided into 100,000-meter squares, as in the UTM grid system.

For the UPS grid system, a secant case of the stereographic projection centered on the pole is used. This use of the projection arranges the scale so that the SFs are constant along the parallels (circles in this case). However, the SFs vary from parallel to parallel. At the pole, the SF is set at 0.994. At approximately 81° latitude, the SF is 1.0. In the vicinity of 80° latitude, the SF increases to 1.0016. This maintains the accuracy of planar calculations at one part in 2,500.

MODIFICATION OF UTM SYSTEM AS USED BY USGS

Together, the UTM and UPS grid systems cover the entire globe at an accuracy of one part in 2,500. On maps that cover just the United States,

however, we don't need such an extensive system. On its quadrangle maps, therefore, the USGS uses a simplified UTM coordinate system. Since all 50 states are in the northern hemisphere, only the zone number (1 through 60) and an easting and northing value are necessary to designate a point.

Within each UTM zone covering part of the United States, the meridian in the zone's center is given a false easting value of 500,000 meters. The equator is given a northing value of 0 meters for the northern hemisphere coordinates and an arbitrary northing value of 10 million meters for the southern hemisphere (if a reference in the southern hemisphere is needed).

When we use this modified system, we observe all the standard conventions for using rectangular coordinate systems. Therefore, as the example in Box 6.D illustrates, there is a direct connection between the coordinates of a point specified in the UTM grid

Box 6.D

Example of UTM (USGS Modification) and SPC Grid Systems on USGS Topographic Maps

Figure 6.12 shows both the 1,000-meter UTM grid lines of Zone 18 and the 10,000-foot SPC grid lines of the Virginia North Zone, as shown on USGS topographic maps. The three different orientations (the graticule, UTM grid, and SPC grid) result from each coordinate grid being tied to a separate map projection and each coordinate grid zone differing from the graticule. In the graticule, all lines are true north-south or east-west. In UTM Zone 18, the meridian of 75°W longitude is the central meridian of the transverse Mercator projection and is the north-south axis about which the UTM grid is arranged. In the SPC Virginia North Zone, the north-south axis of the SPC grid coincides with 78°W30' longitude.

On published topographic maps, positions of grid lines for the two plane coordinate systems are shown only by hard-to-distinguish ticks around the map margins. Point C in Figure 6.12 is the location of the Washington monument in Washington, D.C. You would read grid references of C as follows:

1. **UTM Grid System (USGS modification).** In the UTM (USGS) grid system, the full reference combines the UTM with an easting and a northing related to the central meridian of the zone (500,000 meters E) and the equator (0 meters N). The full UTM grid coordinates to locate the Washington monument, Point C in Figure 6.12 (in a 100-m square), would be 323400 for the easting and 4306200 for the northing. The reference would be UTM Zone 18, 323400 meters east, 4306200 meters north. The initial small digits (3 in the eastings and 43 in the

northings in Figure 6.12) are commonly dropped from a grid reference when working on a single sheet. For instance, the grid reference 234062 would locate Point C within a 100-m square. More precisely, this reference is the location of the southwest corner of that square. Note the similarity between the UTM and the UTM (USGS) grid systems.

2. **Virginia Coordinate System, North Zone.** On a 7 ½ minute, 1:24,000-scale topographic map, 1/100 inch represents 20 feet. The sheet on which the map is printed will stretch and shrink up to ¼ inch with changes in humidity. Thus, trying to read coordinate positions within 1/100 inch is quite useless. Field use is another matter. In the field, state plane coordinates are surveyed to a fraction of a foot and then are fully reported. The Washington monument, Point C in Figure 6.12, is located at approximately 2,416,750 feet east and 448,500 feet north in the Virginia Coordinate System, North Zone. You would state the reference as: Virginia North Zone, 2,416,750 feet east, 448,500 feet north (coordinates rounded to nearest 250 feet).

As you have probably realized, the two systems were devised with each other in mind. The easting is a six-digit number in the UTM (USGS) grid system and a seven-digit number in the SPC system. Likewise, the northing is a seven-digit number in the UTM (USGS) grid system and a six-digit number in the SPC system. These relationships should help prevent confusion about which system is being used, even if the beginning of the reference is lost.

system and the same point expressed using the USGS modification of the UTM grid system.

STATE PLANE COORDINATE (SPC) SYSTEM

In the 1930s, the U.S. Coast and Geodetic Survey (now the U.S. Chart and Geodetic Survey) devised a system of plane rectangular coordinates for each of the 50 states. It did so to provide surveyors with a convenient way to permanently record original land survey monument locations. This grid structure is called the **State Plane Coordinate (SPC)** system. It is based on the transverse Mercator or Lambert's conformal conic projection, tied to locations in the national geodetic survey system.

To keep the unavoidable scale variation to less than one part in 10,000 (four times the accuracy of the UTM system), a state may have two or more overlapping zones. Each of these zones has its own projection and coordinate grid system. The units used are feet, although recently metric equivalents have been published. Tick marks on USGS large-scale topographic maps show locations of the 10,000-foot grids for the SPC zones. (See Figure 6.12).

For SPC zones mapped on the transverse Mercator projection, the relation between coordinate grid and map projection is the same as in the UTM grid system. Scale is constant north-south along the meridian and varies east-west along the parallels. In contrast, when the Lambert conformal conic is used, scale varies north and south of the curved standard parallels of the projection and is constant east-west along those parallels.

To specify a location using SPC notation, you give the state (for example, Virginia), zone name (North Zone), easting (X value), and northing (Y value). The X and Y values are given in feet.

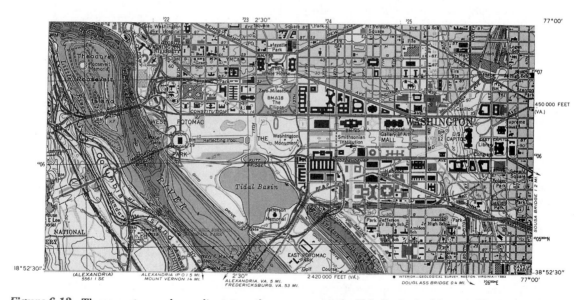

Figure 6.12   Three systems of coordinates as they appear on the U.S. Geological Survey, Washington West, 1:24,000-scale quadrangle. The ticks in the bottom and right margins show the graticule. The dashed tick marks crossing the map margins show the 10,000-foot grid lines of the Virginia North Zone (State Plane Coordinate Grid System), and the black lines show the 1,000-meter UTM grid lines.

PUBLIC LAND SURVEY SYSTEM (PLSS)

In 1785, the U.S. Continental Congress passed legislation setting forth the design of what we now know as the **U.S. Rectangular Land Survey System** or the **Public Land Survey System (PLSS)**. This rectangular land-survey system is different from the UTM and SPC grid systems in two important ways: (1) The basic unit is an acre or an areal unit. (2) The system is defined on the ground, not on a grid superimposed on a map. Today the PLSS covers ¾ of the land area of the 50 states. (See Figure 6.13). Unfortunately for calculation purposes, the system is in U.S. Customary units, and one acre equals 43,560 square feet or ¹⁄₆₄₀ of a square mile. (See Box 6.E).

The PLSS uses a series of origins, as shown in Figure 6.13. A baseline and a principal meridian are defined on the ground. Squares of land six miles on a side are surveyed and marked on the ground. Each six-mile square is labeled by a **township** number north or south of the base line and a **range** number east or west of the principal meridian. Each six-by-six-mile-square township is divided into 36 one-mile-square **sections**. Sections are numbered as shown in Figure 6.14.

Box 6.E

Example of the U.S. Public Land Survey System (PLSS)

Figure 6.E.1 is taken from the 1:24,000-scale USGS topographic quadrangle, Farmersville, Ohio. The village of Farmersville lies primarily in Section 28 of Township 4 North, Range 4 East of the 1st Principal Meridian.

It is easy to see that Section 28 is not a square one mile on a side as the theory of the PLSS would dictate. The original survey marked on the ground was in error. However, by definition Section 28 contains 640 acres. We could locate the church near the center of the village in the PLSS as lying within a 40-acre plot of land designated as the SW ¼ of the NE ¼ of Section 28, T.4.N, R.4.E. of the 1st Principal Meridian. Analyzing this reference theoretically:

T.4.N., R.4.E. designates a 36-square-mile tract of land.

Section 28 designates a one-square-mile of land.

NE ¼ designates a 160-acre square of Section 28.

SW ¼ designates a 40-acre square of the NE ¼.

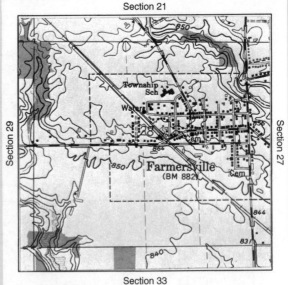

Township 4 North, Range 4 East, 1st Principal Meridian

Figure 6.E.1 A portion taken from the U.S. Geological Survey, Farmersville, OH, 1:24,000-scale topographic map.

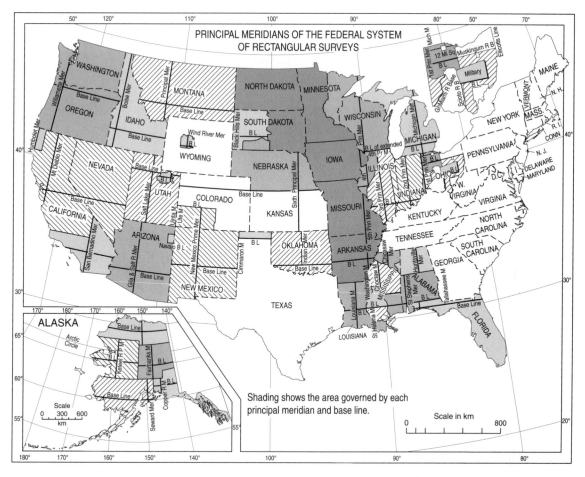

Figure 6.13   Principal meridians and base lines of the U.S. Rectangular Land Survey System.

Within a section, or 640 acres, the land can be subdivided into four quarter sections of 160 acres. Each quarter section can be subdivided into four quarter quarter sections of 40 acres, etc. These parcels can be subdivided further into 10, 2.5, and 0.625 acres.

The above statements represent the system in theory. In practice, the system as defined on the ground was subject to many implementation errors. As a result, sizes of the sections vary considerably.

The PLSS been used for the legal description of property ownership for much of the United States. Because it was initially defined on the ground, and some of the original markers are now missing, the system is also responsible for much litigation over "who owns what?"

The PLSS has left an indelible imprint on the land west of the Appalachians. Here, land use has followed the PLSS grid structure. As a result, roads follow straight lines and intersect at right angles. Building straight roads has increased the cost of road construction and maintenance. Studies show that motorists tend to drive faster on these unswerving roads, leading to more traffic accidents per mile.

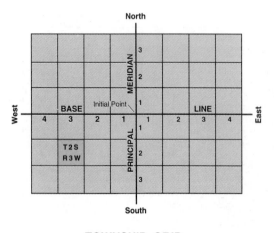

TOWNSHIP GRID

TOWNSHIP 2 SOUTH, RANGE 3 WEST

SECTION 14

Figure 6.14 General diagram of the subdivisions and numbering systems within the U.S. Rectangular Land Survey System.

CONCLUSION

This chapter has emphasized coordinate systems used in the United States. Similar systems exist in other countries. Almost every nation has a variation of the UTM or SPC grid system, although only the United States uses the foot as the basic unit. Likewise, the areal system used in the United States (the PLSS) has been predated and postdated by similar systems in other nations. The first known area referencing system was used by the Etruscan civilization of pre-Roman Italy. Another system was in early use in China, and Canada developed a similar system at the same time as the United States.

The utility of rectangular coordinate systems and the ease of making calculations using grid references in small areas assures that these systems will be in use for many years. Software exists to make the calculations and transformations from latitude and longitude to these coordinate systems. Using this software, we can digitize data in any of these systems to be entered into a database. The data will be compatible with other information referenced in one of these rigorously defined systems.

SELECTED REFERENCES

Claire, C. N., "State Plane Coordinates by Automatic Data Processing," Coast and Geodetic Survey, *Publication 62-4*, U. S. Department of Commerce, Washington D. C., U.S. Government Printing Office, 1968.

Department of the Army, *Grids and Grid References*, TM 5-241-1, Headquarters, Department of the Army, Washington, D. C., 1967.

Mitchell, H. C., and L. G. Simmons, *The State Coordinate Systems*, U. S. Coast and Geodetic Survey, Special Publication No. 235. Washington, D. C.: U. S. Government Printing Office, 1945, revised 1974.

Pattison, W. D., "Beginnings of the American Rectangular Land Survey System, 1784-1800," *Research Paper 50*, Chicago, IL: Department of Geography, University of Chicago, 1957.

Stem, J. E., "State Plane Coordinate System of 1983," *NOAA Manual NOS/NGD # 5*, Charting and Geodetic Services, NOAA, 1989.

Thompson, M. M., *Maps for America*, 3rd ed., Washington, D. C.: U. S. Government Printing Office, 1988.

Wolf, P. R., and R. C. Brinker, *Elementary Survey-ing*, Harper and Row Publishers, 8th ed., New York, 1989 (especially Chapters 21 and 23).

PART III

SOURCES
OF DATA

CHAPTER 7

GROUND SURVEY AND POSITIONING

Study of the shape and size of the earth, called **geodesy**, is the most technically demanding of the mapping sciences. As should be evident from the discussion of basic geodesy in Chapter 4, geodetic specialists, or **geodesists**, need to be highly proficient in math, science, and engineering subjects. They must also have a love for doing careful work, since even small mistakes can lead to big errors when dealing with something the size of the earth.

Fortunately, the earth is so large that its curvature is relatively insignificant at the scale of most human engineering and property division activities. At the local scale, therefore, it usually suffices to treat the earth as if it were flat, but draped with an undulating landform surface (Figure 7.1). The great attractiveness of this flat earth conception is that it lets us avoid the complicated three-dimensional **spherical geometry** that geodesists must use. Instead, we can determine positions using the relatively simple two-dimensional **plane geometry** which we learned in high school.

Thus, rather than hire a geodesist for local surveying jobs, we can get by with a **plane surveyor** (**surveyor** for short). Since the task of plane surveying is much less tedious and demanding than geodetic work, it requires less skilled personnel and can be done far more quickly and inexpensively.

In the remainder of this chapter, we will discuss the principles, instruments, and methods of surveying. For the most part, the topics are relevant to both geodesy and plane surveying. We will also discuss how modern electronic technology is

changing the way surveying is executed. Finally, we will see how surveying concepts have found their way into positioning system technology, which is revolutionizing navigation.

PRINCIPLES OF SURVEYING

The aim of surveying is, basically, to define the location of a point in our environment. Since location is a relative rather than absolute concept, surveyors determine new positions with respect to an existing reference feature.

The **frame of reference** for geodesy consists of such factors as the axis of earth rotation, the equator, a prime meridian, mean sea level, barometric pressure, and the pull of gravity. For most official plane surveys today, geodetic control points, both horizontal and vertical, provide the accepted frame of reference (see "Geodetic Control Networks" in Chapter 4). In the past, however, it was common for plane surveyors to use a local landmark or special survey monument (rock pile, stake, road intersection, etc.) as a point of beginning for local survey activity.

In addition to a frame of reference, surveying is based on the measurement of distance and angles. The way these distance and angle measurements are used in combination establish three fundamental principles of surveying. These principles are straightforward applications of the rules of triangles taken from Euclidean geometry, as we will now see.

ANGLE AND DISTANCE

If two angles and one side of a right triangle are known, the size of the third angle and the length of the other two sides can be determined both graphically and trigonometrically. Thus, a surveyor can determine the position of an unknown point by measuring the direction and distance to it from a known point (Figure 7.2A). If the coordinates of the known point are available, they can be determined for the unknown point as well.

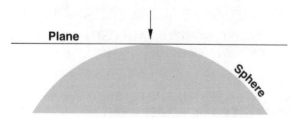

Plane

Sphere

Figure 7.1 A plane is a reasonably good approximation of a sphere in the near vicinity of a point on the curved surface.

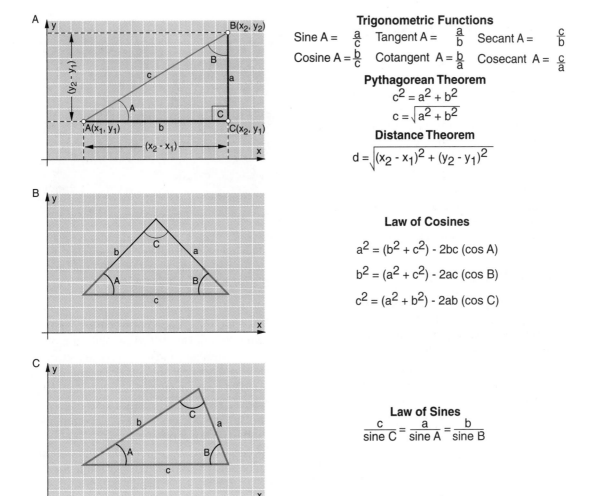

Trigonometric Functions

Sine A = $\dfrac{a}{c}$ Tangent A = $\dfrac{a}{b}$ Secant A = $\dfrac{c}{b}$

Cosine A = $\dfrac{b}{c}$ Cotangent A = $\dfrac{b}{a}$ Cosecant A = $\dfrac{c}{a}$

Pythagorean Theorem
$$c^2 = a^2 + b^2$$
$$c = \sqrt{a^2 + b^2}$$

Distance Theorem
$$d = \sqrt{(x_2 - x_1)^2 + (y_2 - y_1)^2}$$

Law of Cosines

$$a^2 = (b^2 + c^2) - 2bc\,(\cos A)$$

$$b^2 = (a^2 + c^2) - 2ac\,(\cos B)$$

$$c^2 = (a^2 + b^2) - 2ab\,(\cos C)$$

Law of Sines
$$\dfrac{c}{\text{sine } C} = \dfrac{a}{\text{sine } A} = \dfrac{b}{\text{sine } B}$$

Figure 7.2   Known relations between the sides and angles of triangles can sometimes be used to determine the length of lines and the size of angles that are not known.

ANGLE-DISTANCE-ANGLE

If two angles and the included side of a triangle are known, the third angle and other two sides can be determined both graphically and trigonometrically (Figure 7.2B). The known distance is usually called a **baseline**. Thus, a surveyor can determine the position of an unknown point by measuring the angles to it from the endpoints of a known baseline. If coordinates are available for the baseline endpoints, then they can be computed for the unknown point.

DISTANCE-DISTANCE-DISTANCE

If the length of all three sides of a triangle are known, the corner angles can be determined both graphically and trigonometrically (Figure 7.2C). This means that a surveyor can determine the position of an unknown point by measuring the distances to it from the endpoints of a known baseline. If coordinates are available for the baseline endpoints, they can be computed for the unknown point.

MEASUREMENT TECHNOLOGY

The history of surveying is a story of changing technology and methods. Throughout the years, the emergence of a new technology was quickly followed by innovations in the way surveyors made their measurements. There has been a pronounced shift in technology, beginning with manual methods, and progressing in turn through magnetic, mechanical, optical, and electronic technologies. Since the advent of a new technology tends to augment rather than replace older technologies, contemporary measuring instruments used by surveyors tend to be technological hybrids.

MEASURING DISTANCE

For general purposes, visual estimates and step (pace) counting often suffice in determining distance. But surveyors require methods that provide greater precision. **Mechanical aids**, such as measuring rods, flexible chains, and ruled measuring tapes, have all been used. Each of these instruments suffers the problem that someone must physically lay it out in a straight line between the points to be measured. This is easier said than done.

Imagine, for example, the task of stretching a measuring tape between points separated by irregular terrain covered by streams, swamps, lakes, vegetation, buildings, biting insects, and other obstructions. Also, a chain or metal tape will sag if it is stretched between two points without support in the middle. Therefore, the readings must be adjusted to derive the straight-line distance.

Most of the drawbacks of mechanical measuring devices are eliminated with **electronic distance measuring (EDM) instruments**. These tools measure the time it takes a laser beam or radio wave to travel from a transmitter to a receiver (Figure 7.3). With most EDM instruments, the transmitter and receiver are contained in the same unit, requiring that only a reflector be placed at the distant point.

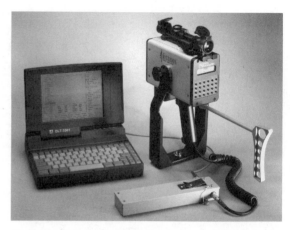

Figure 7.3 Laser beams and radio waves can be used to determine distances between electronic distance measuring instruments with great accuracy. (Courtesy Laser Technology, Inc.)

Since the travel speed of electronic signals is a known constant (the speed of light), the distance between transmitter and receiver or reflector can be easily and accurately computed. The more sophisticated EDM instruments make all calculations automatically, display the result, and store critical data for future use. In some cases, data are telecommunicated to another site for subsequent use.

In addition to accuracy and automation, another benefit of EDM technology is that there is no need to drag a tape over the landscape. EDM instruments do require a unobstructed **line of sight** between transmitter and receiver, however. Surveyors sometimes will circumvent line of sight problems by erecting a temporary tower to hold an EDM device above obstructions. They may even hang an EDM instrument from a helicopter positioned over an especially inaccessible site.

MEASURING DIRECTION

Direction is defined as angular deviation from a baseline. Since the axis of earth rotation provided a fixed frame of reference that could be determined by astronomical observation, the direction to the North Pole was early adopted to serve as a

true north baseline (see Figure 4.7). Due to the great convenience of the magnetic compass, however, it quickly became the instrument of choice for most surveying activity. Unfortunately, the location of the **magnetic north pole** is always changing (see Chapter 4). Currently, it is located more than 1,000 miles south of the true north pole. This means that for most locations, there is significant difference between the true and magnetic north baselines.

The local difference between the true and magnetic north baselines is called **magnetic declination.** Surveyors must use current declination information to convert magnetic readings in the field to true north readings for purposes of official record-keeping. Surveyors must also contend with **compass deviation**, which is the unpredictable error in compass readings caused by such magnetic disturbances as magnetic ore bodies, metal objects, power lines, and thunderstorms.

Early direction measurements were made with mechanical devices which determined angles away from a baseline. The 360 degree gradations on a protractor or compass face provided a useful tool for this purpose. Indeed, once the magnetic compass was understood, it was soon augmented with crude sighting devices.

The breakthrough in direction measurement came with the invention of optical sighting instruments, especially those able to magnify the size of distant objects. By equipping their optics with internal sights, such as the familiar cross-hair, surveyors were able to improve angle-measuring accuracy even further. Surveyors were quick to integrate optics with a compass. By combining optics with a **bubble level** and vertical **degree circle** (a circular protractor ruled in degrees), they could measure vertical angles as well. The resulting instrument was called a **transit**, which evolved into the modern **theodolite** (Figure 7.4). Addition of a digital read-out of angle data in recent years has overcome limitations associated with both making and reading fine gradations on the degree circle.

Since transits use a magnetic compass to establish a baseline, they are subject to disturbances and

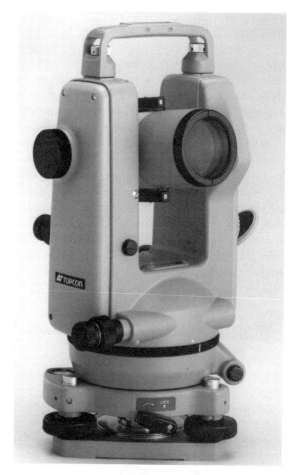

Figure 7.4 By using a transit or theodolite, which has leveling, telescopic, and compass components, surveyors are able to measure both horizontal and vertical angles with accuracy. (Courtesy TOPCON America Corp.)

other problems associated with this magnetic instrument. The invention of the **gyrocompass**, which is able to align a spinning part with the axis of earth rotation, makes it possible to avoid problems that plague the magnetic compass. Unfortunately, a gyroscope requires electrical power to operate and, until recently, has been a bulky instrument. Since both factors limit portability, gyrocompasses have found greater application as navigational aids in ships and airplanes than in surveying.

This situation is likely to change quickly, however, since miniature gyrocompasses that can run on battery power are now being marketed. Units the size of a roll of 35mm film are now available. These small devices are sure to find their way into surveying instruments in the near future.

Still another development in angle measurement is the **radio compass**. This instrument incorporates some form of compass with a receiver and the electronics necessary to determine the direction of radio waves broadcast from a distant location. Radio compasses have been used mostly for navigation purposes in the past. Again, however, miniaturization of electronic components is leading to applications in surveying as well.

TRADITIONAL SURVEY METHODS

Surveying principles can be applied in a variety of ways to accomplish the aim of position finding. The preferred survey method for both two- and three-dimensional position-finding has changed through the years in response to advances in technology. New technologies have also changed the way the different methods are executed, although the same principles still apply.

FINDING HORIZONTAL POSITION

When we think about position, in most cases we have in mind location in a two-dimensional environment. Flat maps encourage this conception of our surroundings. Three surveying methods are also designed to give the horizontal position of points in the environment, as if they were projected vertically onto a single plane. Let's take a look at each of these methods of finding horizontal position.

Traverse

Horizontal surveys had their origin in making careful measurements of the direction and distance

from a known position to an unknown location. The procedure is known as **dead reckoning** or **traversing**. A traverse rarely consists of a single direction and distance measurement. Due to obstructions in the field, it may not be possible to make a single straight line distance measurement between beginning and ending point. The solution is to do the job in a series of legs, each involving a direction and distance measurement. When a traverse ends at a point other than the starting point, it is called an **open traverse** (Figure 7.5A). This type of survey has some drawbacks. Each time a measurement of direction or distance is made, there is risk of making an error, and these errors accumulate with subsequent measurements. Thus, it is routine for surveyors to check the accuracy of their work. Unfortunately, it is difficult to check results of the open traverse method without repeating the whole procedure.

A modification of the open traverse method, called a **closed traverse**, can be used to check the accuracy of a survey. With this procedure, the traverse is extended from the endpoint back to the

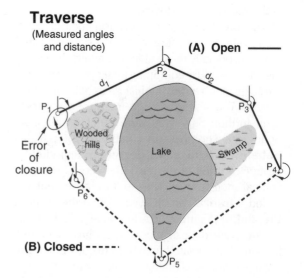

Traverse
(Measured angles and distance)

(A) Open ———

(B) Closed -----

Figure 7.5 The traverse method involves determining the location of an unknown point by making a series of direction and distance measurements.

point of beginning (Figure 7.5B). Due to measurement errors, a closed traverse will rarely "close." But the **error of closure** provides a useful accuracy check, and errors can be redistributed back through the survey legs.

Triangulation

As the development of optics improved angle measuring technology, a method called **triangulation** largely replaced traversing for surveys calling for high accuracy. Triangulation is based on the relationship between the sides and angles of a triangle (Figure 7.6A).

First, we carefully measure a straight **baseline** between two known locations on the ground. Next, from one end, we measure the angle between the baseline and a sightline to the distant third location. Finally, we determine the angle from the opposite end of the baseline to the distant third location. Since the point in question lies on each of the sightlines, it must fall at the point at which the two lines intersect. Thus, starting with a

baseline of known length, we can determine the location of an unknown distant point by measuring angles alone.

Since the two sightlines in the previous example will always intersect at a point, regardless of errors made in one or the other angle measurement, the result appears deceptively accurate. Fortunately, triangulation accuracy is easy to check. If the baseline is segmented or extended to include a third known point, and a third sightline is established to the unknown location, the three sightlines are not likely to intersect at a single point (Figure 7.6B). The intersecting sightlines probably will form a triangle, called the **triangle of error**. The smaller the triangle of error, the greater the accuracy of the angle measurements. Large errors suggest that angle measurements be redone more carefully.

Once the location of a point is determined through triangulation, it can serve as the endpoint of a new baseline, and the process can be repeated many times. This ability to add baselines permits new triangles to be added as well, leading to a **triangulation network** (see Figure 4.12).

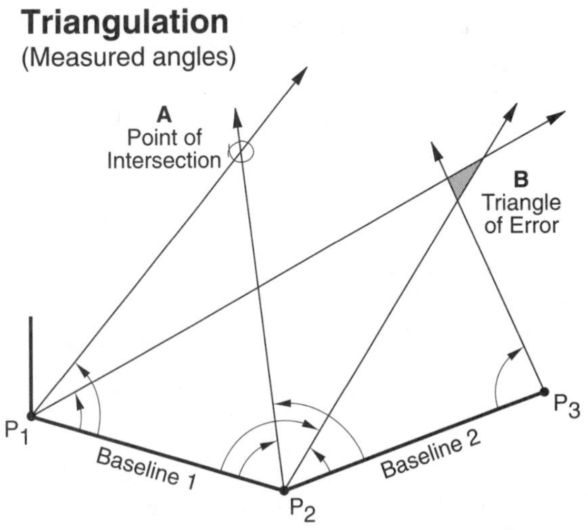

Triangulation
(Measured angles)

Figure 7.6 Starting with a baseline of known length on the ground, the position of an off-baseline point can be determined by triangulation, which involves measuring the angles to the point to be located from the ends of the baseline.

Trilateration

The development of EDM instruments in recent years has greatly improved the accuracy of distance measurements. This greater accuracy has led to an increase in popularity of a survey method called **trilateration**. Trilateration is based on distance measurement alone.

As with triangulation, the first step in trilateration is to establish a baseline (Figure 7.7). The second step is to measure the straight sightline distance from each end of the baseline to the point whose location we wish to determine. The next step is to sweep two arcs, one from each end of the baseline, using the sightline length as each arc's radius. The point at which the two arcs intersect is the location of the position in question, since it is the one point that is the proper distance from each baseline endpoint.

Segmenting or extending the baseline used in trilateration so that a sightline can be made from a

Trilateration
(Measured distance)

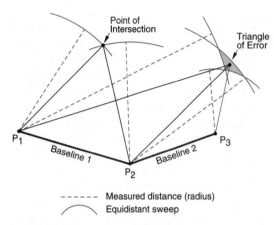

- - - - - Measured distance (radius)
⌒ Equidistant sweep

Figure 7.7   Starting with a baseline of known length on the ground, the position of an off-baseline point can be determined by trilateration, which involves measuring distances to the point to be located from the ends of the baseline.

third known point leads to a triangle of error check on accuracy, as it does with triangulation. Similarly, by adding baselines and making the associated sightline measurements, a trilateration network can be extended across the landscape (see Figure 4.12).

FINDING VERTICAL POSITION

Determining horizontal position suffices for many purposes, such as making planimetric maps. Due to elevation differences between points, however, we often must adjust ground surface measurements to account for the effects of slope. At other times, knowing our vertical position is not a means to an end but our ultimate goal. Surveyors have devised a separate set of methods for finding vertical position. These are generally referred to as **leveling**.

Differential Leveling

Due to its simplicity, an ancient technique called **differential leveling** is still widely used today.

This method involves two instruments. One is a **surveyor's level**, which is a telescopic sighting device equipped with a bubble level. The other device is a **leveling rod**, which is a long stick ruled with fractional gradations.

The survey begins by setting up the level over a control point for which the elevation is known (Figure 7.8A). The leveling rod is held erect by the surveyor's assistant on a line-of-sight position for which the elevation is to be determined. The surveyor then sights on the leveling rod and, compensating for the height of the level itself, determines the height of the second point at which the sightline intersects the ruling. The level can then be set up over the new point for which the elevation is now known, and the process repeated. This is one way a vertical control network is established (see Chapter 4).

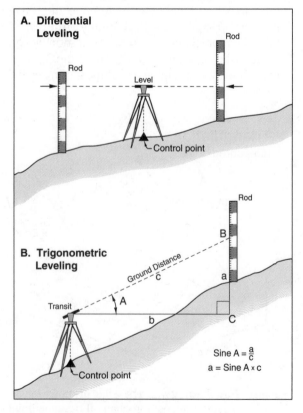

Figure 7.8   The difference in elevation between two locations can be determined using differential leveling (*A*) and trigonometric leveling (*B*).

Trigonometric Leveling

Differences in elevation are also determined using **trigonometric leveling**. In this case, a transit or theodolite is used in place of a level, because a means is needed for measuring vertical angles. With the transit set up over a vertical control point, and a leveling rod held erect on the point whose elevation is to be determined, the surveyor first measures the vertical angle between the horizon (or other level line) and the sightline (Figure 7.8B). The sightline must be focused on the leveling rod at the same height off the ground as the height of the sighting device. The surveyor next measures the ground distance between the known and unknown points.

Trigonometric relations associated with a right triangle are then used to determine both the elevation difference and the planimetric distance between the known and unknown points. As with differential leveling, the process can be repeated over and over to extend the leveling line across the landscape.

AUTOMATED SURVEY SYSTEMS

As we saw in the previous sections, traditional survey methods are laborious and time-consuming. They are also costly. As you might expect, electronic technology has had a great impact on these traditional methods. On one hand, familiar manual procedures have been integrated and automated. On the other, new approaches have been introduced that extend traditional surveying principles into the realm of electronic navigation systems. Let's look at each of these developments.

TOTAL-STATION INSTRUMENTS

The instrument of choice for the modern surveyor integrates an electronic digital theodolite, a electronic distance measuring instrument, and a computer into a single unit. The resulting hybrid instrument is called a **total station** or, less commonly, an **electronic tacheometer** (Figure 7.9).

Figure 7.9 A total-station surveying instrument integrates an electronic theodolite, an electronic distance measuring device, and a computer into a single functional unit. (Courtesy Nikon, Inc.)

A total station automatically measures and displays distance and direction data (both horizontal and vertical angles), and transmits the results to its computer. On-board software then makes the necessary adjustments to reduce slope distances to their horizontal and vertical components. If station coordinates (x,y,z) and a backsight azimuth are entered into the instrument, the coordinates of the sighted point can be computed and displayed.

A total-station instrument is the central component in modern **field-to-finish** surveying systems. Data from the total station are down-loaded into a laboratory computer that possesses data analysis, editing, and plotting capabilities. By combining total station data with information held in existing spatial databases, you have the power to produce a variety of finished maps.

ELECTRONIC POSITIONING

Traditional survey methods are being challenged by several new methods of determining location. Although these methods use longstanding ground survey principles, automated electronic technology has greatly increased the speed of implementation. Since rapid location determination is required in navigation, the term **positioning** has replaced the term survey when referring to these methods. Surveyors can also benefit from positioning systems, however.

Satellite Methods

Electronic technology has led to significant advances in the speed and accuracy of traditional ground survey methods. But the potential of these methods is still limited by the density of control points, earth curvature, and terrain obstructions. These limitations are now being overcome by carrying radio transmitters into space aboard artificial satellites.

The most promising satellite-based system at this time is one supported by the U.S. Department of Defense (DOD). It is called the **NAVSTAR Global Positioning System (GPS)**.* It became fully operational in 1994. Its constellation consists of 24 orbiting satellites (21 plus three operating spares). The satellites are positioned in six evenly spaced orbital planes. Thus, from any point on earth, four to seven satellites will usually be visible at all times (Figure 7.10). Each satellite broadcasts precise time and position data concerning its orbital location.

Figure 7.10 With the aid of the network of GPS satellites you can use a portable GPS receiver to determine your latitude, longitude, and elevation with accuracies of a few feet or less in latitude, longitude, and elevation. (Image of GPS receiver courtesy Magellan Systems Corp.)

The positional accuracy achieved with GPS receivers depends on a number of things. One factor is the level of service supported by the military custodians of the system. DOD now plans to provide a **standard position service (SPS)** and a **precise positioning service (PPS)**. This dual service is called **selective availability**. The accuracy of degraded SPS positioning, which is planned for general public use, falls in the range of 100 meters (about the length of a city block) for most places and times. This is significantly less than the 20-meter accuracy (about the width of a house) achievable with PPS. There is also the risk that DOD will terminate signals open to the public altogether in times of national security concern.

GPS positioning accuracy will also be influenced by such factors as the quality and number of receivers used, the number of satellites contacted, and the time taken to make a fix. The use of multiple receivers is of special interest to surveyors because it can greatly increase positioning accuracy, as well as overcome the potential problem of selective availability. In this multiple-receiver ap-

---

*When the former Soviet Union broke up in 1991, it was in the process of building a global satellite positioning system called GLONASS. Although the GLONASS constellation is not fully configured, it is being used internationally. Unfortunately, the future of this GPS counterpart is uncertain.

proach, called **relative** or **differential** positioning, a fixed-position receiver is located at a control point, and a mobile unit is carried around in the field. Since the fixed receiver is used to transmit local calibration data to the mobile unit, line-of-sight between the two receivers is not necessary. Even with the current incomplete satellite configuration, surveyors using differential GPS techniques can under ideal conditions achieve accuracies in the range of a few centimeters.

Inertial Methods

Electronics has also breathed new life into the ancient traverse method, giving rise to a system called **inertial positioning**. As the name implies, inertial positioning is based on information concerning a vehicle's distance and direction of travel.

Inertial principles are currently being adapted to surveying applications in ground vehicles. Distance is measured in wheel revolutions, and direction is measured using a gyrocompass. More sophisticated three-dimensional systems use accelerometers to measure movement with respect to the x, y, and z axes. With both approaches, a computer is used to make the calculations necessary to determine current position. Inertial positioning is most accurate when carried out over relatively smooth, firm terrain, such as a road surface. Since wheel slippage in off-road situations leads to errors in distance measurement, inertial survey is best limited to relatively accessible points.

SIMULTANEOUS 3-D POSITIONING

Of the various surveying methods, GPS technology is special in that it permits simultaneous determinations of position in three dimensions. This factor, coupled with the relatively high speed at which locational fixes can be made, has opened GPS to a host of navigation applications.

Ships and airplanes, but also ground vehicles, are rapidly shifting to navigational aids that take advantage of GPS technology. In many cases, these navigational devices are able to track the location of the vehicle on a map displayed on a computer monitor. The appropriate map is generally retrieved from some form of disk storage, using the locational information determined by the GPS receiver. The process can be reversed, however, so that the positional data being gathered in the field with a GPS receiver are used as input to a computer mapping system. This lets us map as we go.

These few examples only hint of the impact GPS technology will have on environmental data collection, navigation, and mapping. The changes to come will be revolutionary for cartography. GPS technology promises to change how maps are made and used. It also promises to change what gets mapped and who makes and uses maps.

SELECTED REFERENCES

Buckner, R. B., *Land Survey Review Manual*, Rancho Cordova, CA: Landmark Enterprises, 1991.

Day, N. B., "Private Utility Surveying Engineering: Total Station Experience," *Journal of Surveying Engineering*, 116 (1990): 163-178.

Gerdan, G. P., "Efficient Surveying with the Global Positioning System," *Surveying and Land Information Systems*, 52 (1992): 34–40.

Hurn, J., *GPS: A Guide to the Next Utility*, Sunnyvale, CA: Trimble Navigation, 1989.

Leick, A., *GPS Satellite Surveying*, New York: John Wiley & Sons, 1990

Onsrud, H. J. and D. W. Cook, *GIS for Practicing Surveyors: A Compendium*, Bethesda, MD: American Congress on Surveying and Mapping, 1990.

Reilly, J. P., "P. O. B. 1990 Transit/Theodolite Survey," *P. O. B.*, 16 (1990): 20–31.

Teskey, W. F., "Trigonometric Leveling in Precise Engineering Surveys," *Surveying and Land Information Systems*, 52 (1992): 46–53.

Wolf, P. R. and R. C. Brinker, *Elementary Surveying*, 8th ed., New York: Harper Collins Publishers, 1989.

CHAPTER 8

REMOTE SENSING DATA COLLECTION

*I*n past centuries, the production of maps depicting the earth's surface was hampered by lack of detailed sources of geographic information. Widespread and accurate topographic mapping at large scales only became possible in the 1930s when photographic images obtained by what we now call **remote sensing** began to be taken from aircraft. Remote sensing quickly replaced laborious and inexact data collection by plane table surveying and field sketching.

Remote sensing is the process of collecting, storing, and extracting environmental information from images of the ground acquired by devices not in direct physical contact with the features being studied. Human vision is, of course, a remote sensing process, since we continuously collect and mentally interpret electromagnetic energy from our surroundings. However, our inexact storage and recall of such mental images, our non-vertical ground level or even mountain-top view of the world, and the tiny portion of the electromagnetic spectrum to which our eyes are sensitive make our built-in form of remote sensing less than ideal for mapping. We need images of the same map scale obtained and permanently recorded by remote sensing devices carried high above the ground by aircraft or space vehicles.

Since the 1930s, remote sensing has expanded from aerial photography to encompass imagery of the land surface and ocean floor collected by electronic sensors sensitive to a wider range of electromagnetic energy as well as sonic waves. Platforms carrying sensing devices include orbiting satellites, ocean research ships and submarines, and aircraft ranging from high-altitude reconnaissance jets to remotely controlled drones.

Remote sensing has also been integrated with electronic data processing. Today when a person sits before a computer and begins to call forth data in map format, much of this information may come from databases that were originally obtained by remote sensing techniques. These recent advances in remote sensing not only have given cartographers new tools for creating and updating traditional topographic maps, but have allowed us to map in detail a multitude of new environmental phenomena.

We will begin our study of remote sensing and its use in cartography by looking at the characteristics of electromagnetic energy and sonic waves. We will look at how they interact with ground or ocean features to produce the reflected or emitted energy recorded by sensing devices. This information sets the stage for our examination of different imaging devices, the forms of imagery they produce, and the platforms that carry remote sensing instruments. With these aspects of remote sensing in mind, we can explore the cartographic uses of imagery detailed in Chapter 12.

SOURCES OF ENERGY

Remote sensing devices detect and record **radiant energy**. Radiant energy emitted by or reflected from ground features is transmitted to the sensing instrument in the form of waves. Remote sensing of land surface features is based on detection of radiant energy called **electromagnetic radiation**, or **EMR**.

Electromagnetic radiation from the sun reaches the ocean bottom only in shallow coastal areas. Thus, land-based remote sensing methods are possible only in these areas. However, EMR-based remote sensing of tiny portions of the ocean bottom has been accomplished by artificially illuminating the bottom with floodlamps aboard submarine vehicles. Large sections of the ocean floor have been mapped using a second form of radiant energy—sound waves transmitting **acoustical energy**. Let's look at both sources of energy, beginning with EMR.

ELECTROMAGNETIC RADIATION

Most remote sensing instruments detect and record electromagnetic radiation. Visible light is but one example of EMR, since all objects with temperatures above absolute zero (0° Kelvin, -459.7° Fahrenheit or -273.2° Celsius)* continuously emit a broad range of EMR wavelengths and frequencies.

Wavelength (λ) is the distance between successive wave peaks. Frequency (υ) is the number of wave peaks passing a given point in one second (see Figure 8.1). Wavelength and frequency are inversely related, according to the simple equation λ = cυ, where c is the constant velocity (3×10^8 m/sec) of electromagnetic energy in a vacuum such as outer space.

Although traveling in a wave-like manner, electromagnetic energy is made up of tiny energy packets or particles called **photons**. The energy in a photon is directly proportional to its frequency and hence inversely proportional to wavelength.

The range of possible EMR wavelengths and frequencies, termed the **electromagnetic spectrum**, is immense (Figure 8.2). The electromagnetic spectrum has been divided into energy regions with component wavelengths which are similar in some respect.

Energy regions range from short-wavelength, high-energy gamma rays to long-wavelength, low-energy radio waves. Between these extremes lie the X-ray, ultraviolet (UV), visible, infrared (IR), and microwave regions. Each region is in turn subdivided into bands, such as the spectral color bands of the visible region.

Remote sensing instruments currently operate in all energy regions except radio, X-ray, and gamma ray. The narrow visible region—0.4 to 0.7 micrometers (μm)—is of particular importance not only for human vision, but also because it is ideal in terms of the amount of energy available for detection.

Remote sensing instruments detect and record the energy of photons in the band or bands to which the sensor is sensitive. The amount of energy detected depends on the inherent energy of

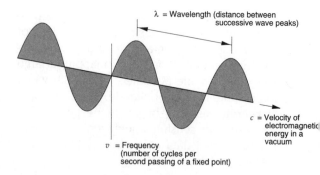

Figure 8.1 An electromagnetic wave.

the photons in the band and on the number of photons reaching the detector during the short time interval of energy collection. The number of photons (energy intensity) reaching the detector varies according to the amount of energy emitted by the illumination source, the amount of energy absorbed by the atmosphere, and the degree to which ground objects reflect and emit energy.

Emission by Illumination Source

We are well aware that warmer objects emit more energy than cooler objects. Physicists further describe this basic physical observation by finding equations that predict the amount of energy that would be emitted across the spectrum at a certain temperature by a **blackbody** surface.

A blackbody is a theoretical object that perfectly absorbs and emits all energy striking its surface. The solid red blackbody energy curve on the top graph in Figure 8.3 shows that: (a) the total energy emitted increases rapidly as the blackbody surface temperature rises, (b) the amount of energy emitted by a warmer blackbody is greater at all wavelengths, and (c) the wavelength of the peak energy emission decreases at a constant rate as the blackbody temperature rises.

No real object can be a blackbody, but the sun comes close, as the solid black line just below the blackbody curve for the sun's 5800°K surface temperature shows in Figure 8.3. Solar energy, with peak emission at approximately 0.5μm (yellow

---

*All objects in the universe must emit electromagnetic radiation, since absolute zero is an unreachably low temperature at which all atomic and molecular motion ceases.

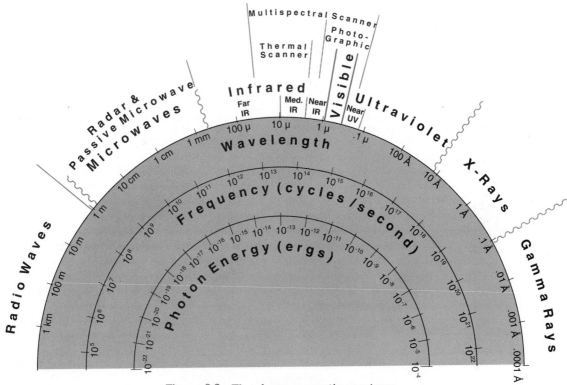

Figure 8.2 The electromagnetic spectrum.

light in the visible spectrum), is the primary source for remote sensing. The earth, with an average temperature of around 300°K, has a peak night-time emission at around 10μm in the middle-infrared band. Many remote sensing instruments are designed to detect and record energy at and near these peak wavelengths.

The blackbody curves also show that sensors operating in the microwave region will have almost no solar or terrestrial energy to detect. Consequently, either energy over large areas must be collected, or the sensor must both emit and detect such long wavelength energy.

Atmospheric Absorption

The portions of the electromagnetic spectrum appropriate for remote sensing are further restricted by atmospheric absorption. This atmospheric ab-

sorption reduces the percentage of energy transmitted through the atmosphere to the ground and back through the atmosphere to the sensing instrument. The shaded areas in the middle graph of Figure 8.3 show the wavelengths in which water vapor, carbon dioxide, ozone, and other molecules selectively absorb energy.

Places of low absorption (high percent transmission), called **atmospheric windows**, are the portions of the spectrum used for remote sensing. In Figure 8.3, we can see the large window in the visible and near-infrared, a place for remote sensing with the human eye, photographic systems, and what are called multispectral scanners.

Photographic devices can record wavelengths from about 0.3 to 1.2μm, or about three times the range of human vision. Electronic scanners or radar detectors must be used to detect longer wavelengths.

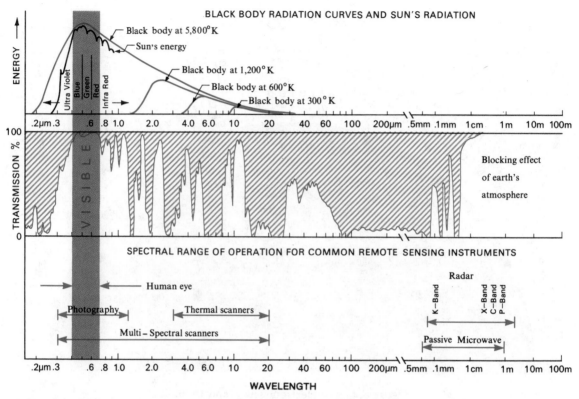

Figure 8.3 The electromagnetic spectrum, showing the spectral regions, energy curves for various blackbody temperatures, and atmospheric windows where remote sensing is possible.

The second major atmospheric window includes the areas of high transmission from 3 to 5 and 8 to 14 μm. This middle (thermal) infrared window is the place for remote sensing by thermal and multispectral scanning instruments. Small windows between 1 mm. and 1 cm. and the high transmission beyond the 1-cm. wavelength allow passive microwave and radar remote sensing, as well as television and radio broadcasting, to be carried out.

Energy Absorption and Reflection by Ground Objects

Ground objects selectively reflect and absorb electromagnetic energy due to differences in the molecular composition of their surfaces. Since this characteristic **spectral response** pattern is predictable, we can use it to identify ground features on remote sensing imagery.

For example, the colors we perceive (see Chapter 19) are due to differences in absorption and reflection at each wavelength within the visible portion of the spectrum. A bright red object is absorbing nearly all incident wavelengths shorter than 0.6μm, while reflecting almost all energy in the 0.6–0.*m range. The absorbed energy at the shorter wavelengths is re-emitted at longer wavelengths invisible to the eye, usually in the middle-infrared.

These differences in reflectance within the visible band also occur in the infrared and microwave bands. Therefore, we can distinguish among objects based on their characteristic patterns of high and low reflectance throughout several energy bands.

Graphing the typical reflectance at each wavelength within one or more bands defines an object's **spectral signature**. Spectral signatures of objects to be identified and mapped by remote sensing are often graphed together (solid lines in Figure 8.4). Thus, portions of the spectrum at which objects differ significantly in reflectance can be selected as the wavelength regions to be used for sensing.

The major problem is that spectral signatures only represent typical or average objects and illumination conditions. Reflectance envelopes, such as the area between the dotted lines for each spectral signature in Figure 8.4, better show the expected reflectance for objects of the same class.

The smoothness of an object's surface also affects the reflection of electromagnetic energy. **Specular** (mirror-like) reflection (Figure 8.5) occurs when surface irregularities are smaller than the wavelength of energy being detected. Most surfaces are far rougher, producing **diffuse** reflection of nearly equal intensity in all directions away from each point on the surface.

In summary, remote sensing systems detect and record electromagnetic energy in several parts of the spectrum, including those invisible to our eyes. Careful selection of regions within spectral bands, based on the spectral signatures of features to be mapped, increases the chance of correct feature

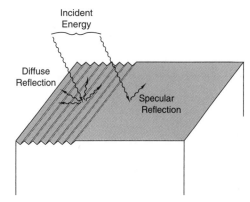

Figure 8.5 Diffuse and specular reflection.

identification. Diffuse reflection of energy from objects with different spectral signatures allows objects to be imaged and distinguished. Images of energy invisible to us provide unusual representations of our surroundings that are important new data sources for mapping our environment. The same can be said of sonic remote sensing devices using acoustical energy.

ACOUSTICAL ENERGY

Remote sensing of the ocean floor that covers three-quarters of the earth's surface is now possible using the acoustical energy of sound waves. Compared to electromagnetic energy, sound waves in water travel at an extremely slow average speed of about 1500 m./sec., or at 1/200,000th the speed of light. However, absorption of sound in water is far less than absorption of electromagnetic energy, which can only be used to remotely sense subsurface features up to 25 m. deep, under perfect conditions.

Sound waves are described by their frequency, which defines their location in the acoustic spectrum (Figure 8.6). Their variation in frequency is from less than 1 to greater than 100,000 hertz (cycles per second).

Remote sensing instruments called **near-surface sonar*** are used to produce medium-scale to

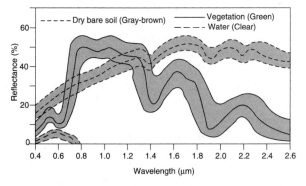

Figure 8.4 Spectral reflectance curves and envelopes for vegetation, soil, and water (adapted from Swain and Davis).

---

*Sonar is an acronym for sound navigation and ranging.

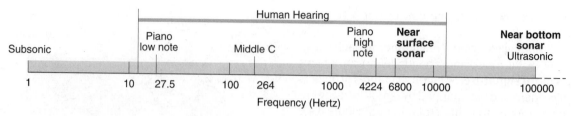

Figure 8.6 The acoustic spectrum.

small-scale images of the ocean floor. They do so by transmitting and recording sound pulses at 6,800 hertz—a fairly low frequency. These sound pulses create a tone we hear as higher than the top note on a piano.

Near-bottom sonar systems are used for large-scale, high-resolution mapping in shallow water. They use a much higher ultrasonic frequency of 100,000 hertz.

ENERGY RECORDING TECHNOLOGY

Images created by a variety of remote sensing devices are used in map compilation and updating. Remote sensing images are also geometrically rectified and mosaicked (see Chapter 12) to form planimetrically correct photomaps.

Passive sensors, such as photography, detect and record solar or terrestrial energy either reflected or emitted by ground objects. **Active sensors**, such as radar or sonar, generate the energy that illuminates the ground and is subsequently reflected from objects back to the sensor. Energy can be recorded both photographically and electronically, and the record may be in hardcopy or digital data file format.

Cartographers should understand the basic operation of major remote sensing instruments and be well acquainted with the forms of imagery each creates. We will begin with photographic remote sensing, then turn to electronic scanning instruments, and finally to active sensors such as side looking radar and sonar.

AERIAL PHOTOGRAPHY

The aerial photograph was the first form of remote sensing used to inventory and map earth features systematically. Various cameras, photographic films, and camera filters are now available for mapping purposes. Although highly sophisticated electronic systems have recently been developed to record images of the environment, the aerial photograph, with its high resolving power, will continue to be a primary form of remote sensing.

Aerial Cameras

Aerial mapping cameras, first developed during World War I, have long provided aerial photographs for topographic map compilation and resource inventories. The standard mapping camera (Figure 8.7) produces individual large-format 9 × 9 inch (23 × 23 cm.) photos. High-precision six-inch or 12-inch focal-length lenses minimize camera-introduced spatial distortions on the photo. Motor-driven film advances are controlled by timing devices called intervalometers. These advance the film in accordance with the flying height and aircraft speed so that a 60-percent overlap is maintained between successive exposures.

Large film format and high-quality, well calibrated lenses make the aerial camera an ideal device for imaging sizeable areas in great detail. But these cameras are expensive, heavy, and require special mounts in aircraft or other platforms. Consequently, lighter and less costly 70-mm. and 35-mm. cameras used in professional photography are often taken into the air and operated manually to obtain vertical photos of small ground areas.

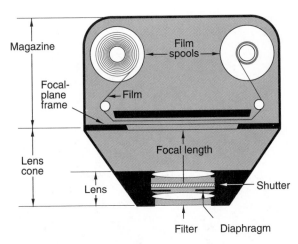

Figure 8.7 A generalized diagram of a mapping camera .

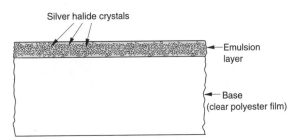

Figure 8.8 Cross-section of black-and-white photographic film.

Photographic Films

The geometric fidelity of aerial photographs depends on the camera and lens design. Photo appearance and resolving power depend primarily on the type of film used.

Photographic films are composed of a transparent plastic base coated on one side with an emulsion of silver salt crystals (Figure 8.8). These crystals are sensitive to electromagnetic energy.

The emulsion is exposed to photons reflected from ground objects and focused by the camera lens. Silver crystals struck by a sufficient number of photons change their chemical properties so as to turn black upon film development. Other crystals exposed below this energy threshold remain in their original state. Photographic emulsions can record energy between 0.3 and 1.2μm, or from the near-UV through the near-IR. Four emulsions— panchromatic, black-and-white infrared, true-color, and color-infrared—have been developed for remote sensing within this spectral range.

Panchromatic film sensitive to the entire 0.3 to 0.7μm near-UV and visible energy range is used most often for black-and-white photography. Unfortunately, haze due to atmospheric scattering is more intense in the shorter wavelength UV and blue spectral regions. To eliminate the loss of image contrast caused by haze, a filter that absorbs these shorter wavelengths is normally placed in front of the camera lens. It is also possible to photograph only selected wavelengths, such as green light, by using filters that block all UV and visible wavelengths except those desired. Several cameras, each with a different filter, can also be aimed identically at the ground and set to simultaneously expose panchromatic films, producing what is called **multiband photography** (Figure 8.9).

The larger the number of photons reflected from an object and focused by the lens to a spot on the film, the darker will be the gray tone on the developed film. This negative image will in turn produce a lighter tone on a photographic print made from the film. Gray tones on the standard paper print (Figure 8.10) record the reflectance from each of the myriad objects imaged.

Human constructions, such as roads and buildings, may be either light or dark on the image, depending on their surface composition (concrete or asphalt). Newly plowed fields, moist soils, and wetlands often are imaged darker than neighboring areas.

The same object may appear light or dark depending on the relative position of the sun and remote sensing device. Highly reflective objects may appear dark if sufficiently smooth so that sunlight is specularly reflected away from the camera lens. On the other hand, a circular white

Figure 8.9 Four 70-mm. Hasselblad cameras mounted for obtaining multiband photography (from Lillesand and Kiefer).

Figure 8.10 A panchromatic aerial photograph of central Washington, D.C.

"sunspot" will sometimes show up in an otherwise dark-toned lake. This spot is caused by specular reflection from the lake surface onto the lens.

Panchromatic photography has many uses in cartography. In addition to being the major data source for topographic mapping, such photos are widely used for mapping environmental features such as soils, croplands, or forests, as well as engineering works such as roads, dams, and reservoirs. Panchromatic photographs are also the most common images used in creating photomaps and orthophotomaps.

Black-and-white infrared film is similar to panchromatic, but with emulsion sensitivity extended into the near-infrared (0.3- to 0.9-μm range). This film is used with a camera filter that absorbs all UV and visible energy, so that only 0.7–0.9μm near-IR energy is recorded. Since highly scattered short wavelength energy has been filtered out, black-and-white infrared photography is very effective in penetrating haze. Thus, it can be used successfully on days when ordinary panchromatic or color film is unsatisfactory.

The greater tonal contrast found on typical black-and-white infrared film transparencies and prints (Figure 8.11) makes them useful for certain map compilation problems. For example, they are widely used in forest inventory and mapping, since different types of forest cover can be distinguished due to the distinct tonal variations on the images. Because water absorbs infrared energy more than visible light, water often registers as a dark gray or black against a much lighter background. This makes infrared photography an excellent data source for mapping the edges of rivers, lakes, and other hydrographic features.

True-color film was developed in the 1930s. At first, because of its high cost and low resolution, it

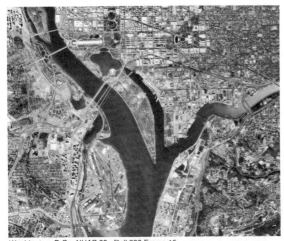

Washington, D.C. - NHAP 80 - Roll 330 Frame 15

Figure 8.11 Black-and-white infrared aerial photograph of central Washington, D.C.

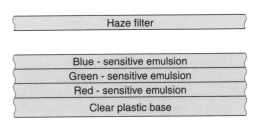

Figure 8.12 Cross-section showing true-color film emulsion layers and UV-absorbing haze filter used to sharpen the image.

was rarely used in mapping. Over the last few decades, however, cost reductions, increases in film speed, higher image quality, and the inherently greater information content of color images have made color film far more attractive for remote sensing than it used to be.

True-color film is similar to panchromatic, except that separate emulsion layers sensitive to blue, green, and red visible energy are present (Figure 8.12). These layers divide the visible band into three components, as in the trichromatic theory of human vision (see Chapter 19). Thus, when we develop the film, we see a realistic multicolor image.

The advantage of true-color film lies in our ability to distinguish thousands of different colors but only a limited number of gray tones (see Color Figure 8.1). Since many objects have characteristic identifying colors, image interpretation is often easier and more accurate when we see objects colored realistically. Subtle color differences may also reveal the condition of natural features, such as the stage of crop maturity. True-color photography has proven to be an indispensable tool for crop monitoring, surface water qual-

ity assessment, soil type delineation, and certain types of geologic mapping.

Infrared-color film differs from true-color in that the three emulsion layers are sensitive to green, red, and near-infrared energy instead of blue, green, and red (Figure 8.13). The three emulsions are also sensitive to blue energy, and a yellow (blue absorbing) filter must be used to correctly expose each layer and to sharpen the image by filtering short wavelength energy. This film, sometimes called **camouflage detection film**, was developed during World War II to assist military image interpreters in distinguishing artificial camouflage from healthy vegetation. This was possible because healthy plant leaves reflect highly in the near-infrared, while camouflage does not. Healthy vegetation and other high infrared reflecting objects appear bright red on film transparencies or prints, features that predominantly reflect red energy appear green, and green reflecting objects will be seen in blue.

Infrared-color images, such as Color Figure 8.2, have many cartographic applications in addition to mapping military equipment and positions. They are ideal for delineating areas of vegetation damage due to diseases, insects, flooding, or other causes. Water typically appears black, making the land-water boundary easy to map. There are also distinctive color differences between natural vegetation and human features that make infrared-color film well suited to mapping land use in urban areas.

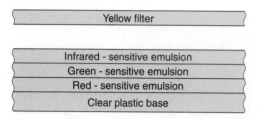

Figure 8.13 Cross-section showing infrared-color film emulsion layers and yellow filter used to absorb blue light to which all layers are sensitive.

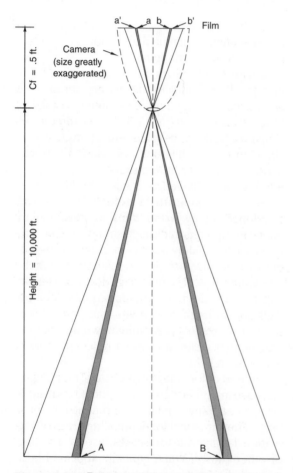

Figure 8.14 Because of the perspective projection of an aerial photograph, the locations of images on the film (and print) are determined by their vertical positions and their distances from the point directly beneath the camera.

Aerial Photograph Geometry

An aerial photograph is a perspective projection of the land surface (Figure 8.14). Assuming no camera lens distortion, no shrinking or swelling of the photographic film or paper, and insignificant earth curvature over the area photographed, the scale across a truly vertical photograph of perfectly flat terrain will be constant. This means that the positions of features on the photo will exactly match those on a planimetrically correct map. In this case, the photo scale may be computed from the simple equation $RF = Cf/H$, where RF is the representative fraction, Cf is the camera focal length, and H is the flying height. For example, a photo obtained by a camera with a six-inch, focal-length lens from a flying height of 10,000 ft. will be at a scale of 0.5/10,000, or 1/20,000.

These ideal conditions rarely exist, however. For one thing, the ground is irregular in height. Hills, ravines, and other landform features introduce local scale variations because of the perspective view recorded by the camera. Higher features will be displaced away from the center of the photo, whereas lower features will be displaced toward the center. This **relief displacement** becomes more pronounced the farther the feature is from the photo center.

For another thing, most images are tilted slightly from vertical. This tilt is due to aircraft roll at the time of exposure.

Because of these local scale variations and image tilting, we can't merely trace information from the photo directly onto planimetric maps. Rather, we must use special photogrammetric instruments (described in Chapter 12) to transfer photo information to planimetrically accurate maps.

ELECTRONIC IMAGING DEVICES

Photographic film emulsions used in remote sensing can only be made sensitive to energy in the 0.3 to 1.2-μm range.* In all other regions of the electromagnetic spectrum, we must use electronic detection and recording instruments.

During World War II, experiments were conducted into the design and use of electronic sensors. This early work was largely concerned with detecting military targets by recording the middle (thermal) infrared energy they emitted both during the day and at night. Electronic imaging devices have now been developed for most parts of the spectrum from the visible through middle-IR wavelengths.

The three forms of imaging instruments currently in use are termed **matrix array, pushbroom**, and **whiskbroom**. These terms roughly describe the way these instruments create images. Let's take a look at each one.

Matrix Array Instruments

Matrix array instruments operate by focusing radiation from ground features onto a two-dimensional array of **charge-coupled device (CCD)** detectors (Figure 8.15). Each energy detector in the array is so small (around $20\mu m \times 20\mu m$) that a $1,000 \times 1,000$ array would just cover your thumb. Each detector measures the intensity of radiation striking its surface, converts the intensity to a digital number, and transmits this number to a large hard disk in the control computer. Thus, the image is simultaneously recorded by all detectors and instantly transmitted to the storage device.

The geometric properties of the image captured by a matrix array instrument are identical to the properties of a standard aerial photograph. This is because a camera lens focuses the ground radiation on the flat CCD surface. There, the entire image is simultaneously recorded. Relief displacement will occur radially away from the center of the image, and adjacent overlapping images will have measurable parallax differences.

Pushbroom Instruments

Pushbroom instruments, also called linear array cameras, detect and record **scan lines** perpen-

---

*Photographic films are used to record X-rays and individual gamma rays, but these very high energy sources are not used in remote sensing.

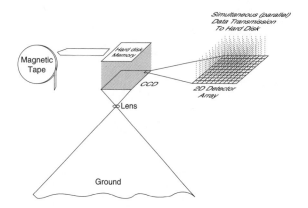

Figure 8.15 Basic components of a matrix array detection device.

dicular to the flightline. To do so, they use a sensor composed of thousands of tiny CCD energy detectors arranged in a linear array (Figure 8.16).

Each detector produces voltage signals proportional to the intensity of photons striking its surface. Voltage signals from all detectors are simultaneously sampled and converted into an array of digital numbers representing one scan line. These numbers drive a photographic film recording device, and may be stored on tape or disk for further analysis. The entire image is built by moving the pushbroom scanner along the flightline at the correct speed so that (in theory) no gaps or overlaps in ground coverage occur between adjacent scan lines.

The geometry of pushbroom scanned images is more complex than photographic or matrix array imagery, and it is often impossible to completely correct the geometric distortions in the scanned image. Geometric distortions on the scanned image are introduced by **platform perturbations** along the flightline as well as by the scanning instrument. Unwanted aircraft motions along the flightline make it practically impossible to collect a long series of perfectly parallel scan lines. Consequently, scanned images may be unsuitable for precise cartographic work. We can, however, manually adjust thematic data from these images to match a planimetrically correct base map.

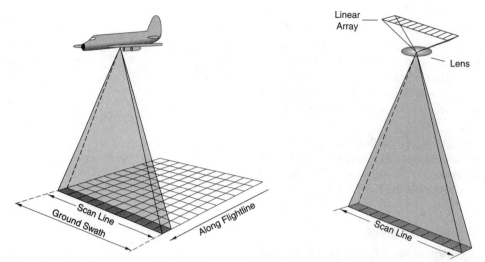

Figure 8.16 Pushbroom instruments detect energy along a strip perpendicular to the flight direction of the sensor platform. The linear array of detectors simultaneously records information for each strip. The rate of recording is synchronized with the aircraft speed to create a continuous image.

Whiskbroom Instruments

Instead of instantly recording an entire image the way a matrix array does, whiskbroom instruments use a single detector to record the energy received in each desired wavelength band. As the remote sensing platform (aircraft or satellite) travels along its flightline, synchronized rotating mirrors in the scanner collect energy coming from a narrow scan line perpendicular to the flightline. The scanner then transmits the collected energy to an electronic detector (Figure 8.17). The electronic photodetector converts different electromagnetic energy intensities into voltages proportional to energy intensity. The continuous voltage signal in turn modulates a point light source that exposes photographic film or paper to produce an image of the scan line. The rotating mirror is synchronized with the aircraft or satellite speed, so that each new scan line is recorded precisely adjacent to the previous line.

The continuous voltage signal from the detector may also be sampled at a constant time interval. Each interval along the scan line defines a **picture element** or **pixel**. The sampled voltage in each

interval is converted to a digital number representing the average reflectance within the pixel. These digital values may be stored on magnetic tape and later hand-carried or electronically transmitted to the ground for digital image processing into image format.

The scanner's **instantaneous field of view (IFOV)** is the product of the instrument's angular field of view and the flying height. The IFOV determines the ground width of the scan line and hence the ground resolution.

Scanner images typically have a lower ground resolution than aerial photography. This is because the IFOV width on the image is considerably larger than the average diameter of silver salt crystals on aerial photographic film.

Like pushbroom instruments, whiskbroom devices are subject to platform perturbations that degrade image geometry. **Tangential scale distortion** further degrades the image so that the scale along the scan line is not constant. This distortion is caused by the scanner mirror rotating at a constant angular speed, sweeping over progressively larger ground distances toward the edges of each scan line (Figure 8.18).

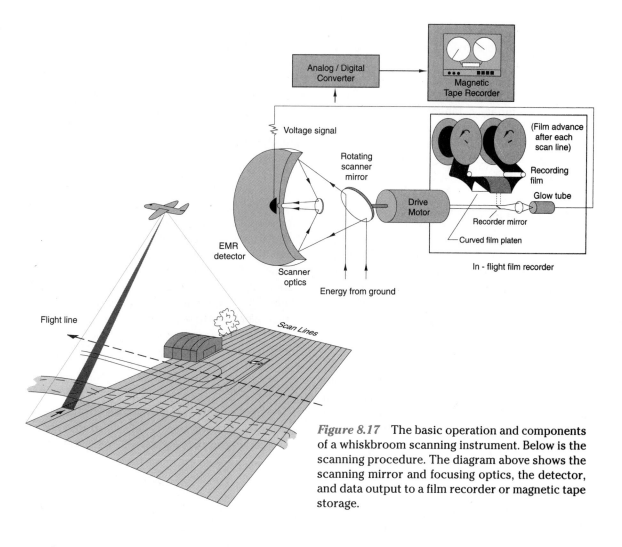

Figure 8.17 The basic operation and components of a whiskbroom scanning instrument. Below is the scanning procedure. The diagram above shows the scanning mirror and focusing optics, the detector, and data output to a film recorder or magnetic tape storage.

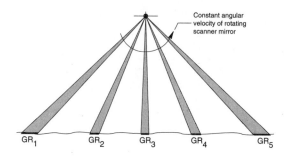

Resulting variations in ground resolution element

Figure 8.18 Tangential scale distortion on whiskbroom scanned images is introduced by the constant angular velocity of the scanner mirror. Ground resolution progressively increases away from the flightline path.

Although some image recorders can be programmed to compensate for tangential distortion, most record onto film revolving on a drum at a constant speed, producing an image that appears progressively compressed toward each edge (Figure 8.19). Fortunately, tangential distortion decreases rapidly as the flying height increases and as the scanner's angular field of view is decreased. Therefore, images from orbiting satellites are little affected.

FORMS OF ELECTRONIC IMAGING

We can use the different electronic imaging devices to create a variety of images. All of these images are made up of pixels. Each instrument produces a characteristic image geometry, however. Let's look more closely at the images produced by these different instruments.

Digital Cameras

Digital cameras using matrix array detectors are the newest form of electronic imaging. A typical digital camera uses a tiny lens to focus ground radiation onto a 1,024 × 1,024 detector array etched into a silicon chip less than one square inch (6.25

cm^2) in area. The microscopically sized detectors allow high spatial resolution images to be collected. Typical spatial resolutions range from 1.5 to 9 feet (0.5 to 3 meters) for aircraft flying heights from 3,000 to 18,000 feet (1,000 to 6,000 m.)

The newest digital cameras contain several matrix arrays that may be adjusted to record different spectral bands within the visible and infrared spectrum. Geometrically identical multispectral images of the blue, green, red, and near-IR regions may be recorded simultaneously, for example. Images such as Figure 8.20 are created using digital image processing methods (see Chapter 12). The images look like standard aerial photographs and have the same image geometry. The overlap area on two adjacent images may be viewed stereoscopically and used to map ground topography. A multitude of other ground features, ranging from land cover types to stream channels, may be interpreted from the image using manual or digital interpretation techniques.

Videography

The detection and recording of electromagnetic energy as video images is called **videography**. A full videographic remote sensing system con-

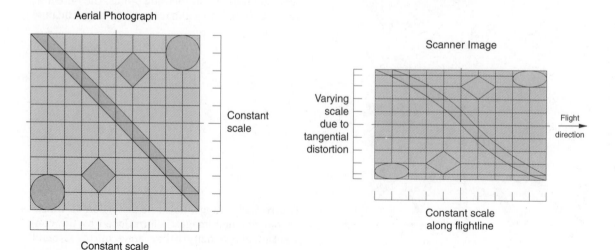

Figure 8.19 Tangential scale distortion compresses the whiskbroom scanner image at both edges.

Figure 8.20 Digital camera image of central Kalispel, Montana, in the green band. (Courtesy Positive Systems, Inc.)

sists of a video camera, a video recording device, a television monitor for image viewing, a photographic hardcopy device, and a "frame grabber" to convert video images into digital images (Figure 8.21).

Video images for remote sensing have been collected by a variety of devices, including **standard video and vidicon cameras**. Early video imaging in the United States was with a NTSC standard television camera (485 lines per frame) sensitive to visible and near-IR imagery.

More recent multispectral cameras use video tubes sensitive to different bands in the visible and near-IR. A typical multispectral design uses geometrically aligned and electronically synchronized cameras with video tubes sensitive to the blue, green, red, and near-IR bands.

A recent development is a single-lens color IR video camera. Energy coming through the camera lens is split into its green, red, and near-IR components. Each of these components is directed to a video tube sensitive to the same band. The video signals from any of these NTSC standard cameras may be stored on videocassette recorder (VCR) tapes in VHS or BETA formats. European standard camera signals can be stored similarly.

Vidicon cameras have been used in remote sensing since the early 1970s. A vidicon camera focuses an image temporarily on a metallic photo-

conductive plate. The image is then quickly scanned by an electron beam recording device. The electronic signals from the recording device are used to produce images on a cathode ray tube (CRT) monitor or to drive a photographic recorder that creates a permanent record on film. Vidicon and photographic images share the same image geometry, since the image is instantaneously focused on a flat energy-detecting surface in both cases.

Video images such as Color Figure 8.3 have several advantages in addition to those just mentioned. Images may be viewed as they are formed and require no film processing. Videotaped imagery may also be played back immediately and written over with new imagery many times, reducing the cost of acquisition.

These advantages must be weighed against the fact that video images have lower ground resolution and longer exposure times than does photography. The balance for cartography seems favorable, however. Video imagery, particularly "frame grabbed" digital video images, are beginning to be used in map compilation projects. These include the mapping of land use changes, crop growth changes, and similar projects that previously relied on aerial photography.

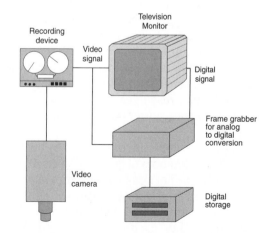

Figure 8.21 Videographic image collection system components.

Thermal-Infrared Scanning

Thermal scanners are whiskbroom instruments that detect middle-IR energy in the 3–14μm range. Many scanners are designed to direct the energy collected to two detectors, sensitive to the 3–5μm and 8–14μm spectral bands. The 8–14μm detector is sensitive to the temperature of the scanning device as well as to energy from the ground. Therefore, the entire instrument must be cooled with liquid nitrogen to an extremely low temperature and enclosed in a heat-proof box. Scanners operating in the 3–5μm band are cheaper to construct and operate, since they must be cooled only to around −200°C.

Since thermal scanners do not operate in the visible region, they may be used to record the nearly equal mixture of reflected and emitted thermal IR energy during the day, or emitted energy alone at night. This day-night sensing capability is an advantage over photography, but the resolving of landscape detail is inferior. One reason for this inferiority is the coarse ground resolution associated with the sensor IFOV. Another reason is that the small temperature differences among objects decrease image contrast.

The 3–5μm and 8–14μm bands correspond to peak energy emission from objects in the 300° to 700°C and −60° to 90°C temperature ranges. Consequently, the 3–5μm detector is well suited to recording locations of very hot features such as forest fires and fresh lava. Images from the 8–14μm detector, tuned to normal earth temperatures, are used to map thermal features such as warm water discharges from power plants, surface geothermal resources, the mixing of cold and warm ocean currents, and underground steam lines.

Image interpretation is complicated by heating and cooling differences of ground objects during the day and night. For instance, the temperature of a lake remains fairly stable; while the surrounding land rapidly warms up during the day and cools down at night. Thus, the lake may appear dark on a daytime thermal image and light on a nighttime image. Similar day-night tonal reversals are used to map moisture-related phenomena such as soil moisture differences.

Another image interpretation problem stems from differences in thermal energy emission among similar features made of different materials. These differences, known as **object emissivity differences**, are illustrated in Figure 8.22. For instance, two roofs at the same surface temperature may appear as dark and light tones on the image. The reason is that one roof is made of tar and one of aluminum. Tar has a greater ability than aluminum to emit thermal energy.

Multispectral Scanning

Multispectral scanners simultaneously record electromagnetic energy in several narrow spectral bands. Most multispectral scanners are similar in design to the whiskbroom scanner described above, but with a prism-like separation of detected energy into narrow bands, called **channels.** Each band is recorded by a separate energy detector tuned to the channel (Figure 8.23). Four-channel multispectral scanners are common, and devices with up to 24 channels have been tested.

Figure 8.22 Thermal scanner image of an industrial area in Millersburg, Oregon.

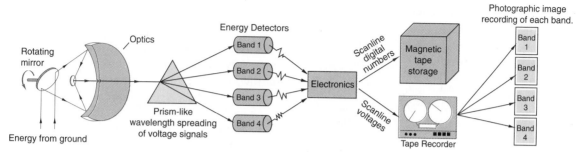

Figure 8.23 Multispectral scanning system components.

Pushbroom multispectral scanners have also been developed. These operate by splitting detected electromagnetic energy into narrow bands. The energy in each band is then directed to a separate linear array of detectors sensitive to those wavelengths. At present, scanners with up to five channels within the 0.4 to 1.1μm range are in operation.

Multispectral scanner images, such as Figure 8.24, have proven useful in many mapping projects. You can view combinations of images of the best spectral bands and thus identify features based on their spectral signatures. Examples include crops, natural vegetation types, soil moisture variations, geological features, and land use classes. You can distinguish and delineate such surface features by studying tonal differences on appropriate sets of multispectral images.

Passive Microwave Sensing

Passive microwave sensors detect and record energy in the 0.1 to 25 cm. band of the microwave region. Here, atmospheric absorption is minimal.

Passive microwave sensors are similar to whiskbroom scanners except that detection of microwave energy requires an antenna device called a **microwave radiometer.** The microwave radiometer, with its narrow-beam antenna, is attached to the scanning device that sweeps along a scan line perpendicular to the flightline. Energy detected is amplified, converted to strings of digital numbers proportional to energy intensity, and stored on magnetic tape. These digital images are often displayed on computer monitors by assigning gray tones or colors to the range of numbers. The digital data are also used to drive a photographic recorder that creates monochrome or color images on film or paper.

Microwave energy emitted by ground objects and transmitted through the first few meters of the subsurface can be detected at night. During daylight hours, the energy detected includes a reflected solar component. In either case, the total energy is very slight, and large areas must be

Figure 8.24 Aircraft multispectral scanner images of central Detroit (courtesy ERIM).

sensed to gather enough energy for a meaningful recording. Consequently, passive microwave images have a low spatial resolution relative to thermal and multispectral imagery.

Passive microwave images are used to map soil moisture, since part of the recorded energy originates in the shallow subsurface, and microwave emission is affected by moisture. Other cartographic applications include mapping ocean surface conditions, inventorying the water content of snowfields, delineating the location of ice-water boundaries, and distinguishing among geologic strata.

SIDE LOOKING AIRBORNE RADAR

The small amount of microwave energy emitted by the sun and earth makes **active** remote sensing possible. An active microwave sensor first emits an energy pulse in the 0.4 to 12 inch (1 to 30 cm.) wavelength range, and then measures the amount of energy reflected back from the ground. We have, of course, just described the basic operating principle behind **radar.***

The first radar developed by the British during World War II was of the **Plan Position Indicator (PPI)** type. A rotating antenna emitted microwave pulses during its full 360° scan. Back-reflected energy was recorded as spots of light on a cathode ray tube. Both the intensity and round trip travel time of all returned energy from a pulse was measured, allowing highly reflecting objects such as storm clouds, aircraft, and ships to be located in terms of their direction and distance from the PPI instrument. Unfortunately, this polar coordinate system form of remote sensing lacks the spatial resolution required for most cartographic applications.

Side Looking Airborne Radar (SLAR) was developed after World War II as a high-resolution remote sensing system capable of producing detailed photo-like images of large ground areas. There are actually two forms of SLAR: real aperture and synthetic aperture radar (SAR).

---

*Radar is an acronym for range detection and ranging.

Real Aperture Radar

A **real aperture radar** image is created by the back-reflected energy from thousands of short microwave pulses. Each pulse is focused in a narrow, fan-shaped beam perpendicular to the flightline on one side of the aircraft (Figure 8.25).

As each pulse sweeps across the ground, variable amounts of energy are reflected back to the radar antenna-receiver device. The farther the object is from the aircraft, the longer the time interval between the pulse emission and returned energy reception. Since the microwave energy travels at the speed of light, the interval is directly proportional to the object's distance from the aircraft, called the **slant range** distance.

When the aircraft has moved forward one pulse width, another pulse is emitted and the detection process is repeated. The entire radar image is created by recording the variable energy returned from many thousands of successive pulses. The returned energy forms one scan line on the image, with the distance along each imaged scan line being equal to the slant range. Each scan line is photographically recorded by a narrow light beam that sweeps across the film, its intensity at any instant controlled by the strength of the returned energy. The entire image is created by stepping the film forward the width of one scan line between exposures.

Synthetic Aperture Radar

A more planimetrically correct and higher resolution image is created by **synthetic aperture radar (SAR)**. Here a much shorter antenna emits and receives returned energy from ground objects.

The details of SAR image formation are beyond the scope of this book, but we will briefly describe the basic principle. Travel time and shifts in the frequency of returned energy (due to aircraft motion) are recorded on an intermediate film. This process is similar to the classical hologram. Laser light is transmitted through this film to construct the final SAR image. The time-frequency data may also be recorded digitally so that a numerical

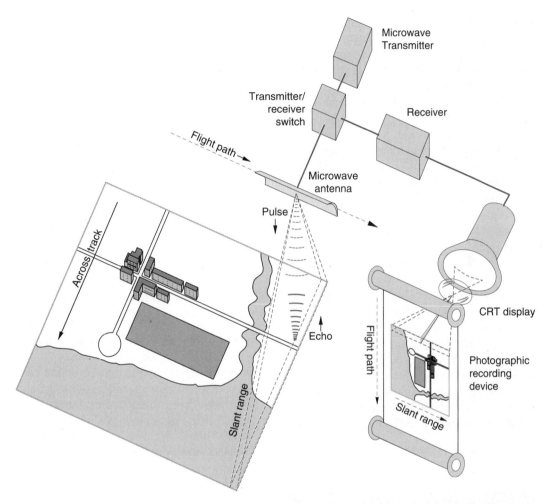

Figure 8.25 Components and operation of a real-aperture SLAR system (adapted from Goodyear Aerospace Corp.)

version of the intermediate image can be transmitted electronically to a ground processing facility.

The differences in returned energy intensity create different gray tones on the SLAR image. These differences are caused by a number of factors: energy wavelength, ground material composition, the topographic position of features relative to the sensor antenna, and the radar depression angle.

Most SLAR systems detect *Ka*-band (0.8 to 2.4 cm.), *X*-band (2.4 to 3.8 cm.), or *L*-band (15 to 30 cm.) microwave energy. All of these allow day and night imaging and will penetrate haze, dust, clouds, and most forms of precipitation. The longer wavelengths are better able to penetrate heavy rain, snow, and hail, at the cost of poorer resolving of ground detail.

Surface properties influencing the amount of back-reflected energy include roughness relative to energy wavelength, the amount of moisture present in each feature, and the inherent ability of objects of conduct microwave energy (their dielectric property). Smooth concrete and calm water often appear black on the image due to specular reflection away from the antenna, whereas rocky areas such as gravel bars near streams often appear light toned (Figure 8.26).

Microwave energy cannot penetrate massive or dense features such as buildings or hills. Building walls and adjacent smooth flat surfaces such as empty parking lots sometimes form microwave **corner reflectors**, causing a white spot on the image marking the building side. Because of the low angle at which many hills are "illuminated" by low-flying aircraft, shadows similar to those cast by hills just before sunset are often formed on the image (Figure 8.27). These shadows enhance the three-dimensional impression of relief and often make subtle landforms stand out clearly. On the other hand, large radar shadows often obscure a significant percentage of the landscape, including many important objects that must be mapped.

The depression angle between the antenna and ground object also affects radar shadows (see Figure 8.27). Given two identical hills, the smaller the depression angle, the greater the radar shadow. This means that the farther the feature is from the aircraft, the greater will be its radar shadow.

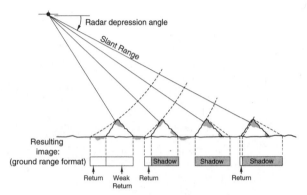

Figure 8.27 Effects of terrain relief in causing displacement on SLAR images (adapted from Lillesand and Kiefer).

SLAR Image Geometry

SLAR image geometry differs considerably from aerial photography and scanner imagery. SLAR and scanner images are similar only in that the image scale in the flightline direction will be the same if a constant aircraft velocity is maintained. The distortion in image scale perpendicular to the flightline is fundamentally different on SLAR images. This is because the scanner records the intensity of returned energy at progressively greater slant range distances from the aircraft, rather than recording planimetrically correct ground range

Figure 8.26 Portion of a SLAR mosaic at 1/250,000 scale of central Washington, D.C. (Courtesy EROS Data Center).

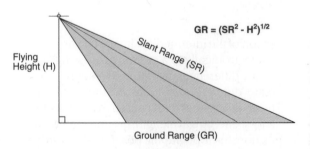

Figure 8.28 Relationship between slant range, ground range, and flying height, assuming flat terrain and minimal earth curvature.

distance. In Figure 8.28, we can see the simple relationship between the slant range (SR), the ground range (GR), and the flying height (H):

$$GR = (SR^2 - H^2)^{1/2}$$

Since a constant flying height should be maintained during image collection, the ground and slant range will differ progressively less as the slant range increases. This means that the portions of the image farthest from the flightline are the most planimetrically correct, exactly the opposite of aerial photography and scanning systems.

SIDESCAN IMAGING SONARS

Oceanographers obtain SLAR-like images of the ocean floor for bathymetric mapping using **sidescan imaging sonars**. The key instrument in the sonar system, other than the survey ship itself, is a **transducer**, which is towed underwater behind the ship (Figure 8.29).

A transducer is like an SLAR antenna in that it both emits pulses of acoustical energy (sound pulses) and records the energy reflected back by bottom features. The energy returned during data collection is converted into a continuous electrical signal. This signal is transmitted over a cable link to the recording device on the research ship. The recorded signals are converted to digital values stored on magnetic tape, or they may drive an image recorder that produces paper images.

Near-surface sonars use transducers towed close to the surface to create images of large sections of the ocean floor at a low spatial resolution. With current equipment, swath widths up to 20 miles (32 km.) on each side of the transducer can be imaged using a 6,800-hertz pulse that is not absorbed significantly by seawater.

Near-bottom sonars operate in shallow water. They place the transducer close to the ocean floor so that high-resolution images of small areas can be obtained. A narrow swath, one mile (1.6 km.) or less on each side of the transducer, is imaged using short ultrasonic pulses at a frequency of around 100,000 hertz.

Sidescan sonar images called **sonographs** (Figure 8.30) are similar to SLAR images. Ocean floor topography plays the dominant role in image appearance. As with SLAR, foreslopes facing the transducer reflect a greater proportion of the energy pulse and are much lighter on the image than

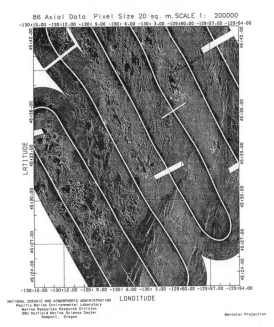

Figure 8.30 Sonograph off the Oregon coast. Note the similarity in appearance to SLAR images. (Courtesy Pacific Marine Environmental Laboratory).

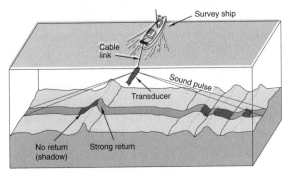

Figure 8.29 Sidescan imaging sonar system components (adapted from Avery and Berlin).

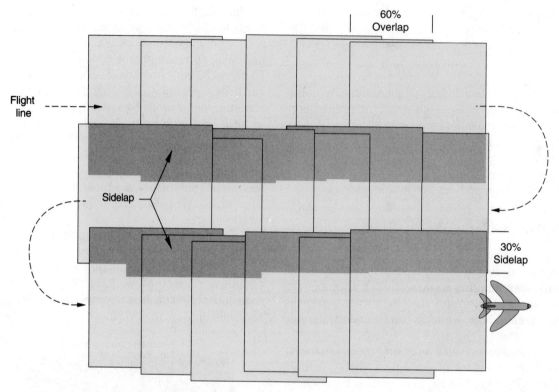

Figure 8.31 A zigzag pattern of adjacent flight lines is flown over a mapping project area.

backslopes. Small depression angles produce large acoustic shadows in hilly ocean floor terrains. Smooth regions of the floor covered with sand and mud cause specular reflections of the sonic pulse away from the transducer, producing dark areas on the image. Rougher bottom materials such as undersea boulder fields reflect sound diffusely, giving light-toned image areas.

REMOTE SENSING PLATFORMS

Platforms are the vehicles that carry remote sensing instruments above the land or water surface. Fixed-wing aircraft have been the traditional platform, but spacecraft have assumed equal importance in the last few decades. Let's look at both, beginning with aircraft.

AIRCRAFT REMOTE SENSING

Fixed-wing aircraft may be fitted with a special mount for holding remote sensing devices in a vertical orientation. Such aircraft have been used to acquire the vast numbers of large-scale photographs and electronic images required for topographic mapping and for large-scale thematic maps. Aircraft have carried all the electronic imaging devices described in the previous sections, but the traditional aircraft sensor in the United States is the six-inch, focal-length aerial mapping camera. This camera exposes 9 × 9 inch (23 × 23 cm.) frames on rolls of black-and-white or color film.

Black-and-white photography for topographic mapping is acquired by flying over the area to be mapped in the zigzag pattern shown in Figure 8.31. Typically, the planes fly at altitudes of 10,000 to 12,000 feet (3,000 to 4,000 m.) The resulting vertical photographs have scales from 1:20,000 to 1:24,000. They cover a ground area of approximately nine square miles (23 sq. km.)

Adjacent flightlines are spaced so that there is a 30 percent sidelap. There is also a 60 percent overlap between successive exposures. Thus (in theory at least), all ground features will appear on at least two photographs. This duplication is a requirement for viewing photos stereoscopically* and for compiling topographic maps using stereoplotting instruments (see Chapter 12).

The most ambitious project aimed at providing a standardized collection of large-scale aerial photographs for topographic mapping and other purposes is the **National Aerial Photography Program (NAPP)**, which began in 1987. Infrared-color photographs covering the entire 48 conterminous United States are obtained by aircraft flying six-inch, focal-length mapping cameras at 20,000 feet (6000 m.) Numerous rolls of 1:40,000-scale, 9 × 9 inch, positive film transparencies have been collected.

Photographs covering larger ground areas at lower spatial resolution—but with less relief distortion—are routinely obtained by cameras carried on **high-altitude aircraft**. One example is the U-2 spy plane of the 1950s and '60s, now operated by the National Aeronautics and Space Administration (NASA). These special planes provide panchromatic and infrared-color 9 × 9 inch transparencies exposed at altitudes ranging from five to 25 miles (8 to 40 km.)

The most ambitious acquisition project was the U.S. Geological Survey's **National High-Altitude Photography (NHAP)** program from 1980 to 1987. The conterminous United States were photographed from an altitude of 40,000 feet (12,000 m.) using a six-inch, focal-length camera to give 1:80,000 black-and-white panchromatic

photos, and an 8.25-inch, focal-length camera to give 1:58,000-scale infrared-color photos. Photographs such as Figure 8.32 have been used by cartographers involved in projects such as producing medium-scale photomaps, revising medium-scale topographic maps, mapping land use across large regions, and outlining natural resources at the county and state level.

Many larger-scale photographs and electronic images are now routinely obtained by sensing devices mounted on low-flying, slow-moving, fixed-winged aircraft including ultralights, as well as helicopters and even the Goodyear blimp. These large-scale images covering small ground areas often contain the level of detail required for site-specific mapping projects such as forest stand inventories. Their major drawback is that relief displacement may be extreme at the photo edges. Consequently, it may be difficult to make the corrections needed to match a distorted image to a planimetrically correct map.

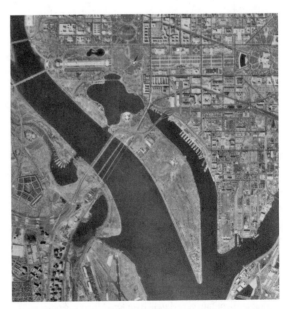

Figure 8.32   Black and white print of a high-altitude infrared-color photograph of Washington, D.C. (Courtesy EROS Data Center).

---

*Stereoscopic viewing involves looking at the same area imaged on two adjacent photos, using devices that allow one photo to be seen with the left eye and the other with the right eye. This causes our mind to form a semi-realistic three-dimensional image of the landscape.

SPACE REMOTE SENSING

Outer space is a natural vantage point for remote sensing instruments collecting lower-resolution, smaller-scale images of the earth.* Space platforms offer several advantages, including minimal relief distortion, long-term repetitive coverage of the entire earth, consistent image scale, and excellent image geometry due to the nearly perfect flightlines followed by space vehicles.

These advantages must be weighed against several drawbacks. For one thing, the low ground resolution is unsuitable for some applications. Also, since energy must travel through the entire atmosphere to the sensor, there is a great deal of energy absorption and scattering. As a result, image clarity and contrast are often severely degraded. In addition, image data are occasionally lost during transmission from ground-controlled satellites to ground receiving stations. Despite these disadvantages, many thousands of photographs and electronic images have been taken from space vehicles over the last three decades and used for mapping purposes.

Human Space Flight Remote Sensing

Remote sensing during human space flights by U.S. astronauts began on February 20, 1962, when John Glenn took some tourist-like 35-mm. color film negatives during his three orbits of the earth. The **Mercury, Gemini,** and **Apollo** missions that followed were aimed at putting humans on the moon, but several forms of remote sensing were tested on Gemini and Apollo missions as a sideline activity.

For example, Gemini 4 astronauts were the first to use 70-mm. Hasselblad cameras to acquire nearly vertical, overlapping true-color photographs. These photographs were used for geological interpreta-

---

*We say this fully recognizing the abilities of military spy satellites to obtain very detailed images of ground objects. The object-specific and classified nature of this type of remote sensing makes such imagery available only to a very limited number of military cartographers.

tion and mapping in several arid regions, including the southwestern United States. The Apollo 9 mission was noted for the first multiband photographs taken by four 70-mm. cameras. These cameras were equipped with filters used to obtain black-and-white photos in the green, red, and two different near-IR bands.

Skylab Imagery The U.S. **Skylab** vehicle operated in a low, 270-mile (435-km.) orbit from May, 1973, until February, 1974. Skylab was equipped with three major remote sensing instruments. One sensor was a single-lens, six-inch, focal-length, multiband camera. This camera split visible and near-IR energy coming through the lens into six bands that exposed different 70-mm. films. Green and red filtered panchromatic, two black-and-white infrared, true-color, and infrared-color films were processed on the ground.

The second instrument was a high-resolution, 18-inch focal-length **Earth Terrain Camera** that could be loaded with panchromatic, true-color, or infrared-color film. Under optimal atmospheric conditions, a 10-m ground resolution was achieved.

The third major sensing device was a 13-channel multispectral scanner that produced relatively poor-quality images. Channels ranged in wavelength from 0.4 to 2.35 and 10.2 to 12.5μm (visible through thermal-infrared).

Multiband camera photographs covered a ground area of approximately 100×100 miles (160×160 km.) The higher-resolution Earth Terrain Camera photos covered 68×68 miles (109×109 km.)

Vertical or near-vertical photographs of numerous research areas between 50°S and 50°N were taken during Skylab's six-month life. These photos were the focus of several mapping-related studies, including the tracing of fault lines in New Zealand and the delineation of seasonal snow cover in Yellowstone National Park.

Soyuzkarta Photography Cosmonauts of the former USSR have been and continue to be involved in taking hand-held camera photographs of the earth and in operating more sophisticated remote sensing instruments. Photography used in

Soviet topographic and thematic map compilation was obtained from remote sensing devices on the *Salyut* and *Mir* orbiting space stations, as well as from the *Resource F1* and *F2* low-orbit automated space vehicles.

Photographs from the Soyuzkarta program were taken using what are called the KFA-1000, MK-4, and KATE-200 cameras. Many of these photos, such as Color Figure 8.4, cover a 75 × 75 mile (120 × 120 km.) area with a purported ground resolution of 15 feet (5 m.). The full 12 × 12 inch (30.5 × 30.5 cm.) photographs are taken from an altitude of 171 miles (275 km.) through a 39.4-inch, focal-length lens, giving an average photo scale of 1:275,000. A two-emulsion "spectrozonal" film, sensitive to green through near-IR wavelengths, or a high-resolution panchromatic film is returned to earth for processing.

Space Shuttle Experiments In addition to launching and retrieving satellites, United States Space Shuttle crews on several missions have collected extensive remote sensor imagery of the earth from two major systems. Most imagery is from the **Shuttle Imaging Radar (SIR)** system, operated on missions in 1981 (SIR-A) and 1984 (SIR-B).

The SIR-A instrument was an *L*-band (9-inch or 23-cm. wavelength) synthetic aperture radar. SIR-B was an upgraded SIR-A system with a selectable initial depression angle. This feature allowed a ground resolution anywhere from 66 to 330 feet (20 to 100 m.), as well as the imaging of ground swaths from 15 to 37 miles (25 to 60 km.) wide, depending on the orbital altitude. This altitude varied between 140 and 224 miles (225 and 360 km.) SIR-C, a new *X*-, *C*-, and *L*-band SAR system, was used on two shuttle missions in 1992 and 1993.

SIR images are useful in identifying geological and terrain features such as faults, folds, and drainage patterns. Vegetation differences, soil moisture variations, and ocean wave patterns also stand out clearly. An unexpected advantage is the penetration of dry wind-blown sand by radar pulses from space, allowing us to map such subsurface features as ancient river channels in the Sahara.

Shuttle crews also experimented with the **Large Format Camera (LFC)**, first operated during a 1984 mission. The LFC is a 12-inch, focal-length, high-precision mapping camera that produces 9 × 18 inch (23 × 46 cm.) photographs. These photographs cover a 140 by 280 mile (225 by 450 km.) ground area from a 186 mile (300 km.) altitude. Ground resolutions from 50 to 100 feet (15 to 30 m.) have been achieved with the panchromatic, true-color, and infrared-color films tested.

Large Format Cameras have produced overlapping photographs such as the section seen in Figure 8.33. Such overlapping photos can be used in compiling topographic maps as large as 1:50,000 in scale, using stereoplotting instruments.

Ground-Controlled Earth Resources and Weather Satellites

Satellites remotely controlled from ground stations have recorded the earth's surface and atmosphere for scientific and environmental monitoring purposes since the early 1960s. Although a great number of images have been transmitted to ground processing facilities by weather satellites, earth resources satellite images and data have been used more in projects involving environmental mapping.

Figure 8.33 Large Format Camera photograph of central Boston. (Courtesy Thomas Loveland).

The five U.S. **Landsat** satellites have produced the most widely used imagery and digital image data, but images and digital data from the French **SPOT** satellite are rapidly gaining in popularity. After we discuss the Landsat satellites, we'll take a look at SPOT and weather satellites.

Landsat 1-3 Between 1972 and 1984, five *Landsat* satellites were launched into orbit at an altitude of 570 miles (917-km.) *Landsat 1, 2,* and *3* were identical in platform design and orbital characteristics, but carried slightly different sensing instruments (see Figure 8.34 for a detailed comparison of the five satellites). Each satellite circled the earth in a near-polar orbit* 14 times a day and repeated its orbital pattern every 18 days.

The vast majority of *Landsat 1-3* images were recorded by the **Multispectral Scanner (MSS)** on each satellite. This was a whiskbroom instrument with a 260-ft. (79-m.). IFOV that swept a 115- mile (185 km.) scan line. The continuous electronic signal from each scan line was converted into approximately 3,200 digital pixel values. Each rhombus-shaped image (due to the near-polar orbit) was composed of around 2,400 scan lines, so that a 7.5 megabyte file held an image of one of four spectral bands. Visible and near-IR energy was split into green (0.5–0.6μm), red (0.6–0.7μm), and two near-IR bands (0.7–0.8 and 0.8–1.1μm). Detected energy in each band was directed to a detector sensitive to the same wavelength range. The *Landsat 3* MSS was also equipped with a thermal-infrared band detector that failed soon after launch.

Return Beam Vidicon (RBV) cameras were also aboard *Landsat 1-3*, but only the two cameras on *Landsat 3* operated successfully. Each camera produced 100-foot (30-m.) ground resolution images of the green to near-IR spectral band covering a 61 × 61 mile (98 × 98 km.) ground area. The two cameras were aimed so as to overlap by 8 miles (13 km.), resulting in a 114-mile (183-km.) ground swath imaged on each orbit.

Landsat 4-5 The last two *Landsat* satellites were launched in 1982 and 1984, and *Landsat 5* was still operating as of 1994. These second-generation satellites are significantly different in appearance and sensor design from their three predecessors. A circular, near-polar orbit was retained, but at a lower altitude of approximately 435 miles (700 km.) The orbital period was slightly lengthened to 14.5 orbits per day, which increased the repeat visit interval to every 16 days (Figure 8.35).

The most important similarity with *Landsat 1-3* is the retention of the MSS sensor, primarily in order to maintain the data continuity required in studies of long-term environmental changes. However, images and digital data from the new **Thematic Mapper (TM)** sensor are now used most often.

The TM is also a whiskbroom multispectral scanning device, but with a 100-ft. (30-m.) IFOV for six of its seven channels. The blue, green, and red visible bands form three channels, along with one near-IR, two middle-IR, and a 400-ft. (120-m.) IFOV thermal-IR channel (see Figure 8.34 for details). Each 115 × 115 mile (185 × 185 km.) TM scene is composed of around 6,000 scan lines, with 6,000 pixels per scan line. This means that in each scene there are almost 36 megabytes of pixel values per channel, except for the far fewer 2.25 megabytes in the thermal-IR channel.

Color Figure 8.5 illustrates several of the many possible images that may be created from Landsat digital data. Note that both MSS and TM green, red, and near-IR bands may be displayed in blue, green, and red to produce the equivalent of an infrared-color photograph. An approximation of true-color photography is also possible using the blue, green, and red TM bands displayed accordingly. Thousands of other color combinations, many strange to the eye but good for emphasizing otherwise hidden detail, can be displayed and printed.

---

*A near-polar orbit is one in which the satellite passes close to, but not over, the North and South Poles. Most earth resources satellites have an orbital inclination of around 82° from the equator which, when coupled with the correct orbital period, makes them sun-synchronous. This means that the satellite always passes over a particular location at the same local time.

	Landsat 1-3	**Landsat 4-5**
Orbital Characteristics		
Altitude	570mi (917km)	435mi (700km)
Period	103 min	99 min
Orbits per day	14	14.5
Return Period	18 days	16 days

Sensor Characteristics

Multispectral Scanner

	IFOV	79m		IFOV	79m	

Spectral Bands	Band #	Range	Band #	Range
	4	0.5-0.6 micron	1	0.5-0.6 micron
	5	0.6-0.7	2	0.6-0.7
	6	0.7-0.8	3	0.7-0.8
	7	0.8-1.1	4	0.8-1.1
	8	10.4-12.6		

Thematic Mapper

Spectral Bands		Band #	Range	IFOV
	_____	1	0.45-0.52	30m
	_____	2	0.52-0.60	30m
	_____	3	0.63-0.69	30m
	_____	4	0.76-0.90	30m
	_____	5	1.55-1.75	30m
	_____	6	10.4-12.5	120m
	_____	7	2.08-2.35	30m
	_____			

Return Beam Vidicon

Landsat 1-2

	IFOV	80m	

Spectral Bands	Band #	Range	
	1	0.48-0.57	_____
	2	0.58-0.68	_____
	3	0.69-0.83	_____
			_____

Landsat 3

IFOV	30m	
Spectral Range	0.51-0.75	_____
		_____

Figure 8.34
Comparison of Landsat sensing devices.

Entire books could be written on the cartographic applications of *Landsat* imagery, but we can only touch on a few key uses. First, we must point out that the Return Beam Vidicon (RBV) sensor was originally intended to be the primary instrument in terms of cartographic applications due to its higher geometric fidelity. However, almost all cartographic products, with the exception of several state or regional image maps made from mosaicked RBV scenes, have been created from

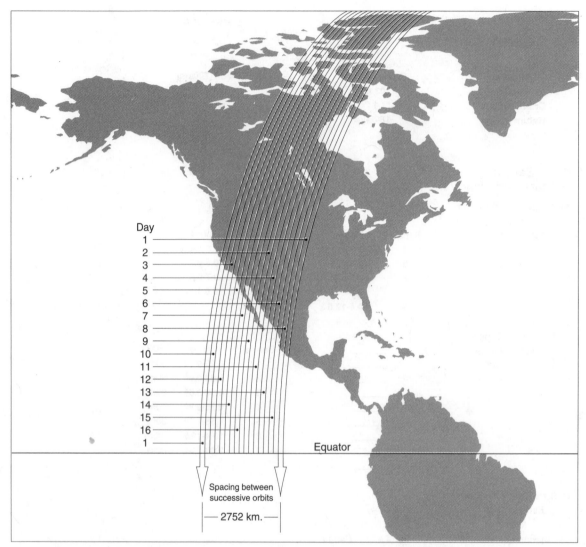

Day
1
2
3
4
5
6
7
8
9
10
11
12
13
14
15
16
1

Equator

Spacing between
successive orbits

— 2752 km. —

Figure 8.35 Timing of adjacent *Landsat 4* and *5* orbital tracks. Adjacent swaths are imaged seven days apart. This orbital pattern repeats every 16 days (adapted from NASA diagram).

MSS or TM images and digital data. Previously mentioned technical difficulties with the *Landsat 1* and *2* RBV sensors, and the low image contrast found on many Landsat 3 RBV scenes contributed to the low usage. Users, including cartographers, also immediately recognized the advantages of the MSS system, especially the ability to digitally analyze and classify multispectral data.

The transition from MSS to TM image use is a matter of users seeing the TM's advantages. These pluses include higher spatial resolution and increased spectral information. Also, it is easier to geometrically rectify a TM scene to a standard map projection (such as the UTM) than it is to rectify an MSS image.

The following are key applications of both MSS

and TM images and digital data (although the TM products are far more likely to be used in all but change detection studies):

- land cover and land use mapping at medium to small scales
- mapping seasonal and other cyclical phenomena such as the annual cycle of vegetation growth or snowmelt within a large county or small state
- delineating rock and mineral types for surficial geological mapping
- identifying different landforms and their characteristic drainage patterns
- mapping non-cyclical changes in regional environmental characteristics, such as loss of farmland due to suburban growth, or the destruction of temperate rainforests due to excessive timber harvesting.

SPOT The French **Systemé Probatoire d'Observation de la Terre (SPOT)** began in 1986 with the launch of the *SPOT-1* satellite into a near-polar, 517-mile (832-km.) orbit. *SPOT-1* is the first earth resources satellite to use pushbroom sensors.

The sensing system consists of two identical **High Resolution Visible (HRV)** instruments that may be operated independently or simultaneously. The pushbroom scanner in each instrument is a 6,000-element linear array detector that records a 37-mile (60-km.) scan line on the ground, resulting in 37 × 37 mile (60 × 60 km.) scenes. When operated simultaneously, the two sensors portray adjacent scenes with a two-mile (three-km.) sidelap. This results in a total ground swath of 73 miles (117 km.) being recorded (see Figure 8.36).

Each HRV sensor may be operated in single-image (0.51–0.9µm) panchromatic mode with a 30-ft. (10-m.) IFOV, the smallest of any non-military electronic satellite scanning system. A multispectral mode (Color Figure 8.6) may also be selected. This mode produces three 60-ft. (20-m.) resolution images in the green (0.51–0.59), red (0.61–0.68), and near-IR (0.79–0.89µm) bands. These may be combined to create an image similar in appearance to infrared-color photography.

SPOT-1 passes over the same location every 26 days, and vertical images may be obtained at this frequency. However, the HRV is the first satellite sensor that may be pointed obliquely so that features up to 250 miles (400 km.) to either side of the flightline may be imaged. This means that areas of interest may be imaged far more often than every 26 days if this is necessary due to cloud cover or rapid changes in environmental features. Another important advantage of the pointable sensors is that oblique images of the same area acquired on

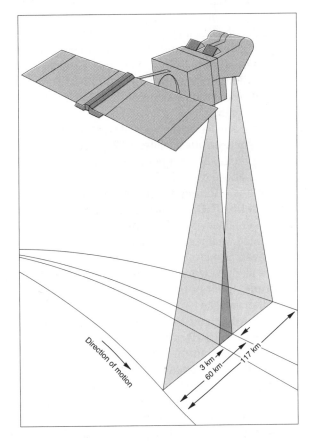

Figure 8.36 *SPOT-1* ground coverage with HRVs recording adjacent swaths.

orbital passes to the east and west of the area allow the two images to be viewed and analyzed stereoscopically.

High image detail, excellent geometric fidelity, and the possibility of stereoscopic viewing make *SPOT* images good cartographic source materials. Topographic maps at 1:100,000 have been compiled, and 1:50,000-scale topographic maps may be revised by stereoscopically viewing oblique image pairs in stereoplotting instruments. Large to medium-scale thematic maps of natural features (such as forests) and cultural features (such as agricultural practices) have been produced from *SPOT* images in several countries.

Weather Satellites We have noted that satellite remote sensing began with the meteorological sensing systems of the early 1960s. These satellites provided the first images of the earth acquired on a repetitive basis, and made possible detailed analyses of global cloud patterns. Ground features, when visible, were imaged poorly at a coarse spatial resolution, a situation that has persisted until recently.

There are two basic types of weather satellites. **Geo-synchronous** satellites in 22,300-mile (35,800-km.) altitude, 24-hour period orbits appear to be parked above a particular point on the equator. The **Geostationary Operational Environmental Satellite (GOES)** that produces low-resolution continental views every 15 minutes brings us the animated weather image map seen by millions on the evening news. Cartographers, however, are more likely to work with digital data from lower-altitude, near-polar orbit **sun-synchronous** weather satellites.

The **Advanced Very High Resolution Radiometer (AVHRR)** is the multispectral scanning instrument found on current U.S. National Oceanic and Atmospheric Administration (NOAA) near-polar orbiting satellites. From an altitude of 518 miles (833 km.), the AVHRR scans an approximately 1,740-mile (2,800-km.) swath on each orbital pass. Each scan line is composed of around 2,500 pixels, each representing a 0.7×0.7 mile (1.1 × 1.1 km.) ground area. Image data are collected from either four or five bands, including the green-red, near-IR, 3–5μm thermal-IR, and 10–12μm thermal-IR.

The best known cartographic application of AVHRR imagery is the mapping of monthly variations in vegetation health and growth across the earth. Image maps such as Color Figure 8.7 have been created by digital image processing techniques to show the extent of healthy vegetation and to monitor areas suffering drought, increased desertification, and loss of forest canopy.

AVAILABILITY OF REMOTE SENSOR DATA

An often overlooked problem in remote sensing data collection is how to determine what images are currently available and how to order existing or new imagery. There are three general categories of imagery availability in the United States and other nations: (1) imagery in the public domain, (2) commercially available imagery, and (3) restricted use imagery.

PUBLIC DOMAIN IMAGERY

In the United States, non-military aerial photographs, satellite images, and digital image data obtained by governmentally operated sensing systems are in the "public domain." This means that any citizen may purchase these non-copyrighted products for a modest fee that covers the cost of duplication and handling.

Although a large number of federal agencies (as well as many local, regional, and state offices) hold aerial photography and other imagery of parts of the United States, the best place to begin a search for coverage of any area is the nearest office of the National Cartographic Information Center (NCIC). The addresses of the NCIC offices are listed in Appendix E. The NCIC is a part of the U.S.

Geological Survey (USGS) that helps individuals locate aerial photographs, satellite images, maps, and other spatial data. NCIC staff do not order images, except for USGS products, but rather put people in contact with the agency holding the desired imagery.

Each NCIC office maintains computer records of agency holdings of aerial photographs, orthophotoquads, images from Landsat and human space flights, as well as traditional topographic, land use, slope and other thematic maps, digital terrain tapes, geodetic control data, and numerous other map-related items. These records include holdings of the U.S. Geological Survey and other agencies such as the U.S. Forest Service, Bureau of Land Management, Bureau of Reclamation, Census Bureau, Central Intelligence Agency, National Oceanic and Atmospheric Administration, Corps of Engineers, Federal Highway Administration, Federal Power Administration, Tennessee Valley Authority, Mississippi River Commission, International Boundary Commission, Library of Congress, Agricultural Stabilization and Conservation Service, National Archives and Records Service, National Aeronautics and Space Administration, and the Defense Mapping Agency. Addresses of several of these agencies with significant imagery holdings are given in Appendix E.

NCIC offices also operate the **Aerial Photography Summary Record System (APSRS)** developed by the USGS. This helps users determine if aerial photographs exist for a particular location, whether they meet the user's requirements, and where copies may be obtained. To do this, the APSRS contains information about cloud cover on images, types of film used, date and time of acquisition, and the agency that holds the imagery.

COMMERCIALLY AVAILABLE IMAGERY

The acquisition of large-scale aerial photography, thermal and multispectral scanner imagery, and SLAR images taken from fixed-wing aircraft, helicopters, and other low-altitude platforms has long

been a commercial as well as governmental enterprise in the United States. Numerous aerial survey companies provide imagery to businesses, university researchers, and governmental agencies. These companies may be difficult to contact, but the American Society of Photogrammetry and Remote Sensing (ASPRS) and its regional sections keep up-to-date address lists.

Satellite remote sensing in the 1960s and '70s was a government affair. Many leaders viewed the images obtained by weather satellites, *Landsat*, and other systems as a national resource to be subsidized heavily so that any taxpayer could obtain imagery and digital data at a minimal cost. This political philosophy changed in the 1980s, as evidenced by the 1984 legislation making *Landsat* a joint governmental-commercial venture. In 1985, the Earth Observation Satellite Company (EOSAT) took over Landsat operations. A sharp rise in the cost of images and digital data soon followed.

SPOT, in contrast, is the first completely commercial satellite system. Consequently, imagery is obtained upon the request of a paying customer. The systematic world-wide coverage provided by Landsat as a public service was never envisioned for *SPOT*. The *SPOT* corporation also retains the copyright on all imagery through a licensing contract that purposefully inhibits its shared use by cartographers and other researchers.

RESTRICTED USE IMAGERY

Imagery collected by classified instruments operated by the United States, Russia, and other military establishments is, of course, not available to the general public. The situation can change, however. Indeed, several of the remote sensing systems described in this chapter were once classified. They only become publically available when the sensing instruments and imagery were so outdated that they no longer presented a security risk. In many nations, this security restriction extends to aerial photography and other large-scale imagery, severely limiting the use of remote sensing imagery in monitoring the nation's environment.

SELECTED REFERENCES

Allan, T. D. (ed.), *Satellite Microwave Remote Sensing*, New York: Halsted Press, 1983.

Avery, T. E. and G. L. Berlin, *Fundamentals of Remote Sensing and Airphoto Interpretation*, 5th ed., New York: Macmillan Publishing Co., 1992.

Chevrel, M., M. Courtois, and G. Weill, "The SPOT Satellite Remote Sensing Mission," *Photogrammetric Engineering and Remote Sensing*, 47, No. 8 (1981) 1163-71.

Colwell, R. N. (ed.), *Manual of Remote Sensing*, 2nd ed., Falls Church, Va.: American Society of Photogrammetry, 1983.

Elachi, C. "Radar Images of the Earth from Space," *Scientific American,* 247: 54-61, 1987.

Estes, J. E., J. R. Jensen, and D. S. Simonett, "Impact of Remote Sensing on U. S. Geography," *Remote Sensing of Environment*, 10 1980.

Harger, R. O., *Synthetic Aperture Radar Systems, Theory and Design*, New York: Academic Press, 1980.

Lillesand, T. M. and R. W. Kiefer, *Remote Sensing and Image Interpretation,* 3rd ed., New York: John Wiley & Sons, 1994.

Lintz, J., Jr., and D. S. Simonett (eds.), *Remote Sensing of the Environment*, Reading, MA: Addison-Wesley, 1976.

Pittenger, R. F. "Exploring and Mapping the Seafloor," *National Geographic*, 177: 61A, 1989.

Sabins, F. F., Jr., *Remote Sensing Principles and Interpretation*, 2nd ed., New York: W.H. Freeman & Co., 1987.

Skolnick, M. I., *Introduction to Radar Systems*, 2nd ed., New York: McGraw-Hill, 1980.

Slater, P. N., *Remote Sensing: Optics and Optical Systems,* Reading, MA: Addison-Wesley, 1980.

Swain, P. H., and S. M. Davis (eds.), *Remote Sensing: The Quantitative Approach,* New York: McGraw-Hill, 1978.

Townshend, I. R., *The Spatial Resolving Power of Earth Resources Satellites: A Review,* NASA Technical Memorandum 82020, Greenbelt, MD: Goddard Space Flight Center, 1980.

Ulaby, F. T., R. K. Moore, and A. K. Fung, *Microwave Remote Sensing: Active and Passive, Vol. I.: Microwave Remote Sensing Fundamentals and Radiometry*, Reading, MA: Addison-Wesley, 1981.

U. S. Department of Interior, Geological Survey, *LANDSAT Data User's Handbook*, rev. ed. Arlington, VA: USGS Branch of Distribution, 1979.

Wolfe, W. L., and G. J. Zissis (eds.), *The Infrared Handbook*, Washington, D.C.: U.S. Government Printing Office, 1978.

CHAPTER 9

CENSUS AND
SAMPLING

*G*round survey and remote sensing methods are not appropriate for all our data gathering needs. It is not practical to use ground survey to locate some features because of their high mobility, short life span, or large number. Obstructions may block electromagnetic energy, preventing other features from being captured on photographs or electronic images. And, of course, intangible phenomena elude survey or remote sensing methods. In all these cases, cartographers use primary data gathered by census or spatial sampling procedures.

POPULATION ENUMERATIONS

Discrete features, such as houses or people, can be counted. The aim of a **census** is to identify and record all members of a population. The U.S Census Bureau's Decennial Census of Population and Housing is an example of an attempt to count all people within a country. In this case, census forms are sent to all known residential addresses. Attempts are also made to identify homeless individuals. Census workers make follow-up visits if forms are completed and returned.

Although a full enumeration of a population would seem to be the ideal form of data gathering, it does have some disadvantages. A census is time-consuming, laborious, and costly when a population is large and widely dispersed. The task of counting all people in the United States every 10 years, for example, requires years of preparation, thousands of workers, and many millions of dollars.

Errors also occur in population enumerations for many reasons. People sometimes respond dishonestly, fearing that the information will not be kept confidential. Due to the variety of addresses people use, a full accounting of residences is difficult to document. Homeless people are especially elusive and often suspicious of census workers' intentions. Census field workers are paid at a rate near the local minimum wage, which is not conducive to recruiting the most skilled and dedicated employees for this critical task.

The problems associated with census-taking mean that the resulting data have to be treated with appropriate care. Except for small, accessible populations, there is no guarantee that census totals are accurate. For specific uses, even small errors may be a problem. But, lacking the resources to gather better information, we have to learn to factor likely census errors into our spatial decision-making and regional comparisons.

GEOCODING

When people collect census data, they must first choose a method of data recording. Although feature attribute information is sufficient for some users of census data, this is not true in the case of mapping. For cartographic purposes, it is essential that a locational reference be associated with the feature attribute information. This practice of attaching locational information to census data is referred to as **geocoding**. Methods of geocoding vary, depending on whether the position of individual population entities is preserved, or whether population elements are aggregated within census region boundaries and reported only as totals.

ENTITY FOCUS

The most accurate way to geocode census data is to identify the location as well as the attributes of each population entity. We can do so in two ways.

In one approach, called **vector geocoding**, we use a coordinate reference system to define the location of point, line, and area features (Figure 9.1). The basic coding unit is a point, which is defined by a coordinate pair (two dimensions) or triplet (three dimensions). Line features are treated as a series of points, and area features are treated as lines that close on themselves.

Terrestrial coordinates (latitude-longitude, UTM, SPC, etc.) are preferred over local coordinates (cartesian) for two-dimensional geocoding. This is because local coordinates are difficult to relate to the earth's surface (see Chapters 4 and 6). Sets of

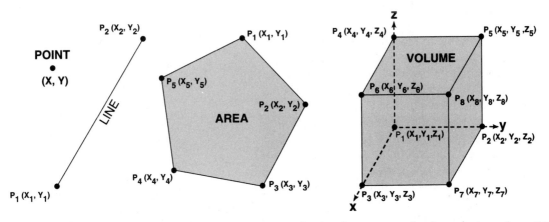

Figure 9.1 Coordinates are used to define the location of point, line, area, and volume features in vector geocoding.

data not readily related to positions on the earth are difficult to merge with other data. Thus, when you want to make a map using data gathered according to different local reference systems, or by a mixture of terrestrial and local reference systems, your merging task will be more complicated than it would be if you only had to deal with terrestrial reference systems.

A third dimension is often added in geocoding with terrestrial coordinates. In this case elevation above mean sea level is most commonly chosen to define the third dimension (see Chapter 27). By using three-dimensional geocoding we can describe the surface characteristics of features such as a mineral deposit, a lake basin, or a toxic plume in the atmosphere.

The alternative to vector geocoding with its use of coordinates is **raster geocoding**, which is based on matrix or grid notation (see Chapter 8). In this case, a position is defined by its row and column cell, which is called a pixel (Figure 9.2). Technically speaking, a pixel defines an area, not a point. In practice, however, if pixels are significantly smaller than the spatial entities they represent, they provide a satisfactory locational reference.

As you will see in the next chapter, you can think of vector and raster geocoding as being

locational aliases. They represent different but equivalent ways to define the position of environmental features. We can readily convert locations in one system to locations in the other. Thus, cartographers commonly practice **rasterization**, the conversion of vector data to raster form, and **vectorization**, the conversion of raster data to vector form.

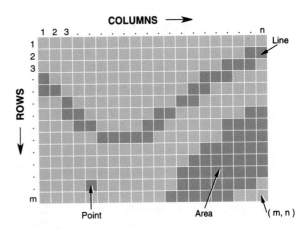

Figure 9.2 Pixels are used to define the location of geographic entities in raster geocoding.

AGGREGATION

Rather than preserve the location of population entities, we may **aggregate** census information within the boundaries of data collection regions. We may do so to reduce costs, since aggregation is by far the less expensive of the two methods. We may also do so for reasons of confidentiality. To protect the identity of individual population elements and to encourage honest responses, it is sometimes desirable to guarantee anonymity.

Aggregating data raises a variety of mapping concerns, however. Maps based on region totals tend to be more abstract and prone to inaccuracy than those based on feature position data. To understand why this is so, we have to look more closely at the relation between population distributions in the environment and the way their characteristics are captured and reported by the aggregation method.

Census Regions

If a population were dispersed uniformly, one mosaic of data collection regions would give as representative a picture of the distribution as the next. Distributions are rarely uniform, however. Furthermore, census regions often vary widely in shape, size, and orientation (Figure 9.3).

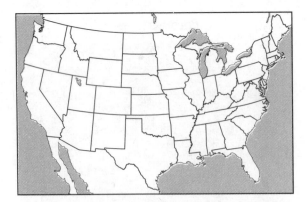

Figure 9.3 When census data for the United States are aggregated by state, the wide variations in shape, size, and orientation of the individual states can introduce significant biases on subsequent maps.

Typically, census regions are defined by political boundaries. Countries, states, provinces, counties, townships, and cities are widely used for data aggregation. Government census organizations often define additional units, such as enumeration districts, census tracts, and blocks. Boundaries for these special census regions generally follow political borders when possible, but also are aligned with prominent natural and cultural landmarks. Natural resource specialists tend to rely on natural landmarks, such as watersheds or floral/faunal communities, for defining their census regions.

Census regions do not have to be political units, of course. Data increasingly are being aggregated by administrative regions such as postal zip codes or telephone area codes. These non-political census regions are especially popular when gathering data of special commercial interest. Marketing and sales firms, for example, rely heavily on these data aggregations.

Data may also be aggregated by units whose boundaries define natural regions in the environment. Watersheds, for instance, have become particularly popular for all sorts of environmental census taking. Climate zones, soil classes, ecological units, and similar regions fall in this natural class of census regions as well.

Relation to Population Distribution

The problem with aggregating census data by a region of any sort is that it is difficult to devise a regional scheme that does not give a biased picture of a population (Figure 9.4). In theory, region boundaries could be defined to achieve a good match between data collection units and variation in a population's distribution. The drawback of this approach is that each distribution would require a different mosaic of data collection regions.

But the extra effort and cost required to come up with an unbiased data collection scheme for a given population often cannot be justified to those funding the activity. Instead, convenient regions, such as countries and states, are used for all manner of data aggregations, regardless of the effect

REGION TOTALS MASK DIVERSITY
AND SUGGESTED UNIFORMITY

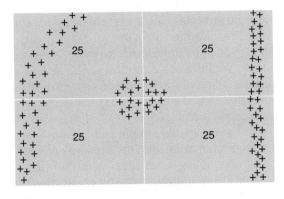

+ Population element

Figure 9.4 When a census region mosaic is randomly superimposed on the landscape, the region totals can mask the true nature of the population distribution.

CENTERS OF AREA AND POPULATION
VARY IN "REPRESENTATIVENESS"

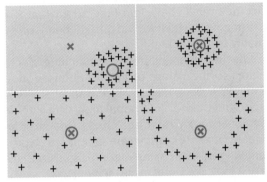

✕ Center of area   ◯ Center of population
+ Population element

Figure 9.5 The reporting of census data by region centroids facilitates data processing and mapping, but can lead to a misleading impression of population distribution.

this choice may have on the quality of resulting data. Maps based on these data will also suffer.

Reporting Method

Aggregated data are generally reported as census region totals. Thus, the information available for mapping is a count by region. This information is commonly represented with value-by-area mapping techniques as well (see Chapter 26).

But we can also report census totals at region **centroids** (Figure 9.5). When we do so, we have several options. We may choose the **center of area**, which is the balance point for the census region shape. The center of area bears no relation to the distribution of population within the region. If the population distribution is uniform, the center of area will provide an unbiased representation. Otherwise, use of the center of area can be misleading.

To make the centroid more responsive to the distribution of population, we may choose the **center of population** to report census data. For point features, the center of population is deter-

mined by averaging the x and y coordinates of the individual population. For this reason, center of population is also called the **bivariate mean**. This centroid can only be computed, of course, if the positions of individual features were recorded at the time of data collection.

SPATIAL SAMPLING

Continuous phenomena, such as temperature or barometric pressure, must be measured rather than counted. It is not practical to collect complete data on these phenomena, since that would entail making observations at all possible locations. Clearly, the census methods discussed previously in this chapter are inappropriate for these phenomena. Even for those features that can be counted, such as trees, a full population enumeration is often not practical.

When a census is not appropriate or too costly, the alternative is a **spatial sample**. The aim is to make observations at a limited number of carefully chosen locations that are representative of a distribution.* From these data, we can then predict the overall character of the population. The accuracy with which we can do so depends on the quality of the sample.

Effective sampling involves matching several factors to the nature of the distribution under scrutiny. A poor match will thwart attempts to reconstruct the overall population with any degree of accuracy (Morrison, 1971). The most important factors to consider are sample size, type of sampling units, and sampling unit dispersion. Let's take a look at each of these elements.

SIZE OF SAMPLE

Although sample size is usually the first factor considered, it is less important than some of the other sampling ingredients discussed in the following sections. The key is to make each sample location provide as much insight into the nature of a distribution as possible. A small number of well chosen sample points can be more revealing than a large number of sample locations chosen haphazardly.

From the literature of inferential statistics, most students learn that a sample size of 30 is about the ideal. However, this 30-sample size has little relevance for cartographers, whose job it is, using limited sample data, to make a map that best describes a distribution. A better guideline is to remember that appropriate sample size is directly related to a distribution's variability. A sample of 20 with a relatively uniform distribution may yield the same level of accuracy as a sample of 500 with a highly variable distribution. (Also see "Relating Sample to Distribution" later in this chapter).

---

*The aim in sampling is to find a strategy that permits us to use the smallest number of observations to achieve a given level of accuracy. Since the strategy becomes less important as the number of observations increases, the choice of strategy is the key to efficient sampling.

SAMPLING UNITS

Common spatial sampling units are points, lines (called **transects**), and areas (called **quadrats**). These are illustrated in Figure 9.6. Some liberty is taken in defining these terms, however. A transect sample, for instance, may consist of counting all features falling within 100 feet on either side of the line. The size and shape of area units is also open to broad definition.

The type of spatial sample units chosen is relatively unimportant within the broader context of the other sampling parameters. It is far more important that the sampling units used are placed so that they generate the most information. There are a number of ways to disperse point, line, and area samples throughout a region.

DISPERSION OF SAMPLING UNITS

The most critical sampling parameter is the way sampling units are **dispersed**. It is convenient to define sample scatter in terms of deviation from a random dispersion. Thus, a spatial sample may be **randomly** dispersed, more **clustered** than random, or more **uniform** than random (Figure 9.7).

Most students learn in their inferential statistics training that a random sample is preferred. In spatial terms, a random sample is one in which each location has the same chance to be chosen, and the choice of one location in no way changes the probability of selecting other locations to com-

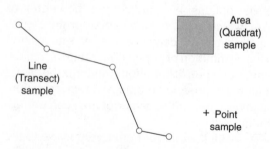

Figure 9.6 Spatial sampling is performed using either point, line, or area units, or some combination of these geometric forms.

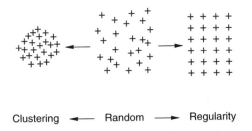

Clustering ◄——— Random ———► Regularity

Figure 9.7 Sample scatter is usually described in terms of deviation from randomness, with clustering at one extreme and uniformity at the other.

plete the sample. Sampling of this type is "blind" in the sense that all prior knowledge about a distribution is ignored. As a result, a random strategy makes for a relatively inefficient sampling method. Although the sampling process is random, the scatter of the resulting sample usually is not. Some areas are likely to be over-represented by the sample, while other areas are not represented at all. For this reason, samples with a random scatter as a rule lead to the least accurate maps

Clustered samples also tend to be inefficient, because concentrating sample sites in a few locations means that other regions are not adequately sampled. When no prior information is available about a distribution, it is preferable that a sample scatter tend toward uniformity rather than toward clustering. The reason is that a uniform sample is not likely to miss any major distribution characteristics. In fact, studies such as Morrison's (1971) show that samples that are somewhat more uniform than random yield the most accurate maps. An exception might occur if variation in a distribution exhibited the same regularity as the sample scatter.

RELATING SAMPLE TO DISTRIBUTION

We saw in the previous section that even a blind approach to dispersing sample sites can be improved by establishing a more regular than random scatter. Still greater sampling efficiency can be achieved if the sample scatter is carefully

matched to the distribution being sampled. This can be done by portioning sample sites out to locations in terms of their distributional importance, and by using a sampling interval that is sufficiently small to capture distributional features of particular interest.

Stratified Sampling

A strategy called **stratified** sampling is widely practiced in national polls and ratings surveys. The method can be remarkably efficient. In U.S. national presidential elections, for example, a stratified sample of 1,500 people is usually sufficient to predict voting behavior within one or two percentage points of the final outcome. Stratified sampling serves mapping purposes equally well.

The logic of stratified spatial sampling is, first, to divide a region into relatively homogeneous subregions considered to bear some relation to the phenomenon being sampled (Figure 9.8). The sub-regions are called **strata**, which gives the procedure its name.

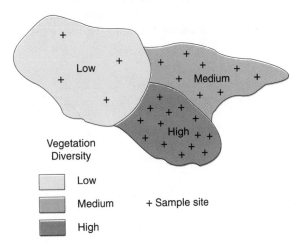

Figure 9.8 Sampling based on stratification is highly efficient, because it takes advantage of what is already known about the distribution being sampled.

The second step is to assign sample sites to each sub-region in proportion to what each area is thought to contribute to the overall picture. Obviously, in the lack of complete information, stratification can only be approximate. But even sketchy knowledge of a distribution vastly improves sampling accuracy when compared to a blind approach.

Sampling Theorem

Although stratified sampling can reduce the sample size needed to achieve a given level of sampling accuracy, we still must determine the minimum spacing between sample sites. This **sampling interval** is addressed by a rule called the **sampling theorem**. This guideline, borrowed from communications theory, states that a sampling interval should be less than half the size of target features in a distribution (Figure 9.9). A larger interval will not let you reconstruct the sampled population and, therefore, will lead to a distorted impression of the distribution.

The sampling theorem provides a useful starting point for those planning to conduct a spatial sample. It suggests that the first step in sampling is to decide the spatial resolution that needs to be achieved in order to capture features of particular interest. For someone wanting to use existing sample data or to evaluate the quality of a map based on sample data, the sampling theorem provides a useful evaluation tool.

PASSIVE DATA COLLECTION

Data captured through census and sampling have traditionally required direct action. Data collectors actively identified, recorded, and reported their data. For centuries, the reporting was done in tabular and cartographic forms. Since the advent of electronic computers, however, an increasing proportion of users want their data in digital form. Unfortunately, the conversion of existing tabular and graphic data to digital records is laborious, time-consuming, and costly (see Chapter 11).

Electronic technology has contributed to growing interest in collecting data from the field in digital form using automated methods. Once such **passive data collection systems** are established, data are sensed, transmitted, received, and processed without need for direct human intervention. This approach is ideal when monitoring movement past a fixed position, as in highway traffic or wildlife migration studies. Passive systems are particularly useful when data must be gathered repeatedly at a large number of locations, or at remote locations.

Transmitters are also being used in passive data collection systems. In biotelemetry studies, for example, census taking involves attaching transmitters to animals, which automatically signal their position to data recording devices. Similarly, traffic flow and density can be monitored with the aid of automated data recorders and vehicles fixed with transmitters.

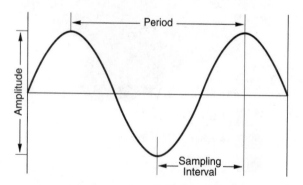

Figure 9.9 According to the sampling theorem, the sampling interval needs to be less than half the size of target features in a distribution.

SELECTED REFERENCES

Berry, B. J. L., *et. al.* "Geographical Ordering of Information: New Opportunities," *The Professional Geographer*, 16: 39-43, 1964.

Dalton, R., and others, *Sampling Techniques in Geography*, London: George Philip and Son, 1975.

Daugherty, R., *Data Collection*, Oxford: Oxford University Press, 1974.

Dixon, C., and B. Leach, *Sampling Methods for Geographical Research*, Norwich, England: Geo Abstracts, University of East Anglia, 1977.

Dueker, K. J. "Urban Geocoding," *Annals of the Association of American Geographers*, 64: 318-325, 1974.

Holmes, J. H. "Problems in Location Sampling," *Annals of the Association of American Geographers*, 57: 757-780, 1967.

Holmes, J. H. "The Theory of Plane Sampling and its Applications in Geographical Research," *Economic Geography*, 46: 379-392, 1970.

Long, C. M., *Understanding Census Data*, Sherman Oaks, California: Western Economic Research Company, 1983.

Lewis, S. "Can You Trust the Census?," *Planning (APA)*, 57: 14-18, 1991.

MacEachren, A. M., and J. V. Davidson, "Sampling and Isometric Mapping of Continuous Geographical Surfaces," *The American Cartographer*, 14: 299-320, 1987.

Marx, R. W. "Automating Census Geography for 1990," *American Demographics*, 5: 30-33, 1983.

Masser, I., and M. Blakemore (eds.), *Handling Geographical Information: Methodology and Potential Applications*, New York: John Wiley & Sons, 1991.

Morrison, J. L., *Method-Produced Error in Isarithmic Mapping*, Washington, D.C.: American Congress on Surveying and Mapping, 1971.

Rhind, D. "Mapping Census Data," in *A Census User's Handbook*, D. Rhind (ed.), New York: Methuen, 1983, 171-198.

Sheskin, I. M., *Survey Research for Geographers*, Washington, D.C.: Association of American Geographers, 1985.

Stoddard, R. H., *Field Techniques and Research Methods in Geography*, Dubuque, Iowa: Kendall/Hunt Publishing Company, 1982.

Trainer, T. F. "Fully Automated Cartography: A Major Transition at the Census Bureau," *Cartography and Geographic Information Systems*, 17: 27-38, 1990.

CHAPTER 10

DATA MODELS FOR DIGITAL CARTOGRAPHIC INFORMATION

The keystone of computer mapping lies in having a set of information in the computer that represents, in digital form, the information that could be shown on a map. In this case, cartography is viewed as an information transfer process centered about a spatial database. This database can be considered, in itself, a multifaceted model of geographic reality.

The database includes the **coordinate data** required to draw a graphic product. It also contains information about the **features** on the map (for example, number of lanes of a highway) and their relationships to each other (for example, the fact that the railroad tracks lie between the stream and the highway).

In the context of geographic information systems (GIS), features are the sum of our interpretations of geographic phenomena. Buildings, bridges, roads, streams, grassland, and counties are examples of features. Spatial data sets are a collection of locational and non-locational information about selected features, attributes of the features, and relationships between the features.

Spatial data models that specify the sets of components and relationships among components must describe both the locational and non-locational aspects of the features of interest. This collection of information forms the heart of a **digital cartographic database**. Once the information is in the database, computers provide the capabilities to handle geographic data in ways that were previously impossible.

Spatial data is often used as an all-encompassing term that includes standardized products such as digital cartographic data, remotely sensed imagery, and census tract descriptions, as well as more specialized data sets such as seismic profiles, distribution of relics in an archeological site, or migration statistics. Taken literally, spatial data could refer to any piece of information related to a location. Such a sweeping definition is too unwieldy for our purposes. Thus, when we refer to spatial data in this book, we mean to restrict ourselves to **geographical data**.

The term "geographical data" differs from "spatial data" only in the sense that geographical data refers specifically to features that describe the earth's surface. When we gather data directly from the environment in digital form, we refer to it as **digital geographic data.** However, if geographical data are first used to make a map, and then the map is subsequently digitized, the result is called **digital cartographic data.** This is a simplifying (some might contend restricting) condition applied to the set of geographic data. The cartographic origin of the data often causes temporal or three-dimensional aspects of a feature not to be represented. Digital cartographic data is the type of geographical data most often considered when digital cartographic data sets are discussed. The remainder of this discussion pertains to this type of information and the ways to add more geographic meaning to it.

MODELING GEOGRAPHIC REALITY

A digital spatial data set represents a certain model of geographic reality. For you to make sense of the data, you must understand the conceptual model underlying the data (Box 10.A). The process of creating a representation of geographic reality is referred to as conceptual modeling*.

A conceptual data model is an abstraction device that allows us to capture and represent certain aspects of the real world. The conceptual data model is in the middle of a set of levels of abstraction that run from reality to a physical data model (Teorey and Fry, 1982; Peuquet, 1984). These levels are shown in Table 10.1.

Every GIS has its own conceptual data model that manifests itself in a unique physical data model. To export a data set, the internal physical

---

*The term conceptual modeling is variously referred to as semantic data modeling, knowledge representation, or semantic network description. The various terms have evolved from work in artificial intelligence and database management research (Winston, 1984, or Brodie and Mylopoulus, 1986).

Box 10.A

Geographical Concepts
within Computer Science Constructs

To be implemented on a computer, a geographer's conceptual model of space and time needs to be reconciled with the conceptual models, developed by computer scientists, that underlie various computer application environments. To the computer scientist, an information system usually stores, retrieves, and manipulates information about some portion of the real world. An information system is built out of a database that conforms to some data model and applications programs that manipulate these data (Borgida, 1986).

Using the terminology of conceptual modeling, the world is described in terms of conceptual objects or entities, which have associated descriptions, and are related to each other in meaningful ways. To express these ideas, researchers have developed a modeling framework that includes the concepts listed below (adapted from Borgida):

- A model is described as a collection of object descriptions.
- Objects in the model correspond to entities in the world.
- Objects are grouped into classes.
- Every object has zero or more attributes which relate it to other objects.
- Classes are also objects. Classes, therefore, are instances of other classes, sometimes called "metaclasses" and may have their own attributes.
- Objects have relationships with other objects. The relationships can also have attributes.
- Attributes can have multiple values.

These modeling concepts have been used in the development of most of the advanced spatial data models.

A data model provides a syntactic and semantic basis for techniques used to support the design and use of databases. Data model concepts are similar to the knowledge representation concepts used in the artificial intelligence field (Brodie, 1986). The three most important data models, historically, have been the hierarchial, network, and relational data models (see Ullman, 1982 for descriptions). Today most commercial database products are based on the relational data model first described by Codd (1970). These relational database management systems are typically used to manage attribute data in a GIS.

Commercial relational database systems (for example, DB2, ORACLE, and INGRES), although useful for handling the attributes of features, are not good at managing vector and raster coordinate data. During the last several years, a variety of next-generation DBMSs have been built, including IRIS (Wilkinson, 1990), ORION (Kim, 1990), POSTGRES (Stonebraker, 1990), and Starburst (Haas, 1990). Sometimes these systems are referred to as object-oriented databases or extended relational database systems (Cattell, 1991). The more general of these systems appear to be usable, at least to some extent, for point, vector, and text data. However, none is yet adequate for the full range of needed capabilities.

Table 10.1
Levels of Data Abstraction

Level of Abstraction	Definition
Reality	The total phenomena as they actually exist.
Conceptual Data Model	The sets of components and the relationships among the components pertaining to the specific phenomena thought to be relevant to anticipated needs. A data model is independent of specific systems or data structures that organize and manage the data.
Logical Data Model or Data Structure	The logical organization of the components of a data model and the manner in which relationships among components are to be explicitly defined.
Physical Data Model or File Structure	A set of rules that specify the machine implementation of a data structure within various computing system environments.

data model is converted to a specific file structure on a given type of media (such as tape or disk). To import a data set, the process in reversed. The following sections describe the basic types of **data structures** used with digital cartographic data.

VECTOR DATA STRUCTURES

The real world of geographical variation is infinitely complex and often uncertain, but must be represented digitally in a discrete, deterministic manner. In one method of describing geographic reality, characterized by the information shown on maps, we think of the world as a space populated by features of various shapes and kinds—points, lines, and areas. These are the elements of the vector data model. Features have attributes that serve to distinguish them from each other. Any location in the space may be empty, or occupied by one or more features. These features may exist in one combined database, or may be separated according to a theme or variable into a number of layers or classes. The terminology used to describe these features is given in Table 10.2. Figure 10.1 shows an example map with point, linear, and areal features as separate layers or files, as well as combined into one layer.

Table 10.2
Terminology for Describing Geographic Features

Feature Type	Example	Digital Database Term (and Synonyms*)
point feature	benchmark	node (vertex, 0-cell)
linear feature	road centerline	arc (edge, 1-cell)
areal feature	lake/pond	polygon (face, 2-cell)

*Synonymous terms exist from geometry and topology and are often seen in the literature. Synonyms that have been used for vertex include: point, node, 0-cell, junction, and 0-simplex. Synonyms for edge include: line, arc, 1-cell, chain, and 1-simplex. Synonyms for face include: polygon, 2-cell, area, and 2-simplex. Books by Bondy and Murty (1976), Giblin (1977), and Lefschetz (1975) provide additional information on graph theory and topology. Corbett (1979) gives some insights on the topological principles used in cartography.

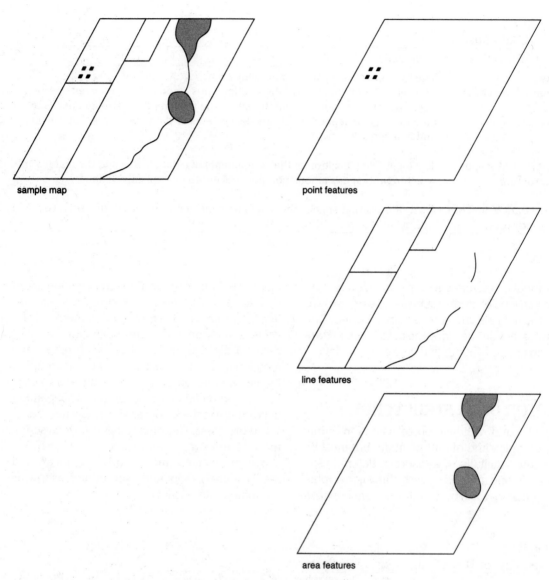

Figure 10.1 **Example map with point, line, and area feature data layers.**

GEOMETRIC COMPONENTS

Mathematically, a map can be treated as a graph. The elements of that graph can be defined using graph theory. Box 10.B describes the basics of graph theory as it relates to cartography. The elements of a map are subdivided into point, line, and area features. Nodes are used to represent point features, such as a water well. Nodes also represent the beginning or ending of every linear feature in the graph and occur at the intersection of linear features. Arcs are ordered sets of points that describe the location and shape of a linear feature such as a road centerline or the boundary of an

---

Box 10.B
Mathematical Foundation for Spatial Objects

The objects used to represent the spatial components of vector data models are formally described using terminology from graph theory. The data models use a directed, planar graph (more precisely a pseudograph) composed of vertexes, edges, and faces to describe spatial objects. See, for example, Harary (1969) for an explanation of these terms and the concepts of graph theory.

Graph G consists of a non-empty set $V(G)$ of *vertexes*, where V is a finite set of points in real vector space R^2 with one point per vertex,

together with a prescribed set $E(G)$ of ordered pairs of (not necessarily distinct) vertexes of G. Each pair $\mathbf{e} = (\mathbf{u,v})$ of vertexes in E is an *edge* of G, and $\mathbf{e}$ is said to join $\mathbf{u}$ and $\mathbf{v}$. The graph is a *plane graph* in that the edges intersect only at their vertexes (the edges are simple curves). The *degree* of a vertex v_i in graph G, denoted deg v_i, is the number of edges incident with v_i. The regions defined by the plane graph are called its *faces*, the unbounded region being called the exterior face.

---

areal feature. Polygons are portions of the map that are bounded by arcs. These bounded regions include areal features such as lakes and counties.

The basic geometry of the point, linear, and areal features as well as attribute data about them can be collected and manipulated by software used for computer-aided design (CAD) work. Data formats supporting only geometric and attribute data include such CAD formats as DXF* (developed by AutoDesk and used by many CAD programs), and IGDS (developed by Intergraph). Both these formats are in the public domain.

TOPOLOGICAL COMPONENTS

The relationships between the nodes, arcs, and polygons are called **topological** relationships. They are shown in Table 10.3. In simple terms, the topological relationships are that arcs start and end only at nodes, and that polygons are enclosed by a series of arcs. Attention is given to neighboring geographical entities when coding arcs. For instance, the topological description of each arc includes not

Table 10.3
Topological Relationships Between Spatial Objects

Object Class	Relationship	Object Class
node	——within———> <——contains———	polygon
node	——bounds———> <——bounded by——	arc
arc	——within———> <——contains———	polygon
arc	——bounds———> <——bounded by——	polygon

only the starting and ending bounding nodes but also the left and right designations of the cobounding regions. In this way, an arc is **directed**. That is, the first coordinate of the arc is that of the start node, and the last coordinate is that of the end node. As we move along the arc from the start node to the end node, there will be a polygon on the left and another on the right. Each polygon record might also include the identification of any interior islands or, if the polygon is itself an interior island, the identification of the single exterior polygon.

This data structure not only accommodates the natural pattern of mapped features efficiently, but

---

*See Appendix B for a description of the DXF File Format.

also provides the information needed to facilitate automated editing and manipulation of the database. Figure 10.2 illustrates these components and lists the types of information that are encoded in a computer to represent these data. In this example, each graphic element has associated with it a set of descriptive, or attribute, data.

Software designed as a geographic information system (GIS) can usually collect and exploit the topological data. Topologically structured data formats include the U.S. Geological Survey's Digital Line Graph (DLG) format (U.S. Geological Survey, 1990), and the Bureau of the Census' TIGER (Topologically Integrated Geographic Encoding and Referencing) data (Marx, 1990). The TIGER data supersedes Census's Geographic Base File/ Dual Independent Map Encoding (GBF/DIME) data, also a topologically structured data set. Commercial GIS vendors, such as ESRI and Intergraph, have proprietary topological data structures.

FEATURE COMPONENTS

Another way to organize spatial data is to remove the attribute information from the polygon, arc, and node components and place it with the feature. The features then describe the various components of which they are composed. Table 10.4 illustrates this feature-based structure.

Features can also be composed of other features (these are termed complex features). This element

Table 10.4
Description of Map Features Shown in Figure 10.2

Features		
ID#	Feature Type	Composed of
1	Park	Polygon 2
2	Lake	Polygon 6
3	River	Arcs 12, 15
4	Road	Arcs 13, 14
5	Shoreline	Arc 16
6	Park Boundary	Arc 11
7	Map Border	Arcs 1–10
8	Houses	Nodes 12, 13
9	Bridge	Node 14

of the design allows you ready access to descriptions of features contained in a cartographic database. For example, parts of the feature "Potomac River"—such as the Georgetown Channel, Tidal Basin, and Washington Channel—are all included in a description of the Potomac River. A number of systems are being developed to support this **feature-based data model**. The U.S. Geological Survey, with Digital Line Graph—Enhanced (DLG-E), and the France's Institute Geographic National, with EDIGéO (Electronic Data Interchange in the field of Geographic Information), are among the national mapping agencies collecting feature-based data (Guptill, 1990, and Salgé and Sclafer, 1989).

RASTER DATA STRUCTURES

Instead of describing discrete geographic objects, we can represent geographical variation by recording the locational pattern of one variable over the study area. When the spatial interval of this sampling is regular (for example, rectangular), the resulting matrix of observations is called a raster data structure (Box 10.C). If there are n variables, then n separate items of information are available for each and every point in the area. Digital elevation data and digital images are typically stored in this structure. Other types of information, such as categories of land cover, could also be encoded in a raster data structure. An example of this is shown in Color Figure 12.5.

The cells of the raster data are uniformly spaced. The geographic location of each cell is implicit in its location in the array. A location is specified for the origin of the matrix. The uniform offsets associated with each cell can then be used to determine the specific location of any given cell.

The use of a raster data structure in cartography is appealing for several reasons. First of all, the data structure is easy to understand. Cartographers have used grid systems in many aspects of traditional cartography, and using a fine mesh grid to encode and store spatial information in some ways is merely an extension of traditional practice. Second, the storage and manipulation

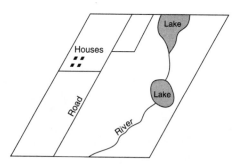

Figure 10.2 A sample map is represented by its component node, arc, and polygon components.

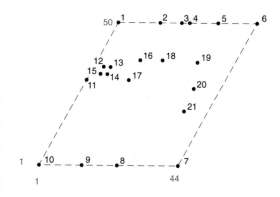

Nodes

ID #	Attribute	X,Y Coordinates
1	nul	1,50
2	nul	14,50
3	nul	21,50
4	nul	23,50
5	nul	32,50
6	nul	44,50
-	-	-
-	-	-
-	-	-
14	house	6,32
15	house	4,32
16	nul	14,37
17	nul	14,30
18	nul	21,37
19	nul	32,36
20	nul	35,27
21	nul	36,20

Arcs

ID #	Attribute	Start Node	End Node	X,Y Coordinates
1	map border	1	2	1,50...14,50
2	map border	2	3	14,50...21,50
3	map border	3	4	21,50...23,50
4	map border	4	5	23,50...32,50
5	map border	5	6	32,50...44,50
6	map border	6	7	44,50...44,1
-	-	-	-	-
-	-	-	-	-
-	-	-	-	-
14	road	17	16	14,30...14,37
15	road	16	2	14,37...14,50
16	road	16	18	14,37...21,37
17	road	18	3	21,37...21,50
18	river	8	21	25,1...36,20
19	shoreline	21	20	36,20...35,27
20	shoreline	20	21	35,27...36,20
21	river	20	19	35,27...32,36
22	river	19	5	32,36...32,50
23	river	19	4	32,36...23,50

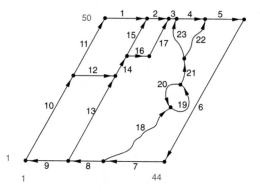

Polygons

ID #	Attribute	Bounding lines
1	map exterior	1,2,3,4,5,6,7,8,9,10,11
2	nul	1,15,14,12,11
3	nul	12,13,9,10
4	nul	2,17,16,15
5	nul	3,23,21,20,18,8,13,14,16,17
6	nul	5,6,7,18,19,21,22
7	lake	19,20
8	lake	4,22,23

Box 10.C
Tessellating the Plane

While raster data sets usually are composed of square grid cells, there is no requirement that this be so. The cells can be triangular, rectangular, or hexagonal (Figure 10.C.1). The type of cell used may be dictated by a collection instrument or chosen because of its mathematical properties. For example, the pixels in Landsat multispectral scanner data images are rectangular, with dimensions of 79×56 meters (see Chapter 8). The hexagonal gridding scheme has the useful property that the center of each cell is an equal distance from the center of all of its neighbors. In fact, the raster cells can be any uniform shape that completely covers a plane without gaps. Geometric figures that do so are said to **tessellate the plane**. The shapes can be quite complex. The figure on the right, below, is a "Koch island" constructed from a fractal curve. It has the unique property that the combination of seven small islands produces an island with the identical shape. This property of self-similarity allows for an infinite nesting of these contorted hexagonal-like shapes.

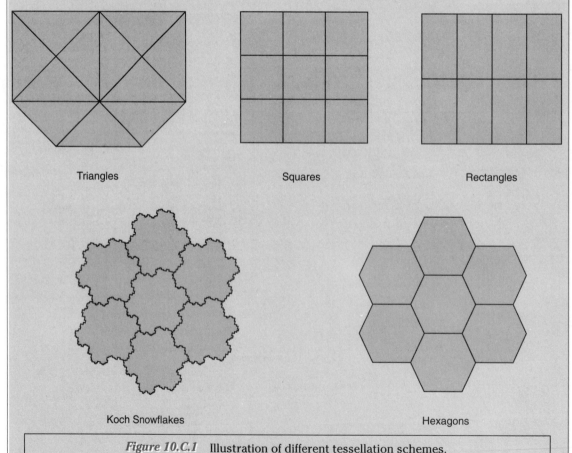

Triangles   Squares   Rectangles

Koch Snowflakes   Hexagons

Figure 10.C.1   Illustration of different tessellation schemes.

of raster data in a computer is straightforward. Multi-dimensional arrays are supported in many computer languages, and numerous algorithms exist for manipulating arrays of spatial information. Third, computer display devices—from color video displays to high resolution film recorders—almost exclusively use raster display technology*. Finally, a number of cartographic data sources (most notably digital satellite and aircraft imagery) are being acquired directly in raster format. The cells of raster data (particularly raster data of images) are commonly referred to as **pixels** (derived from the term picture element).

MODELING SURFACES

While topographic surfaces can be modeled by vector data such as digitized contour lines and spot elevations, or by raster data in the form of an array of heights (a digital elevation model), there is a third method that is often used to model a surface. This method is called a **triangulated irregular network** or **TIN**. TIN models consist of a series of non-overlapping triangular polygons (each defining a planar surface) that completely cover the topographic surface (see Petrie and Kennie, 1991, for more details).

Each vertex of the triangle is encoded with its location and has a height associated with it, as shown in Figure 10.3. The spacing of the elevation points is irregular, resulting in a collection of triangles of different sizes and shapes. Where data points are close together, perhaps in areas of rapid change in elevation, the triangles are tightly packed. Where data points are sparse, such as in flat areas, the triangles increase in size. You can see the result

---

*Only x,y plotting devices (for example, pen plotters) use vector coordinates directly. In the 1960s and 1970s, raster displays were of low quality and resolution. At that time, only devices that directly displayed vectors (whether on plotting film, paper, or the face of a CRT) could provide the high-resolution line work needed for cartographic purposes.

in Figure 10.4, which compares a contour and TIN representation of a surface.

In a GIS, having elevation data in the form of a TIN lets you efficiently calculate slope and aspect data. It also allows you to generate contour, shaded relief, or perspective view portrayals of the data. Figure 10.5 compares shaded relief presentations generated from a TIN with an orthophoto of the same area.

MODELING SOLIDS

So far our discussions have dealt with data on the surface of a plane. However, certain types of data require a more complete modeling of the third dimension. Solid bodies, such as geologic structures (ore bodies, petroleum reservoirs) need to be encoded not only with x,y locations, but also with z or depth information. The surface models mentioned earlier encode z values, but only one z value is permitted for each location. This results in what is termed a 2.5-D model. In true solid or 3-D modeling, multiple z values can exist at a given x,y location.

One way to model such data is to extend the raster data model an additional dimension. Thus the 2-D squares become 3-D cubes as shown in Figure 10.6. The cubes are often termed **voxels** (for volume elements). Each voxel is encoded with attribute data (such as rock type). The geolocation of any given voxel is implied by its location in the x,y,z array and the size of the cubes. The voxel representation allows us to model such things as hollow spheres, with a varying thickness of the shell, or solids with holes through them. Figure 10.7 shows a complicated geologic structure modeled using voxels.

Another way to model solids is to use a technique called **boundary representation** (BR). This is an extension of the vector data model to 3-D. A solid object is defined by its bounding surface. Just as one-dimensional lines bound two-dimensional areas, 2-D planar areas are located in three-dimensional space (as is done with the facets of a TIN) and are used to bound a 3-D solid. The various

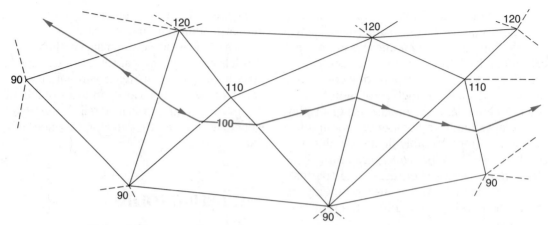

Figure 10.3 Structure of a triangulated irregular network or TIN.

planar areas bound solids, much as the facets bound the shape of a jewel. Figure 10.8 illustrates this method.

The voxel representation is commonly used in earth science applications in a GIS environment. Dynamic processes, such as movement of pollutants through an aquifer, can also be modeled using voxels. The BR approach is often associated with CAD/CAM systems. In such systems, multiple 2-D views of an object, such as a building, can be combined to form a 3-D model.

FEATURE DESCRIPTIONS

Spatial data models, in addition to providing locational data, must handle all the non-locational information used to describe features in a database. For users to understand the nature of the data, there must be a complete specification of the features being described. A comprehensive specification includes the definition of the features as well as definitions of their attributes and attribute values. Other information required for a complete understanding includes: delineation guidelines, data extraction rules, and representation rules.

Understanding how the locational component of a feature is represented is of little use unless you understand what is being described. What is really meant by terms such as "forest," "marshland," "residential land," or "bridge"? For example, you could define a bridge as a built-up structure on or near the surface of the earth. Or you could refine the definition and say that a bridge is a structure erected over a depression or obstacle to carry traffic. Thus, the feature "bridge" is an element of a set of phenomena (an "erected structure"), with the common attributes of function ("to carry traffic") and location. It also has the common relationship of spanning another feature ("over a depression or obstacle"). For good map communication to occur, the definitions of the people who produce the data should match the expectations of those who use the data.

To help solve this definition problem, data producers can use a feature data specification. Such a specification provides a detailed definition of each feature and how it is modeled. This technique is used by national mapping agencies in the United States, France, and Germany. For each feature in a spatial database, a domain of attributes is defined. Each attribute is assigned a value, taken from the domain of values specified by the attribute definition authority. In addition to the definitions, guidance is given on how each feature is delineated, when it is collected, and how it is represented in the spatial data model. This type of information is presented for the U.S. Geological Survey's Digital Line Graph—Enhanced (DLG-E) data using the template shown in Table 10.5.

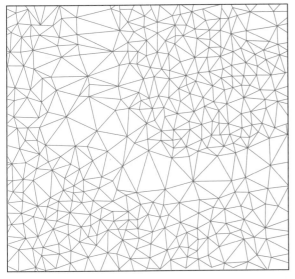

Figure 10.4 Comparison of a contour and TIN representation of a surface.

For an example of how this template is used, see Box 10.D. Here, a complete specification of the feature "bridge" is provided. The specification includes information that traditionally would be found in topographic map compilation instructions and standards for graphic symbolization and publication. This set of information clearly defines the feature being modeled. As we shall see in

Figure 10.5 TIN-based shaded relief and an orthophoto.

Chapter 13, this level of information far exceeds the amount and type of data contained in most digital cartographic databases. However, it has become evident that this information is necessary to make maximum use of digital cartographic data in either a GIS or automated cartographic production system.

MODELING RELATIONSHIPS

Specific relationships (see Box 10.A), termed semantic relationships, are used to express interac-

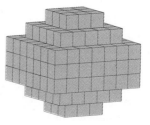

Figure 10.6 Simple voxel representation.

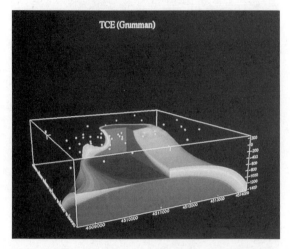

Figure 10.7 A geologic structure modeled using voxels.

tions that occur between features that are otherwise difficult to represent either with locational data, topological relationships, or feature descriptions. Such relationships can include the following:

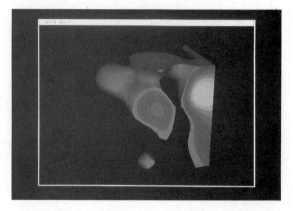

Figure 10.8 Boundary representation of solids.

- **to flow**, the relationship that occurs among the two-dimensional features composing a river system
- **to connect**, the relationship that occurs (at junctions) among the transportation features composing a road network

Table 10.5
DLG-E Feature Data Specification Template

FEATURE   definition as documented in DLG-E domain of features.

ATTRIBUTE/ATTRIBUTE VALUE LIST
    Attribute Value   definitions of attributes and values
        (those common to more than one feature are generic definitions)

DELINEATION   (ground truth)—what the feature looks like on the ground.

REPRESENTATION RULES
    Feature Object Representation Rules   spatial objects that can be used to represent the feature
    Composition Rules   under these conditions, feature is composed of these spatial objects
    Rules for Relationships with Other Feature Objects   relationships that this feature may have with other feature objects

DATA EXTRACTION RULES
    Capture Condition   criteria for determining when a feature is to be captured for inclusion in the data base. This is independent of source.
    Attribute Information   instructions on how to value various attributes of the feature
    Representation Conditions   specific definition rules for a feature instance
    Source Interpretation Guidelines   criteria for extracting ground truth, for those situations that are ambiguous, from various sources (image, field, graphic, DLG-3)

· **to bound** or **to be bounded**, the relationship that occurs between the perimeter of a city and its city limits

· **to be composed of**, the relationship that occurs between a designated route of travel and the transportation features that compose it, or of the stream, and pond features that comprise a watercourse

· **to be above**, the relationship that occurs between the features in an overpass or underpass of a freeway interchange.

The intent of all of these relationships is to model the interactions between features sufficiently to support GIS applications and automated cartographic product generation requirements. These relationships are illustrated in Figure 10.9.

MODELING TIME

Like maps, digital cartographic databases change over time and must be continually revised. There are two ways to handle the temporal dimension associated with spatial data. The simplest way is to maintain multiple versions of a database, akin to keeping a historical map file. When changes occur, you can produce a new edition of the database, release it with the current data, and place the previous edition in a historical archive. Thus, for the maps shown in Figure 10.10, you would have two sets of files—the first with 1970 data, the second for 1989 data. This temporal layering is common with raster data sets. Satellite data sets from many different time periods will be referenced

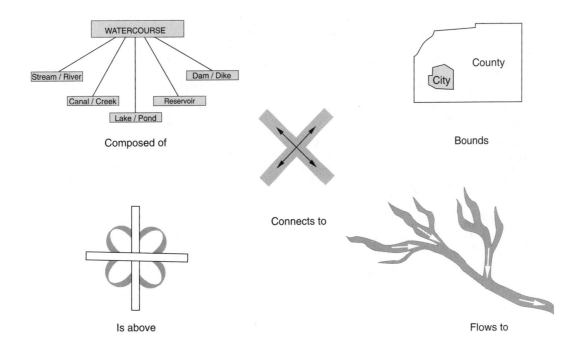

Figure 10.9 Illustrations of semantic relationships.

Box 10.D
DLG-E Feature Data Specification for "Bridge"

The feature data specification provides a detailed definition of each feature and how it is represented in a digital database. Here we provide one example for the feature "bridge." A typical map portrayal of a bridge is shown in Figure 10.D.1.

Bridge - A structure spanning and providing passage over a waterway, railroad, or other obstacle.

Attribute/Attribute Value List

Cover status   **Existence of a cover.**
Covered
Not Covered

Name   **The proper name, specific term, or expression by which the feature is known.**

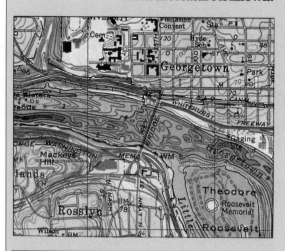

Figure 10.D.1 Depiction of "Key Bridge" on a 1:24,000-scale USGS topographic map.

(Alphanumeric)
Unspecified

Deck Status   Existence of multiple decks
Double Decked
Not Decked
Unspecified

Delineation

The limit of BRIDGE is the extent of the span as defined by the edges of the deck and the end abutments.

Representation Rules

Feature Object Representation Rules
BRIDGE is represented as a 0-dimensional object, a 1-dimensional object, or 2-dimensional object.

Composition Rules
If the 0-dimensional object that represents BRIDGE shares a coordinate pair with any arc, then the 0-dimensional object that represents BRIDGE is composed of a node; otherwise it is composed of a point.

The 1-dimensional object that represents BRIDGE is always composed of a set of one or more arcs.

The 2-dimensional object that represents BRIDGE is always composed of a set of one or more polygons.

Rules for Relationships with Other Feature Objects
None.

continued on next page

Box 10.D (continued)

Data Extraction at 1:24,000 scale

Capture Conditions

Characteristics that affect the decision to capture BRIDGE:

Size

Name

Design

If BRIDGE is _ 240 ft. along the longest axis, OR

If BRIDGE is < 240 ft. along the longest axis and is named, OR

If BRIDGE is < 240 ft. along the longest axis and is a covered bridge, a footbridge, or a drawbridge, then capture.

Attribute Information

For all BRIDGES,

Deck Status=Double Decked or Not Decked.

Representation Conditions

If BRIDGE is captured from graphic, and it is a covered bridge, and is shown without wing ticks, and crosses a 1-dimensional feature, then BRIDGE is represented as a 0-dimensional object.

If BRIDGE is < 125 ft. along the shortest dimension and does not meet the representation conditions for a 0-dimensional object, then BRIDGE is represented as a 1-dimensional object.

If BRIDGE is _ 125 ft. along the shortest dimension and does not meet the representation conditions for a 0-dimensional object, then BRIDGE is represented as a 2-dimensional object.

Source Interpretation Guidelines

All sources

If BRIDGE meets capture conditions and carries ROAD, TRAIL, RAILWAY, or CANAL/DITCH, then capture both BRIDGE and ROAD, TRAIL, RAILWAY, or CANAL/DITCH.

If BRIDGE does not meet capture conditions and carries ROAD or RAILWAY over another ROAD or RAILWAY, then capture ROAD, or RAILWAY, and UNDERPASS.

If BRIDGE does not meet capture conditions and carries ROAD, TRAIL, or RAILWAY over STREAM/RIVER or CANAL/DITCH, then capture only ROAD, TRAIL, OR RAILWAY.

If BRIDGE does not meet capture conditions, and carries CANAL/DITCH over another CANAL/DITCH or STREAM/RIVER, then capture CANAL/DITCH and UNDERPASS to allow definition of the relationship between CANAL/DITCH and the feature over which it passes.

Graphic

BRIDGES over double line drains, symbolized without bridge wing ticks, will be captured from shoreline to shoreline.

BRIDGES symbolized with bridge wing ticks will be captured from wing tick to wing tick.

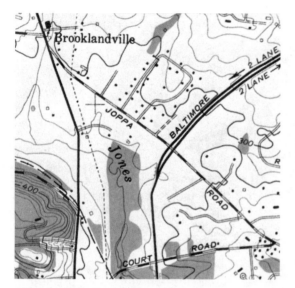 

Figure 10.10   Maps illustrating temporal change from 1970–1989.

to a uniform geographic grid. Each collection of data will be referenced to the common grid and form another temporal layer in the database. For example, the USGS produces bi-weekly composites of AVHRR data for the contermi-nous United States. One CD-ROM holds about eight of these bi-weekly time slices.

A more comprehensive, yet more complex, strategy is to maintain one database (Langran, 1992). Every feature in this database must have time information associated with it. This pro-cess is fairly straightforward for information in raster form. For each cell, you maintain a vari-able-length list of attributes, noting any change in attribution and the time at which the change occurred. For feature-based vector data sets, the process is more complicated. For each fea-ture, you must encode the following informa-tion into the database:

- when the feature was first incorporated into the database

- if the feature presently exists
- when the feature ceased to exist (for example, when a pond was filled in)
- when the feature was recreated, perhaps in a different form (for example, when a road was rerouted).

Thus, you can trace the existence and modifica-tion of any feature. Such a sequence is shown in Figure 10.11.

SELECTED REFERENCES

Bondy, J. A., and U. S. R. Murty, *Graph Theory with Applications*, New York: North Holland, 1976.

Borgida, A., "Conceptual Modeling of Informa-tion Systems," in *On Knowledge Base Man-agement Systems*, Brodie, M. J., and Mylopoulos, J., eds., New York: Springer-Verlag, 1986, pp. 461–469.

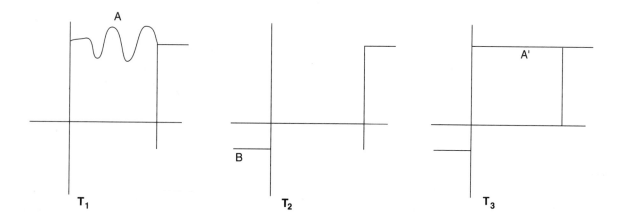

Figure 10.11   Feature tracking over time. At time T2, feature A has ceased to exist, and feature B has been added. At time T3, feature A has been recreated as A' with a different shape than at time T1.

Brodie, M. J., and J. Mylopoulos, (eds.), *On Knowledge Base Management Systems*, New York: Springer-Verlag, 1986.

Cattell, R. G. G., *Object Data Management*, Reading: Addison-Wesley Publishing Co., 1991.

Codd, E. F., "A Relational Model for Large Shared Data Banks," *Communications of the ACM*, 13,6 (1970): 377–387.

Corbett, J. P., *Topological Principles in Cartography*, Technical Paper 48, Washington DC: U. S. Department of Commerce, 1979.

Giblin, P. J., *Graphs, Surfaces, and Homology*, New York: Halsted Press, 1977.

Guptill, S.C., (ed.), *An Enhanced Digital Line Graph Design*, U. S. Geological Survey Circular 1048, Reston, VA: U. S. Geological Survey, 1990.

Haas, L. *et al.*, "Starburst Mid-Flight: As the Dust Clears," *IEEE Transactions on Knowledge and Data Engineering*, March, 1990.

Harary, F., *Graph Theory*, Reading, MA: Addison-Wesley, 1969.

Kim, W. *et al.*, "Architecture of the ORION Next-Generation Database System," *IEEE Transactions on Knowledge and Data Engineering*, March, 1990.

Langran, G., *Time in Geographic Information Systems*, London: Taylor and Francis Ltd., 1992.

Lefschetz, S., *Applications of Algebraic Topology*, New York: Springer-Verlag, 1975.

Marx, R. W., ed., "The Census Bureau's TIGER System," *Cartography and Geographic Information Systems*, 17,1 (1990): 17–113.

Petrie, G., and T. J. M. Kennie, *Terrain Modelling in Surveying and Civil Engineering*, New York: McGraw-Hill, Inc., 1991.

Peuquet, D. J., "A Conceptual Framework and Comparison of Spatial Data Models," *Cartographica*, 21,4 (1984): 66–113.

Salgé, F., and M. Sclafer, "A Geographic Data Model Based on HBDS Concepts: The IGN Cartographic Data Base Model," *Proceedings, Auto Carto 9*, Bethesda, MD: American Congress on Surveying and Mapping, (1989): 110–117.

Stonebraker, M., *et.al.*, "The Implementation of POSTGRES," *IEEE Transactions on Knowledge and Data Engineering*, March, 1990.

Teorey, T. J. and J. P. Fry, *Design of Database Structures*, Englewood Cliffs, NJ: Prentice Hall, 1982.

Ullman, J., *Principles of Database and Knowledge-Based Systems,* Vol. 1 and 2, Rockville, MD: Computer Science Press, 1989.

U.S. Geological Survey, *Digital Line Graphs from 1:24,000-Scale Maps, Data Users Guide 1,* Reston, VA: U.S. Geological Survey, 1990.

Wilkinson, K., *et al.,* "The IRIS Architecture and Implementation," *IEEE Transactions on Knowledge and Data Engineering,* March, 1990.

Winston, P.H., *Artificial Intelligence,* 2nd Edition, Reading, MA: Addison-Wesley, 1984, pp. 251–289.

MAP DIGITIZING

*C*omputer technology provides a modern alternative to traditional manual and photomechanical mapmaking methods. Digital methods generally provide greater mapping flexibility than analog methods. But to gain this freedom, you must have access to suitable digital records.

A large number of computer-compatible databases are already available (see Chapter 13). But there will still be times when existing digital records are inadequate to serve your compilation needs. At such times, you may decide to create the necessary files yourself (or have someone create them for you). Or you may decide to forget about them altogether and rely instead on a nondigital compilation procedure. Both options have their place.

Before beginning a digitizing project, you must weigh the costs and benefits against those of alternative compilation methods. Unless a data file is to be used a number of times, the costs of producing it may be difficult to justify. If, after considering the alternatives, you still feel that digital compilation is the best approach, then you must decide how to go about it. There are several possible methods. You can accomplish analog-to-digital conversion of maps by hand or with the help of image scanners. In either case, the digital records can be structured in either a vector or a raster format for use in GIS or cartographic production software.

COLLECTING THE GEOMETRY

Throughout their history, maps have consisted of graphic analog images made up of point, line, and area symbols (plus lettering), which have been produced by various manual drafting or scribing techniques, and by sticking up preprinted symbols. Regardless of the approach taken, however, the elements of the component artwork have a definite graphic form. As we saw in Chapter 10, a vector data structure preserves the form of point, linear, and areal features. As a result of this "object" focus, vector data structures are functionally similar to the standard way of going about the mapping process.

Vector digitizing was initially done by hand. A sheet of transparent graph paper was overlaid on an existing map, and the coordinate values were scaled from the graph paper. The results were entered on a coding sheet, and the coded data were typed into a computer terminal. As you might expect, the process was slow, tedious, and expensive. As the number of coordinates to be recorded increased, the labor became overwhelming. Furthermore, due to the nature of this digitizing process, numerous errors tended to creep into the digital records. For all these reasons, manual digitizing with graph paper should be avoided if possible. The cost of digitizing equipment has fallen so much that it can be justified for all but the smallest projects. Today, virtually all vector digitizing is done with the aid of automated devices. Although all these machines involve human operators, some are more labor intensive than others.

MANUAL DIGITIZERS

Digitizing can be accomplished with devices as simple as a mouse connected to a personal computer. But much more sophisticated equipment is used both to preserve the accuracy of the information in the original map and to speed up data collection. The most widely used digitizers are manually operated devices called digitizing tables or tablets (depending on their size and configuration).

The digitizer has a number of components, including a coordinate table or digitizing surface (equivalent to electronic graph paper), a movable cursor usually containing a cross-hair and keypad, and a computer interface. The operator moves the cursor by hand over material that has been mounted to the digitizing surface. The equipment's electronics sense the cursor's position and translate the trace of the cross-hair over the image into arbitrary *x-y* coordinates (Figure 11.1). These table coordinates are automatically recorded in a computer file.*

---

*The exact format of this file is determined by the software package being used with the digitizer. These files tend to be product-specific and not interchangeable. This lack of interchangeability has spurred the development of data exchange standards (see Chapter 13).

Figure 11.1 Manual digitizing with magnifying cursor and 16-button keypad.

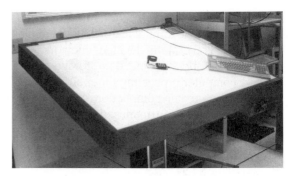

Figure 11.3 Large backlit digitizing table.

Digitizing tablets vary greatly in size and accuracy. Small digitizers with a surface of 8" by 11", with a coordinate accuracy of 0.01", might cost several hundred dollars (Figure 11.2). A large (42" by 60") backlit digitizing surface, with an absolute accuracy of 0.001", will cost 10 to 20 times more (Figure 11.3). Often these precision digitizing surfaces are integrated into the design of a computer graphics workstation (Figure 11.4).

Most manual digitizers can be operated in two modes: point and line. In **point mode**, coordinates are only recorded when the operator gives a special signal, such as clicking a certain button on the cursor. This mode is used to record point

features. It can also be effective in digitizing linear features if only a few points are involved (for example, if the lines are made up of straight-line segments or are geometrically rather simple). This has the additional effect of minimizing the size of the resulting data file.

In **line mode** (often called **stream mode**), the machine is preset to record coordinates automatically at given time or distance intervals. These increments, in conjunction with the scale of the source map, determine the spatial resolution of the resulting digital record. The tighter the curves and the more the detail in the original linework,

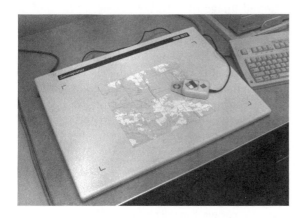

Figure 11.2 Small digitizing tablet.

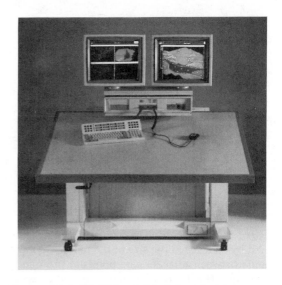

Figure 11.4 Digitizing table integrated into computer graphics workstation.

the greater the digitizer resolution needed to capture or preserve the form of features in the digital file (Figure 11.5). More detailed linework also requires more operator time, data storage capacity, and computer processing time, all of which translate into greater compilation costs. This is why complicated map manuscripts are usually digitized using scanning technology.

SCANNING SYSTEMS

The purpose of scanning systems is to replace the tedious (and error-prone) task of tracing linework on maps with an automated method. Some early automated systems actually tried to automate the tracing function, using a laser beam to follow a line segment. However, such efforts have all been replaced by various forms of raster scanners. The scanning devices collect a digital image of the map in raster form. Scanners on the market today differ

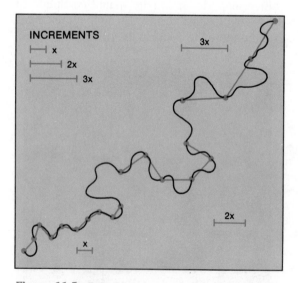

Figure 11.5 Resolution in manual digitizing is defined by changes in the size of the minimum increments between adjacent data points in the *x* and *y* directions. Greater resolution is achieved by specifying finer increments.

in the mechanism used to perform the scanning, the size of document that can be scanned, and the complexity and cost of the system.

Large-Format Drum Scanners

Large-format drum scanners are high-precision, high-cost devices that can scan large pieces of paper or film without cutting. The drum scanner consists of a scanning head and a large cylindrical drum. The map is placed on the drum, which is about a meter long and a meter wide. Some scanners recognize the presence of a line or symbol only in black and white, while others recognize colors. In both types, the scanner optics, which act like a high-powered microscope, divide the area under view into small unit areas. These subdivisions of the map image are called pixels (see Chapter 10). Each pixel represents an area as small as 25 microns in diameter. (The period at the end of this sentence is about 400 microns in diameter).

The finer the resolution, the more scanning time (measured in square inches or centimeters of image processed per minute) and the more data handling, of course. Changing the spot size from 50 microns to 25 microns will quadruple the number of pixels. Thus, the resolution sought should take into consideration subsequent compilation needs. A full matrix of raster data is collected even if the only feature on the map image is a sparse river system. This can result in very large data files. The scanned image of a 1:24,000-scale, 7-1/2 minute USGS quadrangle may be 20,000 pixels wide and 24,000 pixels high, requiring over 480 megabytes of storage.*

The scanning process produces a digital picture of a map. In some cases, particularly when a graphic product is the *only* product desired, this may be sufficient. Map restoration and reproduc-

---

*This assumes that the data file is not compressed or reduced in size. Typically, however, such files are compressed, reducing storage requirements by an order of magnitude or more. Data compression will be discussed in Chapter 15.

Figure 11.6 Portion of Yorktown, Virginia, map, restored and reprinted using map scanning technology.

tion projects are good examples. Figure 11.6 shows a section of a map that was digitally restored and reprinted from a damaged original copy.

If the information is to be used in other than a purely graphic fashion, however, it must be processed still further to complete the encoding of data. Raster data commonly are **vectorized**. In this case, pixels describing a map feature are connected and converted to a series of *x,y* coordinates to describe nodes, arcs, and polygons (see Box 11.A). As we shall see in subsequent sections, the data are then transferred to other equipment or software. Here, geometric and topological checks are performed, and attributes such as road classification and elevation are added.

Electronic Video Systems

An electronic video system can be used in place of a raster scanner for collecting raster map images.

Components of such a system normally include a video (that is, television) camera, a raster cathode ray tube monitor, and a processing unit to make the analog-to-digital conversion. Images can be digitized from either reflection or transmission imagery. (Transmission imagery requires backlighting).

The results may be displayed directly on a raster monitor, or transferred to disk or magnetic tape for future processing. Up to 256 shades of gray can be differentiated in a black-and-white image, so the system is far more sensitive than the human eye.

Electronic video systems are far less expensive than precision raster scanners, but they cannot achieve the extremely high resolution of the more costly equipment. You must weigh these factors against your mapping requirements. The data files collected by the video system can be processed by software similar to that used with data from the raster scanner.

Document Scanners

Document scanners, such as those designed for scanning artwork into desktop publishing systems, can also be used to scan maps. At present, these devices offer relatively high resolution (300 dots per inch or 0.033 mm. spot size to 400 dpi/0.025 mm. spot) at a relatively low cost. Both black-and-white (256 shades of gray) and color (up to 16.7 million color levels) documents can be scanned.* However, the documents are limited to page-size drawings (8 1/2 × 11 inch or A4). You can ease this size limitation by scanning a large document in pieces. You can then use computer software to digitally merge the pieces back together. The files produced by the scanner can be used by various CAD software packages for further processing, including vectorization (see Box 11.A).

---

*Care must be taken when scanning documents that have been screened or half-toned (see Chapter 30) for printing. Instead of detecting a solid line, or a color tint, the scanner will detect each of the half-tone *dots*. The resulting file will be a perfectly acceptable image, but will be impossible to process through vectorizing software.

Box 11.A
The Automatic Digitizing Process

The **raster scanner** (Figure 11.A.1) is a device that converts maps (or any drawing or picture) into digital form. It does so by ascertaining the color of the portion of the map that is in the field of view of its optics. This so-called spot size is very small—0.05 mm. in the scanner shown here. The drum on which the map is mounted rotates, and readings are taken every 0.05 mm. When one rotation is completed, the scanning head moves 0.05 mm. to the left and repeats the process. The resulting digital image of the entire map, which takes about 90 minutes to generate, is kept in a computer file for further processing. Advances in scanner technology are continuously reducing the amount of time necessary to perform the scanning.

Both continuous tone and line (binary) images can be scan-digitized. In the case of continuous tone photos, a range of numbers (say from 0 to 255) is used to represent the tonal density values. With line maps, quantizing generally takes a binary form, with 0's signifying empty cells and 1's representing occupied cells. Some automated scanners work from transmission source material, while others work from reflection copy. Some require miniature images (such as 35 mm. film), while others work with large-format images (up to a meter square). Some scanners, such as the one shown, use color recognition techniques so that even multicolored originals can be scan digitized into separate numerical records for each color. The operator calibrates the equipment by pointing the scanner head at each of the source map's colors. The scanner can then recognize up to 12 colors per scan automatically, making it ideal for digitizing a multicolored input map such as a USGS topographic sheet.

In the digital image (Figure 11.A.2), the original linework seen by the scanner is shown in black. Each line is between seven

Figure 11.A.1 Large-format raster drum scanner.

continued on next page

Box 11A. (continued)

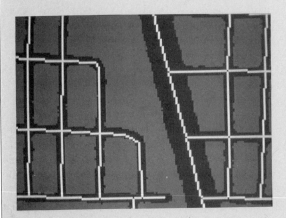

Figure 11.A.2   Raster scanned image of map linework. Original lines are the wide black lines. The centerline is shown in white.

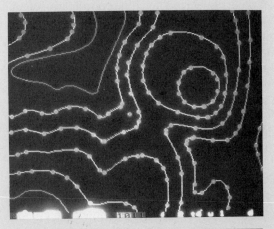

Figure 11.A.3   Dots show the series of x,y coordinate pairs that describe the raster linework shown in the background.

and 15 pixels wide (175 to 375 microns). These lines must be **vectorized**. The process to do this is as follows. The black lines are thinned to white lines one pixel wide (25 microns). These centerline pixels are converted by processing software to a series of x,y coordinate pairs (as illustrated by the dots on the contour lines in Figure 11.A.3). Depending on the processing software, this process may be entirely automated or, as shown in Figure 11.A.4, performed with the assistance of an operator, who will trace the vector centerline on the computer screen to fall within the scanned original line. Such a tracing may be helpful, for example, if there is a very wide line on the original. In such a case, the processing software may have trouble unambiguously locating the centerline.

Figure 11.A.4   Interactive editing of scanned map linework.

ESTABLISHING A GEOGRAPHIC FRAME OF REFERENCE

Vector digitizers and raster scanners both code data with respect to a machine coordinate system. A location on the table or drum is set as 0,0 and all subsequent measures are given with respect to that point. Yet, knowing how many inches a location is in the x and y direction from an arbitrarily established origin on a vector digitizer table, or how many rows and columns a pixel is away from the beginning point (row 1, column 1) in a scan-digitized matrix, is insufficient information for many compilation purposes (Figure 11.7).

A prerequisite of digital compilation is that the digital records representing all source materials be coded with a common reference system. This usually means converting local machine coordinates to a common geographical reference system such as latitude-longitude. Existing cartographic databases in most cases will already have had their data converted to standard geographical coordinates. In addition to latitude-longitude, State Plane Coordinates, Universal Transverse Mercator coordinates and the United States Public Land Survey parcel descriptions have all been used for this purpose (see Chapter 6). However, a user may wish to convert between geographic reference systems, say from State Plane to UTM.

Fortunately, the software appropriate for handling this coordinate transformation chore is available as part of many commercial GIS packages as well as in public domain software.* Computer routines for map coordinate conversion take advantage of the fact that a map projection establishes a systematic mathematical relationship between ground and map locations. Thus, just as

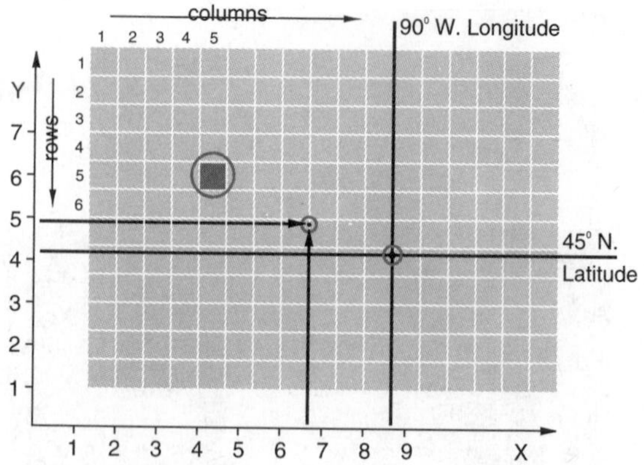

Figure 11.7 Machine reference systems inherent in digitizer table and raster scanner records need to be converted to a common geographical reference system if subsequent mapping flexibility is to be facilitated.

you can move freely from ground reference to map reference (as you would for plotting a graphic from a map database), you can reverse the process and move from map reference to ground reference.

All you need to do is supply pertinent information as input to the computer. First, you must give information about key locations on the source map sheet (for example, several corners or internal tick marks). You must also give the position of the map sheet within its general projection framework (for example, with respect to origin, central meridians or parallels, and so on). You provide this information along with the coordinate file of geographical features, and the computer program does the rest. The result is a new file that is coded with respect to the desired geographical reference base.

---

*The General Coordinate Transformation Package (GCTP) has been developed by the USGS and the Chart and Geodetic Survey (C&GS). It provides forward and inverse transformations between 22 commonly used projection systems (USGS, 1990).

ADDING ATTRIBUTES

Once you have coded the geometry of the map features, you must add attribute information. This attribute data may be quite extensive (as shown in the example in Chapter 10), or relatively simple.

You may input simple sets of attribute data, such as feature type = road, at the time of digitizing. To do so, you use codes inserted via the cursor buttons, or select attributes from a menu located elsewhere on the digitizing surface (see Figure 11.8). The amount of information you can easily encode via this method is rather limited.

However, it is more common for features to have extensive attribute records. Very often the attribute data are held in a relational database management system (RDBMS). You can perform the data input via a "fill-in-the-blank" interface to the RDBMS. You can use the resulting **feature attribute table** for a wide variety of spatial query processing. This process is shown schematically in Figure 11.9.

The ideal situation would be to avoid entirely the manual entry of data already portrayed on a map sheet. Software capabilities to recognize symbolized line work and text in a scanned image and automatically convert them to attribute codes are steadily increasing. Software can, for example, recognize the symbology of a railroad line, derive the vector centerline coordinates, and assign the attribute of "railroad" to the feature. Likewise,

software can recognize different width lines and associate them with various road classes. More advanced systems use **optical character recognition** technology to recognize the numeric values placed on index contours and then attach exact elevation codes to each of the vectorized contour lines. Some software systems can even recognize the names of rivers and roads and link them to the proper features. All these software options hold remarkable promise. Before using any such software, however, you must first define all the required attribute data.

EDITING DATA

Raw digital records as they come from digitizing operations may require considerable processing and manipulation to bring them up to a standard that is cartographically acceptable. To begin with, the digitizing process is usually far from perfect, requiring a variety of cleanup chores before the data can be put to use in compilation. Even with careful planning and execution, things may go wrong in the digitizing operation. Both machine and human errors occur. Additionally, problems with the source material may have been overlooked. In any case, file cleanup and editing are routine chores for the map compiler who is involved with newly created digital records.

The errors that need correction can take several forms. When working with files generated by manually operated vector digitizers, you may find that arcs were crossed by mistake. Also, polygon boundaries may not have closed, features may have been misidentified (such as a river coded as a road), and so forth. Some features may have been digitized twice (usually with slightly different positioning), or they may not have been done at all. You must take care of all these errors, omissions, and deletions during this cleanup and editing stage (Figure 11.10).

Checking the geometry of digitized features usually involves plotting out the digital

Figure 11.8 Encoding attribute data from the digitizing table.

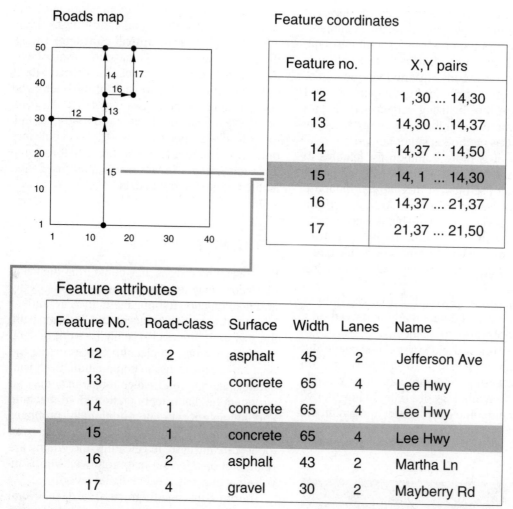

Roads map

Feature coordinates

Feature no.	X,Y pairs
12	1 ,30 ... 14,30
13	14,30 ... 14,37
14	14,37 ... 14,50
15	14, 1 ... 14,30
16	14,37 ... 21,37
17	21,37 ... 21,50

Feature attributes

Feature No.	Road-class	Surface	Width	Lanes	Name
12	2	asphalt	45	2	Jefferson Ave
13	1	concrete	65	4	Lee Hwy
14	1	concrete	65	4	Lee Hwy
15	1	concrete	65	4	Lee Hwy
16	2	asphalt	43	2	Martha Ln
17	4	gravel	30	2	Mayberry Rd

Figure 11.9 Associating feature attributes with geometric elements via a relational database table.

records at the same scale as the original map and overlaying the two. A close examination (often under magnification) of the map and overlay will reveal discrepancies in the digitizing process. Some errors may require redigitizing portions of the map. Other errors may be corrected interactively, comparing the data displayed on the computer screen with the map source, and making adjustments.

Automated topological and attribute checking can identify several types of errors in the digital files. These include arcs that do not connect to other arcs, polygons that are not closed, and polygons with no attributes or with multiple, conflict-

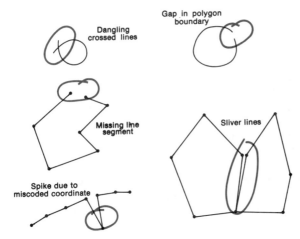

Figure 11.10 Cleanup and editing of digital vector files is usually necessary in order to obtain error-free data. Poor geometry, omissions, and deletions must be taken care of.

Figure 11.11 Errors automatically flagged by editing software for operator correction.

ing attributes. The software can flag these areas and display them on interactive edit screens for you to correct (see Figure 11.11).

This process of editing can take as much time as the basic digitizing and attribute coding. However, its importance cannot be overstated. Correct data, accurately located in a geographic reference system, is essential for creating maps or geographic information systems.

COLLECTING METADATA

Your task of creating a digital database does not end once you collect an error-free file of information. In order to describe the contents adequately, you need to collect and store certain additional information about the data. This additional information, called **metadata**, is prepared by the data producer and consists of a set of structured digital information (Federal Geographic Data Committee, 1993). A set of metadata allows the data user to judge the information's fitness for a given applica-

tion. Some of the items that might be included in a set of metadata are shown in Table 11.1. Many of the standards for **spatial data exchange** require a number of metadata elements. We will discuss these standards in more detail in Chapter 13. This information is also used in **metadata catalogs**, discussed in Chapter 15.

There are several categories of use for spatial metadata. A collection of metadata could accompany a data transfer and serve as documentation. It could also serve as an internal, on-line set of documentation of processing history and file lineage. Finally, metadata could serve as a stand-alone data set for use by spatial data catalogs and indexes.

THE DIGITIZING PROCESS

This chapter has only touched on the major aspects of the map digitizing process. It has been estimated that the process of digitizing represents 60 to 80 percent of the cost of creating a digital cartographic production system or establishing a

Table 11.1

Categories of Spatial Metadata Content

Identification Information	a concise description of the data set
Contact Information	organizational contact to obtain the data
Transfer Information	mechanisms by which the data can be acquired
Status Information	the degree of completeness or availability of the data set
Coordinate System Information	details of the coordinate system and map projection of the data set
Source Information	a description of the origin of a data set
Metadata Reference Information	information about the metadata entry itself
Processing History Information	information about the processing steps that have been performed on the data set
Data Quality Information	a set of measures to provide a user information on which to judge if the quality of the data set is suitable for a planned application
Feature/Attribute Information	an explicit description of the information contained in the data set by providing an exhaustive list of the features, attributes, and attribute values present in the data

geographic information system. The level of effort and expertise required has caused an entire industry, the **map (or data) conversion industry**, to come into being. Each segment of the industry has developed its own processes and procedures (largely dependent on the type and scale of information being collected). However, the concepts presented here are used in all digitizing applications.

SELECTED REFERENCES

Federal Geographic Data Committee, *Draft Content Standards for Spatial Metadata*, Reston, VA: U. S. Geological Survey, 1993.

U. S. Geological Survey, *GCTP—General Cartographic Transformation Package Software Documentation*, National Mapping Program Technical Instructions, Reston, VA: U. S. Geological Survey, 1990.

White, M. S., "Technical Requirements and Standards for a Multi-Purpose Geographic Data System", *The American Cartographer*, 2,1 (1984): 15–26.

PART IV

DATA
PROCESSING

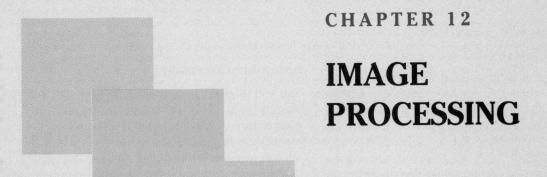

CHAPTER 12

IMAGE PROCESSING

*I*n Chapter 8, we examined the variety of photo-graphic and electronic imagery created by remote sensing devices operated on aircraft and space vehicles. Many of the images seen as illustrations in that chapter were not the original products generated by the sensing system but, rather, were derived from the original image data using **image processing** techniques. Image processing refers to manual and digital techniques used to improve image geometry, enhance image appearance, iden-tify features in an image, and extract selected information from an image.

When applied to images of the environment produced through remote sensing, image process-ing is a powerful cartographic tool. In this chapter, we will first examine manual image processing for cartographic purposes, and then we will turn to the role of digital image processing in mapping.

MANUAL METHODS OF IMAGE INTERPRETATION

Manual image processing techniques have served the needs of cartographers for many years. These procedures range widely in technical sophistica-tion. We will consider a selection of these methods in this section, progressing from lesser to greater complexity in the process.

SINGLE IMAGE

An image poses two problems for the cartogra-pher. We must first determine what we are see-ing. We call this process of feature recognition or identification **image interpretation**. Our sec-ond problem is that we usually want to deter-mine the location as well as the identity of fea-tures. To find the location of a feature on the ground from an image, we must relate the **image geometry** to the geometry of the ground scene. Let's look first at image interpretation and then at image geometry.

Image Interpretation

You will often want to interpret images. A good way to do so is to use **image interpretation elements**. For black-and-white images, these ele-ments include the size, shape, tone, texture, and relative positions of identified entities. The addi-tional element of hue difference will often help you identify features on color images.

A typical problem, for example, might be to interpret residential land use on an image. To do so, you might use clues such as the size and shape of houses and streets, the color and size of yards, and relative locations of houses, yards, and streets.

You may also find it helpful to use **interpreta-tion templates** or **keys**, which have been pre-pared for many features. These keys contain ex-amples of features in typical surroundings and written descriptions of how features look on differ-ent types of film. With some keys, you identify features through an elimination process. You make a series of choices between two or more alterna-tives until only one possibility remains. Keys for electronic scanner and SLAR images are also avail-able for selected features.

Image Geometry

You may consider an aerial photograph to be a planimetrically correct representation of the envi-ronment if: (1) the terrain imaged is flat and level, (2) a small enough area is imaged so that distortion due to the earth's curvature is negligible, and (3) the photo was obtained using a distortion-free camera lens that was aimed vertically at the point directly below the camera. Under these perfect imaging conditions, you may faithfully outline the interpreted features of interest on a translucent sheet laid over the photograph.

If the aerial photo is at the same scale as the base map, you may register the translucent compilation sheet with the base map to serve as a guide for inking or scribing (see Chapter 31). You may also digitize these lines using manual tracing or map scanning methods (Chapter 10) into a cartographic database compatible with computer mapping and geographic information systems.

In most cases, the photo will be different in scale from the base map. You may make the required scale adjustment by placing the photo in an optical projector. With such a projector, you can project the image of the photographic transparency or print over the base map and enlarge or reduce it until the two match exactly (Figure 12.1). You can trace interpreted features directly onto the base map or a translucent cover sheet.

With more sophisticated projectors, you can optically remove distortions on the photograph due to aircraft tilt (or systematic stretching on scanner and SLAR images). With the **Zoom Transferscope**™, for example, you view the projected image through binocular optics. Then you can optically scale, rotate, and linearly stretch the image in any direction until you make the best fit with the base map (Figure 12.2). If images have been stretched in a less regular manner, you can match them with the base map in a piece-wise fashion, although feature outlines may not match along the edges of each image segment.

Hills, valleys, and other landform features introduce irregular **relief displacements** (see Chapter 8) on the aerial photograph (and other forms of imagery) that cannot be corrected with optical projectors. Nevertheless, relief displacements are valuable because they allow us to determine the elevation of all features.

Positional displacements due to differences in elevation occur radially away from the photo **nadir**. The nadir is the point on the photograph at which a vertical line extended from the center of the camera lens would touch the film. On a truly vertical photograph, the nadir coincides with the photo's **principal point**, or geometric center (Figure 12.3). Straight lines connecting opposite **fiducial marks** intersect at the principal point. The amount of relief displacement (r_d) is computed from the following equation relating the local elevation (h), the radial distance from the principal point (d), and the flying altitude (H):

$$r_d = \frac{h \times d}{H} \quad \text{or} \quad h = \frac{r_d \times h}{D}$$

A comparison of similar triangles in Figure 12.4 illustrates the geometrical basis for this equation. We see that photographs of the same area taken at lower altitudes contain more relief displacement than those from higher altitudes. Although high-altitude and satellite photographs contain far less distortion due to relief differences, low-altitude photographs are more useful for determining elevations, particularly if the overlapping area on two adjacent photos can be viewed stereoscopically.

STEREOSCOPIC IMAGES

You perceive depth, the relative closeness and farness of objects, in a variety of ways. With only one eye, you must rely on the relative sizes of objects, clarity of object detail, and on whether one object appears in front of another. When you see an object with both eyes, each eye sends a separate message to your brain, creating **stereoscopic vision**. Because your eyes are separated, each sees an object (if it is within approximately 1,800 feet or

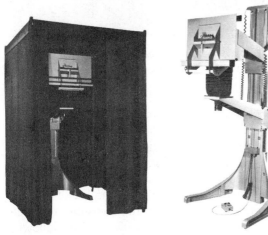

Curtain enclosure

Figure 12.1 Optical projector used to enlarge or reduce an aerial photograph or other image print. The new image is projected onto a planimetrically correct base map (tracing table not shown). (Courtesy Map-o-Graph Co.)

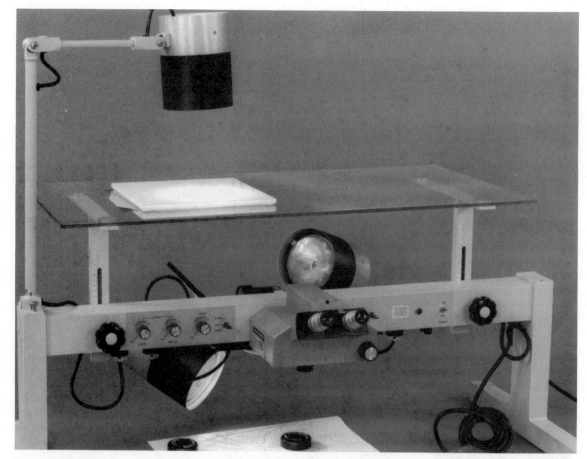

Figure 12.2 Zoom Transferscope™ used to optically enlarge, rotate, and stretch an image (placed on glass stand) so that features may be traced onto a base map (under the instrument). (Courtesy Image Interpretation Systems, Inc., Rochester, New York).

600 m. of you) from a slightly different angle. The subsequent slight variations in the two messages to your brain cause the sensation of object depth.

Parallax

Relief displacement on aerial photographs produces **parallax**, the apparent change in the position of an object due to a change in the point of observation. You can measure the amount of parallax, called the **parallax difference**, on two overlapping photographs aligned as in Figure 12.5. The principal points and conjugate principal points

(location of the principal point of the overlapping photograph) are marked on both photographs. The photographs are aligned so that the four points fall on a straight line. The average distance *b* between principal points and conjugate principal points is then measured. The parallax difference is the difference between the measured distances *xx'* and *yy'* and can be used in the following formula to determine the difference in elevation, Δ*e*, between the two points:

$$\Delta e = H/b \times \text{parallax difference}$$

where *H* is the flying altitude. If point *x* or *y* is a

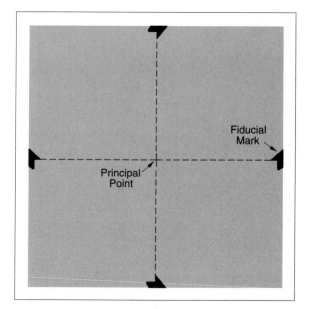

Figure 12.3 Fiducial marks are located in the focal plane of the camera. They may be etched on the surface of the glass, which is in contact with the film. Or, in the case of the open-type focal plane, they are projections of metal that extend into the negative area. In either case, fiducial marks appear on the aerial photo film.

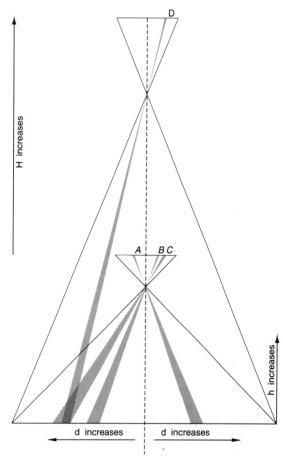

Figure 12.4 As the distance from the center, *d*, of the photograph increases, and/or as the height of the object, *h*, increases, the amount of displacement, r_d (shown in colored tint and measured on the photograph) increases. As the altitude of the camera, *H*, increases, displacement decreases.

triangulation station of known elevation, the true elevations of other points may be computed.

The apparent changes in position on two photographs also allow you to view their overlapping areas and see a three-dimensional image of the landscape. By viewing an object on one photograph with your left eye and the same object on the adjacent photograph with your right eye, you are in effect viewing the object from the two aerial camera positions at which the photos were taken. Your brain is able to simultaneously integrate the parallax differences for all objects into a three-dimensional image. If both photographs are truly vertical, objects in your three-dimensional image will be in their true planimetric position.

Stereoscopes

A **stereoscope** is an instrument that makes stereoscopic image viewing possible. It does so by allowing your left eye to see only the overlap area on one photo and your right eye to only see the overlap area on the second photo.

Several kinds of stereoscopes are manufactured. The two most commonly used are the folding lens

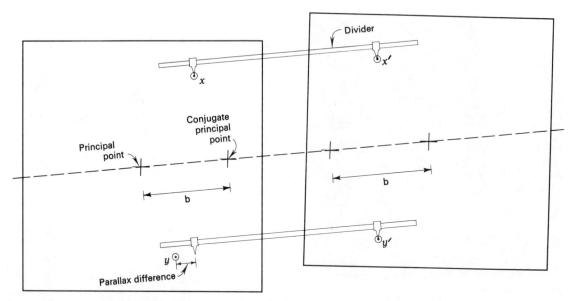

Figure 12.5 Graphically measuring parallax difference. The locations of one of the points on the two photographs is indicated by x and x', and the second point by y and y'. If the lines connecting the principal point to the conjugate principal point on each photograph are placed along a straight line, then the difference $xx'–yy'$ is the parallax difference.

type (Figure 12.6) and the mirror type (Figure 12.7). The folding lens type is lighter and more portable, while the mirror type permits you to view a larger area without moving the instrument.

Both types are used in stereoscopic image interpretation, but both suffer in that you cannot directly mark planimetrically correct feature outlines directly onto a base map. To do this, you need a more sophisticated instrument such as the **Stereo Zoom Transferscope**™ (Figure 12.8). With this instrument, you view the overlap area on the two photos stereoscopically while optically adjusting the scale and orientation of each until your three-dimensional image of the landscape is aligned as well as possible with the base map, which you also see through the binocular viewing lenses. You may then use a pencil to trace the features of interest directly onto the base map in their approximately correct planimetric positions.

Stereoplotters

Stereo Zoom Transferscopes and similar instruments allow you to sketch planimetric features, such as houses and roads, onto a base map. But if you wish to plot such features and contour lines precisely, you will need a **stereoplotter.**

Figure 12.6 Folding lens (pocket) stereoscope. (Courtesy Carl Zeiss, Inc.)

Figure 12.7 Mirror stereoscope with parallax bar used to measure parallax difference. (Courtesy TOPCON Instrument Corp.)

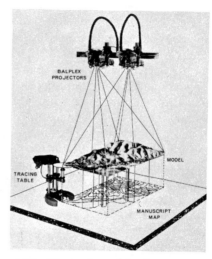

Figure 12.9 Illustration of the basic operation of the Balplex Plotter used for compiling topographic maps. (Courtesy Bausch & Lomb Optical Co.)

Modern stereoplotters are complex computer controlled instruments, but the best way to understand the principles behind stereoplotting is to look at the original optical projector devices, such as the Balplex plotter developed in the 1930s (Figure 12.9).

You can think of optical plotters as operating by reversing the light rays that formed the camera picture and, by projection, producing the image of a three-dimensional model of the terrain on a mapping table. The overlapping portions of two photographs are copied onto glass transparencies that are placed in the two projectors located at the top of the instrument. One transparency is projected downward onto the mapping table through

Figure 12.8 Stereo Zoom Transferscope™ used to optically scale, rotate, and linearly stretch stereoscopically viewed photographs so that they will match a base map upon which interpreted features are traced. (Courtesy Image Interpretation Systems, Inc., Rochester, New York).

a red filter, while the other is projected through a cyan (the complementary color to red) filter. This produces a superimposed pair of images called an **anaglyph.** The cartographer wears special glasses with red and cyan lenses, which allow the left projected image to be seen only with the left eye and the right image only with the right eye. These separate images are mentally fused into a three-dimensional view of the terrain called a **stereo model**.

The two projectors are oriented so that the projected images overlap and the stereo model fits horizontal control points plotted at the desired map scale on the mapping table. This is done in a two-part procedure called **interior and exterior orientation**.

Interior orientation is performed first. This involves moving and rotating the two projectors along their x, y, and z axes until the projected images duplicate the relative position and orientation of the camera at the instant each was obtained.

Exterior orientation follows. Here the mapping table is raised or lowered and tilted until control points marked on the photos match those plotted on the table. When absolute orientation has been achieved, the stereo model is a planimetrically correct representation of the landscape.

The three-dimensional terrain model that we see may now be used to delineate planimetric features and plot contour lines. This is done by tracing with a "**floating dot**" visible on the stereo model. A small tracing table with a white **platen** (circular disc) on top has a tiny illuminated hole in its center. The platen "disappears" into the stereo model, in a manner similar to the apparent disappearance of a projector screen when a slide is projected onto its white surface. The tiny illuminated hole, however, appears to be a dot floating in space at an elevation corresponding with the height of the platen above the mapping table. The platen, and hence the dot, can be raised or lowered a known distance as indicated on its micrometer scale, which in turn is calibrated to known elevations on the stereo model.

The floating dot is adjusted to the desired elevation, and then is moved so as to appear in contact with the terrain. A pencil in the tracing table directly below the illuminated hole is lowered to touch the compilation sheet on the mapping table, and the tracing table is slowly moved, keeping the dot in contact with the terrain. Contours are traced in this manner, adjusting the floating dot for each new contour elevation. Planimetric features such as roads are plotted from the model by tracing the path of each while continually adjusting the floating dot so it is always in contact with the land surface.

Modern **analytical stereoplotting instruments** do not physically project the two photographs but, rather, allow you to view them separately through the left and right eyepieces of a binocular viewing station (Figure 12.10). Interior and exterior orientation is accomplished analytically, which means that the *x,y* coordinates of positions on the two photo transparencies are mathematically transformed to match ground coordinates and elevations.

To perform this transformation, the cartographer enters the camera focal length and additional interior calibration data into the computer that drives the stereoplotter. Next, the cartographer digitizes the coordinates of the fiducial marks seen on the model, as well as several ground control points. As lines are digitized on the stereo model with a cursor device, their digitizer coordinates are mathematically transformed to ground coordinates. Current systems output the ground coordinates of all features to a color monitor, where they may be reviewed and edited prior to storage on disk or magnetic tape for later use in a computer mapping program or GIS.

DIGITAL IMAGE PROCESSING

Today cartographers routinely use **digital image processing systems** to geometrically rectify images, prepare contrast enhanced multicolor image maps, create thematic maps by extracting information from image data, and simplify such thematic maps. Before they can accomplish these

Figure 12.10 Leica SD3000 Universal analytical photogrammetric workstation. (Courtesy Leica, Inc.)

tasks, all imagery must be in raster format. In raster format, each image is defined by a matrix of **pixels** (picture elements). Electronic scanner images typically are collected in raster format, but aerial photography and most SLAR images must be converted to pixels using a scanning digitizer (see Chapter 10). Digital image processing, then, deals with mathematical procedures employed to rectify, enhance, classify, and simplify digital imagery in raster form.

IMAGE RECTIFICATION

Original digital images typically must be rectified to match a particular map projection surface or grid coordinate system laid over the projection surface.* Image maps normally require rectification, as do thematic maps generated from image data. Digitally rectified high-altitude photography, satellite imagery, and some types of SLAR images will be planimetrically correct **orthoimages**, due to the minimal relief distortion across the image.

Image rectification is a **coordinate transformation** and image **resampling** problem. The coordinate transformation problem is similar to converting digitizer coordinates to UTM or State Plane coordinates (see Chapter 6). Our rectification example will be the transformation of a tiny Landsat MSS image segment into 100m × 100m square pixels aligned with the UTM grid system (Figure 12.11). The task is to determine which of the old MSS pixels fall within each new rectified pixel, and to assign a new digital reflectance value accordingly. The following linear coordinate transformation (affine) equations are commonly used to carry out the transformation:

$$col = a + bE + cN \qquad row = d + eE + fN$$

where col and row are the (column, row) coordinates of the MSS pixel center, E and N are the

corresponding UTM grid coordinates, and *a-f* are the transformation coefficients. The six transformation coefficients are determined through the simultaneous solution of three or more such linear equations using control points. The control points required are the UTM coordinates of at least three MSS pixel centers (x_1-x_3 in Figure 12.15). The matrix method for solving these equations is outlined in Box 12.A. If more than the minimum three control points are used, a least squares solution results. This allows the accuracy of the coordinates to be assessed statistically, and the least accurately defined control points to be identified.

The above linear equations remove linear distortions only between coordinate systems. Performing more complex transformations requires higher order polynomial equations, which, in turn, require a greater number of control points. An alternative is a **piece-wise** solution, in which linear equations are determined for a number of sub-areas. The problem here is that coordinates may not match well along sub-area boundaries.

You may perform image **resampling** in several ways, but the simplest is **nearest neighbor**

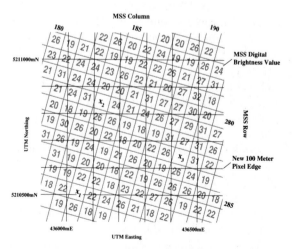

Figure 12.11 Geometric rectification of Landsat MSS pixels to a new image with 100 × 100m square pixels aligned with the UTM grid. x_1-x_3 are control points used in the affine coordinate transformation.

---

*This requirement is so common that Landsat and SPOT images geometrically corrected to the UTM, Lambert conformal conic, or other map projection surfaces may be purchased.

Box 12.A

Matrix Solution to Linear Digital Image Rectification Using Three Control Points

Three Equations to Solve Simultaneously

$$\text{col}_1 = a + bE_1 + cN_1 \qquad \text{row}_1 = d + eE_1 + fN_1$$
$$\text{col}_2 = a + bE_2 + cN_2 \qquad \text{row}_2 = d + eE_2 + fN_2$$
$$\text{col}_3 = a + bE_3 + cN_3 \qquad \text{row}_3 = d + eE_3 + fN_3$$

Matrix Form of Equations

$$\begin{vmatrix} \text{col}_1 \\ \text{col}_2 \\ \text{col}_3 \end{vmatrix} = \begin{vmatrix} 1 & E_1 & N_1 \\ 1 & E_2 & N_2 \\ 1 & E_3 & N_3 \end{vmatrix} \cdot \begin{vmatrix} a \\ b \\ c \end{vmatrix} \qquad \begin{vmatrix} \text{row}_1 \\ \text{row}_2 \\ \text{row}_3 \end{vmatrix} = \begin{vmatrix} 1 & E_1 & N_1 \\ 1 & E_2 & N_2 \\ 1 & E_3 & N_3 \end{vmatrix} \cdot \begin{vmatrix} d \\ e \\ f \end{vmatrix}$$

Matrix Solution

$$\begin{vmatrix} a \\ b \\ c \end{vmatrix} = \begin{vmatrix} 1 & E_1 & N_1 \\ 1 & E_2 & N_2 \\ 1 & E_3 & N_3 \end{vmatrix}^{-1} \cdot \begin{vmatrix} \text{col}_1 \\ \text{col}_2 \\ \text{col}_3 \end{vmatrix} \qquad \begin{vmatrix} d \\ e \\ f \end{vmatrix} = \begin{vmatrix} 1 & E_1 & N_1 \\ 1 & E_2 & N_2 \\ 1 & E_3 & N_3 \end{vmatrix}^{-1} \cdot \begin{vmatrix} \text{row}_1 \\ \text{row}_2 \\ \text{row}_3 \end{vmatrix}$$

col_x and row_x are the column,row coordinates of the MSS pixel center.

E_x and N_x are the corresponding UTM grid coordinates.

resampling. With this approach, you select the digital reflectance value of the MSS pixel center closest to the center of the new pixel. In our example, the new pixel center (solid black dot in Figure 12.11) at (436050,5210850) corresponds to a MSS column, row position of (182.9,287.3), which is rounded to (183,287) to give the nearest MSS pixel center. You would thus assign the new pixel a reflectance value of 28.

A more complex resampling technique uses the distance weighted average of the nearest four MSS pixel reflectance values. This form of resampling, called **bilinear interpolation,** is essentially identical to the distance weighted surface interpolation method described in Chapter 26.

The effect of image rectification is illustrated in Figure 12.12, which shows a UTM rectified Thematic Mapper (TM) image of central Washington, D.C. The rectified image is a satellite orthoimage that would serve well as the base for a satellite image map.

IMAGE ENHANCEMENT

You can use digital image enhancement to improve the appearance of image data displayed on color monitors or written to hardcopy devices such as printers and film recorders. Electronic images and film or paper copies of original image data often are low in contrast, with many features of interest very hard to identify. The problem is that the graytone or color assignment system in the display device is not tuned well to the range of digital reflectance values in the image. The film recorder that produced the Washington, D.C., image in Figure 12.13, for example, assigned printed graytones ranging from black to white to a 0–255 digital reflectance value range. The reflectance values in the image only ranged from 10 to 107. The result is a low-contrast, dark gray print which is in need of **contrast enhancement.**

Figure 12.12 A UTM rectified Thematic Mapper (TM) image of central Washington, D.C.

Contrast Enhancement

The easiest way to increase image contrast is to rescale the original reflectance value range to match the numerical range used in the image display or recording device. To do this, you can use the equation:

$$D = (R\text{-}R_{min})/(R_{max}\text{-}R_{min}) * (D_{max}\text{-}D_{min}) + D_{min}$$

where

D = digital value sent to the output device
D_{min} = minimum digital value used by the output device
D_{max} = maximum digital value used by the output device
R = original pixel reflectance value
R_{min} = minimum reflectance value in the original image
R_{max} = maximum reflectance value in the original image

In our Washington, D.C., image, $D_{min} = 0$, $D_{max} = 255$, $R_{min} = 10$, and $R_{max} = 107$, meaning that the equation simplifies to $D = (R\text{-}10)*255/97$. The effect of this equation is to linearly stretch the histogram of original image reflectance values to com-

Figure 12.13 Original unenhanced Landsat TM band 4 image of central Washington, D.C.

pletely cover the display device range (Figure 12.14). A higher-contrast image such as Figure 12.15 normally results, unless a few pixels have extremely low or high reflectance values. In this case, you can still carry out contrast enhancement by arbitrarily assigning the lowest n percent of reflectance values a new value of 0, and the highest n percent a value of 255. The remaining reflectance values are now scaled from 1 to 254 using the above equation but with $D_{min} = 1$ and $D_{max} = 254$. R_{min} and R_{max} are the minimum and maximum of the remaining reflectance values.

The image produced by this **percentage contrast stretch** is generally much higher in contrast. The same is true of multicolor images, such as Color Figure 12.1. Here each of the three bands used to control the red, green, and blue electron gun intensities (or the yellow, magenta, and cyan separations for color printing) is separately stretched.

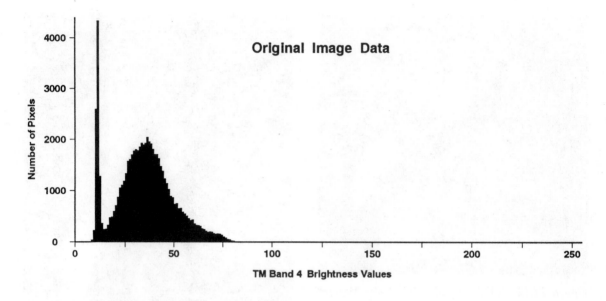

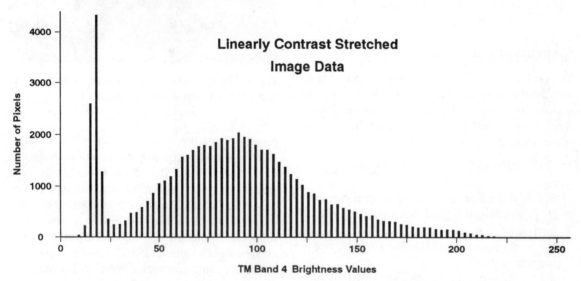

Figure 12.14 Histograms illustrating the linear contrast stretch.

Spatial Filtering

You can also increase image contrast by passing a **spatial filter** across the image. Such a filter emphasizes differences in reflectance values between adjacent pixels. Places on the image having large reflectance differences over short distances are said to have a "high spatial frequency." Areas of uniform or slowly changing reflectance have a "low spatial frequency."

A **high-pass local filter** is used to emphasize high-frequency areas or features, creating what is

Figure 12.15 Linearly contrast stretched image of the data displayed in Figure 12.13.

called an **edge enhancement** effect. The filtering process involves sweeping a small matrix, called a **window**, across each image scan line.

The 3 × 3 window in Figure 12.16 consists of a center cell with a **weighting factor** of 9 surrounded by eight cells with weighting factors of -1. We can envision laying this commonly used high-pass filter over the upper left corner of the digital image segment, so that its center cell overlays pixel (1,1) on the image.

The filter operates by multiplying the reflectance values under the window by the corresponding weighting factor. In other words, multiply the reflectance value under the center window cell by 9 and the eight surrounding reflectance values by -1. Sum these nine products to find the digital number for location (1,1) in the new high-frequency enhanced image.

The window is now shifted one pixel to the right, centered over image location (2,1). Repeat the computations to find the new digital number for position (2,1) in the new image. Doing this for all interior pixels in the image produces a high-frequency enhanced image, such as Figure 12.17.

This new image is higher in contrast than the old one. Also notice the increased ability to discern linear features such as roads and large buildings.

You can use a **low-pass** filter to reduce image contrast on image maps, creating a **smoothing** effect. You might want to smooth an image, for example, if you wish to use it as background for a thematic map showing land use or other environmental features.

The most common low-pass filter is a 3 × 3 window with a weighting factor of ⅑ in each cell. Each new pixel value is given the average value of the nine pixels in the window. The result is a lower-contrast, somewhat blurred image such as Figure 12.18.

Density Slicing

You may also add contrast to a single-band image by assigning a sequence of different colors to the range of reflectance values found in the image. Such a procedure is called **density slicing**.

The Washington, D.C., image in Color Figure 12.2 is density sliced using a 10-class color scheme to accentuate the low and high near-IR values in the data set. Color assignments can be made according to any of the quantitative data classification methods described in Chapter 26. The equal frequency method was used in this example. Thus, the lowest 10 percent of reflectance values in the image are assigned to the first (dark blue) class, and so on. We have thereby created an isarithmic map of near-IR values colored using the layer tinting method (see Chapter 21).

Enhancement by Image Combination

Combining digital data from different sensors can significantly improve image contrast and enhance features of interest. Color Figure 12.3, for example, combines the landscape appearance information inherent in Thematic Mapper data with the detailed terrain depiction found on SLAR images. The following image processing procedures are required to produce such composite images:

High-pass Filter
Moving Window

$$\begin{array}{|c|c|c|}\hline -1 & -1 & -1 \\\hline -1 & 9 & -1 \\\hline -1 & -1 & -1 \\\hline\end{array}$$

Original Brightness Values

Image Column

Image Row	0	1	2	3	4	5	6	7	8	9	10	11	12 ...
0	41	45	45	44	45	45	39	38	42	40	44	57	57
1	40	45	43	41	43	42	36	38	41	42	47	49	39
2	39	44	44	42	40	40	43	47	46	47	46	40	37
3	41	43	44	39	39	43	42	41	46	49	47	47	41
4	38	43	41	41	43	43	44	42	42	44	44	43	38
5	35	40	39	37	43	40	36	32	34	37	31	27	36
6	38	38	36	34	35	35	32	35	36	34	35	36	39
7	38	39	35	36	39	39	36	35	33	37	40	41	42
8	37	38	39	39	39	42	40	36	37	41	41	44	45
9	39	38	40	40	38	40	43	42	41	41	45	47	45
10	33	37	38	43	43	44	44	39	38	40	41	42	38
11	38	42	38	42	46	46	40	37	39	42	44	44	43
12	42	37	36	41	39	38	38	36	35	41	42	44	48

High-pass Filtered Brightness Values

Image Column

Image Row	0	1	2	3	4	5	6	7	8	9	10	11	12 ...
0	31	60	53	45	56	71	30	27	59	18	20	119	145
1	26	64	37	23	48	47	-8	10	29	25	58	74	9
2	18	57	55	45	31	32	58	90	63	59	46	7	21
3	44	53	59	17	20	53	35	17	56	79	63	87	57
4	26	66	43	44	62	57	77	61	53	66	71	76	32
5	7	52	41	21	79	49	21	-13	4	33	-21	-59	15
6	41	42	26	6	12	15	0	41	47	23	32	33	40
7	39	52	16	28	52	53	30	30	6	36	51	47	48
8	27	37	46	45	38	64	47	17	27	54	33	50	60
9	59	41	48	41	12	27	60	60	55	45	68	82	72
10	-7	27	22	62	48	56	65	27	21	29	24	31	5
11	38	79	26	54	78	82	38	24	43	58	60	54	50
12	72	26	10	52	24	19	28	19	7	56	44	42	65

Figure 12.16 High-pass filtering using a 3×3 moving window passed through the image in a raster scan manner.

Figure 12.17 High-pass filtered image of central Washington, D.C., TM image data displayed in Figure 12.13.

Figure 12.18 Low-pass filtered image of central Washington, D.C., TM image data displayed in Figure 12.13.

1. The SLAR image must be scan digitized and the resulting digital image must be geometrically rectified to match the Thematic Mapper image.

2. Reflectance data from three Thematic Mapper bands (1, 2, and 3 in this example) are contrast enhanced and then assigned to the blue, green, and red axes of the RGB color system (see Chapter 19).

3. The RGB coordinates for all pixels are transformed into Hue, Lightness, Saturation (HLS) color model values. Hue and Saturation carry the color information contained in the RGB coordinates.

4. The digital SLAR data are rescaled to match the numerical range of the Lightness dimension (normally 0 to 1), and then are substituted for the Thematic Mapper Lightness values. A high-contrast terrain image has thus been substituted for the lower-contrast TM Lightness component.

5. The new composite image is displayed on a color monitor operating in the HLS color system, or the HLS values are converted back to RGB coordinates for display and hardcopy production on standard equipment.

Other contrast and terrain enhanced image maps have been produced in this manner, including Thematic Mapper imagery merged with digital relief shading. Digitized NHAP high-altitude aerial photography has also been combined with Thematic Mapper imagery to create 1:24,000-scale image maps.

THEMATIC INFORMATION EXTRACTION

The extraction of thematic information from digital image data is a very important image processing activity. There are two general approaches to compiling thematic maps through image processing.

Manual Image Classification

The first feature-extraction approach is a combination of manual interpretation and digital image display. You use a mouse, cursor, or electronic pen to trace the outlines of interpreted features from the image displayed on the color monitor (Figure 12.19). Lines appear to be drawn on the image as they are digitized, and are also stored in a data file for later editing and inclusion in a cartographic or GIS database. This method, of course, is a computerized version of tracing class boundaries from a single aerial photograph or any other hardcopy image.

Some image processing systems also display a stereoscopically viewable image, on which you may trace class boundaries directly. This works best if you can geometrically rectify the 3D image to match control point or digital elevation model data.

Digital Image Classification

The second approach to thematic map compilation is to have the image processing software classify each pixel, based on the reflectance value in each spectral band. Before this is possible, two conditions must be fulfilled: (1) the map classes must have sufficiently distinct spectral signatures, and (2) the digital data for pixels of each class must mirror each spectral signature.

Cartographers are increasingly called upon to perform two major forms of digital image classification. We call these two categories supervised and unsupervised classification.

Figure 12.19 Cartographer outlining land use classes on a high-resolution color monitor that is part of an image processing workstation. (Courtesy ERDAS, Inc.)

Using **supervised classification**, you determine the digital reflectance characteristics of pixels known to be of each feature class. Then you use this information to classify the entire image. The land use map of Washington, D.C., shown in Color Figure 12.4 was created using the simple form of supervised classification outlined below:

a. Display the digital image on a color monitor. Outline small **training areas** known to be of each feature class. Identify training areas with the help of **ancillary information** such as larger-scale images, topographic maps, and thematic maps of the area that show the distribution of related features.

b. Collect digital reflectance values for the pixels falling within the training areas for each class. Determine statistics for this sample, such as the **mean** reflectance and reflectance **range**, for each band. Next, produce **histograms** for all bands. These histograms show the frequency distribution of reflectance values for each class. Several training area histograms used in classifying the Washington, D.C., image are shown in Figure 12.20.

c. For each band, find upper and lower reflectance value **bounds** for each class. To do so, examine the histograms for each class, and select the range of reflectance values for each band that best defines the class. Each range must be wide enough to include the expected variations in reflectance, yet be narrow enough to minimize overlapping class bounds. One way to select the range is to simultaneously examine two bands, with training area reflectance values for both bands plotted on a two-dimensional **scatter diagram** such as Figure 12.21. You can then determine class limits by drawing non-overlapping boxes outlining the cluster of similar reflectance values for each class. Extend the edges of each box to the scatter diagram axis, where you can read the lower and upper bounds for the two bands.*

d. You can now classify the entire image by seeing which box each pixel falls into. For

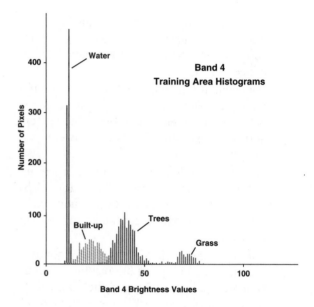

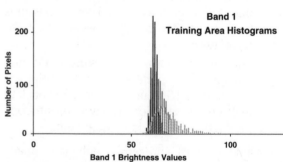

Figure 12.20 Histograms of training area digital reflectance data used to classify the central Washington, D.C., image in Color Figure 12.4.

example, a pixel with reflectance values of 60 and 45 in bands 1 and 4 falls into the "trees" class box, and is classified accordingly.* *Pixels not falling into any box remain unclassified or are placed in an "unknown" category.

Unsupervised classification is not based on training areas but, rather, on finding inherent clusters of similar reflectance values. The computational details of cluster definition are beyond the scope of this book, but the basic procedure can be explained for two-band unsupervised clas-

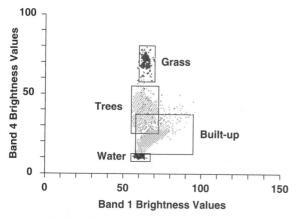

Figure 12.21 Scatter diagram of training area reflectance data for the central Washington, D.C., land use map. Boxes outline the upper and lower bounds for each class.

sification. We begin by examining a scatter diagram of band 1 and 2 reflectance values for all pixels in the area to be classified (Figure 12.22). The idea is to find on the diagram clusters of similar reflectance values, called **spectral classes**, that represent different ground features. When all clusters have been identified and outlined, the pixels falling in each cluster are displayed in the same color. The multicolor thematic map that results from displaying all clusters (see Color Figure 12.5) must now be examined to determine what ground feature class corresponds to each spectral class. This class assignment requires the same ancillary information used in supervised classification to define training areas.

Neither supervised nor unsupervised classification must be based on image data alone. Ancillary data, such as a digital elevation model geometrically registered to the image, may be used as

another band. Incorporating such data may significantly improve classification accuracy, particularly if the different feature classes are closely linked to the ancillary information. Vegetation types often occur within certain elevation ranges, for example. This "hybrid" type of classification is often carried out in a GIS (see Chapter 15), in which the remote sensor data form only one part of the total database used in classification.

DIGITAL IMAGE SIMPLIFICATION

There are several cases in which you may want to digitally simplify classified images. You may, for example, want to generate a smaller-scale generalized thematic map from a larger-scale classified image. Or a classified image may contain a scattering of single unclassified pixels or pixels of a class different from their neighbors. In either case, you achieve simplification by changing the pixel's class to that of the majority of its neighbors. You can do so by using a small local filter, such as the 3 x 3 window overlaid on the thematic map segment illustrated in Figure 12.23.

The simplification filter works by finding the center pixel's class, and then determining the classes of the eight surrounding pixels. If a majority of the surrounding pixels are of a different class,

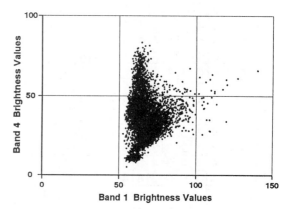

Figure 12.22 Scatter diagram of reflectance values for all pixels in the central Washington, D.C., image area. Notice the inherent pixel clusters.

---

*Three and higher dimensional boxes can be envisioned as additional bands are employed in the classification.

**This classification procedure is identical in concept to using the Boolean .and., .or. operations in a GIS to classify digitized map data in raster or vector format.

the center pixel is changed to the majority class on the new simplified map. Otherwise, the original class is transferred to the new map.

When the new map is complete (see Color Figure 12.6), you may filter it in the same manner to produce a second simplification iteration. You may continue the iteration process until a minimal number of pixels has been changed.

CARTOGRAPHIC PRODUCTS

Aerial photographs and other remote sensor images are environmental base maps as well as major data sources for the compilation of topographic maps and other traditional cartographic products. For decades, aerial photographs have been assembled into **mosaics** which, in turn, have been overprinted with selected map symbols to produce **photomaps** such as Figure 12.24. Because they can be prepared quickly, photomaps have

been used during military operations as a replacement for unavailable topographic maps. Mosaics also serve well as a base upon which soil, geologic, or vegetation class outlines may be overprinted.

MOSAICS

Mosaics are of two kinds, uncontrolled and controlled. Both kinds use only the central part of each photograph, thereby reducing the geometric distortions due to relief displacements. The central areas are carefully trimmed along features in such a way that the joints between pieces can be easily camouflaged upon assembly.

If the sections of the photographs are laid in place only by visually matching images, the mosaic is referred to as being **uncontrolled**. If horizontal ground control points are first plotted on a map base and if the photographs are positioned so that the control points seen on the photo coincide with the plotted control, the mosaic is said to be

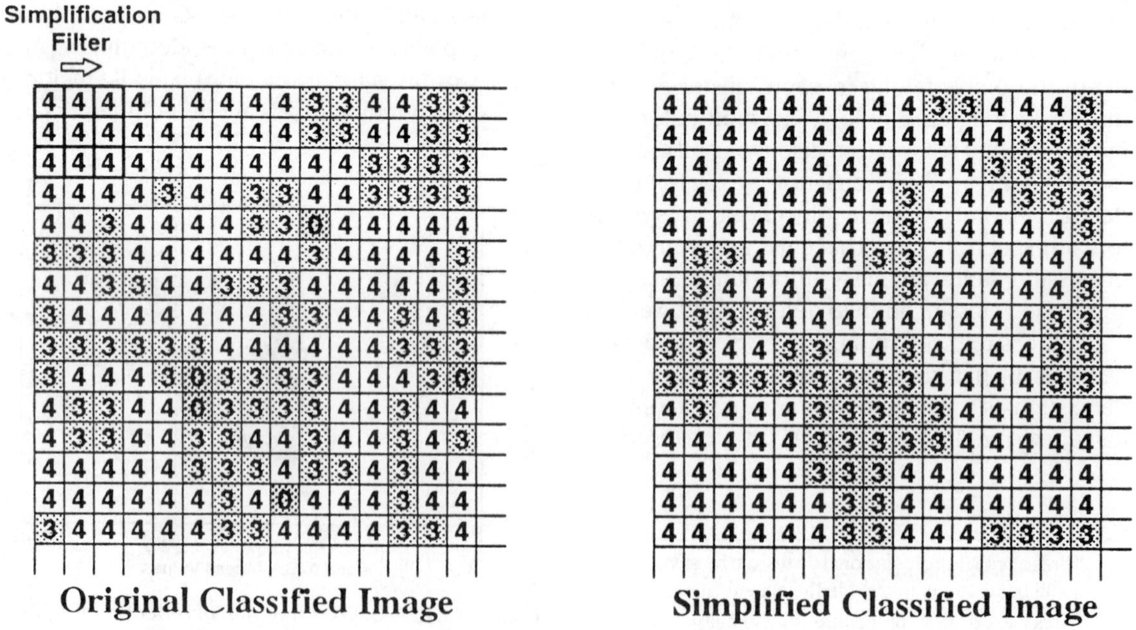

Figure 12.23 Original (*left*) and simplified (*right*) image classification based on a 3 × 3 simplification filter.

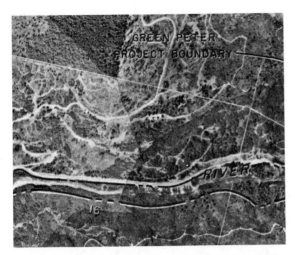

Figure 12.24 Large-scale photomap taken near Sweet Home, Oregon, made from a controlled mosaic. Notice the tonal differences between photos used in the mosaic. (Courtesy U.S. Army Corps of Engineers).

controlled. Although the overall scale of a controlled mosaic is likely to be better, we must remember that none of the relief displacement has been removed from the individual photographs. Consequently, the scale across the mosaic is not constant as on a planimetrically correct map.

Uncontrolled and controlled mosaics have been assembled from SLAR images as well as Landsat and SPOT scenes. Mosaic assembly proceeds as described above, except that on SLAR images the more geometrically correct areas away from the flightline are used. The entire satellite image may be used, due to the low geometric distortion throughout the image.

ORTHOPHOTOS

An **orthophoto** is a planimetrically correct aerial photograph. The perspective geometry on the photograph and the resulting relief displacements can be changed to match a planimetrically correct base map by use of a device called an **orthophotoscope**.

Although there are a variety of instruments that produce orthophotos, most operate on the basic principles underlying the first instrument developed in 1953 (Figure 12.25). The projection of the three-dimensional stereoscopic image produced by a stereoplotter is exposed to a film sheet through a very small opening moved across the model in a scan line manner. As the tiny aperture moves along the narrow scan line strip, the film sheet being exposed remains stationary in its horizontal position but is raised or lowered to keep the aperture "in contact with the surface" of the three-dimensional stereo model. The operation is much the same, then, as tracing correct planimetry from a stereo model by continually adjusting the floating dot to keep it in contact with the terrain surface. After the aperture has moved across the stereo model along one scan line, it is stepped sideways a distance equal to the width of the aperture and the next scan line is exposed.

Current orthophotoscopes also operate by first scanning the entire stereo model using the floating dot method, or by mathematically computing the parallax difference at each point using a

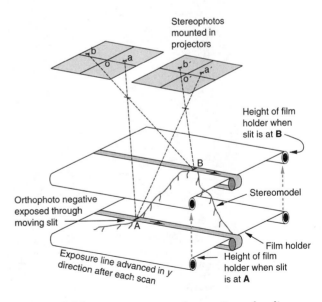

Figure 12.25 Components and operation of a direct optical projection orthophotoscope.

statistical procedure called **cross correlation**. The elevations computed along each scan line from the parallax difference are stored in a digital elevation model data file. This file is then used to raise and lower the aperture as it moves along each scan line in a computer controlled film plotting instrument.

ORTHOPHOTOMAPS

Orthophotos can be assembled to form a mosaic, which, in turn, can be overprinted with map symbols to produce an **orthophotomap** such as Figure 12.26. Photomaps prepared from conventional uncontrolled or controlled mosaics have long been used as planimetric map substitutes, but the relief displacement inherent in the photos limited their usefulness. The orthophotomap has all the advantages of the photographic image and is of consistent scale as well. **Orthophotoquads**, orthophotomaps made to match 1:24,000-scale and other topographic quadrangles, now exist for many parts of the United States.

Figure 12.26 Portion of orthophotoquad of Black Earth, Wisconsin. (Courtesy U.S. Geological Survey).

SELECTED REFERENCES

Batson, R. M., K. Edwards, and E. M. Eliason, "Synthetic Stereo and Landsat Pictures," *Photogrammetric Engineering and Remote Sensing,* 42 (1976): 1279–84.

Chavez, P. S. Jr., "Digital Merging of Landsat TM and Digitized NHAP Data for 1:24,000-Scale Image Mapping," *Photogrammetric Engineering and Remote Sensing,* 52 (1986): 1637–46.

Hartigan, J. A., *Clustering Algorithms,* New York: John Wiley and Sons, 1975.

Hord, R. M., *Digital Image Processing of Remotely Sensed Data,* New York: Academic Press, Inc., 1982.

Jensen, J. R., *Introductory Digital Image Processing,* Englewood Cliffs, NJ: Prentice-Hall, 1986.

Moffitt, F. H. and E. M. Mikhail, *Photogrammetry,* 3rd. ed., New York: Harper & Row, Inc., 1980.

Muller, J. P., *Digital Image Processing in Remote Sensing,* Philadelphia: Taylor & Francis, 1988.

Richards, J. A., D. A. Landgrebe, and P. H. Swain, "A Means for Utilizing Ancillary Information in Multispectral Classification," *Remote Sensing of Environment,* 12 (1982): 463–77.

Shlien, S., "Geometric Correction, Registration and Resampling of LANDSAT Imagery," *Canadian Journal of Remote Sensing,* 5, No. 1 (1979): 74.

Slama, C. C. (ed.), *Manual of Photogrammetry,* 4th ed., Falls Church, VA: American Society of Photogrammetry, 1980.

Wharton, S. W. and B. J. Turner, "ICAP: An Interactive Cluster Analysis Procedure for Analyzing Remotely Sensed Data," *Remote Sensing of Environment,* 12 (1982): 279–93.

CHAPTER 13

DIGITAL
DATABASES

*A*s shown in Chapter 11, collecting digital cartographic data can be a major undertaking. Fortunately, a large amount of cartographic data have been collected by government agencies, quasi-governmental agencies, and private firms, and is available to the public. You will find many of these data sets useful if you want to create customized data sets or graphic cartographic products. Available data sets include base cartographic data, statistical and geographical data, thematic data, digital elevation data, and digital image data.

EARLY CARTOGRAPHIC DATABASES

Digital cartographic data in vector form have been collected since the 1960s. The first data sets were very simple: small-scale maps digitized as strings of coordinates with a descriptive attribute. Five of these early digital databases have been used widely in cartography. Perhaps the first was the **Dahlgren** database, collected in the early 1960s. It is a global database of coastlines and international boundaries digitized mostly from 1:12,000,000-scale maps. The database is small (only 8,300 coordinate points describe the geometry) and provides a very general portrayal of the world. Thus, it is suitable for small-scale (that is half-page or less) drawings of continents or the world.

A database of the United States was developed in the mid-1960s by the Department of Transportation. The **DIMECO** database consists of county and state boundaries along with the coastline. The data were collected from 1:5,000,000-scale maps, and the resulting digital files contain 115,000 points.

World Data Bank I (WDB I) was developed by the Central Intelligence Agency in 1966. World coastlines and international boundaries were digitized from 1:12,000,000-scale maps, resulting in 100,000 points. The U.S. coastline contains 20,000 points.

These three data sets are not being systematically maintained or distributed by the government. However, they are widely available from various universities or as sample data sets with GIS or mapping software. These files can be quite useful, especially as a base for small-scale reference, thematic, or statistical maps. In fact, WDB I was used to plot the map projections shown in Chapter 5.

World Data Bank II (WDB II) was released by the CIA in 1977. It represented a major increase in both data content and geometric detail. The data were digitized from maps that ranged in scale from 1:4,000,000 to 1:1,000,000. The data set contains multiple files, each containing a specific theme of information. These themes include country and state boundaries, coastlines, islands, lakes, rivers, and selected roads and railroads. The resulting files contain over 6 million points. Data for the United States were digitized from 1:3,000,000-scale maps and contain 1.5 million points. Some efforts have been made over the years, both by the public and private sectors, to maintain and enhance WDB II. Numerous geometric and attribute errors have been corrected, and the data have been topologically structured. Versions are available from the U.S. Geological Survey (USGS) as well as from private firms.

By 1982, the USGS had completed its project to digitize the base cartographic information shown on the 1:2,000,000-scale sectional maps of the *National Atlas of the United States of America*. The resulting database contains about 7 million points. The content of the database includes political boundaries (counties and states), transportation (roads and railroads), hydrography (streams and water bodies), federal lands, and populated places. A unique characteristic of this data set is that the features have been ranked from "most significant" to "least significant." Thus, by choosing various combinations of categories and levels of significance, you could compile a county outline map, for example, that was "light" on drainage, but "heavy" on roads. This data set is sold by the USGS on one CD-ROM. The CD-ROM contains the data in the Geological Survey's Digital Line

Graph (DLG) format (USGS, 1990a), as well as in a graphic format with the data expressed as coordinate strings (in latitude, longitude) without topological linkages. Other versions of the data are sold by a number of private firms.

The relative information content of these five data sets is illustrated in Figure 13.1.

BASE CARTOGRAPHIC DATA

Government mapping agencies (both civilian and military) around the world collect base cartographic data, which is the information typically shown on topographic maps. In the United States, the USGS is responsible for producing base carto-

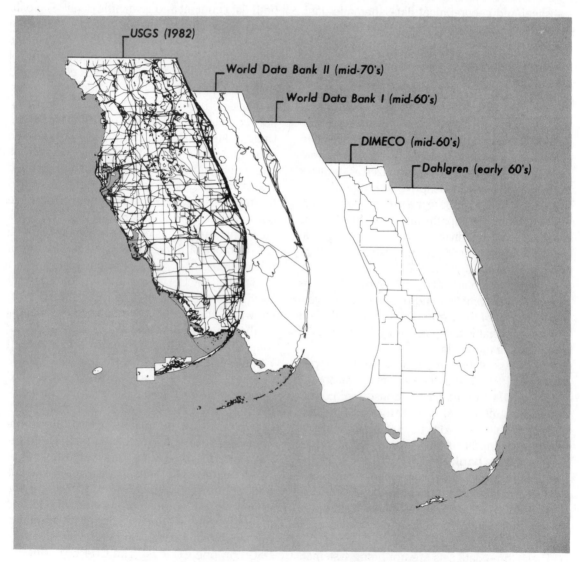

Figure 13.1 Comparison of level of detail of early small-scale databases.

graphic data for the nation; the Defense Mapping Agency (DMA) produces data for areas outside of the U.S. The USGS sells not only its own data sets, but also the publicly available DMA data. The next sections describe the major data products of DMA and USGS.

DEFENSE MAPPING AGENCY DATA

The U.S. Defense Mapping Agency's Digital Chart of the World (DCW) is a comprehensive vector database of the world (Danko, 1991). The 1:1,000,000-scale Operational Navigational Chart (ONC) series was the primary cartographic source material used for constructing the DCW database. The ONC base map information for the U.S. was revised in the early 1980s. However, no other major revisions or enhancements were made to the ONC before digitization. The feature content and portrayal characteristics of the DCW reflect that the ONC series was designed for air navigation. Features that are visually significant from the air or are important for navigation are shown on the ONC, and thus are now in DCW. The ONC is the largest-scale unclassified map series that provides consistent, continuous global coverage of base-map features.

The database is organized into 17 thematic layers, including the following categories of information: elevation contours at 1000-foot intervals (with supplemental contours and spot heights at some locations), roads, railroads, populated places with city outlines, hydrographic features, and international boundaries. The database also contains a worldwide index by place name with more than 100,000 entries. The database is structured in DMA's Vector Product Format (VPF), which is a topological data structure. The DCW product consists of four CD-ROMs, containing 1.5 gigabytes of data, and sophisticated viewing and data retrieval software for use on MS-DOS based personal computers. The database can be accessed directly from the CD-ROMs or can be transferred to a magnetic media. Figure 13.2 shows some of the information content of the DCW.

U.S. GEOLOGICAL SURVEY DATA

The U.S. Geological Survey's Digital Line Graph (DLG) data are digital representations of the information shown on USGS topographic maps. Topographic and planimetric features are derived from cartographic source materials using manual digitizing and scanning methods. The DLG data are classified as large-scale, intermediate-scale, and small-scale.

Large-scale DLG data are derived from USGS 1:20,000-scale, 1:24,000-scale, and 1:25,000-scale 7.5-minute topographic quadrangle maps (USGS, 1990b). Intermediate-scale DLG are derived from 1:100,000-scale 30- by 60-minute quadrangle maps (USGS, 1989). Small-scale DLG are derived from 1:2,000,000-scale *National Atlas* sectional maps, as discussed earlier in this chapter.

Large-scale and intermediate-scale DLG data are collected and sold in several data categories. These include:

- hypsography, including contours and supplementary spot elevations
- hydrography, including streams, water bodies, and wetlands
- boundaries, including state, county, city, and other national and state lands, such as forests and parks
- transportation, including roads and trails, railroads, pipelines and transmission lines, and miscellaneous transportation features
- the U.S. Public Land Survey System, including township, range, and section information.

The DLG data distributed by the USGS are DLG-Level 3 (DLG-3), which means the data contain a full range of attribute codes, have full topological structuring, and have passed a series of quality control checks. Attribute codes are used to describe the characteristics of the feature represented by the DLG node, line, and area elements. Each DLG element has one or more attribute codes composed of a three-digit major code and a four-digit minor code. Consider, for example, the line attribute code 290 5001. The major code—290—

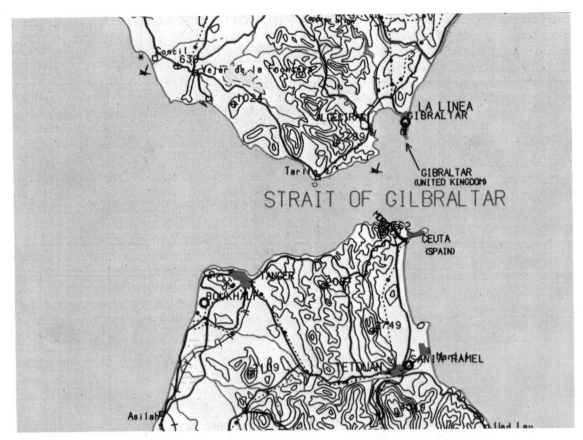

Figure 13.2   Plot of sample DCW data.

tells you that the feature is a road. The minor code—5001—identifies the road as an interstate. The DLG coordinates are referenced to the Universal Transverse Mercator projection system.

The DLG format data are organized into 11 distinct types of computer records. The first four records are header records: (1) file identification and description; (2) accuracy (not used); (3) control point identification; and (4) data category identification. The data records follow the header records and consist of seven types: (1) node and area identification; (2) node-to-line linkage; (3) area-to-line linkage; (4) line identification (also contains line-to-node and line-to-area linkages; (5) coordinate string; (6) attribute code; and (7) text

(not used). The data are sold on magnetic tape and, in some cases, on CD-ROM disks. Figure 13.3 shows the content of some intermediate-scale data. Figure 13.4 shows some large-scale DLG data.

BUREAU OF CENSUS STATISTICAL AND GEOGRAPHICAL DATA

The Bureau of the Census, U.S. Department of Commerce, offers a variety of useful geographic data (Bureau of the Census, 1992). Foremost of these is the **Topologically Integrated Geographic Encoding and Referencing (TIGER)**

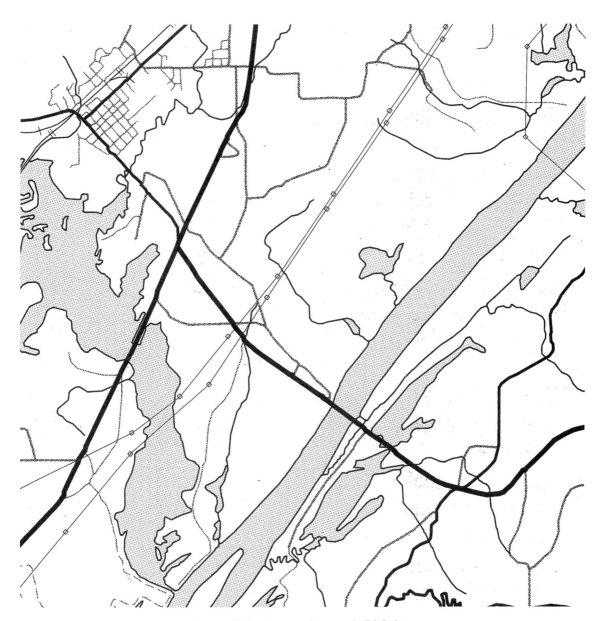

Figure 13.3 Intermediate-scale DLG data.

system. The TIGER database does not contain any statistical data, but provides the boundaries of thousands of legal and statistical areas. Table 13.1 lists the geographic entities of the 1990 Decennial Census. Map features, such as roads, streams, and railroads are identified by their name and coordinates.

The TIGER database includes the range of address numbers located along each side of each street segment for most of the urban and densely settled suburbs of the country. The database also identifies bodies of water and selected landmark features. Figure 13.5 illustrates some of the contents of the

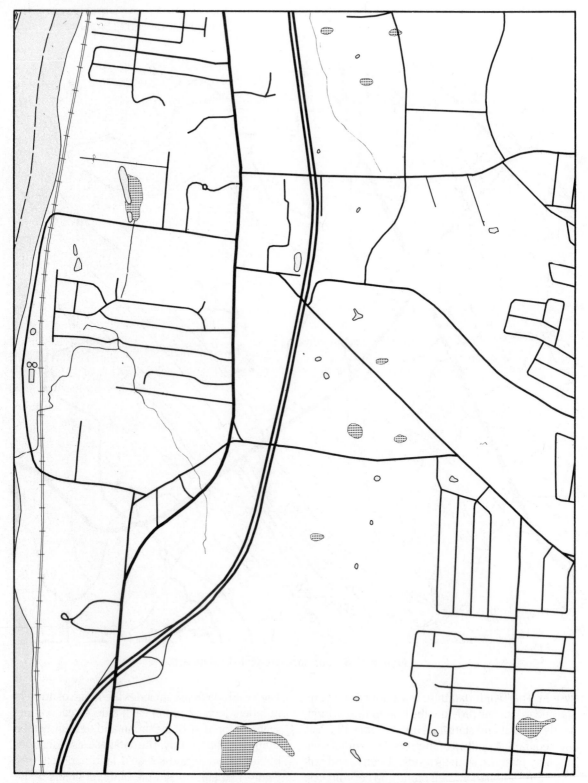

Figure 13.4 Large-scale DLG data.

TIGER database. Most important, it contains the boundaries, names, and numeric codes for all the geographic entities listed in Table 13.1. The codes correspond to those in the summary statistical files produced by the Census Bureau.

The Census Bureau produces four Summary Tape File (STF) series that provide detailed statis- ˙ tics for all levels of Census geography. Table 13.2 lists the content of the 1990 Census of Population and Housing. Figure 13.6 shows the "census block" units for Centrailia, Missouri. The summary statistics for the shaded block, number 433, are shown in Table 13.3. The combination of computerized geographic and statistical data allows you to make a wide variety of cartographic products.

Much of the data in the TIGER database was collected as part of a joint production agreement with USGS. USGS supplied the Census Bureau with computer files containing descriptions of water and transportation features as well as major powerlines and pipelines as shown on the

1:100,000-scale USGS quadrangle maps. (See the discussion of intermediate-scale DLG data above). At the same time, the Census Bureau extended and updated the features in its Geographic Base File/Dual Independent Map Encoding (GBF/DIME) files. These topologically structured data files contained descriptions of the streets and other features, address ranges, and ZIP Codes in 345 highly developed areas. The GBF/DIME files covered less than five percent of the land area but about 60 percent of the population. These files were joined with the USGS data to create TIGER. Figure 13.7 shows the relationship between the USGS DLG data and the TIGER data.

THEMATIC DATA

Thematic data are often considered as special-purpose data sets, as opposed to the general-purpose nature of base cartographic data. The-

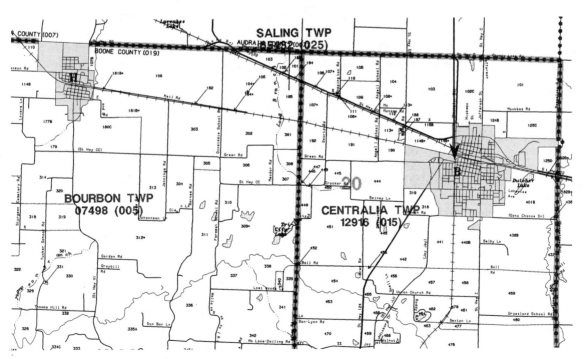

Figure 13.5 Representative content of the TIGER database.

Table 13.1
Census Geographic Units of the 1990 Decennial Census.

Legal/Administrative Entities

United States	1
States and statistically equivalent entities	57
State	50
District of Columbia	1
Outlying areas	6
Counties and statistically equivalent entities	3248
Minor Civil Divisions (MCD's)	30386
Sub-MCD's	145
Incorporated places	19365
Consolidated cities	6
American Indian reservations	310
American Indian entities with trust lands	52
Alaska Native Regional Corporations	12
Congressional Districts	435
Voting districts	148872
School districts	16000*
ZIP Codes	39850*

Statistical Entities

Region	4
Divisions	9
Metropolitan statistical areas (MSA's)	268
Consolidated metropolitan statistical areas (CMSA's)	21
Primary metropolitan statistical areas (PMSA's)	73
Urbanized areas (UA's)	405
Alaska Native village statistical areas (ANVSA's)	217
Tribal jurisdiction statistical areas (TJSA's)	17
Tribal designated statistical areas (TDSA's)	19
County subdivisions	5903
Census county divisions	5581
Unorganized territories	282
Other statistically equivalent entities	40
Census designated places	4423
Census tracts	50690
Block numbering areas	11586
Block groups	229192
Tabulated parts	363047
Blocks	7017427
Traffic analysis zones	200000*

*Estimated

matic data sets are often associated with a particular discipline or activity and are used by specialists in that field. Two such databases deal with soils and wetlands.

SOILS

The Soil Conservation Service (SCS), U.S. Department of Agriculture, uses field mapping methods to construct the soil maps in the Soil Survey Geographic Database (SSURGO) (FDGC, 1992, p. 30). The mapping scales range from 1:12,000 to 1:31,680 and represent the most detailed level of soil mapping done by the Soil Conservation Service. The maps are digitized in vector format, and the data files are available in either the USGS DLG data format or the SCS Geographic Exchange Format. SSURGO data are collected and archived in 7.5-minute quadrangle units. The data are distributed as a complete coverage for a county or area consisting of 10 or more quadrangle units. An example of a detailed soil survey map is shown in Figure 13.8.

SSURGO is linked to a Soil Interpretations Record attribute relational database, which gives the proportionate extent of the component soils and their properties for each map unit. This database includes over 25 soil, physical, and chemical properties for approximately 18,000 soil series in the United States.

WETLANDS

The National Wetland Inventory (NWI) program of the U.S. Fish and Wildlife Service, Department of the Interior, provides information on wetland location and type (FGDC, 1992, p. 53). NWI wetlands data are available in map form at 1:24,000-scale for approximately 75 percent of the conterminous United States. Wetland point, linear, and areal features are mapped and overlaid on USGS 7.5-minute topographic maps. All photo-interpretable wetlands are mapped. In general, the minimum size of the mapped wetland area is from one to three acres, depending on the wetland type. Wetlands are classified according to the Cowardin (1979) wetland classification scheme. An example portion of a wetlands map is shown in Figure 13.9.

Table 13.2
Content of the 1990 Census of Population and Housing

100-percent Component	
(Asked of all persons and housing units)	
Population	Housing
Household relationship	Number of units in structure
Sex	Number of rooms in unit
Race	Tenure - owned or rented
Age	Value of home or monthly rent
Marital status	Congregate housing (meals included in rent)
Hispanic origin	Vacancy characteristics

Sample Component	
(Asked of a portion or sample of the population and housing units)	
Population	Housing
Social characteristics	
Education-enrollment and attainment	Year moved into residence
Place of birth, citizenship and year of entry to U.S.	Number of bedrooms
Ancestry	Plumbing and kitchen facilities
Language spoken at home	Telephone in unit
Migration (residence in 1985)	Vehicles available
Disability	Heating fuel
Fertility	Source of water and method of sewage disposal
Veteran status	Year structure built
	Condominium status
	Farm residence
	Shelter costs including utilities
Economic characteristics	
Labor force	
Occupation, industry, and class of worker	
Place of work and journey to work	
Work experience in 1989	
Income in 1989	
Year last worked	

Digital wetland data are digitized from the 1:24,000-scale wetland maps. The digital data are available by 7.5-minute quadrangle units in a number of formats, including the USGS DLG format.

DIGITAL ELEVATION DATA

Digital elevation model (DEM) data consist of an array of regularly spaced elevations. For the United States, the USGS sells a variety of DEM products,

with various array spacings and geographic coverages (USGS, 1990c). All USGS DEM data are similar in logical data structure. The data are arranged in profiles. Within these profiles, the data are organized from north to south. The profiles themselves are arranged by column from east to west. The data are sold in 7.5-minute, 15-minute, 30-minute, and one-degree units, representing map scales of 1:24,000, 1:62,500, 1:100,000, and 1:250,000.

The 7.5-minute data consist of an array of data spaced at 30-meter intervals and referenced hori-

Table 13.3

Example Summary Statistics for Block 433

Housing Units	45
1 unit detached or attached	41
10 or more units	0
Owner occupied housing units	32
Renter occupied housing units	11
Mean value	74,300
Mean contract rent	149
Persons in occupied housing units	111
One person households	7
Persons	111
White	111
Black	0
American Indian, Eskimo, or Aleut	0
Asian or Pacific Islander	0
Persons of Hispanic Origin	0
Age, persons under 18 years	27
Age, persons over 65 years	16

zontally on the UTM coordinate system (on the North American Datum of 1927). These data cover a 7.5-minute by 7.5-minute area that corresponds to the USGS 7.5-minute topographic quadrangle map series (Figure 13.10). The vertical accuracy of these data is equal to or better than 15 meters. These data have been collected by four different production methods. One source is an automated photogrammetric system (the Gestalt Photo Mapper II) that produces orthophotos as well as digital terrain data. A second source is manual profiling. Here, the operator traces the terrain surface along a straight profile line on a photogrammetric stereomodel. This is done using stereoplotters equipped with encoders recording x, y, and z (vertical) coordinates. A third source is the interpolation of an elevation array from contour lines that have been digitally collected by stereoplotters.

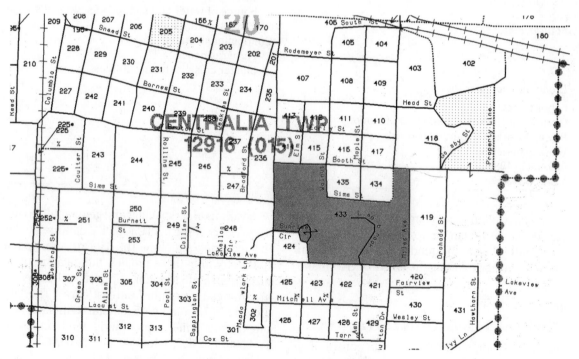

Figure 13.6 Census block units.

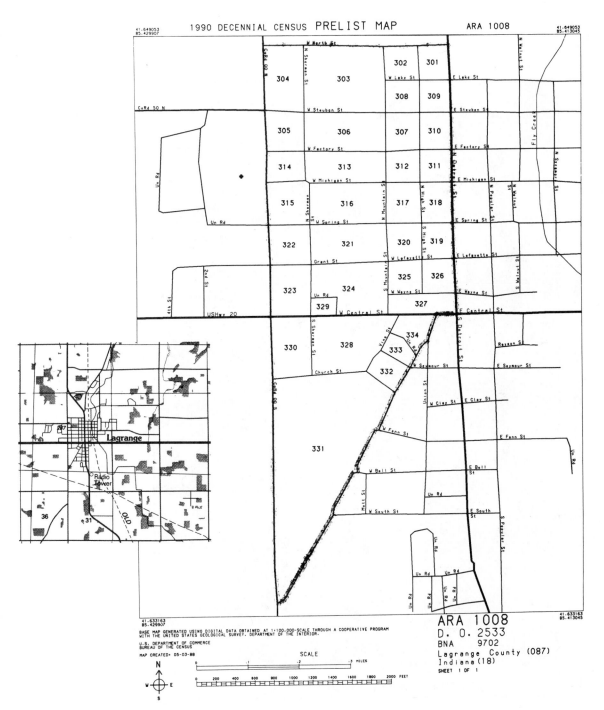

Figure 13.7 Relationship between TIGER and DLG files.

Figure 13.8   Example of a detailed soil survey map in Waushura Co., Wisconsin.

The final source uses interpolation software and DLG data representing the hypsography and hydrography data shown on a 1:24,000-scale map.

The 15-minute DEM data correspond to the USGS 15-minute topographic map series in Alaska. The data are derived from DLG hypsography and hydrography data. The DEMs are referenced to geographic coordinates. (That is, they are aligned with the latitude-longitude grid). The DEMs have a spacing of two arc seconds of latitude by three arc seconds of longitude. The DEM accuracy is equal to or better than one-half of a contour interval of the 15-minute topographic map.

The 30-minute DEM data cover 30-minute by 30-minute areas that correspond to the east or west half of USGS 30-by 60-minute, 1:100,000-scale quadrangle series topographic maps. The data are referenced to geographic coordinates and have a uniform spacing of two arc seconds. These data are derived from DLG data of 1:100,000-scale maps, or are resampled from higher-resolution DEM data (for example, 7.5-minute DEM). The DEM accuracy is equal to or better than one-half of a contour interval of the 30- by 60-minute topographic map.

The one-degree DEM data were originally produced by the U.S. Defense Mapping Agency by

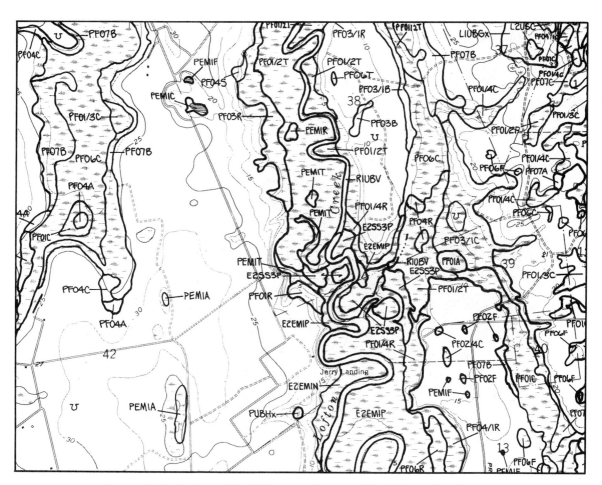

Figure 13.9 Portion of the NWI wetlands map of the Hedges, Florida, area.

interpolating elevations from digitized contours. The data have been repackaged for sale by the USGS. The data are referenced to latitude-longitude (but on the World Geodetic System 1972 Datum). The elevation model has a spacing of three arc seconds, or about 100 m. on the ground in the mid-latitudes. In Alaska the spacing varies with the latitude. The data have been collected and revised using a variety of cartographic and photogrammetric sources. These data have an absolute accuracy of 130 meters horizontally and 30 meters vertically.

DIGITAL IMAGE DATA

Digital imagery is playing an ever increasing role in cartography. Not only is it useful as source material for map compilation, but it can also serve as a dramatic background for various cartographic portrayals.

DIGITAL ORTHOPHOTOS

The USGS has begun a program to produce digital orthophotos. As described in Chapter 12, an orthophoto is an aerial photo that has been differentially rectified to remove the effects of relief displacement. The digital orthophotos sold by the USGS are quadrangle based (four images per 7.5-minute quadrangle), cast on the UTM projection system, and have a pixel size of one meter. The radiometric resolution of the image is eight bits, corresponding to gray scale values of 0-255. The orthophoto extends beyond the edges of the 3.75-minute quarter-quadrangle area to form a rectangle. These "overedge" data allow users to mosaic adjacent images. Metadata records include the photographic source type, date, instrumentation used to create the digital orthophoto, and information on the DEM used to rectify the image. (See Chapters 11 and 15 for information on metadata). Figure 12.26 shows a portion of a digital orthophoto.

Cannonville, Utah

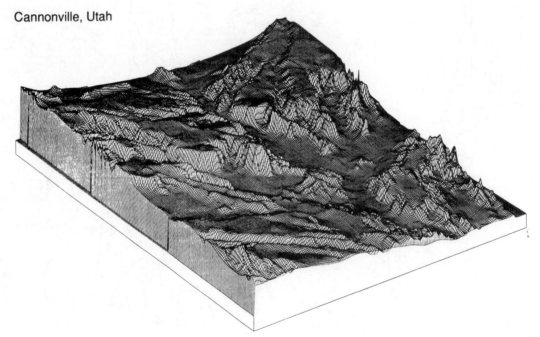

Figure 13.10   DEM plot of Cannonville, Utah.

GOVERNMENT SATELLITE DATA

The U.S. Geological Survey produces and distributes remotely sensed satellite data sets of the United States and the entire earth. The data sets include not only raw satellite imagery, but also satellite image data integrated with other related land data to provide repetitive, synoptic coverage of regional, continental, and global land areas. The USGS acquires, archives, and accesses **AVHRR** and **Landsat** source data to provide periodic, consistent coverage of the earth's land surface on a global scale. (The characteristics of these remote sensing satellites are covered in Chapter 8).

AVHRR

The USGS presently acquires and archives daily coverage of AVHRR one-km. data for much of North America through its real-time reception capability for High Resolution Picture Transmission data. The USGS also acquires and archives significant global land coverage of 1-km. data through capture of the tape-recorded Local Area Coverage data stream of all foreign land coverage recorded by the National Oceanic and Atmospheric Administration (NOAA). USGS AVHRR holdings exceed 20,000 scenes of worldwide coverage (Bailey, 1992).

USGS produces AVHRR 1-km. composited data sets that provide calibrated and geo-registered spectral data and a vegetation greenness index. The compositing procedure relies on selecting pixels from daily AVHRR scenes over a specified number of days to minimize cloud cover over regional and continental land areas. Each composited data set is composed of 10 bands. These bands include the maximum Normalized Difference Vegetation Index greenness value for the compositing period and the corresponding AVHRR bands 1-5. The bands also contain the solar and satellite view angle geometry and the calendar date for each of the composited pixels. The USGS produces data sets for both North and South America on a periodic basis. Data set collection for the entire conterminous United States began in 1990. The USGS

contributes to the development of AVHRR 1-km. data sets for other continents when possible. Production efforts will be coordinated with other processing centers to ensure consistent land data sets on a global scale and to achieve maximum production efficiency through collaboration. The left image in Figure 13.11 shows an AVHRR scene, as received from the satellite, whereas the right image is a reproduction of a colored vegetation classification of composited AVHRR images. Color Figure 8.7 is a greatly reduced example of a portion of one of the composited AVHRR data products.

Landsat MSS

The USGS, as the national archive for Landsat satellite data, records and distributes Landsat Multispectral Scanner (MSS) and Thematic Mapper data. About 800,000 scenes of Landsat data have been archived by USGS. From 1972-78, about 310,000 scenes of MSS data were acquired. Of these, 15 percent have been converted to computer compatible tape (CCT) format for distribution. Since 1979, over 300,000 MSS scenes and over 170,000 TM scenes have been acquired, with about 5 percent of these converted to CCTs. As the result of a 1990 agreement between the Earth Observation Satellite Company and NOAA, restrictions on public availability of Landsat MSS data two years or older were removed. The USGS is now distributing these data at prices reflecting only the cost of reproduction.

These early Landsat scenes provide a valuable base of land observations. They are useful for verifying and refining changes in land use. They are also used to study impacts of global climate change that are initially made on lower-resolution AVHRR one-km. data. The USGS will provide scenes of calibrated, corrected, and co-registered Landsat data as required in multiscale studies of global change.

COMMERCIAL SATELLITE DATA

The Earth Observation Satellite Company (EOSAT) is under contract to the U.S. government to operate

Figure 13.11 Raw AVHRR image on the left as it comes from the satellite, and a vegetation classification of the composited AVHRR images on the right (see also Color Figure 8.7).

the Landsat remote sensing satellites and to market Landsat data. The Landsat satellites collect digital imagery with two types of sensors—the Multispectral Scanner (MSS) and the Thematic Mapper (TM). (These sensors were described in Chapter 8). EOSAT offers data with various levels of geometric correction and orientation to map coordinate systems.

SPOT Image Corporation sells digital data products derived from the SPOT remote sensing satellites (see Chapter 8). The SPOT satellite is capable of collecting stereo data. Thus, SPOT data products include not only imagery but also digital elevation data produced from the stereo satellite data.

DATA EXCHANGE STANDARDS

Dictionary definitions of "exchange" refer to reciprocal giving and receiving. In the context of geographic information systems (GIS), **data exchange** is better characterized as data import and data export. The data exchange process is not typically reciprocal. Rather, users import data purchased from data exporters (providers)—usually government agencies or commercial data sources. In addition to these formal data exchanges, numerous informal exchanges of data between researchers also occur.

The need to receive data from a number of sources and to exchange data between dissimilar

GISs presents many challenges. For users, the challenge is to cope with a heterogeneous data environment. For data exchange to be successful, users need to understand how this heterogeneous environment affects data import and export.

As we saw in Chapter 10, every GIS has its own unique internal data model. Thus, to export a data set, the internal model must be converted to a specific file structure on disk or other media. To import a data set, the process is reversed.

The goal of data exchange is to transfer information to enable understanding of the phenomena being represented. What on the surface may seem a straightforward task has many levels of meaning, particularly when we consider data content issues. Each of these data content issues needs to be understood for the data exchange process to succeed.

As noted in Chapter 10, data structures specify the logical organization of components of a data model. File structures specify how the data structure is implemented within a given computing environment. Documentation of these structures allows computer programs to be written to perform the data exchange. There are three basic design strategies that can be used by data exchange software: (1) direct translation; (2) switchyard conversion (that is, translation to an internal standard, then retranslation); and (3) translation to a neutral format. Let's take a closer look at these three design strategies.

DIRECT TRANSLATION

In **direct translation**, there is a direct conversion of data from one system to another (as if you translated a book from English directly to German). The software converts the data in the internal file structure of system A to the internal file structure of system B (see Figure 13.12). This is an efficient data exchange method. It merely requires understanding how the two systems work.

However, if data exchange between more than two systems is required, the problem becomes more complicated. Add two more systems, C and D, to the

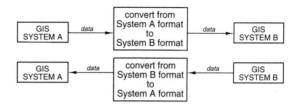

Figure 13.12 Data exchange between GISs using direct translation.

data exchange mix. Now, instead of two programs being required (translate A to B, translate B to A), 12 would be needed. The number of permutations grows rapidly as more systems participate in the exchange. (Exchanges between 10 systems would require 90 direct translation routines!)

The direct conversion scenario provides the most efficient path between two parties exchanging data. A scheme can be devised to translate the set of features, attributes, and relationships used in the first system to those used in the second system. Software can be designed to convert information that is allowed in one data format but not another. Suppose that the first system allows multiple-valued attributes (for example, attribute = building function; attribute values = police station, fire station) and the second does not. The data exchange software can include routines to alter the data organization to fit the specifications of the second system (attribute value = police and fire station). The major disadvantage is that such routines need to be written for every pairwise permutation of systems involved in the data exchange.

SWITCHYARD CONVERSION

An alternative to direct translation is the concept of a "switching yard" conversion. This method avoids the factorial increases in the number of translation programs as more parties are added to the data exchange. (See Figure 13.13).

With **switchyard conversion**, the software translates incoming data to one internal standard. Then, on output, the software translates the data from the internal standard to that of the desired

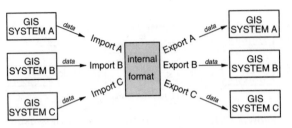

Figure 13.13 Data exchange between GISs using a "data switchyard."

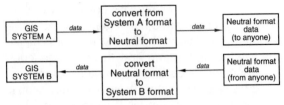

Figure 13.14 Data exchange between GISs using neutral format.

export format. (It's as if you translated a book from English to Japanese, then from Japanese to German). For each party added to the exchange, only two translation routines are needed. Thus, data exchange among 10 systems would require only 20 translation routines, rather than the 90 required in the direct translation scenario.

The "data switchyard" attempts to retain the advantages of direct conversion, while reducing the number of routines needed to convert information. There is a problem, however. The hub of the switchyard—the internal representation of the data in the central conversion system—must be all-inclusive. That is, it must be able to accommodate all the characteristics of the data being imported or exported. The switchyard software, like direct conversion software, can be designed to accommodate differences in data representation in order to place the data into the central internal data set. But if data sets are transformed on input, then the inverse transformation must occur upon output. For example, if you concatenate multiple-valued attributes upon input, then you need to restore the multiple-valued attribute from the concatenated representation on output.

NEUTRAL EXCHANGE FORMAT

The third method is for the parties to agree to use a standardized common (neutral) exchange file structure. (See Figure 13.14). With this approach, system A translates its internal file structure into the standard structure. System B reads the standard structure and converts it into its own internal

file structure. (It's as if you translated a book from English to Esperanto, then Esperanto to German). This **neutral exchange** method has the significant advantage that, in theory at least, only two software routines are required (one to read and one to write the exchange format).

Under special circumstances, there may be problems with this method. Difficulty may occur if the neutral file format is highly flexible, with a number of optional fields and records. In that case, one computer program may not be adequate for encoding (or decoding) data to (or from) the neutral format. Separate routines may be needed to handle variations from government agency Y or GIS vendor Z. The number of routines required for data exchange then reverts to the solution given for the "data switchyard."

As with the central hub in the data switchyard, the neutral file structure must be all-inclusive. Otherwise, it won't be able to handle the variety of data structures (and underlying data models) used in geographic information systems.

Also, the parties of a data exchange must agree on common implementation of the neutral format. Each party must model phenomena in the same way and use the same records and fields. Otherwise, varying implementations or dialects of the standard format may emerge. Each of these would require a separate import and export routine, thus defeating the goal of having only two routines. A standardized neutral format that does not allow variations (or dialects) can be written, but this standard format has a reduced ability to handle variations in data sets being exchanged. Trying to achieve a balance between uniformity and flexibility is a constant problem for people preparing standards. From the point of view of a user receiving data from a number of sources, however, the idea of acquiring the information in the same format is very appealing.

To date, the most widely used "neutral" formats for the exchange of vector data have been fairly rigidly specified data structures. These structures have been developed either by major data producers (for example, the DLG format of the USGS) or by developers of widely used software packages (for example the DXF format of AutoCAD).

A similar situation exists with raster data. The DEM format of the USGS is widely used for elevation data. The computer-compatible-tape formats of Landsat (USGS, 1984) and SPOT (CNES, 1987) are used for digital image data. These de facto standard formats place an emphasis on uniformity and specificity in describing the data rather than providing a flexible envelope capable of holding a wide variety of data descriptions.

Efforts are underway to develop more flexible standardized data exchange formats for digital cartographic data. These efforts are variations on the "neutral file structure" theme, with differing degrees of flexibility within the file structure. The International Cartographic Association Working Group on Digital Cartographic Data Exchange Standards reported that the following countries indicated an active involvement in developing exchange standards: Australia, Canada, China, Federal Republic of Germany, Finland, France, Hungary, Japan, New Zealand, Norway, South Africa, Sweden, Switzerland, United Kingdom, Union of Soviet Socialist Republics, and the United States (Moellering, 1991). In 1992, a formal standard for the exchange of spatial data was approved by the National Institute of Standards and Testing for use by the United States federal government. The Spatial Data Transfer Standard, FIPS 173, is discussed in Box 13.A.

The standardization activities of the United States (reviewed in Box 13.A) illustrate one approach to the standardization question. Members of the computer graphics community have taken an alternate strategy. Their efforts to create a spatial data exchange standard are discussed in Appendix B.

ANALYSIS OF STANDARDS ACTIVITIES

Given the increase of producers and users of geographic information, the drive toward a standard for data exchange is inevitable. The question is: What is the best way to achieve such a standard? Mapping agencies in the United States are leading efforts to establish national spatial data exchange formats. At the same time, the computer graphics industry is leading an international effort to develop a comprehensive data exchange format that might be able to accommodate geographic data as a specific application.

The ultimate measure of success of any standard lies in its acceptance by users. For such acceptance to occur, computer vendors must take an active role. They must develop software tools to encode and decode spatial data between the standard format and their own system format. A concerted effort has been made during the development of the standards mentioned to keep the private sector informed and to use their input to improve the standard. It remains to be seen which is the best path for achieving exchange of spatial data.

Standards are written to accommodate a wide variety of data. As a result, any given implementation (that is, the transfer of a specific type of data, such as contours or census tracts) will use only a subset of the features (or data elements) available in the standard. As noted by the designers of the United Kingdom's National Transfer Format (NTF): "No doubt producers and users who regularly exchange the same type of data will quickly establish an appropriate subset of the total transfer package" (Sowton and Haywood, 1986). The net result is a set of "standard" or "de facto" implementations of the standard itself.

This concept of a standard within a standard will probably be widely supported by GIS vendors. Take the case of SDTS, for example. GIS vendors will most likely support the import and export of data using SDTS. But they may support only the subset of SDTS used by the Geological

Box 13.A
Spatial Data Transfer Standard—FIPS 173

The Spatial Data Transfer Standard (SDTS), after nine years of development, was approved on July 29, 1992, as Federal Information Processing Standard (FIPS) Publication 173. SDTS provides solutions to the problems of transferring the full range of spatial (that is, geographic and cartographic) data. Using FIPS 173, you can exchange vector and raster data of many different types and structures between dissimilar systems. You can also exchange associated attribute data of widely varying models.

The SDTS addresses a number of issues necessary to successful spatial data transfer. These issues (arranged from highest to lowest level) include:

- conceptual modeling (at the highest level)
- definition of feature and attribute terms
- inclusion of metadata and quality reporting
- logical structuring
- details of physical encoding (at the lowest level).

The SDTS consists of three distinct parts. **Part 1** provides logical specifications for spatial data transfer. These specifications include:

- a **conceptual model** of spatial data
- **data quality report** specifications
- detailed logical **transfer format specifications** for SDTS data sets.

Part 2 provides a model for the definition of real-world spatial features, attributes, and attribute values. It includes a standard but working and expandable list with definitions.

Part 3 specifies the byte-level format implementation of the logical specifications in Part 1. It uses ISO/ANSI 8211 (FIPS 123), a general data exchange standard accepted by the International Standards Organization (ISO) and the American National Standards Institute (ANSI).

Knowledge of the quality of data and its fitness for use is an integral part of data transfer. The SDTS recognizes the importance of letting users evaluate the suitability of data for their particular use. Thus, it requires a **data quality report** (as noted in the description of Part 1 above). The report must be part of a transfer, and it must also be available separately so that evaluation can be done without actually acquiring the data.

Part 1, Section 3 specifies five portions of a data quality report. Temporal information should be included in each portion. These five portions are:

1. **Lineage**. This portion describes source and update material (with dates), methods of derivation, transformations, and other processing history.

2. **Positional accuracy**. This portion is concerned with how closely the locational data (encoded coordinate values) represent true locations. Different methods of obtaining measures of positional accuracy are described in Section 3.

3. **Attribute accuracy**. This portion is similar to positional accuracy, except that it is concerned with non-locational descriptive data.

4. **Logical consistency**. This portion discusses the fidelity of encoded relationships in the structure of the spatial data. For example, the degree to which topological relationships have been verified should be reported, when appropriate.

continued on next page

Box 13.A (continued)

5. **Completeness**. This portion includes information about geographic area and subject matter coverage. For example, selection criteria and other mapping rules may be included.

Employing a truth-in-labeling approach, as opposed to setting absolute quality thresholds, the SDTS places responsibilities on both the producer and receiver of data. It is up to the producer to report what quality information is known in each of the five quality areas. (What is known may be nothing, in which case "don't know" is a permissible entry in a quality report). It is up to the receiver to determine whether data are suitable for a given application.

To facilitate the exchange of data, SDTS specifies a series of transfer modules. Each module contains a collection of module records. Each module record contains data fields that have been grouped together because of the function of that information. Transfer module types are grouped in relation to a certain type of data organization. Groups of modules to be transferred are called a **transfer**.

A transfer may consist of a single physical file containing multiple modules. Or it may consist of multiple physical files, each containing one or more modules. Table 13.A.1 shows the transfer modules grouped according to the following: Global Information, Data Quality, Attributes, Spatial Objects, and Graphic Representation.

Table 13.A.1
SDTS Module Categories and Types

Global Information
    Identification
    Catalog
        Directory
        Cross-Reference
        Spatial Domain

Spatial Reference
    Internal Spatial Reference
    External Spatial Reference
    Registration
    Spatial Domain
Data Dictionary
    Definition
    Domain
    Schema
Security
Transfer Statistics
Data Quality
    Lineage
    Positional Accuracy
    Attribute Accuracy
    Logical Consistency
    Completeness
Attributes
    Attribute Primary
    Attribute Secondary
Spatial Objects
    Composite
    Vector Modules
        Point-Node
        Line
        Arc
        Ring
        Polygon
    Raster Modules
        Raster Definition
        Cell
Graphic Representation
    Text Representation
    Line Representation
    Symbol Representation
    Area Fill Representation
    Color Index
    Font Index

The content and size of a transfer determine what modules and fields are present and how they are collected into records and the records into files. SDTS allows the user considerable freedom to structure records and files as needed.

Survey to provide DLG-E data. If other major data sources, such as the Census Bureau, employ a different SDTS encoding scheme, the vendors would probably be driven by market forces to support that as well.

If major data providers would agree on a common implementation of SDTS, third-part data access would be simplified. Agreement on the use of spatial objects and relationships is probably possible. Agreement on common definitions and representations of geographic features is a more serious problem.

It is clear that the time for spatial data standards is here. Standards have many advantages for data collectors, processors, and users, particularly those who need to use data from several sources. A standard transfer specification will facilitate data exchange throughout the public and private communities. Such a specification will also make it easier to display, analyze, and integrate spatial data for a growing number of applications. The availability of information about data quality (such as lineage, completeness, accuracy, and logical consistency) will help users evaluate the fitness of the data for a particular use.

Unfortunately, there is a multitude of standards from which to choose. At least 16 countries are currently establishing their own national spatial data exchange standards. Related standards, such as those from the computer graphics field, can be used to exchange spatial data as well. It seems unlikely that an international exchange standard will emerge in the near future.

Meanwhile, all parties involved in the mapping process must continue to struggle for mutual understanding. They must understand the nature of the data model, data content, data structures, and file structures used to represent spatial phenomena. Only then will successful data exchanges occur.

SELECTED REFERENCES

Bailey, G. B., T. M. Holm, and J. A. Sturdevant, "EOS Land Processes DAAC Science Data Programs," *Proceedings, 1992 ASPRS-ACSM Annual Convention,* Volume 1, Bethesda, MD: American Society for Photogrammetry and Remote Sensing, (1992): 1–10.

Bureau of the Census, *Maps and More—Your Guide to Census Bureau Geography*, Washington DC: U. S. Department of Commerce, 1992.

Centre Spatial de Toulouse, *SPOT Users Handbook, Vol. 1*, Toulouse, Cedex, France, 1989.

Cowardin, L. M., *et al.*, *Classification of Wetlands and Deepwater Habitats of the United States*, Washington, DC: U. S. Government Printing Office, 1979.

Danko, D. M., "The Digital Chart of the World Project," *Proceedings 1991 ACSM-ASPRS Annual Convention,* Volume 2, Bethesda, MD: American Congress on Surveying and Mapping, (1991): 83–93.

Federal Geographic Data Committee, *Manual of Federal Geographic Data Products*, Washington, DC: Office of Information Resources Management, U. S. Environmental Protection Agency, 1992.

FIPS PUB 173, *Spatial Data Transfer Standard (SDTS)*, National Institute of Standards and Technology, Washington, DC: Government Printing Office, 1992.

ISO 8211, *Specification for a Data Descriptive File for Information Interchange*; also FIPS 123, National Institute of Standards and Technology, Washington, DC: Government Printing Office, 1986.

Moellering, H., (ed.), *Spatial Database Transfer*

Standards: Current International Status, New York: Elsevier Applied Science, 1991.

Ordnance Survey, *National Transfer Format, Release 1.1*, Southhampton, UK: Ordnance Survey, 1989.

Sowten, M. and P. Haywood, "National Standards for the Transfer of Digital Map Data", *Auto Carto London, Proceedings, Vol. 1*, London: Royal Institution of Chartered Surveyors, (1986): 298–311.

U. S. Geological Survey, *Landsat 4 Data Users Handbook*, Washington, DC: Government Printing Office, 1984.

U. S. Geological Survey, *Digital Line Graphs from 1:100,000-Scale Maps, Data Users Guide 2*, Reston, VA: U. S. Geological Survey, 1989.

U. S. Geological Survey, *Digital Line Graphs from 1:2,000,000-Scale Maps, Data Users Guide 3*, Reston, VA: U. S. Geological Survey, 1990a.

U. S. Geological Survey, *Digital Line Graphs from 1:24,000-Scale Maps, Data Users Guide 1*, Reston, VA: U. S. Geological Survey, 1990b.

U. S. Geological Survey, *Digital Elevation Models, Data Users Guide 5*, Reston, VA: U. S. Geological Survey, 1990c.

CHAPTER 14

GEOGRAPHIC AND CARTOGRAPHIC DATABASE CONCEPTS

*M*any of cartography's major activities—determining content level, compilation method, data accuracy, and appropriate display methods—are embodied in the concept of scale. These scale issues take on special importance as we move from traditional to digital procedures.

In the past, a map's scale greatly influenced map content and data resolution. The level of environmental detail captured in data intended for mapping was inhibited by concerns about creating a cluttered map. There was no point to having more data detail than could be shown on a map of the specified scale. Efficiency demanded a close match between data detail and map scale.

In contrast, digital databases are scaleless in a theoretical sense. Data can be stored at the level of detail found in the environment. Since these databases are used for other purposes besides mapping, the choices of content and data resolution are not determined solely by map products. When using the database, the cartographer is responsible for choosing the content and resolution of data appropriate for a given map.

In a practical sense, however, scale is still a critical factor with digital databases. Scale is important because of the ways we create digital databases. When digital data are gathered directly from the environment, collection occurs at a level of resolution set by field instruments. When data are digitized from existing maps, the content and resolution of the resulting database are influenced both by the resolution of digitizing instruments and by the effects of cartographic abstraction and production factors.

When cartographers shift from manual to digital operations, they often have problems with scale. They may have trouble realizing that with digital methods, many traditional concepts of scale take on new meaning. To clarify these differences, let's begin by looking at the relation between scale and spatial resolution.

RELATIONSHIP OF SPATIAL RESOLUTION TO SCALE

If cartographers saw a 1:100,000-scale graphic map showing the shorelines contained in World Data Bank II (a cartographic database digitized from 1:3,000,000-scale maps), they would be indignant. They would note that this was an inappropriate use of the data. They would declare that the data lacked the resolution to be properly displayed at such a scale.

But why would they be so outraged? What is this relationship between resolution and scale? How can you tell if you are violating a basic element of cartographic display?

Table 14.1 shows a set of approximate relationships among scale, spatial resolution, and feature detection. These numbers are derived from the notion that the plottable error of a map (the smallest physical mark that a cartographer can accurately draw on a map) provides a limit to which measurement can be made on maps of different scales. (See Box 14.A). The plottable error or resolution of the map is the smallest feature or distance that can be recorded true to scale on a map (assuming the finest line on the map is 0.5 mm. wide). Thus on a 1:50,000-scale map, the resolu-

Table 14.1
Relationships Among Scale, Spatial Resolution, and Feature Detection (After Tobler, 1988)

Scale	Resolution (precision)	Detection (accuracy)
1:1,000,000	500 m	1,000 m
1:500,000	250 m	500 m
1:250,000	125 m	250 m
1:100,000	50 m	100 m
1:50,000	25 m	50 m
1:10,000	5 m	10 m
And in English units		
1:24,000	39 ft	78 ft
1:63,360	104 ft	208 ft

tion is 25 m. (0.5 mm. × 50,000 = 25 m.) Any object of smaller dimensions must be shown symbolically, with its size exaggerated and location displaced. This cartographic principle also is embodied in the U.S. Geological Survey's National Map Accuracy Standard. The end result is that as the

scale of a map is reduced, accuracy is lost and generalization increases.

These relationships give you a feel for the relation between resolution and "optimal" scale of visual presentation. The contents of a database with 50-m. resolution could be plotted at a scale of

Box 14.A
Describing Spatial Precision and Accuracy

Imagine that you are reading the documentation accompanying a digital elevation model. You notice that the data values were given as four-digit integer numbers expressed in meters. You also note that the data were originally collected from 1:100,000-scale maps with a contour interval of 20 meters. You are confused. If the contour interval is 20 m., then why are such elevation values as 3,186 m. in the data set? Are all those numbers meaningful?

To figure out what is going on, you need to understand the concepts of accuracy, precision, and significant figures. You must also remember that cartography involves measuring location and determining position of features on the earth's surface, and that all measurement and positioning processes involve error.

First we must define a few terms with respect to spatial data. Spatial **accuracy** refers to a measure of how close a recorded location comes to its true value. Spatial **precision** is a measure of how exactly a location is specified without any reference to its true value. The precision of a measurement will depend on the measuring (or positioning) process. The precision of a location is implied by the way in which it is written. To indicate precision, we write a number with as many digits as are **significant**. The number of significant figures in the result of a scalar measurement (nominal, ordinal, and scalar measurement are discussed in Chapter 16) is defined as follows:

- The leftmost non-zero digit is the most significant digit.
- If there is no decimal point, the rightmost non-zero digit is the least significant digit.*
- If there is a decimal point, the rightmost digit is the least significant digit, even if it is zero.
- All digits between the least and the most significant digits are counted as significant digits.

Returning to our cartographers dilemma, we now see that the data set has a precision of one meter (as determined by the number of significant digits in the data, as well as the statement in the documentation). That is, elevation values vary in 1-meter increments within the range from 0 to 9999. The data set has an accuracy of 10 meters (calculated by taking one-half the contour interval of the original data source).

---

*Computer storage of integer numbers causes some ambiguity with this rule. For example, the number 5280 is considered by the rule to have only three significant digits, even though if it were in an integer data set (...5173, 5201, 5280, 5307...), you might logically assume that the last digit is significant. In strict mathematical notation, 5280 should be shown as 5280. or 5.280×10^3 to remove the ambiguity. However, these solutions have negative ramifications in computer storage. The **integer** representation of 5280 occupies one-half the storage of the equivalent **floating point** representation of 5.280 E+3. As a result, integer arrays are used for storage of coordinates and elevation values. To remove ambiguity, therefore, a separate statement of the precision of measurement (for example, to the nearest meter) should be given to resolve such questions.

1:1,000,000, but most of the detail would be lost in a clutter of lines. Likewise, the data could be drawn at 1:10,000-scale, but the resulting linework would look overly simplified and stick-like. The same principles apply as those in the display of remotely sensed data, where an optimal visual presentation does not reveal the "blockiness" of the pixels in the imagery.

GEOGRAPHIC AND CARTOGRAPHIC DATABASES

As we have seen, the resolution limits of traditional maps are linked to scale and the ability to depict features true to scale. Beyond those limits, features become generalized, displaced, or further abstracted to allow for a readable, aesthetically pleasing map. These limits do not exist in digital databases. For example, spatial databases easily allow the storage of locational data to a precision of 1 meter, permitting the accurate positioning of feature data without further abstraction. The question now arises: Which type of database is preferable for making maps? Are we better off with a geographic database, containing the "true" geographic location of each feature? Or should we use a cartographic database, containing an abstraction, according to cartographic rules, of that information?

A geographic database contains digital information that has been gathered directly from the environment. The location of features in the horizontal and vertical dimensions can be given as accurately as positioning technology allows. An example would be a network of weather observation stations accurately located with GPS receivers. Or you can think of geographic databases as accurate tracings of the features as shown in a high-resolution aerial photograph (see Figure 14.1).

In contrast to a geographic database, a cartographic database is created by digitizing existing cartographic products. In a cartographic database, the location of certain features may have been

Figure 14.1 Orthophoto with traced geographic features. (courtesy Intergraph Corporation).

deliberately moved. For example, house symbols may have been "pulled back" from a road, which itself has been exaggerated in width (see Figure 14.2). Or a group of seven buildings, if too close together for an uncluttered presentation, might be shown as a representative cluster of four or five building symbols (see Figure 14.3). Parallel features, such as roads, streams, and railroads might be offset from each other. Short road segments might be eliminated entirely.

This difference in compilation rules between geographic and cartographic databases can lead to problems, particularly when cartographic databases are revised or updated from a geographic database or image source. Consider the situation shown in Figure 14.4. In Figure 14.4a, feature A represents the geographic location of a road. Feature B is a displaced cartographic representation

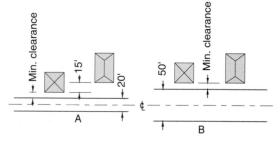

Figure 14.2 Effect of road width symbolization on location of housing symbols.

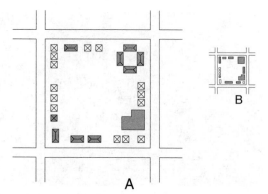

Figure 14.3 Generalization of buildings.

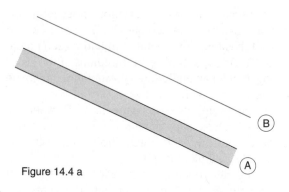

Figure 14.4 a

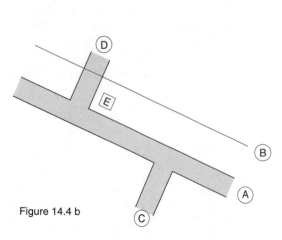

Figure 14.4 b

Figure 14.4 Revision of geographic and cartographic databases.

of the same road. At a later time, depicted in Figure 14.4b, roads C and D and building E are constructed. Cartographers, given the task of updating the cartographic database, are faced with a choice. They can move feature B to its true geographic position (A). Or they can displace and deliberately mislocate new features C, D, and E. There is not a correct or incorrect choice here. The decision must be driven by the underlying intent of those building the database.

HANDLING DATABASES OF VARYING SCALES

Cartographers also face problems when they encounter data sets of different resolutions. Consider the problem of creating a state-wide road database. Suppose that data collected by urban counties have been compiled from 1:25,000-scale maps. At the same time, rural counties have collected their data from 1:100,000-scale maps. What happens to the roads at a rural-urban county line? As shown in Figure 14.5, some roads continue across,

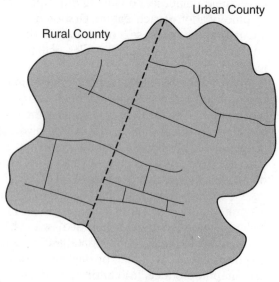

Figure 14.5 Combining data sets of different scale.

others appear to continue but have been displaced at the boundary, and some roads stop altogether. Many ambiguous situations can result from such combinations. These ambiguities can only be resolved with ancillary source materials, imagery, or field work.

The problems are compounded when using data sets not only of different resolutions, but also different themes. Figure 14.6a shows roads and drainage at a scale of 1:24,000. Figure 14.6b shows the same 1:24,000-scale roads, but combined with drainage information from a data set digitized from 1:100,000-scale maps. Because of either the relative accuracy or displacement rules, the stream has shifted from the west to the east side of the road. Similar inconsistencies can result from combining elevation data and drainage and can cause such anomalous conditions as water running uphill, or non-level lakes or ponds.

As a result, the cartographer should take a great deal of care in combining data sets of varying resolution and accuracy. Any anomalies that result must be resolved before the data can be used for GIS analyses or cartographic production. The actions taken should be described as part of a **data quality report** and included in the **metadata** (discussed in Chapter 15) associated with the data set. This report will help subsequent users judge that data set's fitness for use.

WHY HAVE CARTOGRAPHIC DATABASES?

The previous discussion highlights some of the problems associated with cartographic databases. This is not meant to imply that they are not useful. Let us review some of the reasons cartographic databases are of value.

Cost of Creation Collecting digital data from an existing cartographic source is far cheaper than compiling data at high accuracy and precision levels from high-resolution imagery or field surveying.

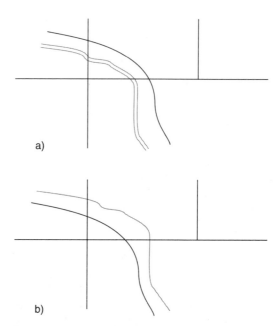

Figure 14.6 Combining data of different scales and themes.

Appropriateness of Use Some GIS analyses are conducted at national or regional scales. For example, a highway routing analysis between Atlanta and San Francisco may require only the interstate highways. More detailed data sets require more storage and longer times to process, perhaps to the point of making the problem totally unwieldy.

Lack of Alternative Data Sources Some information may have been recorded only on maps. This is particularly true of historical data portrayed on old maps.

Graphic Output Graphic output is a key component of any GIS or automated cartographic production system. The fact that the output is being generated by computer does not negate the need to follow the cartographic principles developed to ensure readable graphic output. Generalized cartographic output from a detailed geographic database may be difficult or impossible to produce without a large amount of processing and manual intervention.

SCALE LEVELS

We can apply only a limited range of magnification and reduction to a data set and still have features remain legible and aesthetically pleasing. Without intervention, this range is about 2X. That is, a 1:24,000-scale cartographic database could be displayed from a range of 1:10,000 to 1:50,000 and still be visually pleasing. Beyond that range, we must apply generalization operations to the data being reduced in scale. These procedures are discussed in Chapter 24. At present, only some of these generalization techniques can be totally automated.

In the case of enlargements, we can perform visual enhancements such as connecting coordinate points with curves rather than straight-line segments to present a smooth shape. This approach is illustrated in Figure 14.7. Although the result has a pleasant appearance, it lacks the detail needed for accurate line placement compared to maps initially compiled at this larger scale.

Given these factors, it is not surprising that databases are being developed at various resolutions or scale levels. At the largest scale or highest resolution is the geographic database. Here, the location of features is not influenced by cartographic issues. The resolution of this database will vary according to the mission of the agency doing

the collection and is ultimately determined by the resolution of the source material (either imagery or surveying data).

For a national mapping agency, the data resolution might be on the order of 10–15 m.; for a county 0.1–1.0 m. Once past these resolution levels, the geographic database transitions into a cartographic database. The second level of data resolution is a reduction of four to 10 times that of the previous level. The third level is four to 10 times the resolution of the second level. Table 14.2 shows the scale levels of digital databases of various national mapping agencies.

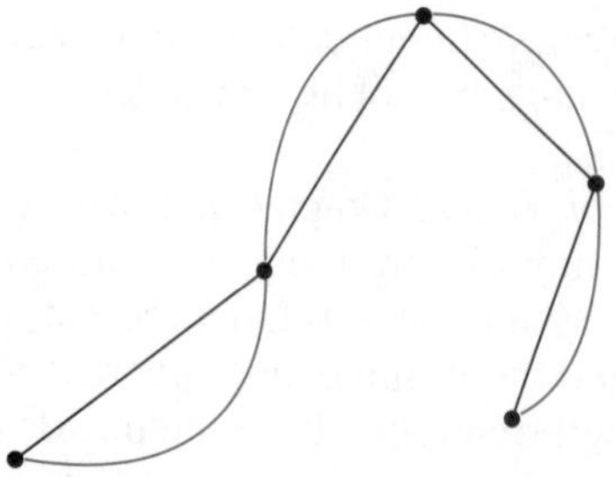

LARGE-SCALE DATA

As the resolution of digital data sets increases, the scope of information grows beyond that commonly considered in cartography. Two special applications of digital databases are **land information systems (LIS)** and **automated mapping/facilities management (AM/FM)** (Federal Geodetic Control Committee, 1989). In some cases, the data needs of these activities are consolidated and termed a **multipurpose cadastre** (National Research Council, 1983). The typical content of a multipurpose cadastre is shown in Table 14.3. The information is usually digitized from large-scale (for example, 1:600-scale or 1:1,200-scale) maps,

Figure 14.7 Replacing straight-line segments with curved segments to approve appearance.

Table 14.2
Scales of Various Digital Databases

USA	United Kingdom	France
	1:1250	
	1:5000	
1:24,000 (USGS)	1:25,000	1:25,000
	1:50,000	
1:100,000 (USGS)		1:100,000
	1:625,000	
1:1,000,000 (DMA)		
1:2,000,000 (USGS)		

Table 14.3
Multipurpose Cadastre—Basic Content

- Parcel boundaries
- Parcel dimensions and area
- Block numbers, lot numbers, parcel identifiers
- Boundaries of political and governmental subdivisions
- Right of way and easement boundaries

Table 14.4
Land Information System Data Content

- Legal land parcel descriptions
- Parcel ownership and sales history
- Land use; land and building improvements
- Land status (leases, permits, license)
- Assessor data

although direct digital input from field surveying instruments can also be used. A typical large-scale parcel map produced from a multipurpose cadastre database is shown in Figure 14.8.

LIS activities are concerned with managing land parcels. Thus, some additional information, shown in Table 14.4, may be needed.

The legal description of parcel boundaries is usually stated in surveying terms (a series of bearing

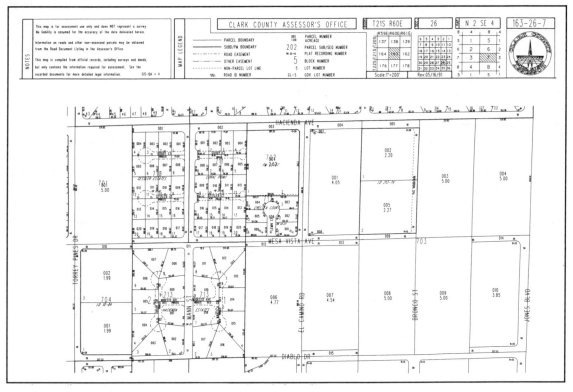

Figure 14.8 Example of a land parcel map. The underlying database was compiled by a combination of survey data and map digitizing.

Table 14.5
Land Survey Data Elements

- Survey number
- Base x-coordinate, y-coordinate
- Bearing/radius 1
- Distance 1
- Bearing/radius 2
- Distance 2, etc...
- Closure (yes/no)

Table 14.6
Information Categories for AM/FM Applications

- Public water system (lines, hydrants, valves)
- Sewer system (lines, manholes, pumping stations)
- Gaslines
- Electric utilities
- Telephone utilities
- Cable TV
- Storm Drains
- Road signage, traffic signals

and distance measurements) rather than as a polygon with x,y coordinates.* A survey record might have data elements as shown in Table 14.5.

AM/FM applications are concerned with describing the public safety or public works of a given area. One set of information content for AM/FM applications is shown in Table 14.6.

Many categories will contain more levels of detailed data than shown in the table. Electric utilities, for example, may maintain data on: distribution and transmission facilities (both above and below ground); sub-station information and diagrams; facility descriptions and operating histories; and customer information.

As you can see in these brief examples, large-scale geographic databases provide a framework for storing many types of spatially referenced data. The information on paper maps, traditionally maintained separately by the city engineer, registrar of deeds, and tax assessor, can be combined and

shared. The potential overlap of these large-scale data sets with the information on traditional large-scale topographic base maps argues for a data-sharing mechanism between national and local mapping agencies.

SELECTED REFERENCES

Federal Geodetic Control Committee, *Multipurpose Land Information Systems—The Guidebook*, Brown, P. M. and D. D. Moyer (eds.), Rockville, MD: National Geodetic Survey, 1989.

Tobler, W., "Resolution, Resampling, and All That," in *Building Databases for Global Science*, Mounsey, H. and R. Tomlinson, (eds.), New York: Taylor and Francis, 1988, pp. 129–137.

National Research Council, Panel on a Multipurpose Cadastre, *Procedures and Standards for a Multipurpose Cadastre*, Washington, DC: National Academy Press, 1983.

---

*The State Plane Coordinate System (SPCS) was devised by surveyors to replace bearing and distance measurements with coordinates. However, this system is little used in land surveying practice. The availability of GPS receivers that can produce SPCS coordinates directly may cause a change in this practice.

CHAPTER 15

MANAGING LARGE DATABASES

As more and more digital cartographic data are collected, greater amounts of computer memory, hard disk, and tertiary storage are needed for the data. Clearly, sophisticated techniques are needed to manage these vast data holdings. Some of these techniques **organize** the data holdings to make them more manageable and accessible. Other techniques **compress** the data so that less storage is needed. In this chapter, we will explore both these approaches, beginning with data organization.

DATA ORGANIZATION

Three methods are commonly used to organize large databases. The first is to break the database up into more tractable pieces—that is, to **partition** or **tile** the data. The second is to build a set of **indexes** to enable rapid access to any portion of the database. Finally, a set of summary information about the data, **metadata**, can be created so that users can browse through this smaller data abstract rather than the entire database. Let's take a closer look at these three methods.

PARTITIONING

Digital spatial databases are partitioned or tiled in order to handle a large amount of information within a set of operational limitations. These limitations include: operational limits in an application software package, size limitations on a magnetic or optical disk, and conventions used in data production or maintenance.

There are two basic strategies to tiling a spatial database. One is to have each partition contain approximately the same amount of information, allowing the geographic extent of each tile to vary. The second is to use a fixed geographic partition and allow the amount of data in the tile to vary. The first approach is often called **adaptive tiling**, and the second **fixed tiling**.

Adaptive Tiling

Spatial databases that have a wide range of feature densities may be good candidates for adaptive tiling. Where the data density (say of a road network) is high, there would be numerous small tiles. Where density is low, there would be fewer, larger tiles. All the tiles must nest within each other.

How are these tiles formed? One approach is a systematic spatial quartering of the study area. In this scheme, the four tiles produced within each quartering are numbered clockwise from one to four, starting with the upper left quadrant. The first two levels of partitioning of a world projection are shown in Figure 15.1. Four levels of partitioning of cell 13 are shown in Figure 15.2.

Once a partitioning scheme has been developed, database subdivision continues until each partition contains the desired amount of data. As an example, assume that we wanted each partition to contain no more than one megabyte of data. Figure 15.3 shows data volumes for partitions in cell 13 after the partitioning shown in Figure 15.2.

Based on these data volumes, cells 131 and 133 should be further subdivided, but cell 134 can be left alone. The final tiles are shown in Figure 15.4, with data volumes in parentheses under the tile identification codes.

Fixed Tiling

Fixed tiling defines a set of partitioned sub-areas into which database content is placed. A fixed tiling scheme can be created without consideration for the size or distribution of the database. Or, the size and shape of the tiles can be chosen after database characteristics have been quantified as much as possible. Tiles do not have to be regular shapes or fixed sizes. They may be tailored to a specific application. For example, the 1990 Census TIGER database is partitioned along county boundaries. Each tile holds one county of data. Counties in California have much more data than those in Alaska.

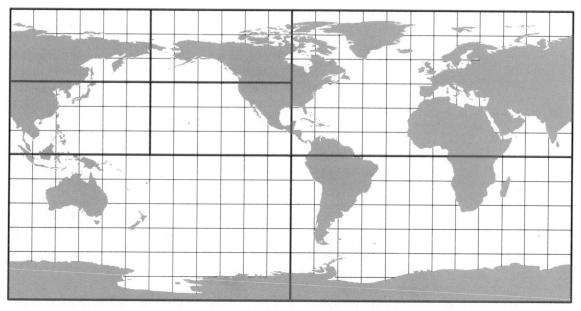

Figure 15.1   Two levels of partitioning of a Plate Carreé world projection.

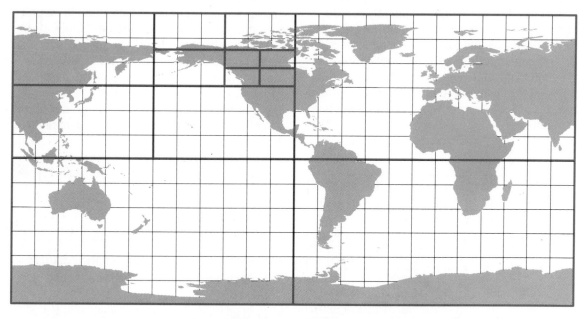

Figure 15.2   Four levels of partitioning.

2,443,303	477,530	661,039
	2,380,837	
	537,942	704,326
759,208	2,425,523	

Figure 15.3 Data volumes for partitions of cell 13.

By far the most common fixed tiling scheme is based on quadrilaterals—a set of systematic partitions along the latitude-longitude grid, very often following the boundaries of various printed map series. The question is what size tile should be used—1 degree by 1 degree, 30 minutes by 30 minutes, etc. If available, sample data sets or data volume estimates can be used to bound the range of data in a cell of a given size. For example, for the Digital Chart of the World, empirical tests determined that cells ranging in size from 3×4 degrees to 4×7 degrees would contain acceptable amounts of data to meet performance and application criteria, such as the time required to retrieve and draw a given set of features. After further evaluation, a tile size of 5×5 degrees was chosen for the final product.

1311	1312	1321	1322
(537,526)	(977,321)	(477,530)	(611,039)
1314	1313	1324	1323
(439,794)	(488,662)	(537,942)	(704,326)
134		1331	1332
(759,208)		(412,338)	(873,188)
		1334	1333
		(533,617)	(606,380)

Figure 15.4 Final set of tiles and data volumes for cell 13.

SPATIAL INDEXES

In database terminology, an **index** is a method to speed access to a particular portion or record in a database. A spatial index speeds retrieval of information about a geographic feature when a **query** or **request** is based on the feature's location. Queries such as those shown in Table 15.1 would benefit from the existence of a spatial index.

The key to executing rapid searches and data retrievals based on location is to reduce the amount of database that must be searched. Tiling the database helps in this data reduction. If each feature has an index field noting the tile in which the feature is contained, then detailed query processing can be restricted to that tile.

Quadtrees

Tiles provide only the first level of restricting a spatial search. Many spatial coding procedures use an extension of the tiling process to create an index. This process is one of **recursive decomposition**—progressively partitioning a fixed region into finer and finer units. The result is a hierarchial data structure termed a **quadtree** (Samet, 1990).

In a quadtree, each area is divided into four parts. Those parts are then further subdivided, much as in the adaptive tiling scheme described earlier in this chapter. In this case, however, decomposition continues to a much finer level. Theoretically, in fact, partitioning proceeds until there is only one spatial feature contained in each quadtree tile. In reality, such refined partitioning is impractical for all but the sparsest data set. Thus,

Table 15.1
Example Spatial Queries

- Find all the features at a given location.
- Find all the features within x meters of a given location.
- Find all the features within a given polygon.
- Find the feature and related attributes of the object being pointed at by a user.

the partitioning process is usually revised so that a quadrant can contain only so many features before it is split (just as in the adaptive tiling discussion). This process is illustrated in Figure 15.5. In this figure, a region is decomposed until each member of a set of cities has been placed in its own quadrant.

The quadtree **address** or **locational code** of any given cell is calculated by subdividing the space into quadrants and assigning codes to each quadrant. For instance, 0, 1, 2, 3 might stand for SW, NW, SE, and NE. There are various ways to assign codes to quadrants. The **Morton coding** scheme (Comeau, 1981) is illustrated in Figure 15.6.

Quadtrees are often used as space-ordering methods. That is, two-dimensional space is mapped into one dimension, creating a linear sequence that can be used as a spatial index. These index values can be calculated directly from x,y coordinates using a technique invented by Italian mathematician Guiseppe Peano in the 19th century. The spatial index values are often called **Peano codes** or, more coyly, **Peano keys** (White, 1986).

Peano began by assuming that a finite coordinate space is a decomposition of a plane to a unit grid. He went on to discover a space-filling curve (called a Peano curve) that traverses every intersection of the unit grid. This curve establishes a relationship between points in one-dimensional

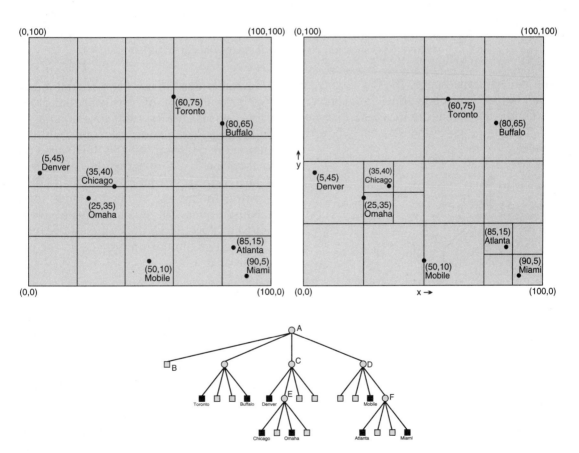

Figure 15.5 Distribution of cities in a quadtree representation. (Source: Samet, 1990, pp. 15, 19).

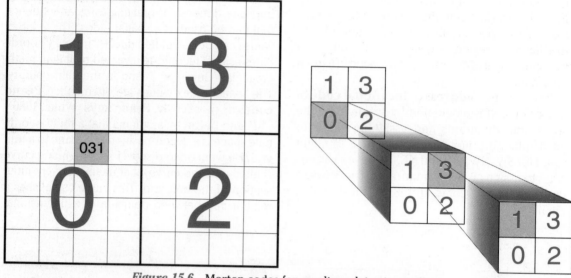

Figure 15.6 Morton codes for quadtree data structures.

and two-dimensional space. The Peano curve is defined by the Peano code, which is the order in which grid intersections are traversed. (See Box 15.A). The line formed by this traversal is shown in Figure 15.7.

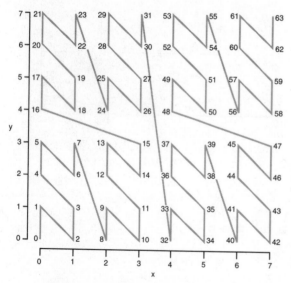

Figure 15.7 Space-filling Peano curve.

Peano codes are used as a spatial index to speed the location of nearby spatial objects. In the Census TIGER system, for example, each node has a Peano code associated with it. To answer the question "What is the nearest node to location x,y?", you merely calculate the Peano code for the query location and find its position in the list of Peano codes for the existing nodes. The Peano codes above and below this position point to the nodes that are close, in two-dimensional space, to the place in question.

R-trees

An **R-tree** is an alternate indexing scheme for spatial objects. The R-tree is a hierarchial tree. Each sub-element of the tree (referred to as a leaf) contains a rectangle and a pointer. The pointer points to a feature having the rectangle as a bounding box. In other words, each spatial feature in the database is placed in a rectangular container, defined by the minimum and maximum x,y coordinates of the feature. This is illustrated in Figure 15.8.

Box 15.A
Calculating Peano Codes

Peano codes can be calculated directly by performing a bit-wise interleaving of the x and y coordinate values of any point. This process is illustrated in Figure 15.A.1.

How are Peano codes related to quadtrees?

Assume that we chose the Morton coding scheme for quadtrees. In this scheme, 0=SW, 1=NW, 2=SE, and 3=NE. Then the Peano code gives the location of the lower left corner of the quadtree quadrant of the same code.

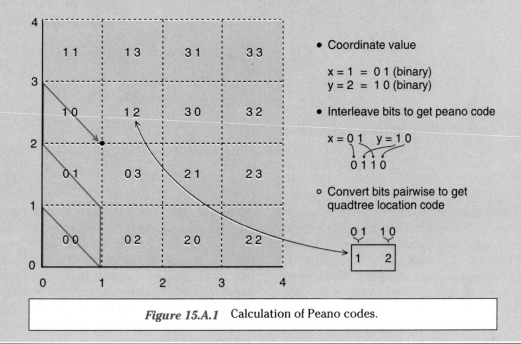

Figure 15.A.1 Calculation of Peano codes.

The collection of rectangles is used as an index to features. This index is particularly useful to retrieve all objects about a certain location. To do so, you find all the boxes that contain the point in question. Then you retrieve the objects in the boxes.

The index is also useful in solving the overlay operation in a GIS. Assume that boxes R1 and R2 represent two maps that are being overlaid. Rather than testing each feature in R1 against each feature in R2, you first find the set of boxes that overlap between R1 and R2. Thus, you can eliminate the

features in boxes R3 and R7 immediately. You need only test the features in boxes R5 and R4 for spatial intersections. See Laurini and Thompson (1992) for a thorough discussion of spatial indexing methods.

METADATA

Large spatial data sets contain a great variety and amount of information. Tiling schemes and spatial indexes can help manage the spatial component

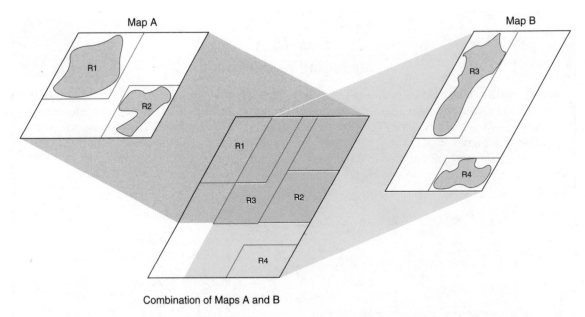

Figure 15.8 Example of an R-tree structure and the geometric forms it represents.

of the data. But other means are needed to provide summary information on and access to the database content. Using spatial data depends on being able to locate the data and to understand the data's characteristics and quality. **Metadata**, or "data about data," are a key to this ability. They give information on a data set's characteristics and history. They also identify organizations to contact to obtain a data set.

Standardized metadata elements would provide a way to document data sets within an organization. Standardized metadata would also help users understand the contents of data sets they receive from others. (See Chapter 13 for more on standardization).

Metadata components could include a set of data elements that accomplish the following:

- identify the data
- identify the custodians and access conditions for the data
- describe the projection, status, content, lineage and processing, and quality of the data.

Metadata could be gathered for each tile in a partitioned database as well as for the collection as a whole. The metadata could be collected in a data catalog to help users evaluate existing data. Metadata could also be provided in a data transfer to help users adapt the data to their applications. Figure 15.9 illustrates the range of usage of metadata, moving from general catalog to specific data transfer functions.

Box 15.B illustrates organization and content of metadata for a data set. From this example it is clear that the collection of comprehensive amounts of metadata is an enormous operation. Ideally all of the metadata elements would be recorded automatically by GIS software during the process of data collection. However this is not yet common practice. Indeed some elements of metadata, such as those related to data quality, require further research to develop the quantitative descriptive measures to be recorded in the metadata. Nonetheless, metadata catalogs are the keys to identifying the desirable data sets contained in large data archives.

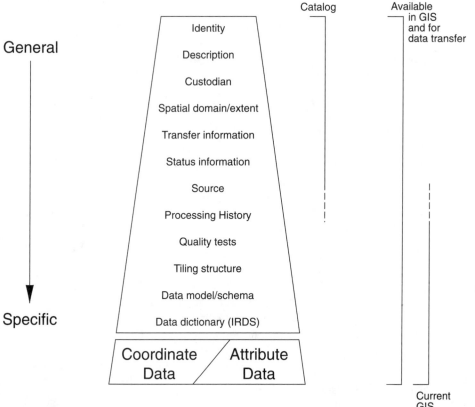

Figure 15.9   Range of metadata usage.

DATA COMPRESSION

Another way to deal with the problem of massive amounts of data is to use data compression schemes to reduce the physical size of a data file. There are numerous algorithms for performing compression. Some are completely **invertible**. That is, the original data are reproduced exactly upon decompression. These are called **lossless compression** techniques.

Other algorithms offer greater compression factors (they produce smaller data sets), but cannot exactly reproduce the original data when decompressed. These are called **lossey compression** algorithms. For image data, some data loss might be acceptable in order to achieve a greater level of compression. Some of the more common compression schemes are discussed here.

RUN-LENGTH ENCODING

Run-length encoding (RLE) is a lossless compression technique used with raster data sets. In RLE, an array of data is replaced by a series of variable-length records, each representing a row in the matrix. This record contains pairs of values. The first number in the pair represents the data value stored in the array. The second number in the pair represents the consecutive number of times that value is repeated in a part of the row. Figure 15.10 illustrates this principle. Data sets that contain large homogeneous regions will achieve greater compression than heterogeneous data sets.

Box 15.B
Elements of Spatial Metadata Content

Below is an example of metadata for a data set. This example illustrates the organization and content for metadata.

Identification Information

Data Set Identity	Hydrologic unit map, intermediate scale
Identification Code	Not applicable
Theme Keywords	Hydrology, basin, hydrologic unit, watershed
Use Restrictions	None, public domain
Access Restrictions	None
Spatial Data Structure	Vector
Data Set Description	Hydrologic units as defined by the federal government, collected and digitized at 1:250,000-scale with 1:100,000-scale insets in several western states. Hydrologic units have been appended into a single data set that comprises the conterminous 48 states, removing map edges in the process. Polygon and boundary attributes are present for site analysis and basin boundary symbolization.
Bounding Coordinates	21.4346,-126.0221 49.23,-67.000 NAS
Geographic Keywords	Conterminous United States
Browse Graphic	None

Contact Information

Contact Type	Custodian
Contact Organization	U.S. Geological Survey, Water Resources Division
Contact Person	Ms. Jane Doe
Contact Mailing Address	444 National Center, Reston, VA 22092
Contact Telephone	(703) 555-1212
Contact Fax Number	(703) 555-1234
Contact Electronic Mail	jdoe@usgs.gov
Contact Instructions	Contact for technical information via e-mail or regular mail.
Contact Liability	Data are in public domain. Custodian does not assume any liability.

Transfer Information

Transfer Format	DLG-3 Optional
Transfer Mode	Online/Offline
Transfer Size	40 MB
Transfer Instructions	Data are available through WAIS software and anonymous ftp from Internet site 130.11.51.171. Tape requests are filled at cost of duplication.
Fees	Fees range from $50 to $500 depending on tape format and media. Contact for detailed pricing.

Status Information

Data Set Status	Available
Release Date	19921001
Maintenance and Update Frequency	As needed

Coordinate System Information

Horizontal Coordinate System	Albers Equal-Area Conic

continued on next page

Box 15.B (continued)

Coordinate Type	Map projection
Coordinate Units	Meters
Origin Latitude	Not applicable
Origin Longitude	-96.0
Latitude of Standard Parallel One	29.5
Latitude of Standard Parallel Two	45.5
False Easting	Not applicable
False Northing	Not applicable
Central Azimuth	Not applicable
Central Scale Factor	Not applicable
Datum Name	North American Datum 1927
Ellipsoid Name	Clarke 1866
Ellipsoid Semimajor Axis	6378206.4
Ellipsoid Reciprocal Flattening	294.9786982
Vertical Coordinate System Name	Not applicable
Vertical Coordinate Type	Not applicable
Vertical Datum Name	Not applicable
Vertical Datum Type	Not applicable

Source Information

Source Descriptor	Land use and land cover digital data from 1:250,0000- and 1:100,0000- scale maps—GIRAS series data from USGS National Mapping Division,
Source Citation	U.S. Geological Survey, 1990, land use and land cover digital data from 1:250,000- and 1:100,000- scale maps, data users guide 4.

U.S. Geological Survey, 1982, codes for the identification of hydrologic units in the United States and the Caribbean outlying areas: U.S. Geological Survey Circular 878-A.

Source Data Resolution	10 m.

Metadata Reference Information

Metadata Date	19930105
Metadata Contact	Jane Doe, U.S. Geological Survey

Processing History Information

Process Used	Original materials were USGS/NMD quadrangle based digital files in the UTM projection. These files were processed as follows: 1. Performed an affine transformation between the internal map coordinates (to nearest 10 meters) and true UTM coordinates for each quadrangle. 2. Eliminated bounding neatline from each quadrangle and replaced with geometrically correct neatline. Extended under-shoots to intersect new neatline using 500 meter maximum extension. Clipped overshoots to neatline. 3. Re-estab-

continued on next page

Box 15.B *(continued)*

lished topology of polygon and line features using 2 meter feature-feature tolerance, clipped all overshoot features. 4. Joined all quadrangle data sets together, preserving line and polygon features and attribution. 5. Removed map edge lines where hydrologic unit identities (HUC) were the same on either side of a line.

Process Parameters	See process description.
Process Date	19920801

Data Quality Information

Horizontal Positional Accuracy	+/-150 meters
Horizontal Positional Accuracy Explanation	Deductive estimate, map materials at 1:250,000-scale do not comply with National Map Accuracy Standards.
Vertical Positional Accuracy	Not applicable

Vertical Positional Accuracy

Explanation	Not applicable
Thematic Accuracy	Greater than 90%
Thematic Accuracy Explanation	Value derived from comparison with 1:2M scale source values
Logical Consistency	Data set is verified topologically-structured polygon and line data with nodes at all intersections.
Completeness	All hydrologic units identified by U.S. Geological Survey are included in this digital map product. Several hydrologic units may be composed of multiple, non-contiguous polygons.

Feature/Attribute Information

Feature Label	Hydrologic unit codes
Feature Definition	Hydrologic units delimit catchment basins for stream segments.
Feature Definition Source	U.S. Geological Survey Circular 878-A
Attribute Label	Area
Attribute Definition	Area measured in equal area meters
Attribute Definition Source	Algorithm used by software
Attribute Feature Association	Area associated with each hydrologic unit
Attribute Domain Value	Positive real numbers

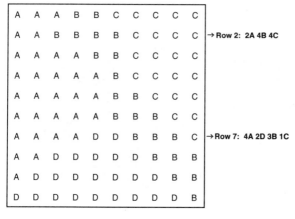

Figure 15.10 Example of run-length encoding.

QUADTREE ENCODING

In addition to acting as spatial indexes, quadtrees can also be used to perform a lossless compression of raster data sets. They are particularly effective in compressing areal or regional data represented as binary (black-and-white) images or bit planes. We achieve compression because not all elements of an array need to be stored. Instead, we only store the location codes of the quadrants that are totally occupied (that is, bits that are "black" or "on"). This **quadtree encoding** technique is shown in Figure 15.11.

JPEG COMPRESSION

During the last decade, the Joint Photographic Experts Group (JPEG) was organized to establish standards for easy exchange of graphic data, including data compressed using lossey techniques (Pennebaker and Mitchell, 1993). The resulting methods have been widely applied, particularly to image data such as digital orthophotos.

The JPEG specification performs compression using *either* a **lossless** (and reversible) differential pulse code modulation (DPCM) *or* a

lossey discrete cosine transform (DCT). The lossey DCT compression provides large amounts of data compression with a minimal introduction of visible distortion. Some of this compression is achieved by reducing the number of gray levels in an image or quantizing the image. The quantization step has the effect of **increasing redundancy** in image data. Such increased redundancy is useful in areas with little variation in data values. Examples of such areas are homogeneous fields or water bodies. With increased redundancy in the information, the compression can achieve higher compression ratios. In areas with a wide range in data values, such as at field edges and other higher-contrast areas, quantization is reduced or eliminated, thus preserving photointerpretation edge detail. Figure 15.12 shows a set of images subjected to various levels of **JPEG compression**.

FREEMAN CHAIN CODES

Freeman chain codes are used with vector data sets (Freeman, 1980). A line, instead of being represented by a series of *x,y* coordinates, is represented by a starting location. This starting location is followed by a sequence of numerical codes. Each of these codes represents the direction of a step to be moved along a line, as shown in Figure 15.13.

These chain sequences could themselves be run-length encoded for even greater compression. Since only a limited number of vector azimuths is permitted and vector length is constant, Freeman coding may result in the *loss* of data.

OTHER COMPRESSION SCHEMES

If you are familiar with personal computers, you may be aware of data compression software which works with *any* type of data file. Programs such as PKZip and LHA analyze the byte structure of the file and use sophisticated encoding schemes to

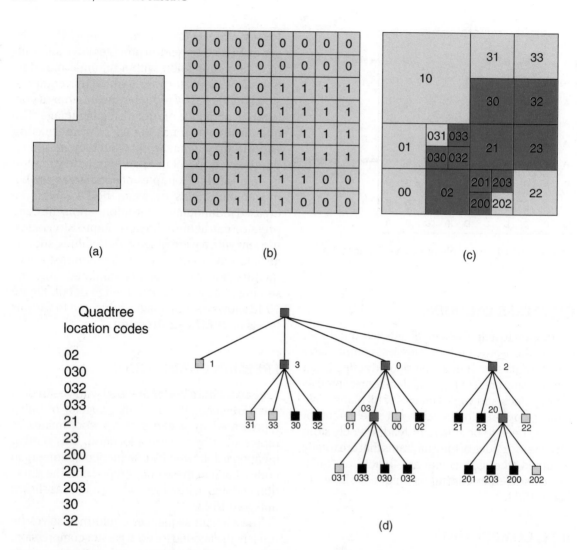

Figure 15.11 Quadtree encoding of a region. (a) region, (b) binary image, (c) quadtree blocks covering region, and (d) the corresponding quadtree. The list of quadtree codes contains 11 entries, whereas the binary image contains 64 entries (based on Samet, 1990, p. 58).

Original

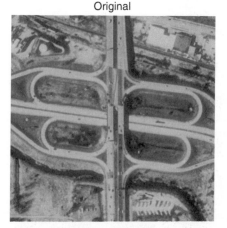

10X Compression

20X Compression

Figure 15.12 Images processed with JPEG compression. Original image (top); 10 X compression (middle); 20 X compression (bottom).

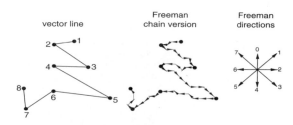

Freeman Code = 6563232436576532332367676565 57107

Figure 15.13 Freeman chain encoding.

remove "wasted" space from a data file. The decompression software reproduces the original exactly. Unlike the methods discussed earlier, these programs do not alter the logical way in which the spatial data are encoded.

SELECTED REFERENCES

Comeau, M.A., *A Coordinate Reference System for Spatial Data Processing*, Technical Bulletin No. 3, Ottawa, Ontario, Canada: Canada Land Data Systems Division, 1981.

Mandelbrot, B.B., *The Fractal Geometry of Nature*, New York: W.H. Freeman and Co., 1983.

Laurini, R., and D. Thompson, *Fundamentals of Spatial Information Systems*, San Diego, CA: Academic Press Inc., 1992.

Pennebaker, W.B., and J.L. Mitchell, *JPEG: Still Image Data Compression Standard*, New York: Van Nostrand Reinhold, 1993.

Freeman, H., "Analysis and Manipulation of Lineal Map Data", in *Map Data Processing*, Freeman, H. and Pieroni, G., eds., New York: Academic Press, Inc., 1980, pp 151–168.

Samet, H., *Applications of Spatial Data Structures*, Reading, MA: Addison-Wesley, 1990.

White, M., *N-Trees: Large Ordered Indexes for Multi-Dimensional Space*, Report of the Statistical Research Division, Washington, DC: U.S. Bureau of the Census, 1986.

CHAPTER 16

DATA MEASUREMENT AND BASIC STATISTICAL PROCESSING

*E*arthly phenomena occur in bewildering numbers and intricate arrays. One way to make sense of this confusing complexity is to make maps. An advantage that maps have over unprocessed photographs or remotely sensed images is that they generalize the information to be communicated. An image as it comes from a camera or sensor simply reports what is sensed. To make maps, cartographers must select features to be portrayed, process (classify, simplify, and exaggerate) information about the features, and symbolize this information in order to create an effective portrayal. These important processes, when correctly combined, help map users conceptualize the information and visualize the represented features.

As we saw in Chapter 10, the modeling of geographic reality involves using variables of many types. These include point, line, area, volume, time, vector, and raster data. In order to deal logically with all the options open to cartographers, we shall begin this chapter by discussing scales to use when measuring geographic features. We will then introduce common statistical measures appropriate for different types of information used in cartography.

THE NATURE OF GEOGRAPHIC PHENOMENA

Anything that is anywhere, whether it is material such as a road, or immaterial such as religious adherence, is a geographical or spatial feature. It has location as an attribute and therefore can be mapped.

Because geographical variables are so diverse and complex, we must understand their essential nature. Foremost are their locational relationships, or their **geographical ordering**. Geographical ordering is an inherent attribute of all spatial features. Equally important is a systematic approach to describing geographical features. Without a logical way to deal with qualitative and quantitative characteristics of geographical features, there would be nothing but confusion.

We tend not to be very systematic in the way we deal with geographical features. Depending on our mapping purpose, we may use different dimensional symbols to map the same feature. For example, we may conceive of New York City as a point (in contrast to Philadelphia), or as an area (a particular administrative region, as contrasted with an adjacent region, Westchester County), or as a "volume" of humanity. The same intellectual conceptualization is not necessary when we deal with images. Some geographical arrays are **discrete**. That is, they are composed of individual items at particular locations. Intervening areas are empty of that feature or have a value of zero for the attribute value. Such would be the case, for example, with individual houses or industrial plants.

In contrast, other distributions are **continuous** in that no location is empty. Temperature values on the earth's surface or categories of land cover exist everywhere. Thus, they are examples of continuous features.

Spatial data that intrinsically are discrete can be transformed conceptually into continuous data. People, for example, are discrete. But we can relate numbers of people to the areas they occupy by applying the density concept (number of persons per square kilometer). The ratio thus becomes continuous. All areas and the points contained within them must have some density value even if the value is zero.

The surface of a geographical distribution is either smooth or stepped. With **smooth** phenomena, the differences from place to place are transitional rather than abrupt. For example, atmospheric pressure varies gradually from place to place, and the pressure between two places close together will be intermediate. On the other hand, some phenomena (such as tax rates reported by states) change abruptly at boundaries and create a **stepped** surface.

MEASUREMENT OF GEOGRAPHICAL VARIABLES

When dealing cartographically with geographical features, it is necessary to determine their locations. That provides the geographical ordering attribute required for the fundamental function of the map. The location attribute alone is not enough, however. It is also necessary to differentiate among other feature attributes. A map that shows locations of rivers, roads, boundaries, and railways by lines but fails to show which is which would not be very useful.

For cartography, the most efficient method of describing observed characteristics and specifying categories within a set of variables is one that involves four levels of measurement. The four levels of measurement, in increasing order of descriptive richness, are: nominal, ordinal, interval, and ratio.

NOMINAL SCALES OF MEASUREMENT

We use **nominal** scales when we distinguish among features only on the basis of **qualitative** considerations. With nominal distinctions, there is no implication of a quantitative relationship. We are only saying that feature *A* is of a different class than feature *B*. See Figure 16.1.

For instance, we might differentiate a bench mark from a spring or land from water. We might distinguish a river from a road, or a maritime air mass from a continental air mass. All these are examples of nominal differentiation.

Although we can conceive of a particular geographical volume on a nominal scale, we cannot map it as a volume without using a higher level of measurement (ordinal, interval, or ratio). Because the ordinary map is only two-dimensional, a volume without any quantitative attribute can only be treated as a point, line, or area feature. For example, a volume of foreign-born population might be mapped as existing in a city (point data) or in an area (area data); a lake or an air mass would have to be mapped as occurring over an area (area data); and a volume of freight moved over a given railroad line could be mapped as a line.

ORDINAL SCALES OF MEASUREMENT

Ordinal scales involve differentiation by class, but they also differentiate within a class of features on the basis of rank according to some **quantitative** measure. Only rank is involved; that is, attribute values are ordered from lowest to highest without any definition of numerical values. (See Figure 16.2). For example, we can differentiate major ports from minor ports or intensive from extensive agriculture. Or we can distinguish among small, medium, and large cities. Such ordinal scales

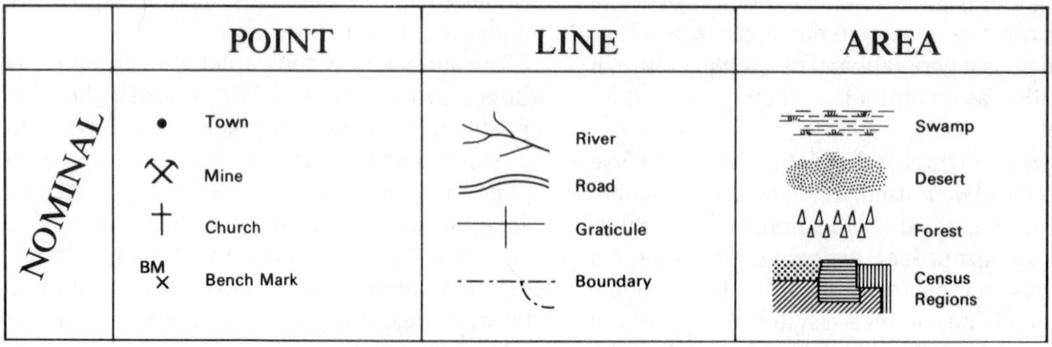

	POINT	LINE	AREA
NOMINAL	• Town ⚒ Mine † Church BM× Bench Mark	River Road Graticule Boundary	Swamp Desert Forest Census Regions

Figure 16.1 Examples of differentiation of point, line, and area features on a nominal scale of measurement.

let map readers know that some instances of a variable are larger or smaller, more or less important, younger or older than others. But they do not indicate any specific magnitude of difference.

INTERVAL SCALES OF MEASUREMENT

Interval scales add information about distance between ranks to the description of kind and rank. To employ an interval scale, we must use some kind of standard unit (which may be quite arbitrary) and then express the amount of difference in terms of that unit. For example, we can differentiate among temperatures by using a standard unit, the degree (Celsius or Fahrenheit). Or we can distinguish among elevations above an arbitrary datum, mean sea level.

Although interval scales of attributes referencing points, lines, areas, and volumes provide more information than nominal and ordinal scales, we must be careful not to infer more than is warranted by the nature of the unit and the scale to which it applies. For example, we should not say that 40°F. is twice as warm as 20°F., because 0°F. is arbitrary. It is not a true number meaning absence of temperature.

RATIO SCALES OF MEASUREMENT

A **ratio** scale is a further refinement of an interval scale. It provides magnitudes that are intrinsically meaningful. To do so, it uses a scale which begins at a non-arbitrary zero point (where zero indicates absence of the feature). Depth of snow or precipitation, populations of cities, and tons of freight are all measured on ratio scales. Most measures pertaining to weight, length, area, and volume are developed on ratio scales.

Also popular in cartography is range-grading of ratio scale measurements. (See Chapter 25 for more on range-grading). Range-grading acts much like interval scale measurements. Often, for instance, we group features by attribute values into classes such as family incomes of "less than $10,000," "$10,000 to $30,000," "$30,000 to $50,000," and "over $50,000."

We must be careful not to assume that a family in the "$30,000 to $50,000" income bracket earns five times as much as a family in the "less than $10,000" bracket. The reason is that a family in the lower bracket may have an income of $9,999, while a family in the upper bracket may have an income of $30,001. Clearly, the latter is not five times larger than the former.

It is important for cartographers to realize that the four measurement levels—nominal, ordinal, interval, and ratio—are **nested in one direction only**. In other words, data available at higher measurement levels can be reduced or generalized to lower levels, but the reverse is not possible. We can't take data available at lower measurement levels and portray them at higher levels.

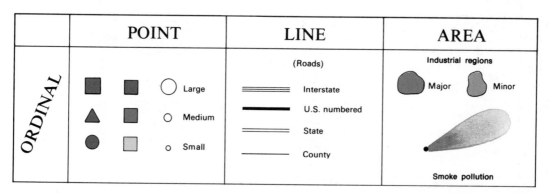

Figure 16.2 Examples of differentiation of point, line, and area features on an ordinal scale of measurement.

Doing so would require additional information about a feature, and that information was never collected in the first place.

Thus, ratio data about a geographic phenomenon may be portrayed on a map as ordinal or even nominal data. Such a reduction in measurement level is a proper form of generalization (see Chapter 24). Conversely, data measured on a nominal scale should not be mapped with symbols that connote ordinally scaled data.

Data suitable for mapping are available at all measurement levels. The cartographer's task is to decide how to represent these data. For mapping purposes, the data can be represented at or below—but not above—their measurement level.

From the point of view of mapmaking, there is no difference between the symbolization of geographical data categorized on interval and ratio scales. (See Figure 16.3). In both instances, a range is being displayed. In terms of visualization, it is immaterial whether the scale begins at an arbitrary or non-arbitrary zero. The map user, however, must be very careful when interpreting such maps, since the two scales of measurement allow different inferences to be made.

BASIC STATISTICAL CONCEPTS AND PROCESSES

To meet the needs of sophisticated map users, cartographers often find it necessary to manipulate raw data prior to mapping. Statistical techniques serve this purpose. Already a large number of maps portray statistical information. Given current trends, we can expect an even greater proportion of statistical maps in the future.

In this book, we can't detail all the statistical techniques available. However, every cartographer must be familiar with basic statistical concepts. We will, therefore, give a brief rundown of important statistical basics.

Before mapping statistical data, the cartographer may have to deal with a variety of problems concerning the attribute values of selected features. We can think of this as a pre-map data manipulation stage in the cartographic process.

When you obtain statistical data from different sources, for example, you need to equate them so that they represent **comparable values**. For instance, different countries use different units of measure (such as metric, long or short tons, U.S.

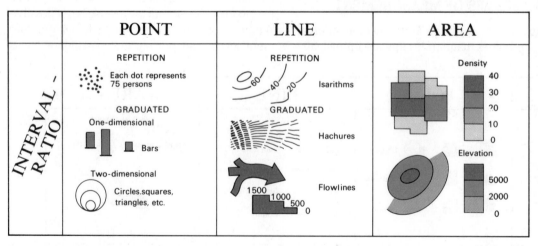

Figure 16.3 Examples of differentiation of point, line, and area features on an interval or ratio scale of measurement.

gallons or Imperial gallons, hectares or acres, and so on). If you were preparing a map of energy reserves, it would not be sufficient to change only the volume or weight units to comparable values. You would also need to bring the figures into conformity on the basis of their energy ratings (such as British thermal units or BTUs).

Or imagine that you were preparing a continental map to reveal relationships among temperatures, latitudes, and air masses. You would need to be sure that local effects of elevation were removed from the reported figures. To do so, you would need to ascertain the altitude of each weather station and then convert each temperature value to its sea-level equivalent.

After data have been made comparable, the next step may be to convert them to mappable data. Such statistics as averages, ratios, densities, potentials, variation, and correlation may be calculated. Clearly, the cartographer must understand these statistical processes and their effects on the data before using statistically processed data to make a map.

ABSOLUTE AND DERIVED DATA

All maps fall into one of two classes: They portray either (1) absolute qualities or quantities, or (2) derived qualities or quantities. Examples of the first class are "raw data" maps showing land use categories, production of goods, elevations above sea level, and the like. Qualities or quantities are observed concerning a single class of features. These attributes are expressed on the map in absolute terms according to some measurement scale.

The second group of maps portray values derived from raw data. These **derived values** express some kind of summarization or relationship between features. Examples include number of people per square kilometer, average July temperatures, per capita income, or categories of similar ground slope. This group includes four classes of relationships: averages, ratios, densities, and potentials. Let's look at each of these classes, beginning with averages.

Averages

Averages are the most common derived value. They are often called **measures of central tendency**, because a selected or calculated quality or quantity is used to characterize a series of qualities or quantities. There are many kinds of averages, but only three are commonly used in cartography.

The first of these commonly used averages is the **arithmetic mean**. Most maps of climate, income, or production are based on arithmetic means derived by reducing large amounts of statistical data. The equation for calculating the arithmetic mean is:

$$\bar{Z} = \sum_{i=1}^{n} Z_i / N$$

where ΣZ_i is the summation of all the Z attribute values, and N is the number of instances of Z. The arithmetic mean is appropriate for ratio-scaled measurements.

The mean sometimes should be **areally weighted**. For example, suppose we wish to prepare a map of the United States showing value of farmland per acre by state. Say that we plan to use data reported by county averages. But what if some counties in a state are much smaller than others? Then to give equal weight to all values would seriously bias the average for that state. Therefore, whenever the attribute values (Z) in a distribution are in any way related to areal extent, they must be areally weighted. The best way to do so is to multiply each Z value by the area of the region to which it refers, sum these products, and divide the sum by the total area. The general expression for any areally weighted mean is, therefore,

$$\bar{Z} = \sum_{i=1}^{n} aZ_i / A$$

in which ΣaZ_i represents the sum of the products of each Z attribute value multiplied by its area, and A is the total area (that is, Σa). The areally weighted mean is also called the **geographic mean**. Arithmetic means may be weighted by factors other than area.

The second commonly used average is the **median**. If we rank (arrange in numerical order) all the attribute values of a set of similar features measured on an ordinal, interval, or ratio scale from lowest to highest, the median is the attribute value in the middle. That is, half the values will be higher and half lower than the median value.

Let's use the previous example of farmland value per acre by state, obtained from data reported by counties. For each state, the county attribute values would be ranked without regard for the areas of the counties, and the midpoint value would be taken as the representative median value.

As before, if counties within a state varied greatly in size, the median should be areally weighted. When this is done, the **geographic median** is that value below which and above which half the total area occurs. The method of calculation for areal weighting requires that the values be ranked, and that the area associated with each value be known. Then, beginning at the bottom of the ranked array, we progressively sum the areas associated with each observation. The geographic median value is the attribute value of the instance (for example, county) whose associated area, when it is added to the accumulating areal sum of counties of lower rank, makes that sum equal to half the total area of the state.

The third commonly used average is the **mode**. It is the value that occurs most frequently in a distribution. The mode is used as the basis for maps that portray a region's major type of phenomena (such as land use, soil character, vegetation cover, or linguistic areas). Determining the mode is often quite simple for attributes such as the predominant fuel used at generating plants or the party gaining the most votes in an election.

Determining the modal quality of an area can be very elusive, however, unless the distribution being mapped is visible. For areas, the **area modal class** is defined as that class which occupies the greatest proportion of an area, not the greatest frequency of occurrence. Even then, it is possible that two or more categories equally comprise an areal unit or that variation within a unit area is so

great that less than 25 percent of the area occurs in any single class. The smaller the unit areas, the less likely this is to be a problem.

Each of the three measures of central tendency will in some instances provide mapping data that truly represent the character of the distribution. In other cases, the distribution may be such than an average is not very representative. There are indexes of variation that help to describe how well the arithmetic mean, the median, and the mode characterize a distribution. These will be described later.

RATIOS

The second class of derived quantities includes **ratios** or **rates, proportions**, and **percentages**. With this class, we measure something per unit of something else, or we single out some element of the data and compare it to the whole. Maps that show percentage of rainy days, proportion of all cattle that are beef cattle, or rate of growth or decline of some phenomenon are examples. In this group, the numerical value mapped will ordinarily be the result of one of the following basic kinds of operations:

$$\text{a ratio or rate} = n_a/n_b$$

$$\text{a proportion} = n_a/N$$

$$\text{a percentage} = n_a/N \times 100$$

where n_a is the number of feature instances in one category, n_b the number in another category, and N the total of all categories.

Sometimes such statistics can themselves take on the characteristics of a spatial distribution. For example, calculation of per-capita income by county can be mapped as a continuous income surface for a state.

Cartographers design maps of derived quantities to show place-to-place variations in the relationship mapped. Often they map percentages, ratios, and rates on the basis of enumeration units such as counties or states. The problem is that map readers might assume that such values are dispersed uniformly within each region. Thus, if the data used to derive values are not evenly distributed within each region, the mapped data may be quite misleading.

Another problem is that derived values can mask the nature of available raw data. For instance, to map County A as having tractors on 100 percent of farms and County B as having only 50 percent of farms with tractors hides the fact that County A may have only one farm, while County B has 500 farms.

Quantities that are not comparable should never be made the basis for a ratio. For example, we ought not even calculate—let alone map—the number of tractors per farm by dividing the total number of tractors by the number of farms in a county unless the farm sizes and practices are relatively comparable. Common sense will usually dictate ways to insure comparability.

Densities

A third class of derived quantities is called **densities**. We use these measures when our major concern is the relative geographical crowding or sparseness of discrete phenomena. Examples are maps of number of people (or trees, cows, or whatever) per square kilometer, or average spacing between phenomena such as service stations or mail collection points. Density is computed by

$$D = N/A$$

where N is the total number of feature instances occurring in an enumeration unit (for example, a county), and A is the area of the unit.

The average spacing between phenomena, another way of looking at density, is computed by

$$S = 1.0746 \sqrt{A/N}$$

where S is the average spacing of features or the mean distance between them (expressed in the same units used for A when the elements of N are assumed to be equidistant). If we assume a square arrangement, the **interneighbor interval**, as it is

called, is the square root of the reciprocal $\left(\sqrt{1/D} \right)$ of the population density (D) expressed in linear units of the measure used to calculate density.

Density, because it includes area in its calculation, is more closely related to the land than other averages and ratios. For this reason, we can derive several useful measures from density figures. For example, 5,000 feature instances in an area of 100 square kilometers results in a density of 50 per square kilometer. If arranged hexagonally, each instance would be about 150 meters from its neighbors. The interneighbor interval would be 141 meters.

In some circumstances, a density value derived by relating total population of a statistical unit to its total area is not as significant as one which expresses the ratio between feature instances and the part of the area related to the function of interest. For example, density of rural people to cultivated land may be more useful than the simple total population to total area density figure. If data are available, we can easily relate population to cultivated land or to any spatial characterization that is important to the objective of our analysis.

When working with densities and average spacings, cartographers are limited in the detail they can show. The main limiting factor is the size of statistical units (such as townships, counties, or states) for which enumerations have been made. The larger the units, the less useful the map will usually be.

In many cases, cartographers must supplement the initial data with other sources in order to present a useful visualization of a distribution. A systematic way to do so is to estimate the density of a portion of the enumeration district by using supplementary data. It is then necessary to adjust the value for the other portion so that the modifications are consistent with the original data. This procedure is called **estimating the density of parts** and is explained in Box 16.A.

Potentials

A fourth derived quantity is termed **potential**. Maps of potential assume that individuals comprising a distribution (such as people or prices) interact or influence one another. The degree of interaction is *directly* proportional to the magnitudes of the phenomena and *inversely* proportional to the distance between their locations. Because this assumption is derived from the physi-

cal laws governing the gravitational attraction of inanimate masses, it is called a **gravity concept**. It has been applied to a variety of economic and cultural elements.

A potential's value at a given point is the sum of the influence of all other points on the given point,

Box 16.A

Estimating the Density of Parts

Assume an enumeration district with a known density, D, of 90 (Figure 16.A.1). Assume further that the district is divided in two portions, m, comprising 0.7 of the total area, and n, making up 0.3 of the area. If the density of area m is estimated to be 15, then the density of area n must be calculated so that the two regions together have a density of 90.

The basic equation is

$$D_n = D/(1-a_m) - D_m a_m/ (1-a_m)$$

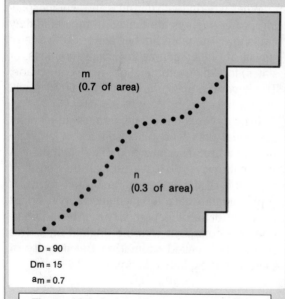

D = 90
Dm = 15
am = 0.7

Figure 16.A.1   A hypothetical enumeration district divided in two portions, *m* and *n*, by the dotted line.

in which D_n = the density in area n (see Figure 16.A.1)

D = average density of the area as a whole (number of units/area)

D_m = estimated density in part *m* of area

a_m = the fraction of the total area comprised in *m*

$1 - a_m$ = the fraction of the total area comprised in *n*

D_m and a_m are estimated approximately. Sometimes if we study supplementary information about neighboring areas, we can gain clues to the value we should assign to D_m. For example, other maps may show a similar type of distribution prevailing over D_m. It would be reasonable, therefore, to assign to D_m a density comparable to that value from another map.

Having assigned estimated but consistent densities to the two parts of an area, we may then again divide one or each of these parts into two subdivisions and work out densities for those two subdivisions. We can apply this method in the mapping of any phenomena for which statistics are available only by larger units. But the resulting map's quality depends on the validity of the estimates of D_m. Figure 16.A.2 illustrates the refinement that can be made.

continued on next page

Box 16.A *(continued)*

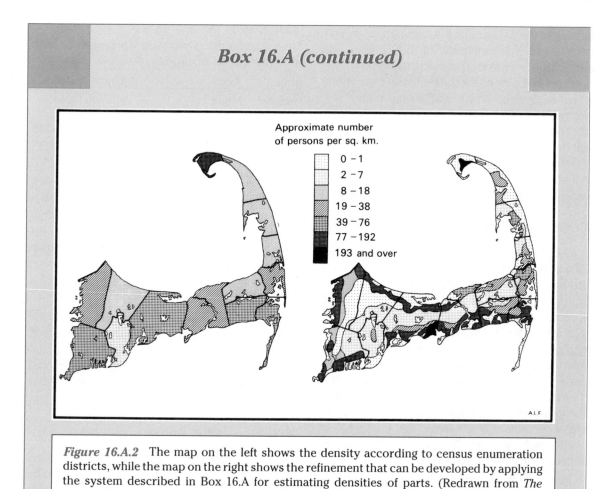

**Approximate number
of persons per sq. km.**

	0 – 1
	2 – 7
	8 – 18
	19 – 38
	39 – 76
	77 – 192
	193 and over

A.L.F.

Figure 16.A.2 The map on the left shows the density according to census enumeration districts, while the map on the right shows the refinement that can be developed by applying the system described in Box 16.A for estimating densities of parts. (Redrawn from *The Geographical Review*, published by the American Geographical Society of New York).

plus its influence on itself. The potential P of place i of feature Z will be

$$P_i = Z_i + \sum_{i=1}^{n} \frac{Z_j}{D_{ij}}$$

where Z_j is the value of Z at each place involved, and D_{ij} is the distance between place i and j. Preparing a potential map requires that this summation be repeated for each place. This is a task for a computer, since if j is only 50, for example, more than 2,500 calculations will be required. The resulting map values will be in units of Z at particular points.

Often we limit the number of values used in computing a potential to a local neighborhood of points.

INDEXES OF VARIATION

Averages are probably the most common way of summarizing data in cartography. However, an average is a generalization, and you cannot "tell by looking" whether it is a good or poor summary of the generalized information. Accordingly, it is the cartographer's responsibility to (1) understand the character of the distribution being mapped, and (2) make the map user aware of the critical characteristics of the data so that the summarization will not mislead. The remainder of this chapter will address the first task by discussing the appropriate averages and describing the indexes of variation, which suggest how valid the averages are. The second responsibility involves simplification, classification, and symbolization procedures, which will be discussed in Chapters 24, 25, and 26.

Table 16.1 lists the appropriate averages and indexes of variation with which to summarize information, depending on the scale of measurement.

Mode—Variation Ratio

In a set of attribute values measured nominally, each is distinguished qualitatively but not quantitatively. When we map unsummarized absolute values, such as actual land use or kinds of road surfaces (gravel, black top, concrete), the actual facts are mapped. On the other hand, when the map's purpose or scale requires it, the mode (the class that occurs most frequently) is mapped. The mode can be either the most frequently occurring class in a set of attribute values, or, if it is an area phenomenon, it can be the class that occupies the largest proportion of the area.

The **variation ratio (v)** indicates how representative of the distribution the mode is. It provides an index of the proportion of nonmodal cases; v varies from a value of near 1, which indicates that nearly all units differ from the modal unit, to 0, which indicates the occurrence of only the modal class. The nearer v is to zero, therefore, the better the quality of the mode as a summarizing statement. The variation ratio is calculated by

$$V = 1 - f_{modal}/N$$

where f_{modal} is the frequency or number of feature instances in the modal class, and N is the total of all feature instances.

Land type (upland, lowland, hills, swamps) is an example of features scaled nominally in which the areal extent of the features is mapped. Mapped categories are determined on the basis of the predominant or modal class in a series of areas. In this case, the frequency of each feature is the

TABLE 16.1

Scaling Systems: Their Appropriate Averages and Indexes of Variation

Scale	Average	Index of Variation
Nominal	Mode	Variation Ratio
Ordinal	Median	Decile Range
Interval	Arithmetic Mean	Standard Deviation
Ratio	Arithmetic Mean	Standard Deviation

proportion of the total area it occupies in each mapping unit. The modal class is, therefore, the category occupying more area than any other category. When we are concerned with areal frequency, the calculation of variation ratio must be weighted by area and becomes

$$V = 1 - a_{modal}/A$$

where a_{modal} is the area of the unit occupied by the modal category, and A is the total area of the unit. The areally weighted variation ratio indicates the proportion of the unit occupied by the nonmodal categories. The nearer it is to zero, the better that area mode represents the unit.

Median—Decile Range

We have seen that the median is the attribute value midway in an array of ranked variables. It can be used as a descriptive measure for summarizing attribute values ranked ordinally or for ranked data on interval and ratio scales.

An appropriate descriptive statistic to evaluate dispersion among attribute values ranked on an ordinal scale is a quantile range. A **quantile** is obtained by dividing a distribution into equal segments; **quartiles** divide it into four categories, **deciles** into 10 categories, and so on, to **centiles** (or **percentiles**) which divide it into 100 categories.

Let's use the **decile range** as an example. The decile range (d) is the range of ranked values between those in which the first (d_1) and ninth (d_9) deciles occur; in effect, $d = d_9 - d_1$. The first decile is the rank below which 10 percent of attribute values occur. The ninth decile is the rank above which 10 percent occur. Thus, the value d indicates the dispersion of the middle 80 percent of feature instances. The nearer d is to the total range of ranked values, the poorer the median is as a representative statistic.

When considerable differences occur in attribute value-areal extent relationships, it is appropriate to areally weight the decile range information. To do so, we determine deciles for 10 percent of the total area involved (not 10 percent of the number of feature occurrences, as is normally done). By carefully inspecting the data, we determine whether the ordinary median or the geographic median will provide the more representative average when working with deciles. Once we choose the appropriate measure, we apply it uniformly to all the decile data.

Arithmetic Mean—Standard Deviation

The concepts of mode and median underlie much mapping of qualitative and ranked distributions. But most maps based on averaging attribute values, measured on either an interval or ratio scale, use the arithmetic mean. Calculating a mean, or its areally weighted equivalent, is relatively simple. However, it is also important to evaluate how well the mean serves as a measure of the distribution's "central tendency."

If a series of observations, such as low temperatures each day or number of automobiles passing a point between 4:00 P.M. and 6:00 P.M. daily, fall within a narrow range, then the mean of the series is generally a good summary. Conversely, if the numbers vary greatly (that is, if they are widely dispersed), then the mean is not a good summary. The most useful measure or index of dispersion of data about the mean is a statistic called the **standard deviation** (symbolized as σ).

The standard deviation is based upon an important concept in statistical analysis known as a **normal distribution**. A normal distribution describes the predicted frequency with which various values will occur in a set of repeated observations.

In a normal distribution, we expect values near the arithmetic mean to occur most often. The greater the deviation from the mean, the less often we expect that value to occur. When we plot values' frequency of occurrence on the Y axis against values on the X axis, the result is a bell-shaped curve called the **normal curve** (See Figure 16.4). The standard deviation (σ) is a way of describing the dispersion of values around the mean of a normal distribution. It is calculated by either of the following formulas:

$$\sigma = \sqrt{\frac{\sum_{i=1}^{n}\left(X_i - \overline{X}^2\right)}{N}}$$

$$\sigma = \sqrt{\frac{\sum_{i=1}^{n}X_i^2}{N} - \left(\overline{X}\right)^2}$$

in which σ is the standard deviation, and $(X_i - \overline{X})$ is the difference between each value, X_i, and the mean, $\overline{X}$. The standard deviation is also called the **root mean square**. The square of the standard deviation is known as the **variance**.

When data are in any way related to areal extent, we need to calculate an areally weighted standard deviation. To do so, we use the formula:

$$\sigma = \sqrt{\frac{\sum_{i=1}^{n}aX_i^2}{A} - \left(\frac{\sum_{i=1}^{n}aX_i}{A}\right)^2}$$

in which we obtain ΣaX^2_i by first squaring each X_i value, then multiplying it by the area it represents, and summing the products. The term $(\Sigma aX_i/A)^2$ is merely the square of the geographic mean.

The importance of the standard deviation is that the arithmetic mean plus or minus $l\sigma$ predictably

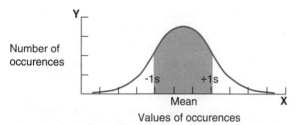

Number of occurences

Mean

Values of occurences

Figure 16.4 A normal curve results from plotting the values of a normal distribution on the X axis against the frequencies of their occurrences on the Y axis. The mean is the value at the peak of the curve. The shaded area under the curve shows the proportion of occurrences included within one standard deviation (plus and minus) of the mean.

includes slightly more than two-thirds (68.27 percent) of all attribute values in a normally-distributed data set. Consequently, the smaller the value of σ, the closer the values cluster around the mean, and the better the mean characterizes the distribution.

The mean and standard deviation are widely used in both descriptive and inferential statistics. Maps of distributions such as rainfall variability, probability of temperature extremes, or potential crop yields depend upon standard deviations. In addition to its descriptive usefulness, the standard deviation is used to set class limits for mapped distributions. It has been suggested that only by using the mean and standard deviation for class limits is it possible to obtain comparability through time for statistics about distributions like birth and death rates, diseases, or income.

It is not uncommon for cartographers to be asked to derive and work with attribute values from information that is already mapped. For example, suppose you are asked to determine average elevation above sea level for a certain region. You are also asked to determine the standard deviation of the elevations in that region. Your source data is an existing contour map. What procedure should you use?

We will illustrate this procedure using a conically shaped island mapped with contours at a 30-meter interval. (See Chapter 26 for more on contours and contour intervals). This island is shown

in Figure 16.5. Contour lines are labeled from 0 to 240. Areas between successive contours are labeled $a_1, \ldots a_9$. In this instance, the geographic frequencies are the relative areas of the occurrences. Consequently, you need to obtain the areas of the spaces between contours. You can express your results in any convenient square units, such as square centimeters. The measurements obtained from the original drawing of Figure 16.5 (before reduction for printing) are shown in the third column of Table 16.2.

Using the areally weighted mean formula given above, the data in Table 16.2 yield

$$z = 9329.25/102.65 = 90.9 \text{ meters}$$

as the areally weighted mean elevation of the island.

Many problems of cartographic analysis include data with **open-end classes**. That is, the top or bottom class may be available only on a "less than" or "more than" basis. This is the case with the series of elevations shown by contours on the island in Figure 16.5. In that figure, the contour of highest value is 240 meters. By definition, the area inside that contour is above (that is,

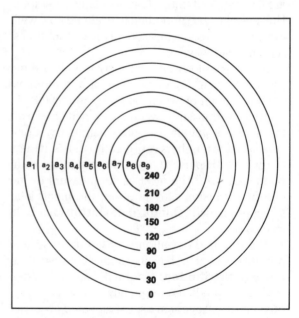

Figure 16.5 A conically shaped island delineated by contours at a 30-m. interval.

Table 16.2
Calculation of the Mean Elevation of the Island Shown in Figure 6.5

Elevation Classes (meters) (z)	Midvalue of Elevation Classes (z_m)	Map Areas of Elevation Classes (cm²) (a)	(az_m)	Cumulated Areas (cm²) ($a_1 + a_2 + a_3...$)
0–30	15	21.42	321.30	21.42
30–60	45	18.84	847.80	40.26
60–90	75	16.45	1233.75	56.71
90–120	105	14.06	1476.30	70.77
120–150	135	11.55	1559.25	82.32
150–180	165	8.84	1458.60	91.16
180–210	195	6.39	1246.05	97.55
210–240	225	3.81	857.25	101.36
240–270	255	1.29	328.95	102.65
Total		102.65	9329.25	

more than) 240 meters, but by exactly how much is not shown. In other cases, the neat line or map edge may cut through a statistical unit area and produce a similar result. When the limits of all classes are not given, we cannot accurately ascertain the mean value. However, judicious estimation and interpolation will usually make significant errors unlikely.

To illustrate calculation of areally weighted standard deviation, the necessary data are shown in Table 16.3. As you can see, class midvalues are used.

$$S = \sqrt{1,261,424.25/102.65 - \left(9329.25/102.65\right)^2}$$

$$S = 63.47\,m$$

equals the areally weighted standard deviation of the elevation of the island.

Thus, you can expect that about two-thirds of the island has an elevation within 63.5 meters of the mean. Or, about two-thirds of the island lies above approximately 27.5 meters (90.9 - 63.47 m.) and below approximately 154.5 meters (90.9 + 63.47 m.); that is, about two-thirds lies between those two elevations.*

These calculations are tedious and subject to error when performed manually. Fortunately, you can use computers to calculate averages and measures of dispersion in a few seconds.

SOME BASIC STATISTICAL RELATIONS

Since the introduction of computers into cartography, the sophistication of statistical techniques cartographers use to process data prior to mapping has increased substantially. Cartographers must understand these statistical techniques and be able to use them wisely.

The concepts of regression and correlation are basic to many statistical procedures cartographers need to use (see Chapters 24, 25, and 26). Regression is used in point or feature elimination algorithms and as the basis for surface-fitting routines in cartographic simplification. The correlation coefficient is often used to indicate adequacy of a surface fit or of a class interval scheme. Correlations are also used in factor analytical schemes which result in classifications of data.

Extensions of linear regression to multiple regression and of correlation to autocorrelation are also important in cartography. Surface-fitting is really a multiple regression technique. Autocorrelation of a distribution is a useful parameter

---

*Since this is only true if the elevation data are *normally distributed*, we have assumed that this is the case with the conical island in our illustration.

TABLE 16.3

Calculation of the Standard Deviation of Elevations on the Island Shown in Figure 16.6

Midvalue of Elevation Class (m) (z_m)	Map (or actual) Area of Class (cm²) (a)	(az_m)	(z_m^2)	(az_m^2)
15	21.42	321.30	255	4819.50
45	18.84	847.80	2025	38151.00
75	16.45	1233.75	5625	92531.25
105	14.06	1476.30	11025	155011.50
135	11.55	1559.25	18225	210498.75
165	8.84	1458.60	27225	240669.00
195	6.39	1246.05	38025	242979.75
225	3.81	857.25	50625	192881.25
255	1.29	328.95	65025	83882.25
Total	102.65	9329.25		1,261,424.25

for cartographers to know when selecting an interpolation algorithm.

In the remainder of this chapter, we review two commonly used statistical techniques: linear regression and correlation. These techniques form the basis for several more advanced statistical data-processing techniques. What follows is a condensed introduction and reference, not a substitute for a book on statistics.

REGRESSION ANALYSIS

Regression can be used to provide solutions to numerous prediction problems and to summarize relationships between quantitative variables. A retail store chain may wish to predict sales volume for a given store based on its location. Or a farmer may wish to predict harvest yields based on climatic factors. In its simplest form, regression analysis includes at least one independent variable (such as a measure of climate) and one dependent variable (crop yield). A simple model can then be established of the type

$$y = a + bx$$

where x is the independent variable, y is the dependent variable, and a and b are constants. You may recognize this as the equation for a straight line in a plane coordinate system, in which a is the point where the line intercepts the y axis and b is the slope of the line.

Figures 16.6 and 16.7 illustrate a calculated linear regression equation for the observations of one independent variable, x (per capita personal income) and one dependent variable, y. In Figure 16.6, y represents per capita educational expenditures. In Figure 16.7, y represents number of first-degree graduates. The areal units are shown in Figure 16.8.

The solution of a linear regression equation is usually based on the **least-squares criterion**. This means that the regression line is situated so as to minimize the sum of squares of deviations between the line and the graphed points. The graphed points represent observed values, while the regression line represents the predicted relationship. (The observed y values are represented in Figures 16.6 and 16.7 as points. The predicted values are read on the y axis for any given x value by moving vertically from the selected x value to the plotted regression line and then horizontally to the y axis to obtain the predicted value). Using this least-squares criterion, the correct values for the constants a and b are given by the following formulas:

$$b = \frac{\sum_{i=1}^{n}\left(x_i - \overline{x}\right)\left(y_i - \overline{y}\right)}{\sum_{i=1}^{n}\left(x_i - \overline{x}\right)^2}$$

$$a = \overline{y} - b\overline{x}$$

where

$$\overline{y} = \frac{\sum_{i=1}^{n}Y_i}{N} \qquad and \qquad \overline{x} = \frac{\sum_{i=1}^{n}x_i}{N}$$

Table 16.4 and Figures 16.6 and 16.7 illustrate this linear regression technique.

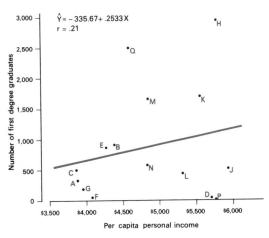

Figure 16.7 A scatter diagram with fitted linear regression line. Letters refer to areal units designated in Figure 16.8. Y refers to the predicted values for the dependent variable, and r refers to the coefficient of correlation. The data values are given in Table 16.4.

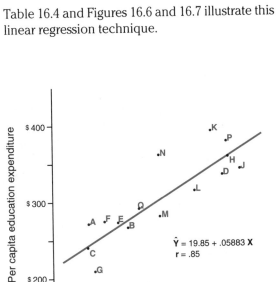

Figure 16.6 A scatter diagram with fitted linear regression line. Letters refer to areal units designated in Figure 16.8. Y refers to the predicted values for the dependent variable, and r refers to the coefficient of correlation. The data values are given in Table 16.4.

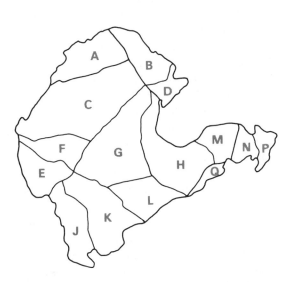

Figure 16.8 A map of a fictitious island showing the areal units to which the data in Table 16.4 refer.

Table 16.4

Data for a Given Year for a Fictitious Island with 15 Civil Districts (data used for Figures 16.6 to 16.10)

Area	Per Capita Personal Income	Per Capita Educational Expenditure	Number of First-Degree Graduates
A	$3882	$273	330
B	4395	266	910
C	3870	240	500
D	5695	333	40
E	4282	273	870
F	4082	276	70
G	3952	210	240
H	5770	357	2920
J	5938	340	530
K	5550	390	1760
L	5304	314	460
M	4840	280	1670
N	4830	360	580
P	5745	376	0
Q	4570	287	2500

Maps may be created from observed values, predicted values, or residual values (the difference between observed and predicted values), as shown in Figures 16.9 and 16.10. To prepare such maps, you should be aware of the assumptions behind this type of analysis. You must also remember that selecting appropriate class intervals for mapping is critical (see Chapter 26).

From a geographic point of view, the computed relationship of x and y does not take spatial position into account unless x or y is a position-related variable (that is, a variable whose measurement inherently contains some element of spatial position, such as latitude or longitude). In practice, most cases of x and y pairs of observations used in regression analyses are not locational in that sense. Instead, they are simply attached to places.

For example, note the regression analysis in Figure 16.6. Per capita personal income is used as an independent variable to predict per capita educational expenditures for the island. Areas A through Q are used as the data collection units. In this case, a pair of values of the dependent variable (per capita educational expenditures) and the independent variable (per capita personal income)

are associated with each area (location). Location itself is not involved in the computations.

Another assumption underlying regression analysis is that the data are normally distributed. Only by making this assumption can we make inferences about the strength of the relationship between variables. Otherwise, we only know the relationship between the specific points used to derive the a and b values of the equation. Only by assuming a normal distribution can we assess the reliability of any prediction made on the basis of the sample data.

Cartographers must be concerned with this fact, since they need to indicate relative reliabilities when mapping predicted values or residuals. For example, compare the maps in Figures 16.9 and 16.10. The predicted maps in each figure alone indicate nothing about their reliability. However, compare the mapped data in their corresponding scatter diagrams (plots of one variable against another) in Figures 16.6 and 16.7. The quality of the prediction mapped in Figure 16.9, as revealed by Figure 16.6, is much greater than the predicted map in Figure 16.10. One way to indicate reliability is to map residuals as well as predicted values, as shown in Figures 16.9 and 16.10. Only the residual map tells us whether the predicted map is of much or little value.

We have discussed only the simplest case of regression analysis. Cartographers must often map the results of multiple linear regressions (that is, regressions that include more than one independent variable) or curvilinear regressions (regressions that involve equations of curved instead of straight lines). Occasionally, ordinal information is incorporated into regression analyses. Although the computations become more involved, the mapping requirements remain the same: to map the predicted, observed, and residual values, and to indicate their reliability.

CORRELATION ANALYSIS

One important indicator of the strength of the linear relationship between two sets of attribute

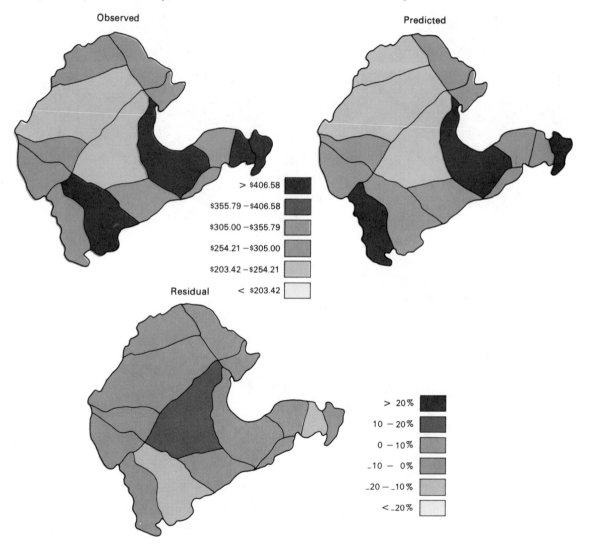

Figure 16.9   Maps of the fictitious island showing observed per capita educational expenditures, predicted per capita educational expenditure based on per capita income, and residuals from the regression shown in Figure 16.6.

values, *y* and *x*, is the **coefficient of correlation**, *r*, between *y* and *x*. The value of *r* can vary between -1 and +1, where *r* = 1 indicates that an increase in *x* is associated with a corresponding increase in *y*, *r* = -1 indicates that an increase in *x* is associated with a corresponding decrease in *y*, and *r* = 0 indicates the absence of a predictable relationship (that is, knowledge of the *x* value gives no predictive information about *y*). The formula for the coefficient of correlation is:

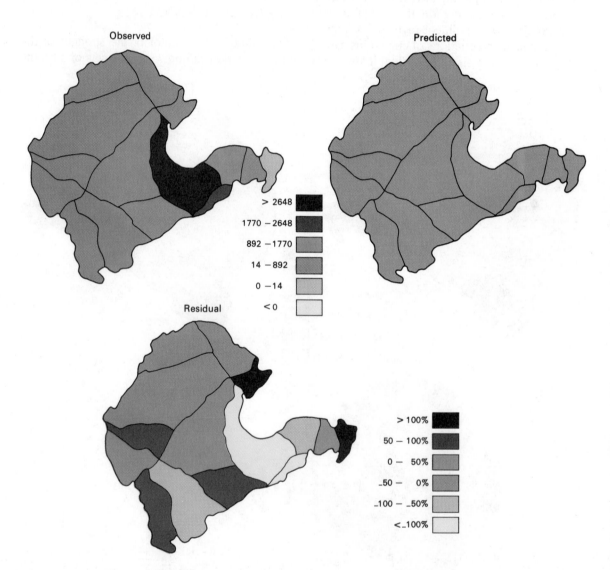

Figure 16.10 Maps of fictitious island showing observed numbers of first-degree graduates, predicted numbers of first-degree graduates based on per capita income, and residuals from the regression shown in Figure 16.7.

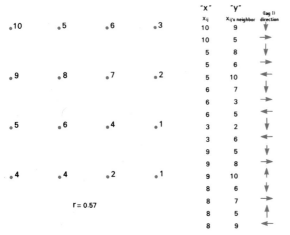

"x"	"y"	
x_{ij}	x_{ij}'s neighbor	(lag l) direction
10	9	↓
10	5	→
5	8	↓
5	6	→
5	10	←
6	7	↓
6	3	→
6	5	←
3	2	↓
3	6	←
9	5	↓
9	8	→
9	10	↑
8	6	↓
8	7	→
8	5	↑
8	9	←

r = 0.57

Figure 16.11   The calculation of an autocorrelation coefficient. The value for *r*, the lag-one autocorrelation coefficient for the data array in this illustration, is obtained by defining *x* and *y* as shown in the right-hand columns. The equation for *r* is given in the text.

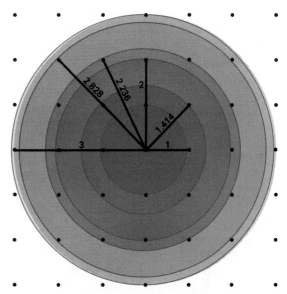

Figure 16.12   For the matrix of data values indicated by points in this illustration, various lags can be defined as illustrated.

$$r = \frac{\sum_{i=1}^{n}\left(x_i - \bar{x}\right)\left(y_i - \bar{y}\right)}{\sqrt{\sum_{i=1}^{n}\left(x_i - \bar{x}\right)^2 \cdot \sum_{i=1}^{n}\left(y_i - \bar{y}\right)^2}}$$

The value *r* itself is a summary measure relating to an entire set of paired observations. Therefore, the cartographer cannot map an *r* = 0.5, for example. It is possible that a separate value of *r* might be calculated for sets of attribute values (for example, number of calls to the automobile club for aid in starting the car, and temperature) for a given location. If similar values for *r* were computed at a series of locations, this set of r values could be mapped.

Perhaps of more importance to the cartographer is the concept of correlation when applied to neighboring values in one data set. This is called **spatial autocorrelation**. A series of coefficients

of autocorrelation can be computed for a single spatially distributed set of feature attribute values.

The position of attribute values in the spatial array is used to determine which observations are

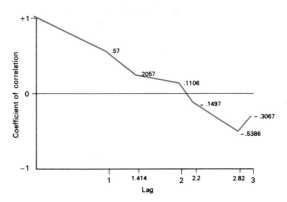

Figure 16.13   The plot of a graph, approximating the autocorrelation function for the data array shown in Figure 16.11 using six different autocorrelation coefficients calculated for increasing lags.

paired to calculate the autocorrelation coefficient (see Figure 16.11). The formula for calculation can be identical to the formula for correlation. Each node value is paired with another node value a certain distance and direction from it (see Figure 16.12). This distance-direction shift is called a **lag** or **lag shift**. When lags are repeated, they are referred to as **first lag** (**lag one**), second lag (lag two), and so on.

The number of possible lags is numerous. We can calculate a value for r for each lag. If lags are systematically scaled in distance and direction, we can obtain a plot of r values against lags (Figure 16.13). This is the **autocorrelation function** for the distribution being analyzed. The importance of this concept for cartographers will become more apparent in the following chapters, when we discuss cartographic generalization and portrayal of surface volumes.

SELECTED REFERENCES

Barber, G. M., *Elementary Statistics for Geographers*, New York and London: The Guilford Press, 1988.

Chang, K., "Measurement Scales in Cartography," *The American Cartographer,* 5 (1978): 57–64.

Clark, W. A. V., and P.L. Hosking, *Statistical Methods for Geographers*, New York: John Wiley and Sons, 1986.

Clarke, K. C., *Analytical and Computer Cartography*, Englewood Cliffs, NJ: Prentice-Hall, 1990.

Court, A., "The Inter-Neighbor Interval," *Yearbook of the Association of Pacific Coast Geographers,* 28 (1966): 180-82.

Davis, J .C., and M. J. McCullagh (eds.), *Display and Analysis of Spatial Data*, NATO Advanced Study Institute, New York: John Wiley & Sons, 1975.

Gregory, S., *Statistical Methods and the Geographer*, 4th ed., London and New York: Longman, 1978.

Lewis, P., *Maps and Statistics*, New York: John Wiley & Sons, 1977.

Shaw, G., and D. Wheeler, *Statistical Techniques in Geographical Analysis*, Chichester and New York: John Wiley & Sons, Ltd., 1985.

Taylor, P. J., *Quantitative Methods in Geography: An Introduction to Spatial Analysis*, Boston: Houghton Mifflin Co., 1977.

Thomas, R. W., and R. J. Huggett, *Modelling in Geography: A Mathematical Approach*, Totowa, NJ: Barnes and Noble, 1980.

Wright, J. K., " 'Crossbreeding' Geographical Quantities," *Geographical Review,* 45 (1955): 52–65.

CHAPTER 17

GEOGRAPHIC INFORMATION SYSTEMS

*G*eographic information systems, or GIS, have arrived on the cartographic scene with tremendous impact during the past few years. What is a GIS? Why should cartographers be interested? Will the cartographic world ever be the same again?

To document the interactions of GIS and cartography, it is necessary to describe the profound effect of **digital** technology on cartography. When using **analog** technology, the map served two functions: (1) as a storage medium for spatial data, and (2) as a medium of communication between humans about spatial relationships. Digital technology has separated these functions into (1) a digital database containing spatial data, and (2) the ability to create maps from that digital data through an automated system. One such system is called a GIS.

Using analog technology, cartographers collected and stored the data and made all the map products. Using digital technology, cartographers still collect and store spatial data and make map products, but an entirely new class of user-cartographers can now make their own spatial communications using cartographers' databases and digital technology. The suite of tools that allows this to occur is GIS.

Many definitions for GIS have been written, and universal agreement on a definition is still evolving. We will use a rather broad definition from the U.S. Federal Interagency Coordinating Committee (1988). A GIS is "a system of computer hardware, software, and procedures designed to support the capture, management, manipulation, analysis, modularity and display of spatially referenced data for solving complex planning and management problems."

GIS is proving to be an empowering set of tools that allows everyone to communicate spatial relationships more easily than ever before. In this chapter, we will explore GIS and all that these new tools offer.

COMPONENTS OF A GIS

There are four factors that are important in a GIS: (1) computer hardware, (2) computer software, (3) data, and (4) people. It is important to consider each of these factors at the outset and to chart their current trends. Although these trends may change in the future, they have been rather stable lately.

HARDWARE

Hardware refers to the computer components that form the physical framework on which the GIS runs and on which manipulations and analyses are performed. Devices for storing data, displaying analyses, and creating output also fall in the hardware category.

When computer-based GIS first became available, hardware was expensive relative to the other factors. In the last few years, however, hardware cost has declined. Today, hardware is a minor portion of the total cost of a mid-level GIS.

A GIS can be operated in a stand-alone environment or in a **distributed processing** environment consisting of a series of personal computers (PCs) or workstations connected by a network. In the past, a GIS was often run on a mainframe or in a minicomputer environment, but the number of these configurations is declining.

In either the stand-alone environment or on a network, the GIS consists of a processing device and an interactive display and editing device. At the low end of the spectrum, the GIS may simply consist of a PC with its associated monitor and keyboard. In the workstation environment, a large digitizing table (see below) and two large monitors may be used. (See Figure 11.4).

The computer processor should have sufficient internal memory and a rather rapid response and instruction execution time. The number of instructions and amount of data are often very large in a GIS application.

The interactive display and editing device is usually a graphics terminal with an alphanumeric keyboard. (See Figure 17.1). Size and screen resolution are important factors to consider when selecting the interactive display and editing device.

In addition to the processor and monitor(s), a complete GIS usually requires some peripheral units. These units may include an input or data-capture device. The input device can be a digitizer or a scanner. (See Chapter 11). Digitizers capture data in two or three dimensions. The two-dimensional device is called a digitizing table or tablet. (See Figures 11.2 and 11.3). It can have a free-standing cursor or a fixed cursor. Digitizing tables first became readily available around 1970. Today, they are available in many sizes, resolutions, repeatabilities, and accuracies. Three-dimensional devices are commonly used for stereocompilation of data from images. (See Chapter 12). These devices are considerably more expensive than the low-end digitizing tablet and require more skill on the part of the operator.

Scanners are now widely available in inexpensive, hand-held versions and are often used by stores to process sales transactions. These hand-held scanners aren't very practical for working

Figure 17.1 A low-end hardware configuration capable of using a small GIS software package.

with map sheets, however. Instead, most cartographic laboratories use page-sized, desk-top scanners. These scanners usually have spatial resolutions from 300 to 600 dpi (dots per inch) and can distinguish up to 256 gray tones. A serious cartographic workplace may also invest in a large flatbed or drum scanner with resolutions measured in thousandths of an inch. (See Chapter 11). As the resolution and accuracy of such scanners increase, cost escalates.

Peripheral hardware may also include an **external data storage device**. Data can be stored in a number of ways: (1) on disk, either hard or floppy, (2) on CD-ROMs or laser disks, (3) on optical disks, or (4) on tape. Which device to use depends on your required storage capacity, access speed, transfer rate, reliability, and portability. For massive digital databases, a jukebox device based on a redundant array of inexpensive disk (RAID) technology is becoming popular. (See Figure 17.2).

A GIS may also include a hardcopy **output device**. Output devices for hardcopy products are of two principal types: printers and plotters. (See Chapter 20). **Printers** can be dot-matrix, laser, xerographic, or light (optical). **Plotters** can be analog or incremental. Plotters can consist of a flat plotting surface or a drum. They can have a pen head, a photohead, or use a scribing tool. The choice is usually based on cost and required resolution of output.

All these hardware options are readily available, and their use has been proven for GIS. A minimum system with no external storage, data-capture, or hardcopy output capabilities can cost under $2,000. A large stand-alone system with all peripherals can cost in excess of $2,000,000. These are 1993 prices. The trend in hardware cost continues to be downward, however.

SOFTWARE

GIS **software** is also readily available in today's marketplace. Several comprehensive software systems, fully supported by their creators and

Figure 17.2 Massive amounts of digital data can be stored and quickly accessed on a high availability disk array (HADA) storage subsystem based on a redundant array of inexpensive disks (RAID), similar to the one pictured here.

decreasing in price relative to the number of functions, speed, and ease of use, are in widespread use in cartographic workplaces. In addition, there are several public domain software systems, as well as systems designed by and for university training. These software packages are inexpensive and operate on small systems. Nonetheless, they provide a large percentage of the functions we expect in a GIS. (See Box 17.A).

Software costs are a constant but increasingly small share of overall GIS cost. A low-end GIS software package contains mainly manipulation and analysis tools. These tools enable users to overlay spatial data layers and perform limited analytical functions. Such software may also allow users to transfer the final edited screen version of the analysis to some form of hardcopy output device. More comprehensive GISs, in addition to providing more functionality for manipulation and

analysis, will offer options for data display and product output. Systems may also include data-capture and data-entry operations.

A fully comprehensive GIS will include software functions to import, capture, and manage data in a database, manipulate and analyze the data, and display and output a set of desirable cartographic products. (See Box 17.B). Examples of specific functions supported by a comprehensive GIS software package will be discussed in the following sections.

DATA

GIS users are awakening to two important facts. First, output from the system is only as valid as data input to the system. Second, the GIS tool itself accounts for only a fraction of total GIS cost. The most expensive part of a GIS is the collection and processing of the data. These facts underscore the vital importance of creating or gaining access to reliable and accurately collected and maintained spatial data.

Every detail that our eyes, ears, and nose can sense is a candidate for conversion to a digital record useful as input to a GIS. The universe of potential data is infinite. Even the simplest of problems, when spatial context is included, involves staggering numbers of digits representing the data.

Prior to the availability of comprehensive GIS, sample data sets were sufficient for two important reasons. First, they were the only data available. Second, though not often admitted, they approached the maximum amount of information that a user could process manually. With the aid of a computer-based GIS, decision makers are asking for much more detailed data to input into an analysis. This not only requires the ability to collect, store, and analyze this larger data set. It also requires that we tag or index the data set so that the system can retrieve it efficiently.

Since the costs associated with data capture, indexing, and storage are enormous, it is mandatory that the data be multipurpose in nature. The data must be usable by many unrelated independent

Box 17.A
Low-End, Inexpensive GIS Software

GISs come in all ranges of functionality and cost. This box introduces two inexpensive "low-end" systems that operate on PCs. The following material is taken primarily from the developer's brochures with little or no editorializing by the authors of this book.

Considering their costs, these two systems offer the cartographer excellent introductions to GIS. Both systems can serve the individual user or small office quite satisfactorily. There are other available inexpensive GISs, perhaps equally useful. When purchasing a GIS, you would be well advised to compare several systems to get one that best matches your requirements.

IDRISI

Under the direction of Dr. R. Eastman, the Graduate School of Geography at Clark University began disseminating IDRISI in 1987. IDRISI is designed to stimulate the education and research uses of GIS and remote sensing. IDRISI is DOS oriented and requires an AT or PS/2 compatible computer with 512K of free RAM, a hard disk, and an EGA, VGA, or 8514/A graphics adaptor. A math co-processor is highly recommended but not required.

IDRISI has a modular-based program design that is easy to learn and use. It consists of over 100 modules or functions, categorized as image processing, geographic analysis, statistical analysis, file exportation/importation, exploration and conversion, and a number of core modules helping the user in project management, display, data entry, attribute data management, and spatial data management. IDRISI accommodates real, integer, byte, and run-length encoded data formats and images up to 32,000 rows by 32,000 columns. It supports full

geo-referencing, and full vector-to-raster conversion. IDRISI includes TOSCA, an interactive digitizing vector editor, and supports any digitizer capable of ASCII output. Import/export utilities exist for Arc/Info, ERDAS, MAP, DLG, DEM, TIFF, and many other formats.

This amount of functionality in a PC grid based GIS for considerably under $1,000 must be considered a "best buy." IDRISI, Version 4.0, is fully supported and maintained by the Graduate School of Geography at Clark University and is available by contacting the School at 950 Main Street, Worcester, MA 01610-1477.

GRASS

The U.S. Army Construction Engineering Research Laboratories (USACERL) developed the Geographic Resources Analysis Support System (GRASS) to provide computer-based management tools to Army environmental planners and land managers. GRASS has many capabilities, including the handling of different data representations, such as raster (or grid cell) data, vector data, point data, and imagery (satellite or aerial photographic) data. Users can also link to a database management system for additional data storage or definition.

A necessary hardware configuration to run GRASS would include a display device capable of 256 simultaneous colors, a processor running UNIX or a similar operating system, at least 8 megabytes of system memory, at least 140 megabytes of disk space (300 to 600 megabytes are recommended), a line printer, graphics library, a ¼ inch or ½ inch tape drive, and a mouse pointing device. Other options include a digitizer for map input, any of several color printers for hardcopy output or pen plotters, *continued on next page*

Box 17.A *(continued)*

and modems or network connections to communicate with other machines.

GRASS is run through the use of standardized command line input, and can be run under the X Window System. There is an internal language that allows users and programmers to create application and demonstration models and to link GRASS with other software packages. Users can input new data through digitization or the use of a scanner, with a screen pointing device, from a floppy disk, or from computer tapes. New data can also be created by selecting data elements from existing files for analysis. Outputs include statistical tables, text files, and maps that can be displayed on a color monitor or printed on several types of hardcopy printers or plotters.

USACERL completed version 4.0 of GRASS in 1991 and distributes it with source code; reference, tutorial, and programmer documenta-

tion; and an extensive sample data set. This public domain system is available from the GRASS Information Center, USACERL, ATTN: CECER-ECA, P.O. Box 9005, Champaign, IL 61826-9005.

Current GRASS workstations include Sun, Intergraph, MacIntosh II, CDC 4000 machines, PC-386s and 486s, DEC, Tektronix 88K, Silicon Graphic's IRIS, Data General, IBM RISC and PS/2, Hewlett Packard, and AT&T 3B2.

GRASS allows Army environmental planners and land managers to analyze, store, update, model, and display landscape data quickly and easily. Data files can be developed for large or small geographic regions at any scale desired within the limits of the original source documents and the hardware's storage capacity. Users can perform analysis and display operations for an entire geographic region or for any user-defined area within this region.

users if the costs are to be justified. Prior to computer-based GIS, data were often collected and processed for a single use. Today the costs of collecting a data set, indexing and storing it, managing it within a computer environment, and using it in a GIS require that data be collected once, correctly, and made available to a wide array of potential users.

The data problem requires that standards be set and met by the data-providing community. Only then can we maximize the benefits of multiple-use data in GIS. Metadata (that is, data about the data contained in a digital file) are required (see Chapter 15). At the same time, it is increasingly important that this metadata information be standardized. Metadata should be accompanied by data quality and data lineage information. In fact, in

terms of actual data storage, more information in a digital file may consist of metadata and data quality reports than data describing geographical features themselves.

Data exchange standards are also needed (see Chapter 13). Unrelated parties must be able to exchange data. It shouldn't matter which computer manufacturer, operating system, or GIS you are using. You should be able to transfer the data quickly and without loss between these systems.

Finally, the importance of data quality must be emphasized. When using a GIS, you must realize that the manipulations, analyses, and output all contain a fixed and measurable amount of potential error. This problem is especially relevant during output, because a GIS can impart an appearance of quality to any data set. Upon output, "bad"

Box 17.B
High-End, Full-Functionality GIS Software

If Box 17.A represents the minimalist stand-alone workstation GIS, the two systems reported in this box represent large comprehensive GISs. These two systems can operate in a networked environment, on a large workstation, or even on a mainframe. These systems, costing thousands of dollars, provide the cartographer with a wide array of tools. An operator of one of these systems can become a highly trained specialist within an organization. The systems are complex and require training and daily use in order to be most productive. The systems are well supported and maintained, and each has a large satisfied international user support group.

ARC/INFO

The Environmental Systems Research Institute, Inc. (ESRI) has created ARC/INFO as its flagship software product. ARC/INFO is a high-end GIS that consists of a complete geo-processing toolbox for the automation, modification, management, analysis, and display of geographic information. The coupling of ARC/INFO software with ArcData databases supports an unlimited variety of applications.

ARC/INFO is the result of over 20 years' development at ESRI. ESRI itself is a software house totally devoted to GIS. The system, ARC/INFO, runs on all four major classes of computers (PCs, engineering workstations, minicomputers, and mainframes). It operates on over 20 different hardware platforms. ARC/INFO is hardware independent, and ESRI prides itself on offering a full suite of software services to meet the specific needs of individual clients. New versions are issued at least once each year, and the annual ESRI users' conference in Palm Springs, California, attracts GIS users from around the globe.

The system itself integrates raster, vector, image, CAD, tabular, surface, and video data. It operates, with PC/ARC, on PCs, but most useful is the full-blown GIS ARC/INFO. The software has full analytic capabilities, including the general classes of geo-processing, contiguity, coincidence, connectivity, surface modeling, logical expression, raster modeling, image processing, coordinate management, and query tools. It also offers a comprehensive cartographic tool kit, including type set quality typefaces. GUI tools are included on a GIS dashboard that allows trained users easy access to operating the GIS.

ARC/INFO is available from the Environmental Systems Research Institute, Inc., 380 New York Street, Redlands, CA 92373. Training, a necessity for the beginning user, is available through ESRI or through a number of other vendors.

MGE—Modular GIS Environment

Intergraph's MGE GIS allows users to define GIS to meet their specific needs. MGE is a set of modular tools capable of handling the entire span of discipline workflows in a common environment. It runs on all models of Intergraph's high-performance RISC-based UNIX workstations and servers. These workstations support all major industry standards for operating systems, networking protocols, and software.

continued on next page

Box 17.B (continued)

Whether a system is used for data collection, management, analysis, visualization, or quality output, MGE offers seamless access to a full suite of solutions. Separate modules are available for specialized applications. MGE modules include the following:

MGE/SX Version 2.0 is the foundation which offers tools for setting up, controlling, and manipulating data. It also serves as an integration point for all of the GIS and mapping applications (for example, efficient 3D data capture, edge matching, feature validation for geometry and attribution, extensive status monitoring, and improved forms interfaces that provide lists of possible inputs to streamline workflows).

MGE Analyst (MGA) Version 2.0 module permits the creation, query, analysis, and display of topologically structured geographic data.

MGE Network Analyst (MGNA) Version 1.0 module allows network creation and analysis.

MGE Grid Analyst (MGGA) Version 1.0 does fast grid cell data analysis through overlay and proximity analyses and zone, cost surface, and optimal path generation.

MGE Projection Manager (MSPM) Version 2.0 integrates data including transformation and projection conversion.

MGE ETI Version 2.0 provides for the input of field survey data.

MGE Map Finisher (MGFN) Version 1.0 permits efficient and easy thematic map compilation.

MGE Map Publisher (MAPPUB) Version 2.0 allows the production of digital film separates for cartographic publishing.

MGE Terrain Modeler (MSM) Version 1.0 creates a digital terrain modeling product for creating and managing elevation models.

MGE Imager (MSI) Version 1.0 is an easy-to-use image processing module for vector and raster data integration.

MGE Dynamic Analyst (Dynamo) brings sophisticated dynamic analysis and presentation of geographic data to the MGE environment. Dynamo extends standard MGE analysis functions by using the latest technology in object data structures and programming techniques.

Although MGE was designed to run on Intergraph hardware systems, it has been ported to other systems. For information, write Intergraph Corporation, Huntsville, AL 35894-001.

---

data are indistinguishable from "good" data. It is crucial, therefore, that a statement of data quality accompany any file of digital data. This statement should give the reliability, precision, accuracy, completeness, and consistency of the data in the digital file.

PEOPLE OR "LIVEWARE"

To operate a GIS, we need more than hardware, software, and data. We must also have **liveware**—that is, the people or **operators** necessary to run the system. Liveware is expensive. GIS education is necessary, if somewhat lacking today. Operators need to be computer literate and understand the functions available in a GIS. They also must know what cartographers have traditionally been taught about analysis and display of spatial data. Few people today are adequately trained in all these areas. Nor is there an abundance of quality training programs available.

Technicians can perform some of the tasks required to operate a GIS. But the person who oversees the entire operation and whose responsibility it is to make the decisions must be professionally trained. Such well-trained professionals are relatively rare today, and most are learning on the job.

We are seeing an increasing number of specialized technicians who can handle the computer system (that is, maintain the system and assure that each hardware component is talking to the others). We are creating a corps of trained digitizer operators, data collectors, data editors, and taggers. We are relying on software houses and universities to provide the software functionality used to train cartographers. But we must add cartographic training to create a group of GIS professionals who understand the overall system and can make efficient use of this technology.

STRUCTURE OF A GEOGRAPHIC INFORMATION SYSTEM

What, then, constitutes the "system" of a GIS? The basic structure requires, at a minimum, a computer and an operator, along with access to software and a database. Ideally, we can think of the four components of a GIS—hardware, software, data, and liveware—as being independent. But all four must be present to some extent to have an operational GIS.

At the simplest level, a person has a PC with a low-end GIS software package. The operator may manually input data and create the output on the computer monitor. At the other extreme, we have a networked set of cartographic workstations online to a large spatial database and are able to access several comprehensive GIS software packages like those described in Box 17.B.

THE PROCESS FLOW OF A GIS

The flow of digital spatial data in a GIS is directed by an operator (person) from an external or internal data storage device to a hardware processor. (See Figure 17.3). The software is resident or addressable by the processor, and the operator selects the functions that need to be performed. This selection is often accomplished through a **graphic user interface** (**GUI**). With a GUI, you use a menu to plan and select the sequence of functions.

When you work with complex GIS analyses, you will find **macros** useful. Whenever you use a series of functions frequently, you can convert that series to a macro. Then, merely by selecting the macro name, you cause that entire series of functions to be performed.

After you have performed the data retrieval and analytical functions, you may interactively design a composition of the output desired. You can then send this output to a peripheral hardcopy output device. If you need a new data layer entry during the analysis, you may use a peripheral data-entry device to capture the needed information and to structure it into the database.

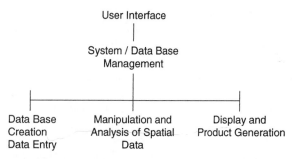

Figure 17.3 A diagram of the process flows through a generic GIS.

STAGES OF MATURITY AND COMPLEXITY

A GIS can be categorized by three levels of maturity and complexity. These levels characterize the stages through which an organization might evolve while implementing a GIS. Because of their unique missions, some organizations may not progress beyond the first or second stage. Most organizations, however, will eventually move from level 1, through level 2, to the most sophisticated system, level 3. The old saying, "You must learn to crawl before you walk," applies to GIS evolution in an organization.

Stage 1

The first level is one of inventory. Often the intended result is to create a consistent, complete data layer. Only simple queries, mostly related to location, are used. Analysis is limited and relates primarily to location or to database structuring.

One example of a GIS in Stage 1 is the 1:100,000-scale digital database of U.S. highways and roads created by the USGS (with input by the U.S. Bureau of the Census) for the 1990 Census enumeration maps. The purpose of this project was to create a positionally correct digital record of all U.S. road and highway centerlines. This data layer is useful to answer queries about the connectivity of the U.S. highway network, to retrieve place-name information, and to prepare census maps that facilitate data collection in the field. Sophisticated analyses were not done. Yet if collection was performed properly, this data layer could be considered multiuse and could be included with other data layers in more sophisticated analyses.

A second example of a GIS in Stage 1 is the capture and archiving of digital records from the Landsat satellite system. The digital data are received, and a number of rectification and enhancement algorithms are performed on the data. The data are then entered into the Landsat archive for multipurpose use. Once again we are dealing with data from one source which constitutes an inventory. The data are processed to allow for simple queries that include the attribute of location.

Stage 2

The second level of maturity adds more complex analytical operations to those performed in Stage 1. In a level-2 GIS, a pattern is sought within a given data layer, using statistical and spatial analytical tools. Alternatively, several layers of data are analyzed to determine relationships. Rather complex queries are possible—for instance, "If a region's soil type is A, and its landform type is B, then what is its erosion potential?" Siting buildings or locating parks are examples of outputs from this level of GIS complexity. The level depends upon multiple layers of data analyzed together to assess locations or patterns having similar sets of attributes.

For example, suppose that your town needs to enhance its water supply. Provided the needed data layers are available, you could query a level-2 system in the following way:

> We would like to locate all land that has a surficial material of class X, is not currently zoned commercial, is within 50 meters of an existing highway, is not within 300 meters of a known pollution source or waste disposal site, and which is not in a protected wilderness area. Further, we would like the land to be available for purchase and within 10 miles of the current service network of the town's water supply.

Figure 17.4 illustrates some of the stages in such a location siting.

Stage 3

The third, most advanced level of GIS use evolves beyond a transaction processing system and beyond a siting capability to become a true decision support system. Models are used with existing data sets to ask "what if?" questions.

More sophisticated modeling techniques allow for routing decisions and for real-time examination of traffic flows and other patterns of activity. They also allow us to compare real-world patterns with the results of modeling. For example, the ability to predict the timing, magnitude, and direction of maximum fallout from a volcanic eruption would aid in evacuating people from potentially affected areas. In the same way, we could predict

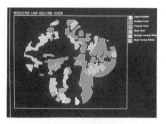

Satisfactory
landuse / landcover

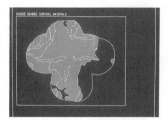

Satisfactory
coarse grained
surficial materials

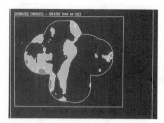

Satisfactory saturated soil
thickness > 12 meters

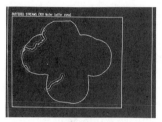

Unsatisfactory
polluted stream
with 100 meter buffer

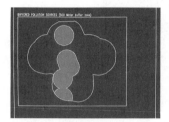

Unsatisfactory
known point
pollution sources
with 500 meter buffer

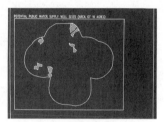

Areas for potential
public water supply
well sites

Question: Where is land suitable for a public water supply well site?

Must currently have satisfactory landuse / landcover
Must be composed of coarse grained surficial materials
Must have saturated soil thickness > 12 meters depth
Must not be within 100 meters of a known polluted stream
Must not be within 500 meters of a known point pollution source

Result: Areas left within source area that would be suitable for
a public water supply well site.

Figure 17.4   A series of data layers and intermediate overlay solutions to a well-siting problem in northern Connecticut.

the spatio-temporal behavior of approaching hurricanes, forest fires, and floods using a level-3 GIS. (See Figure 17.5). This third-stage GIS serves as an indispensable tool in helping humans interact with their environment.

You can expect a level-3 GIS to be larger, more costly, and with more capabilities than a level-2 or level-1 system. There are some exceptions to this rule, however.

Clearly, not all GISs need to evolve to the maturity of a level-3 system. There are legitimate uses for level-1 and level-2 systems. However, any level-1 system that wishes to survive must be aware of the data needs of level-2 and level-3 systems. It is foolhardy to prepare a data layer using a level-1 system that will not be capable of serving a level-2 or level-3 system. This is the essence of the need to create multipurpose data sets.

Figure 17.5 A series of predicted intermediate stages from a model of forest fire spread in Montana. (Courtesy of Dr. K. Clarke).

GIS FUNCTIONALITY

The number of possible functions embedded in a sophisticated GIS is large. Some GISs have over 1,000 functions. Organizing the desired functions is itself difficult. Following Figure 17.3, we have organized a sample of desirable functions into the following five categories: (1) the GUI, (2) system and database management, (3) data entry, edit, and validation, (4) manipulation and analysis, and (5) display and product creation. Let's look at each of these functions, beginning with the GUI.

GRAPHIC USER INTERFACE (GUI) FUNCTIONS

A graphic user interface allows the GIS operator to interact with the specific GIS. It allows commands to be given and serves as intermediary between the person and the system of hardware, software, data, and peripherals. A GUI usually contains menus of functions, menus for answering questions, and help menus. System and function manuals should also be available at the GUI.

A menu of functions consists of a list of options, organized either hierarchically (like a dictionary) or in icon form. In either case, the operator's task is to select a sequence of functions. In more advanced GISs, the GUI may contain several macros, which allow a sequence of functions to be performed by a single input stroke.

Between the GUI and the hardware is an operating system. The operating system for the hardware may perform memory assignment, allow system access, and perform accounting functions. The operating system also maintains command control, directing what happens next in the system. Also present are language compilers, device drivers, disk backup capabilities, and subroutine libraries. In addition, you will sometimes find special communication software (such as local area network support), remote communications using modems or other devices, device emulators, and protocol converters necessary for access to other installations using different operating systems.

SYSTEM AND DATABASE MANAGEMENT FUNCTIONS

The functions usually contained in the system and database management portion of a GIS include routines for: (1) format transformation, (2) geometric transformation, (3) data conflation and merge functions, (4) transaction tracking, and (5) other utility routines. Let's examine these routines.

Format Transformations

A GIS is designed to use data of a given format. Data having a different format must be transformed before being imported into the GIS database. Sometimes this transformation is accomplished through an intermediate format such as the SDTS (see Chapter 13). During this processing, it may be necessary to strip the incoming data of its topology and to recreate that topology. Or, if the incoming data lack topology, it may be advantageous to create new topology for the data file.

You may also wish to transform the data in other ways. If, for example, data in the GIS database are stored by tiles, then any incoming data must be restructured to conform to the existing tiles (see Chapter 15). It may also be necessary to construct a data dictionary for efficient use of incoming data.

Geometric Transformations

The incoming data may be in coordinates from a given map projection or coordinate system. This will certainly be the case if the incoming data are being digitized from a printed map. The data in the database must be held in some locational coordinate system, and you will need to transform the incoming data so that they are compatible with the other data in the database. Similar transformations are required if the incoming data are in arbitrary rectangular coordinates such as the UTM or SPC systems (see Chapter 6). Many databases accept the geographical coordinate (latitude and longitude) system as the preferable system in which to store data.

Sometimes when we bring data into a database, only a few of the features are referenced to latitude and longitude. In extreme cases, no feature is referenced to an earth position; the features are only referenced to one another in a relative sense. When this happens, we can perform a function called **rubber-sheeting** to transform the positions of the data into the desired locational framework. (See Figure 17.6).

To do so, we select a few well-defined features. We add their latitude and longitude position as an attribute to their record. Rubber-sheeting then allows us to use a mathematical procedure to obtain an approximated latitude and longitude for the rest of the features.

Data Conflation and Merge

Often a data layer will exist in the database, and a second version of that data layer will become available from another source. After you import the second file to the database, you may discover that there are differences between the two files.

Rubber-Sheeting

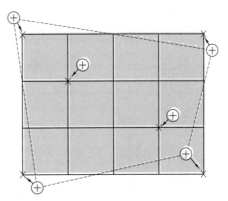

Figure 17.6 Rubber-sheeting force-fits the positions of data points into a desired locational framework.

(The time of the capture of the files may be different, or different constraints in the data-capture function may have been used). Such conflicting data are said to be **conflated**. You will need to reconcile the differences between the two sets of data. (See Figure 17.7). You may be able to do so by interacting directly with the two conflated data layers. Or you may deconflate the data by choosing between the two files, based on available metadata.

Incoming files must also be **merged** with existing data. Data in a database may be held in tiles. As described in Chapter 15, the tile structure may vary between databases. Regardless of the tile structure used in a database, the edges of an incoming file must be joined with existing data files. (See Figure 17.8). Sometimes this **edge-matching** may create almost insolvable problems, and the operator may have to make arbitrary decisions in order to merge the two data files. The time spent correctly edge-matching files usually proves to be cost effective.

Transaction Tracking

Often the operator of a GIS will want a list of all accesses and changes that have been made to a database. Such a file is referred to as a **transaction file**. The file lists every entry into the database,

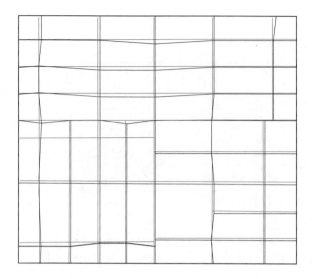

TIGER ——— ETAK ———

Figure 17.7 An example of conflated data from two different sources for highway centerline data. Colored lines are from an older U.S. Bureau of the Census file, and black lines are from a more recent file.

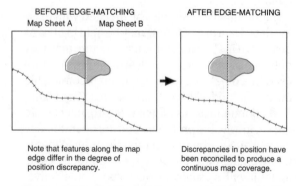

BEFORE EDGE-MATCHING
Map Sheet A   Map Sheet B

AFTER EDGE-MATCHING

Note that features along the map edge differ in the degree of position discrepancy.

Discrepancies in position have been reconciled to produce a continuous map coverage.

Figure 17.8 On the left is an example of a problem arising from merging data digitized from adjacent tiles or map sheets. An edge-matching function is applied to these data files to create the figure on the right.

regardless of whether it is for data retrieval, data input, data change or update, or merely data validation. Such transaction files are important to document when updates to files were done, to charge customers for using data in the database, and to record unauthorized uses or attempts to access the files.

Utility Functions

Additional **utility functions** may be included in the system and database management segment of a GIS. These include **vector-to-raster** and **raster-to-vector** conversion routines. All spatial data are originally recorded digitally in either raster or vector files. Most analysis routines operate on one or the other of these files. Therefore, it is often necessary to convert raster data to vector data, or vice versa, in order to use a particular analysis routine.

Another routine that may be contained in this segment of a GIS is the **scaling** function. Data available at different levels of measurement from one region to another may need to be reduced to a common level for analysis and mapping. Similarly, data available from maps of different scales may have to be generalized to a common scale before being used in a GIS application.

DATA ENTRY, EDIT, AND VALIDATION FUNCTIONS

When the necessary data are not available in the database, you may need to convert a data set into a form that can be ingested by the GIS. It is usually advisable to take time to do the conversion carefully so that the resulting files can be incorporated permanently into the database. Sometimes, however, you will only need to enhance data already in the database with new data about recently developed features or changes of special importance to you.

Several functions may be used to prepare data for use in a GIS. Let's begin by looking at the digitizing function.

Digitizing

Sometimes a data layer must be digitized to be used in GIS analysis. There are several ways to digitize data, depending on the data's form (see Chapter 11). If the information is in photographic or image form, it will probably be digitized as raster data. If the information is in line graphic form (consisting of point, line, and area symbols), it will probably be digitized as vector data. The vector-to-raster and raster-to-vector functions can be used to convert the data layer once it has been digitized. On rare occasions, the operator may draw the input on the screen and then digitize the screen display. Such interactive digitizing is relatively rare when precise data are desired.

Tagging

Once we have digitized data location attributes for features included in the file, we usually need to add other attributes about the features. We also need to specify the metadata that will accompany the data file. This process of adding intelligence to the data file is called **tagging**.

For example, we might encode feature names or labels as an attribute. Or we might specify feature codes that correspond to information in a data dictionary already held in the database. We could add time reference data in the form of an attribute. We also need to identify the source and lineage of source material.

Edit

Edit functions include routines that clean up the data in the file. For example, when digitizing is performed manually or rapidly by machine, or when the source is not of high quality, small remnants of the digitizing procedure itself may remain in the data file. Figure 17.9 illustrates the possible consequences. These effects include **spikes**, **over-**

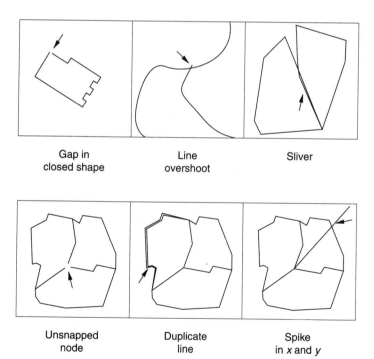

Gap in
closed shape

Line
overshoot

Sliver

Unsnapped
node

Duplicate
line

Spike
in *x* and *y*

Figure 17.9 Examples of problems occurring in digitized data files that result from the digitizing process. Edit functions can be used by an operator to correct these data discrepancies.

shoots, **gaps** (lines that fail to connect), and **slivers** of areas. (These slivers may result when we separately digitize outlines of two adjacent areas that share a common border. The coincident edge may contain slight differences in digital form). The GIS operator can correct these remnants automatically or interactively by using an edit function.

Other edit functions are important as well. When points are connected by straight-line segments, the result may be unwanted sharp angles at the points. It may be desirable to smooth these lines by using arcs or splines between local points (usually three points at a time) to create a smooth flowing line connecting the points. (See Figure 17.10). These arc or spline functions are generalizations in small areas and are extremely useful in characterizing some phenomena. For example, most railroads are smoothly flowing arcs, not sharply angled paths. Edit functions allow the cartographer to input this characteristic into the delineation of a railroad without digitizing its entire course.

Finally, it may be useful to delete or modify a feature's digital record. For example, you might have a string of coordinates representing a road but discover that one of those coordinates is wrong. You can modify this coordinate interactively using the edit function. Likewise, when composing output from the database, you may want simply to delete a given feature or features in order to make the output more legible.

These editing functions can be useful not only in the data-entry segment but also in the manipulation and analysis segment or the display and product creation segment of a GIS. (We will discuss these segments later in the chapter).

Validation

An entirely separate function is the **validation** necessary at data entry. Logical consistency checks may be built into data-entry functions to make sure that blunders do not occur. These logical consistency checks simply apply the logic of the world to the data in the data file. If, for example, the file shows that a road goes under a body of water but no tunnel is encoded at that position, then a validation check will record this inconsistency.

Before

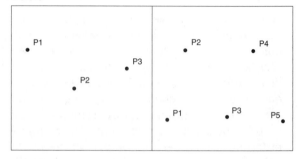

After

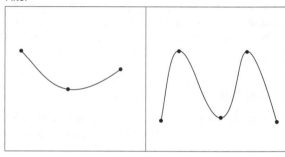

Three point arc   Spline curve

Figure 17.10   Additional edit functions used to create smoothly flowing curves through a series of points.

Topology Creation

The creation of topology in a digitized data set can identify errors or blunders but also can serve as additional information for data analysis. **Topology** defines the geometric relationships among features in a data file. Examples of topological relationships are intersections, adjacencies, insideness or outsideness, betweenness, and so on.

For example, if we have four separately identified non-coincident roads, two running east-west and two running north-south, and four nodes (a special type of point) of intersection, then we must have an area bounded by these four roads. (See Figure 17.11.) Topologically speaking, there are four nodes, four line segments, one inside area, and one or more outside areas. Once this topology is created, network connectivity can be calculated. Attributes can also be attached to nodes, lines, and areas. This information becomes a vital part of the database.

Topological data need to be validated to insure completeness and consistency. For instance, if the centerlines of two segments of highway intersect, you would expect a node to be recorded at that intersection. A validation check would establish a node, if none was recorded. Topological validation checks are routine and very helpful in cleaning a file after initial digitizing.

MANIPULATION AND ANALYSIS FUNCTIONS

This category and the next contain the functions which are the most visually apparent. These are also the areas in which most new GIS developments are occurring. The routines included in these two segments are the digital equivalent of the analog procedures that cartographers have used for 50 years.

Imagine yourself the GIS operator-cartographer sitting at a workstation compiling a map on your monitor. What tools would you like to have at your disposal? Most of these tools exist today in one GIS software package or another. The following description of functions will only be representative, not exhaustive.

First of all, you need a retrieval function so that you can obtain digital files from the database. At times, you may wish to retrieve an entire data layer. At other times, you may want to retrieve only those features or parts of a data layer that satisfy a given set of conditions. Immediately after retrieval, you may wish to modify, update, or change the data in some way. Software to do so is available. Next, you may wish to change the map projection or to rubber-sheet the underlying geometry of the area to achieve a desired effect.

You may want to make choroplethic maps and other displays of classed data, and the wide variety of classification routines outlined in Chapters 24 and 26 are included in most GISs. You can apply these classification functions to the features themselves or to their attributes or attribute values. You may wish to save the classification values as additional attribute values for the features.

Of course you might want to overlay one data layer on another, or maybe even combine several data layers at once. These routines can be arithmetic overlays (in which you perform such operations as addition, subtraction, multiplication, and division). Or they may be logical overlays, in which you find areas that satisfy a set of criteria. You may wish to use Boolean logic (a special logic of great utility in computer data processing, based on and/or operators and true/false values). Or you may want to use the topology imbedded in the data to form new data combinations. Again, you may wish to modify and save these new data-layer combinations.

In addition, a number of neighborhood functions are useful in analyzing spatial data. These include:

- creating a buffer zone, x meters wide, around a point (or node), line, or area (see Figure 17.12)
- searching for all features which have a given attribute and which are located within a certain radius
- locating a point or line within a given polygon
- calculating surface values such as slope, aspect, or gradient (see Figure 17.13)
- segmenting a surface by Thiessen polygons (see Figure 17.14)
- interpolating between points, lines, or within areas

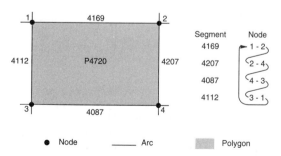

Figure 17.11 Topological structure: four nodes, four lines, and two areas.

- contouring a set of attribute values associated with points
- defining contiguity or adjacency measures
- calculating distances between points by space or time
- defining networks and calculating statistics about the created network
- defining the spread of a phenomenon while considering absolute barriers or permeable barriers
- seeking the direction of flow across a surface defined by a set of attribute values
- calculating the intervisibility at points on a surface (see Figure 17.15)
- defining a window and extracting the data layers of interest only for that window (see Figure 17.16)
- zooming in or out on an area
- panning around an area
- matching edges of data that come from different tiles.

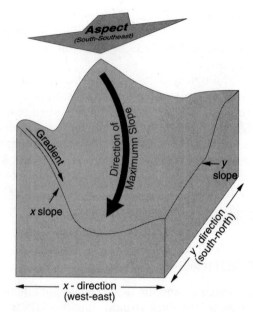

Figure 17.13 Slope, aspect, and gradient are often useful in analyzing spatial data.

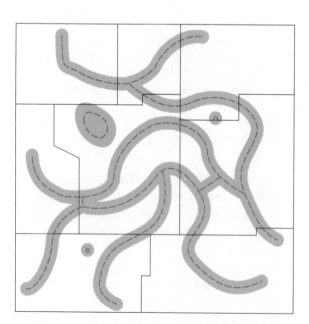

Figure 17.12 Tinted areas delineate buffer zones around existing points, lines, or areas.

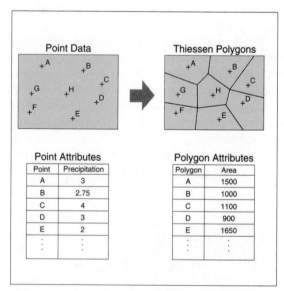

Figure 17.14 Thiessen polygons constructed around a set of points. Each Thiessen polygon encloses all points which are closer to the specified point inside each polygon than they are to any other specified point.

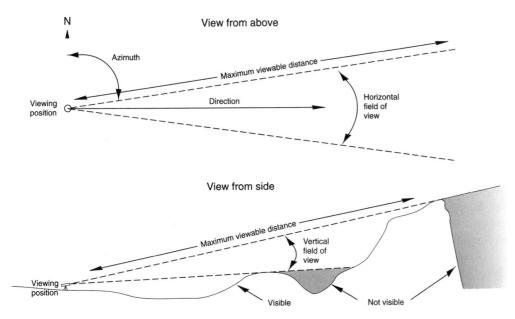

Figure 17.15 Intervisibility on a surface. This concept identifies all locations that are within an unobscured line-of-sight from a specified position.

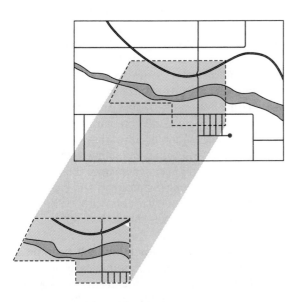

Figure 17.16 Windowing allows the cartographer to specify a user-defined geographic area for further analysis in a GIS.

The list of possible functions seems endless. As noted earlier, there are over 1,000 functions available in some GIS software today.

DISPLAY AND PRODUCT CREATION FUNCTIONS

Many functions could be listed both here and in the previous section. Among them are the capabilities to produce perspective views (see Figure 26.20) of a data layer or a combination of data layers. Quite common today is the desire to drape an image of an area over a digital elevation model of the area, and then to highlight by vector enhancement certain features on the display. Such displays are informative and can be sequenced into a series that approximates a fly-over of the area (see Chapter 29). We can now even produce respectable shaded relief from elevational data using the capabilities of a GIS.

Admittedly, specialized equipment is often used to enhance this process.

Functions are also available to produce the information which you find around the margins of many printed maps. (See Chapter 23). For example, it is easier to place the map's title and legend in a GIS environment than manually, because you have more flexibility to try different positions.

Another useful function is the scale indicator. You can produce a dynamic display of data layers, position a scale bar next to a feature, and measure the feature's distance. You can then save this measure and delete the scale bar from the monitor until you need it again.

Text of all forms can be handled in a GIS. You can store long text explanations in an added file. Or you can reposition or remove names when they obscure other data. Some touch screens allow you merely to touch a point on the screen; when you do so, a text list of attributes associated with that point will appear on the screen.

The number of available text fonts and sizes surpasses the number available in manual cartographic production. This is because in a computer environment some fonts are made scalable. That is, they can be enlarged with no loss of graphic quality. (See Chapter 22). Any point size or fractional point size may be specified for the display of these fonts.

The entire range of graphic symbols traditionally used by cartographers is available. In addition, you can custom design symbols interactively. You can manipulate all the elements of cartographic design (size, shape, value, hue, chroma, orientation, and texture arrangement) to compose an output product.

Finally, we cannot overemphasize the interactive and dynamic display potentials of GIS (see Chapter 29). In the analog cartographic world, it was necessary to specify a product design and then to execute it. Once begun, very little change could be accommodated, and such accommodations were costly. In a GIS, you can change the composition at any time. Also, the chance to interact with information enhances your understanding of it. You can add a data layer and then remove it. You can even flash it on and off while you study the display. The capabilities to zoom in or out on the display, to vary data layers, and to rotate the map in a perspective view add new insights about the map. Users have never experienced such flexibility, and we are only beginning to learn to take full advantage of it.

Dynamic displays or movies of an area shown in perspective are especially poignant visually. The ability to create several perspective views from differing angles and then to replay those views in sequence creates the impression that the display is rotating in front of your eyes on the monitor screen. The next step is to enter the virtual-reality world and to display the earth in still other ways.

THE MEANING OF GIS TO CARTOGRAPHY

GIS is popularizing cartography. These developments are exhilarating as well as stressful. GIS allows everyone to access spatial data and to communicate with these data. This ability may revolutionize our society. We can't predict the magnitude of effects that will occur. Some people are predicting that corporations and organizations will restructure geographically because of GIS. Targeted marketing, delivery truck routing, utility repair, and mortgage lending are already taking advantage of GIS. In car navigation systems, "smart" vehicles (such as buses that navigate by taking advantage of GIS data and electronic positioning technology) and a plethora of programmable moving robots will soon be commonplace. Cartographers should be excited about these possibilities.

The "old" map in new forms will become a principal medium for communication and will provide the basis for an increasing number of societal decisions. Cartographers will have to relinquish some control that they enjoyed when analog technology was used. But we can also view the analog technology period as the "dark ages" of map use. The new empowerment of society that GIS promises using digital technology may prove to be a "renaissance."

In fact, uses of GIS are limited only by our imagination. Increasingly, GIS will create products that meet users' needs better than did the traditional hardcopy printed map or photographic image. Although printed maps and images will be needed in the future, more and more of the work of society will be accomplished with spatial data in a GIS environment. Cartographers are well prepared to lead this revolution in spatial data use made possible by GIS.

SELECTED REFERENCES

Antenucci, J. G., D. Brown, P. L. Croswell, and M. J. Kevany, *Geographic Information Systems: A Guide to the Technology*, New York: Van Nostrand Reinhold, 1991.

Aronoff, S., *Geographic Information Systems: A Management Perspective*, Ottawa: WDL Publications, 1989.

Burrough, P. A., *Principles of Geographic Information Systems for Land Resource Assessment*, Oxford: Clarendon Press, 1986.

Chorley, R. (ed.), *Handling Geographic Information*, Department of the Environment, London: Her Majesty's Stationery Office, 1987.

Clarke, K. C., *Analytical and Computer Cartography*, Englewood Cliffs, NJ: Prentice-Hall, 1990.

Environmental Systems Research Institute, *Understanding GIS: The ARC/INFO Method*, Redlands, CA: ESRI, 1991.

Goodchild, M. F., and K. K. Kemp, (eds.), *Core Curriculum in GIS*, Santa Barbara, CA: National Center for Geographic Information and Analysis, 1990, three volumes.

Guptill, S. C. (ed.), "A Process for Evaluating Geographic Information Systems," Federal Interagency Coordinating Committee on Digital Cartography, U. S. Geological Survey Open File Report 88–105, 1988.

Huxhold, W. E., *An Introduction to Urban Geographic Information Systems*, New York and Oxford: Oxford University Press, 1991.

Maquire, D. J., M. F. Goodchild, and D. W. Rhind, (eds.), *Geographical Information Systems: Principles and Applications*, Harlow, Essex, U. K.: Longman Scientific and Technical, 1991, two volumes.

Onsrud, H. J., and D. W. Cook, *Geographic and Land Information Systems for Practicing Surveyors: A Compendium*, Bethesda, MD: American Congress on Surveying and Mapping, 1990.

Raper, J., *Three Dimensional Applications in Geographic Information Systems*, London: Taylor and Francis, 1989.

Star, J., and J. Estes, *Geographic Information Systems: An Introduction*, Englewood Cliffs, NJ: Prentice-Hall, 1990.

Tomlin, C. D., *Geographic Information Systems and Cartographic Modeling*, Englewood Cliffs, NJ: Prentice-Hall, 1990.

PERCEPTION
AND DESIGN

CARTOGRAPHIC DESIGN

*D*esign is creation. Graphic design produces visual forms. Some maps, such as air photographs and satellite views, are mechanical images. Others are prepared in piecemeal fashion by arranging marks to form a visual representation of selected spatial phenomena. In order to display the data describing the phenomena, we use an almost unlimited variety of graphic signs. By relating graphic characteristics of the marks to attributes of the data, we assign qualitative and quantitative meanings to the signs, and they then become **designated symbols**. By arranging the symbols in the horizontal plane, we endow them with geographical meaning, and the display becomes a map.

In any graphic system of communication, each sign must be clearly distinguishable from every other sign, just as the letters of the alphabet must appear different from one another, so that we do not get mixed up as to the sounds they represent. Furthermore, by the systematic use of graphic similarities and differences among the signs, we can express likenesses and distinctions among the data they symbolize.

If a map presentation is to be effective, the signs must be carefully chosen and fitted together so that they form an integrated whole. The task of map design has much in common with writing. An author—a literary designer—must employ words with due regard for many important structural elements of the written language, such as grammar, syntax, and spelling, in order to produce a first-class written communication. Likewise, the cartographer—a map designer—must pay attention to the principles of graphic communication. Regardless of the positional accuracy or appropriateness of the data, if the map has not been carefully designed it will be a poor map.

In this chapter, we will focus on the **design culture** or environment that sets the context for mapping. We will do this by surveying the principles of graphic communication and map design. We are concerned with the design choices that must be made and the principles that influence these decisions. Three important components of map design—color, pattern, and typography—are unusually complex and will thus be treated separately in the next four chapters.

OBJECTIVES OF MAP DESIGN

Your main objective in designing a map is to evoke in the minds of viewers an environmental image appropriate to the map's purpose. There are a great many graphical ways you could represent geographical data, concepts, and relationships. Your first concern in map design is to assign specific meaning to the distinctive marks you use. Your second concern is to arrange the marks in a total composition that will make the viewer see the result you intend. These two aspects of design are not really separate, of course, since you must select the symbolism with the ultimate objective in mind.

The communication objectives of cartography range along a continuum from general reference maps, such as atlas or topographic maps, to thematic maps, such as maps of population or rainfall. An individual map usually involves some of each aim, but it is worthwhile to contrast the two as they affect map design.

A "pure" **general reference map**, at one end of the continuum, aims to display a variety of geographic information. When designing such a map, you must encode each attribute in a unique way, so that its meaning will not be confused with any other attribute. Your primary aims should be legibility and graphic contrast among symbols. Unless emphasis is specifically desired, no mark should appear more important than another, and no class of marks should dominate. (In reality, such a "pure" general reference map is rare. Usually, some geographic characteristics are thought to be more "important" than others and will consequently be given visual emphasis).

A "pure" **thematic map**, at the other end of the continuum, is concerned with portraying the overall form of a given geographical distribution. It is the structural relationship of each part to the whole that is important. Such a map is a kind of graphic essay dealing with the spatial variations and interrelationships of some geographical distribution. When designing a thematic map, you must be sure that the modulations of marks and symbols you choose work together graphically to evoke the overall form of a distribution. Although spatial images and graphic integration are primary aims, you must also provide an adequate locational reference base.

Clearly, the fundamental design objectives of general reference maps and thematic maps are quite different and lead to distinct graphic treatments. In practice, however, most maps actually combine objectives to some degree. When designing a map, you must recognize how these objectives are combined before proceeding with your design.

FUNCTIONAL DESIGN

Cartography does not qualify as an aesthetic art form like painting, music, fiction, or dance. Unless a map bears strong fidelity to reality, the purpose of mapping will not be served. The functionalism of cartography, together with the limitations imposed by external controls on the mapping process (see later section), put too many constraints on the cartographer to allow "full freedom of expression."

This does not mean that there is no room for creativity in mapping or that there is only one correct cartographic expression. Quite the contrary. There are many graphic media and design possibilities. Skill and artistry have played a significant role in map design throughout the history of cartography.

Taken together, these factors suggest that mapmaking, like architecture, is a mixture of art, science, and technology. The preparation of a map is not a mechanical process like taking a photograph. Instead, it involves purposeful assembling, processing, and generalizing diverse data

and then symbolically displaying them as a meaningful, functional portrayal. That is a highly creative operation, and an important part of it is the development of the graphic design.

Map design is a complex task, since there are almost unlimited options for organizing the visual character of the display. Furthermore, **most design choices are compromises**, since it is normal for "intellectual" and "visual" objectives to be in conflict. If, for example, you wish to develop a strong graphic contrast at a coastline, you would make one side dark and one light. If you wish to indicate greater depth of ocean by deeper shades, then you would make the shallow water light. A shoreline contrast then requires dark land, but that means type and symbols on the land will not be very visible. You can only proceed by making choices after weighing all the pros and cons.

Map design choices involve a combination of intuition and rational choice. Some design combinations will come closer than others to meeting the ultimate aim, which is excellence in communication. Like the study of literary composition, the basic elements of graphic composition lend themselves to systematic analysis, and the principles can be learned.

A basic requirement in graphic design is a willingness to think in visual terms, uninhibited by prejudices resulting from previous experience. The range of imaginative innovation must, of course, be disciplined to some extent. Cartography, like many fields, has developed traditions and conventions. To disregard the powerful forces associated with these traditions and conventions would inconvenience map users, which would in itself be poor design.

SCOPE OF DESIGN

The word "design" is used both as a verb and a noun. To avoid confusion, therefore, we must distinguish between design as process and design as product. Let's look at both these aspects of design in detail.

DESIGN PROCESS

The design process involves a series of operations. In map design, it is convenient to break this sequence into three stages.

In the **first stage**, you draw heavily on imagination and creativity. You think of various graphic possibilities, consider alternate ways you can approach the problem, and visualize different solutions. The term **graphic ideation** is used to describe this intuitive process. During this stage, you decide on the type of map, spatial format (size and shape), basic layout, data to be represented, mapping technique, and so on. The result is a general design plan for your map.

In the **second stage**, you develop a specific graphic plan. You analyze various alternatives and weigh them within the limits of your general plan. You decide on kinds of symbolism, number of classes and class limits, color use, typographical relationships, general line weights, and the like. You make these choices so that the map components will meld into a coherent graphic display. By the time you complete the second stage, you have made all but minor decisions.

During the **third stage**, you prepare detailed specifications for map construction, whether by automated or manual methods. You define all symbols, line weights, screens, colors, lettering sizes, and so on. Preparing these specifications requires a thorough understanding of all the processes involved in map construction.

Digital methods have significantly affected the map design process. When you use computer-based procedures, the design steps are much more integrated, and the design stages are less distinct. Also, when working with a digital system, the final map artwork can be plotted or printed automatically (see Chapters 30 and 31).

The flexibility of software-driven procedures and the ease of making changes on an electronic map display have also made it relatively easy and inexpensive to develop several **design prototypes** before becoming locked into a specific plan. In contrast, when using manual methods it may become too expensive and time-consuming to make a major design change after you get very far into the construction process.

DESIGN RESULT

Due to the complexity of the design environment, any one design challenge will likely have a considerable number of possible solutions. The different solutions will not be equally successful in achieving the mapping goals, however. You must be able to evaluate the design alternatives. Thus, learning how to critique a map's design is an important aspect of becoming a better cartographer.

Our problem is that critical thinking with respect to map design is strongly intuitive and not easy to describe in words. Good design simply "looks" right. It is simple (clear and uncomplicated). Good design is also elegant, and does not look contrived. A map should be aesthetically pleasing, thought-provoking, and communicative.

In evaluating a map, you should focus on the logical as well as the visual aspect of design. You should also evaluate the way the tools and media of design were handled. Good map design is the result of a clear need or purpose, well-developed powers of imagination and visualization, and skill in working with available graphic technology.

PERCEPTUAL CONSIDERATIONS

In order to represent data in a meaningful way, you must be able to identify and discriminate between map symbols. This process is made possible by varying the appearance of symbols. You can do so by systematically adjusting their graphic character.

GRAPHIC ELEMENTS

Point, line, and area marks constitute the primitive building blocks of pictorial representation (Figure 18.1). We call these marks the **basic graphic elements** because they can be used to create all

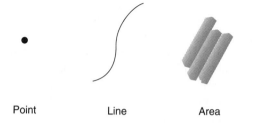

Point   Line   Area

Figure 18.1 A graphic representation is made up through combination and repetition of primitive point, line, and area marks, which are called the basic graphic elements.

visual designs. **Point** marks convey a sense of position and are the most fundamental of the three types of marks. **Line** marks exhibit direction as well as position, and can be thought of as a linear array of points. **Area** marks exhibit extent, direction, and position, and can be thought of as a two-dimensional array of points.

VISUAL VARIABLES

We can make point, line, and area marks on the map appear more or less distinctive and prominent by altering their shape, size, orientation, or color (hue, value, chroma). (See Figure 18.2). We call these graphic variations the **primary visual variables***:

Although there is considerable similarity between our list of visual variables and Bertin's list, they are not completely the same. For example, a systematic variation of both size and spacing com-

---

*The term "visual variable" was introduced in 1967 by Jacques Bertin in *Semiologie graphique*, 2nd ed. (Paris: Gautier-Villars, and Paris-La Haye: Mouton & Cie, 1973). The essentials of that work were condensed in Jacques Bertin, *La graphique et le traitement graphique de l'information* (Paris: Flammarion, 1977), and published in English as *Graphics and Graphic Information Processing*, translated by William J. Berg and Paul Scott (Berlin and New York: Walter de Gruyter & Co., 1981). The second edition of the larger work has been published in English: Jacques Bertin, *The Semiology of Graphics*, translated by William J. Berg (Madison, Wis.: University of Wisconsin Press, 1983).

prise Bertin's visual variable grain (Fr.), which has been translated in English as "texture."

1. **Shape**. Shape is the graphic characteristic provided by the form of a graphic mark. A shape may be regular and geometric, as in the case of a square, circle, or triangle. Shapes may also be irregular, as in the case of a pictographic version of a tree or bridge.

2. **Size**. Marks vary in size when they have different apparent geometric dimensions—length, height, area, volume. Usually the larger a sign, the more important it is thought to be.

3. **Orientation**. Lines and elongated shapes exhibit the visual variable we call orientation. A directional frame of reference is needed to define the orientation of a graphic mark. The map border or the graticule may serve this purpose.

4. **Hue (Color)**. Hue is a very important and complex visual phenomenon, which will be discussed at greater length in the next three chapters. Here suffice it to point out that the common use of the term "color" really refers to hue. When we say things are different colors, we are usually describing characteristics such as blue, green, red, and so on. Since this book is only printed using two inks, Figure 18.2 can only suggest the range of possible hue distinctions. See the color plate insert for better examples of hue differences.

5. **Value (Color)**. As a graphic quality, value refers to the relative lightness or darkness of a mark, whether of black or any other hue. A surface that reflects a noticeable amount of light is said to have a **tone**. Perceptually, a given tonal surface (as physically measured) may appear different under varied viewing circumstances. Consequently, when referring to the sensation of tone, it is better to use the term value, which refers to the perceptual scale of lightness and darkness. In the perceptual scale of values, light is referred to as high value and dark as low value.

6. **Chroma (Color)**. As a graphic quality, chroma refers to the degree to which a hue departs in "colorfulness" from a gray tone of the same value. Chroma can range from a gray hue with no apparent color pigment to a pure hue with no apparent gray. Terms such as saturation, intensity, richness, lightness, and purity are also used to refer to this complex visual variable.

Repetition of basic graphic elements (marks) representing various combinations of these primary visual variables produces a areal graphic effect that we call **pattern**. In turn, a pattern exhibits the characteristics of arrangement, texture (spacing), and orientation (Figure 18.3). These aspects of pattern are made up of component marks that are characterized by the primary visual variables. Thus, we call pattern arrange-

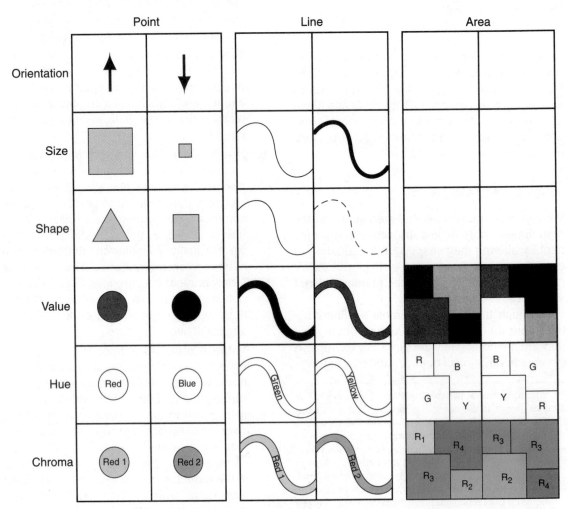

Figure 18.2 The primary visual variables with some examples of their simple application to the classes of symbols.

ment, texture, and orientation the **secondary visual variables**:

1. **Arrangement (Pattern)**. Arrangement refers to the shape and configuration of component marks that make up a pattern. The pattern of points or lines may be random or systematic in structure.

2. **Texture (Pattern)**. Texture refers to the size and spacing of component marks that make up a pattern. A fine texture is produced by a close spacing of small marks. This contrasts with a coarse textured pattern produced by an open arrangement of larger marks. Regular textures are often described as being composed of so many **lines per inch (lpi)** or **lines per centimeter (l/cm)**.

3. **Orientation (Pattern)**. Pattern orientation refers to the directional arrangement of parallel rows of marks as they are positioned with respect to some frame of reference.

CLASSES OF SYMBOLS

The primary and secondary visual variables contain the essential visual ingredients of all maps. Just as there are almost unlimited ways of combining the sounds of speech for audible communication, there are almost no bounds to the ways we can combine the visual variables for pictorial communication. Through these combinations, we create symbols to represent an unlimited variety of environmental features and distributions.

In order to consider the ways you can use symbols, it is helpful to classify them. A three-class scheme of point, line, and area symbols is popular. This would make sense if there were direct correspondence between the geometry of map symbols and the environmental phenomena they depict. But that is not the case. In fact, a powerful tool of cartographic abstraction is to map with fewer geometric dimensions than the

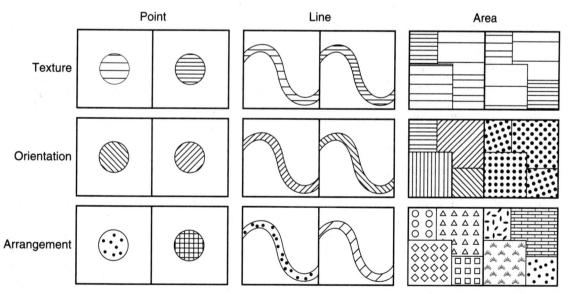

Figure 18.3 The secondary visual variables with some examples of their simple application to the basic graphic elements.

data exhibit. A contour line represents volumetric data, not linear data. Confusion results because this three-way classification focuses on the graphic appearance of a symbol rather than on its spatial meaning.

A better way to classify symbols is to focus on their spatial emphasis. Zero in on what the symbol represents, not what it looks like. This approach yields point-, line-, area-, and volume-emphasizing symbols:

1. **Point-emphasizing symbols**. Point symbols are individual signs, such as dots, triangles, and so on, used to denote a position, the location of a feature, the intensity at a place, or a representative location for spatial summary data. Examples include a coordinate location, a radio tower, a spot height, the centroid of some distribution, or a conceptual volume at a place, such as the population of a city. Even though the mark may cover some map space, it is a point symbol if it refers conceptually to a location.

2. **Line-emphasizing symbols**. Line symbols are individual linear signs used to represent a variety of geographical phenomena. Lines depicting rivers, roads, and political boundaries are common examples. But use of a line symbol does not always mean that the class of feature being represented is linear. For instance, contours are lines used to represent elevation and depth data, from which volumes may be determined.

3. **Area-emphasizing symbols**. Area symbols are markings extending throughout a map area to indicate that the region has some common attribute, such as water, administrative jurisdiction, or some measurable characteristic. When used this way, an area symbol is graphically uniform over the entire area it represents. The homogeneous character of the symbol may result from an even color, or it may consist of uniform repetition (pattern) of point or line symbols.

4. **Volume-emphasizing symbols**. Volume symbols represent the vertical or intensity dimension of a spatial phenomenon through space. In landform mapping, for example, the terrain surface may be symbolized with non-uniform markings that produce value (lightness) variation, as is the case in relief shading (see Chapter 27). The landform surface may also be represented by spot height data (point symbols) or linear surface traces such as profiles or contour lines (line symbols).

Figure 18.4 shows a few examples of the great variety of point-, line-, area- and volume-emphasizing symbols used to portray some kinds of qualitative and quantitative data. The four classes of symbols, along with alpha-numeric annotations (see Chapter 22), make up the basic building blocks of cartographic presentation. But in order to endow the marks with meaning, to make them similar or different from one another, and to make them more or less prominent, you must adjust their appearances by using the visual variables (see *Visual Variables*, discussed earlier).

GRAPHIC COMMUNICATION

Because cartography is a visual medium, it is based on principles of graphic communication that are critical to good visual composition. These principles parallel such rules of literary composition as syntax, paragraphing, organization, and phrasing. But while language and graphics have a common communication objective—clear, unambiguous transfer of information—there are many differences between the two media. One important difference is that graphic displays are made up of visual stimuli. People react to such stimuli quite differently than they do to written and spoken communication, as we shall see in the next section.

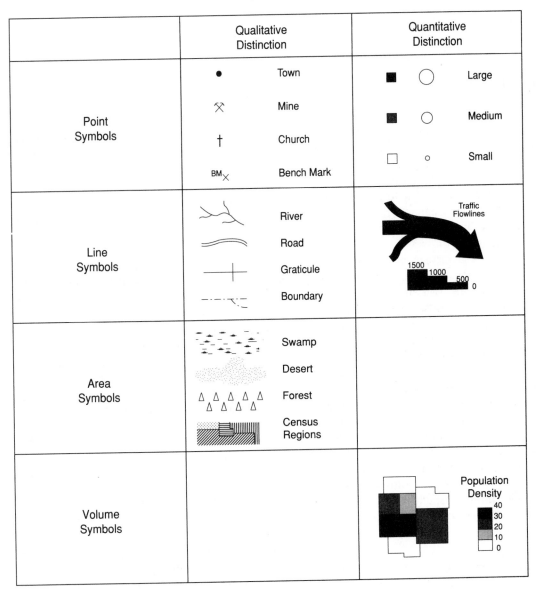

	Qualitative Distinction	Quantitative Distinction
Point Symbols	• Town ⚒ Mine † Church BM✗ Bench Mark	■ ◯ Large ■ ◯ Medium □ ∘ Small
Line Symbols	River Road Graticule Boundary	Traffic Flowlines 1500 1000 500 0
Area Symbols	Swamp Desert Forest Census Regions	
Volume Symbols		Population Density 40 30 20 10 0

Figure 18.4 Some examples of the four classes of symbols (point-, line-, area-, and volume-emphasizing) and how they might be used for a few of the kinds of qualitative and quantitative data.

PERCEPTION OF GRAPHIC COMPLEXES

Through long familiarity, the sounds of language and the written words that encode these sounds tend to be "transparent" to us. We understand their meanings without paying much attention to the actual sounds or the physical appearance of words. This is not true with graphic presentation. Graphic marks tend to be "opaque." We pay a great deal of attention to their appearance and arrangement.

When you read, or when someone speaks to

you, you receive information in serial fashion. That is, the words follow each other in sequence. You are thus programmed to receive the thoughts in a definite order. Graphic communication is quite different. With graphics, you receive visual impressions synoptically (all at once) instead of in sequence. Your perception of every mark on a map is affected by its location and appearance relative to all the other marks. This means that when you make a map, you can't structure your communication sequentially as you do when talking or writing. Instead, you must always think in terms of the map as a whole. Everything on a map is related visually to everything else. If you change one thing, you change everything.

When you view a display of any kind, you process it perceptually. When you hear a series of words or other sounds, you automatically try to make sense of it. You unconsciously do something similar with graphics. Because people tend to reject visual monotony and ambiguity, you will organize the graphic display so that it makes sense visually. You will assign visual meaning to each mark and its attributes, such as size, shape, or hue.

It is inevitable, therefore, that viewers of graphic displays will see them structurally. Some marks will seem more important than others, some shapes will "stand out," some things will appear crowded, some colors will dominate, and so on. If these visual relationships coincide with the cartographer's intentions, effective communication can take place. If not, the map design is likely to fail.

DESIGN PRINCIPLES

There is little agreement among professional designers about what they mean by graphic design. No definitive set of design rules is available to guide the novice. But there is general agreement that certain design qualities are visually significant. We can call these qualities the **principles of graphic design**. For cartographers, the most important of these principles are legibility, visual contrast, figure-ground, and hierarchical structure.

Expert designers take full advantage of each one. Let's look more closely at each of these design principles.

Legibility

Whenever you use graphic symbols, the first rule is to be sure they are easy to read and understand. To do so, you need to choose the proper graphic marks and portray them clearly. The shapes of point symbols must not be confusing. Lines must be clearly distinguishable in form and width. Differences between colors, patterns, and shadings used to differentiate between symbols must be visually distinct.

One important aspect of legibility is size. No matter how much care you put into producing a symbol, it serves no purpose if it is too small to be seen. Even if a symbol is visible, there is still a size limit below which it cannot be identified.

Although there is some disagreement about the exact measure of this size threshold, a practical application sets it as a size that subtends one minute of angle at the eye. This limit sets an absolute minimum, since it assumes perfect vision and perfect conditions of viewing. Because these assumptions are unrealistic, it is wise for the cartographer to establish the minimum size somewhat higher. Thus, two minutes is a more realistic measure for average vision and viewing conditions. Table 18.1 (based on two minutes) is useful in setting minimum values of visibility.

Some types of symbols are inherently more visible than others. The width of line symbols may be reduced considerably, since the length seems to enhance visibility. Similarly, contrasting colors or shapes may increase visibility. But these enhancements have a limited impact on legibility. If a symbol does not fall at or above the sizes given in Table 18.1, it is not likely to be legible. In other words, it might be seen (visible), but it might not be recognized or identified (legible).

Another quality that affects legibility is familiarity. It is easier to recognize something familiar than something new. Thus, you may see a name in a particular place on a map and, even though it is too

Table 18.1
Approximate Minimum Sizes for Legibility of
Point Symbols

International		U.S. Customary	
Viewing Distance	Size (width)	Viewing Distance	Size (width)
50 cm	0.3 mm	18 in.	0.01 in.
2 m	1.15 mm	5 ft.	0.03 in.
5 m	2.9 mm	10 ft.	0.07 in.
10 m	5.8 mm	20 ft.	0.14 in.
15 m	8.7 mm	40 ft.	0.28 in.
20 m	11.6 mm	60 ft.	0.42 in.
25 m	14.5 mm	80 ft.	0.56 in.
30 m	17.4 mm	100 ft.	0.70 in.

small to read, you can tell what it is from its position and general shape.

Visual Contrast

Making symbols large enough to see does not in itself make them legible. An additional graphic principle, that of visual contrast, is necessary. Assuming that each item on a map is large enough to be seen, the way a sign contrasts with its background and adjacent signs determines its visibility.*

No graphic factor is as important as contrast. Contrast is the basis of seeing. The critical eye seems to accept moderate and weak graphic distinctions passively and without enthusiasm, while it relishes greater contrasts. How crisp, clean, and sharp a map looks depends largely on the amount of contrast it contains. Note in Figure 18.5 that the most "interesting" sections are those with considerable contrast.

Do not assume, however, that maximum contrast is automatically desirable. Some map components differ only slightly, and you may want to indicate this lack of contrast with symbols. Also, as Figure 18.5 illustrates, too much contrast can sometimes be unpleasant.

*A study of factors affecting contrast is Michael W. Dobson's "Visual Information Processing During Cartographic Communication," *The Cartographic Journal*, 16, No. 1 (1979): 14–20.

Note, too, that if you vary two signs in two ways (for example, in both shape and size), the contrast between them will be greater than if you vary only one element. If the sign is simple, you may need to vary only one element to achieve the needed contrast. But the more complex the sign is, the more elements you must vary to obtain contrast. Color contrast is also important, as you will see in Chapter 21.

Figure-Ground Organization

Another important aspect of perception is the figure-ground phenomenon. Your eye and mind, working together, react spontaneously to any visual array, whether or not it is familiar. You immediately organize the display into two contrasting perceptual impressions: a **figure** on which your eye settles, and the amorphous **ground** around it. You perceive the figure as a coherent shape or form, with clear outlines, in front of or above its surroundings. You don't confuse this figure with

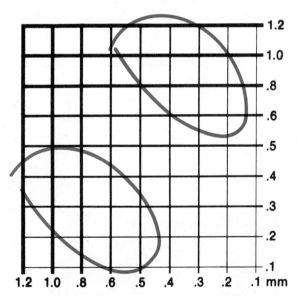

Figure 18.5 Size contrast of lines. Uniformity produces unpleasant monotony. The areas outlined in red appear to be crisper and more interesting than areas with less contrast.

the less distinct, somewhat formless ground. While this spontaneous visual organization is occurring, you also bring your previous experiences into play as you decide what the figure might represent.

Separation of the visual field into figure and ground is automatic. It is not a conscious operation. It is a natural and fundamental characteristic of visual perception and is therefore a primary consideration in graphic map design. Map users must be able to tell land from water and to recognize outlines of towns, islands, and harbors. They should be able to focus immediately on the cartographer's objectives without struggling to decide what they are supposed to see.

A simple illustration of this phenomenon is provided by Figure 18.6A. Here, a simple line, wandering across the visual field, merely separates two areas and is quite ambiguous. Notice how confusing this map is. There is no way to know what it represents, unless you happen to be familiar with the area.

In Figures 18.6B and 18.6C, areas are clearly defined from one another, but it is still unclear which is land and which water. Finally, in Figure

18.6D, the land emerges as the figure. Notice how the cartographer has clarified the map by using land-based names and familiar boundary symbols. The cartographer has also used the graticule to "lower" the water and separated land from water with slight shading. Thus, figure and ground clearly emerge.

As Figure 18.6 makes clear, certain visual stimuli help map readers perceive figure and ground. Let's examine some of the traits necessary to promote good figure-ground perception.

Differentiation must be present in order for one area to emerge as the figure. The figure area must be visually homogeneous. Also, the map as a whole must not be more uniform than the figure. Otherwise, the whole map will become the figure and its surroundings the ground. You can promote differentiation in a variety of ways, such as by using different colors and patterns.

Closed forms, such as islands and countries, are more likely to be seen as figures if shown in their entirety than if only partially shown. If all of the peninsula of Italy were shown in Figure 18.6A, it would be easier for us to see it as figure. In

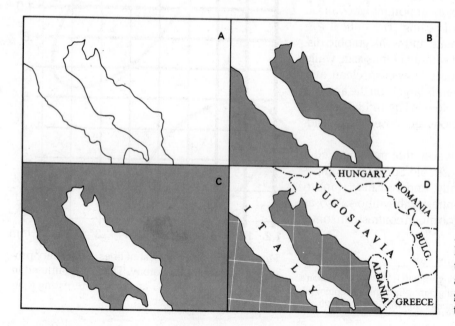

Figure 18.6
Four simple sketch maps to illustrate various aspects of the figure-ground relationship. See the text for explanation.

general, a small, surrounded area tends toward figure.

Familiarity has a great influence on the promotion of figure. We readily fix on a well-known shape. This suggests that if you want users of your map to see an unfamiliar area as figure, you will need to make an extra effort. You will have to apply more graphic devices that promote figure structure than you will with a familiar area.

Lightness (value) difference promotes the emergence of figure. All other things being equal, the darker area tends toward figure. (Of course, all other things aren't always equal. For example, in Figure 18.6D, the lighter area tends toward figure because of the overriding effect of other influences).

Good contour is the graphic equivalent of the term "logical" or "unambiguous." When something appears continuous, symmetrical, or sensible, it will lead toward figure-ground differentiation. Viewers tend to move toward the simplest visual explanation of graphic phenomena, even if this requires them to mentally fill in missing information.

The principle of good contour finds many applications in cartography. In Figure 18.6D, for example, the graticule appears to be continuous beneath the land, thus raising the land area and helping it to become figure. This differentiation of land and water is a common map design strategy. Another example of good contour is the breaking of a line for lettering or other map components. This practice is possible because visual logic will continue the line where it is not actually shown.

Detail promotes figure. In Figure 18.6D, for instance, notice all the detail on the land. In contrast to the simple graticule on the water, the land contains complex lettering and boundaries. These markings help the land to emerge as figure. Details such as city symbols, names, rivers, transportation routes, and relief representation all contribute to figure.

Size is important in differentiating between figure and ground. Smaller areas tend to emerge as figure, with larger areas as ground.

Each time you design a map, you will need to focus on different factors from the above list. Some geographical relationships simply have "bad con-

tour," are large, cannot be closed, and so on. In those cases, you must lean more heavily on other graphic elements to promote good figure-ground perception.

Hierarchical Organization

Due to the complexity of reality, successful map communication requires some kind of internal **graphic structuring**. On a thematic map, for example, the distribution's basic form is more important than the base data against which it is displayed. On a general reference map, classes of roads vary in importance. Or, a map may separate types of soil or vegetation into groups.

As a cartographer, it is your task to separate meaningful characteristics and to portray likenesses, differences, and interrelationships. To do so, you need to set up levels of relative importance. Thus, we can call this process **hierarchical organization**. Using different combinations of the visual variables, you can achieve a **visual layering** of mapped features. Let's look at the three kinds of hierarchical organization you can use.

Stereogrammic With one form of hierarchical organization, you give map users the impression that classes of features lie at different visual levels. The purpose of this hierarchical structuring is to direct readers through the layers of map information that make up the integrated whole. In doing so, you use the figure-ground principles discussed earlier. You direct attention to prominent layers of information, so that those layers become the figure. Less prominent layers then become the ground.

Stereogrammic organization is least developed in general reference mapping. The reason is that we purposefully design general reference maps to serve a variety of purposes. Having one category of features stand out prominently over the rest does not suit the goal of general reference mapping. Thus, we avoid strong visual layering on general reference maps. When people use these maps, then, they can concentrate on a single layer of information (such as roads or rivers) and achieve figure-ground separation **by attention**.

In contrast, stereogrammic organization **by design** is central to thematic cartography. In this case, the cartographer wishes to emphasize one map feature or relationship. See, for example, Figure 18.7. Here, the objective is to show territories that have changed hands in Europe since 1900. The cartographer therefore made those regions appear above the background base data. Thus, the eye focuses on the featured territories and only incidentally sees the geographical setting lying beneath them.

With stereogrammic organization, you may use other design principles in addition to figure-ground differentiation. You may invoke cues related to depth perception, such as superimposition, progression of size, and value progression (all illustrated in Figure 18.8). Differences in chroma may also be useful, since they have some connotation of depth.

Extensional Sometimes you are primarily concerned with ranking line networks or point symbols. For example, you may wish to show road systems made up of several classes of roads or drainage systems with several orders of streams. Or you may wish to show a hierarchy of settlements, such as megalopolis, metropolis, city, town, village, and hamlet.

Extensional organization refers to this ranking of features and, therefore, is usually ordinal. Even though you may use interval or ratio scales for individual symbols, the overall objective is to show relative importance. Thus, you will most often use the visual variables of size, value, and texture. Again, you may sometimes find chroma, or value and chroma in combination, useful as well. Figure 18.9 is an example of extensional organization.

Figure 18.7 An example of stereogrammic hierarchical graphic organization. In (A), all elements lie in the same visual plane. In (B), the land seems to be above the water, and modern boundaries rise above the visual plane of the land.

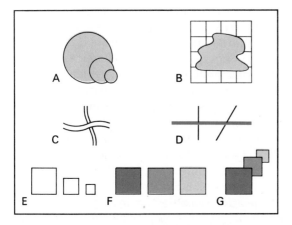

Figure 18.8 Some examples of depth cues that may be useful in stereogrammic organization. (A), (B), (C), and (D) illustrate various kinds of superimposition. (E) illustrates a progression of size, and (F) illustrates a progression of value. Like the graphic elements, depth cues may be used additively, as is done in (G).

Subdivisional   When you want to portray the internal relationships of a hierarchy, it's best to use a subdivisional structure. For example, you might subdivide the land use categories of agricultural and nonagricultural into several related uses. Figure 18.10 is an example of complex subdivisional organization.

When you use subdivisional organization, you are primarily concerned with area symbols. You distinguish between classes and subclasses using differences in color and pattern. The differentiation may be solely qualitative (nominal), or it may be a combination of nominal and quantitative (ordinal, interval, ratio) scaling. Since at least two primary levels of organization are involved, at least one of which incorporates at least two subdivisions, you must focus on using the elements of visual contrast and similarity. Each primary category must have distinctive visual homogeneity. Furthermore, distinctions

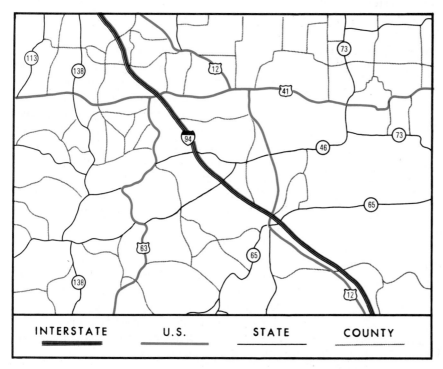

Figure 18.9   An example of extensional hierarchical graphic organization in which a set of roads is graded according to relative importance.

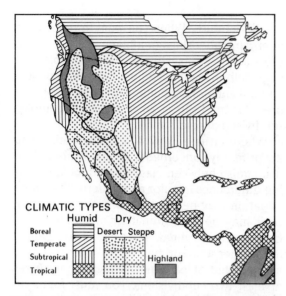

Figure 18.10 An example of subdivisional hierarchical organization in which the primary division is between humid and dry climates, with a secondary subdivision based on temperature, and a tertiary subdivision based on desert versus steppe.

must be greater between major categories than between subdivisions of each primary class.

CONTROLS ON MAP DESIGN

Map design does not take place in a vacuum, of course. A host of **controls**, or external forces, influence the design process. Before you begin a map project, think about the constraints you are likely to face. Such thought may save you from an unexpected design problem later on.

Let's take a look at these controls on graphic design. You may find it useful to consider each one in relation to a specific map.

PURPOSE

The purpose for which a map is made is the essential determinant of its final form. You should fit all your design decisions as much as possible to the map's purpose.

The mapping purpose has two aspects. The **substantive objective** relates to the information a map is to include. Of particular importance is the breadth of purpose. General reference maps are inherently multipurpose. Thematic maps, on the other hand, are meant to focus on overall form. In thematic mapping, the greater the number of identifiable objectives, the harder it is to incorporate them all successfully. Because maps tend to be difficult and costly to make, there is a strong tendency to combine many objectives in one map. Often, the result is a poor map.

The **affective objective** in map design relates to the map's total look. While the substantive objective concerns what is to be mapped, the affective objective concerns how this information is portrayed. Should the map appear dark or light, crowded or open, precise or approximate? The affective objective also relates to the general subjective feeling to be conveyed, such as graceful, bold, traditional, or modern.

REALITY

Every region or distribution has a certain geographical character, over which the cartographer has little control. No matter what you do, for instance, Chile will remain long and narrow. The New York megalopolis area will remain a great complex of transportation facilities. Some landforms are extremely rugged; soil variations are often very complex; and areas of recent glaciation tend to have enormous numbers of lakes. There's no way you can change these geographical realities, even though they may make your design task more difficult.

The constraints of reality will sometimes make it hard for you to organize things hierarchically as you might like. The features you wish to emphasize may be smaller than ones you want to fade into the background. A significant item may occur inconveniently in a faraway corner of the map. A color convention, such as blue water, may prevent the color series you would like to use.

These states of reality, and many more like them, place serious constraints on your graphic design possibilities. You must anticipate them in the beginning so that they do not later interfere with your plans.

AVAILABLE DATA

In theory, map purpose should determine what gets mapped. In practice, however, map design is also affected by the nature of available data. Different data sources require different design strategies. The amount of data must be sufficient to support the map purpose. In addition, the data's quality in terms of positional accuracy and timeliness must be compatible with the map purpose.

Maps which give an impression of accuracy or completeness greater than warranted by the source material are deceptive. Thus, you must convey a sense of the character of data used in mapping. Poor-quality information should be mapped in less detail and at smaller scales, or it should be mapped with less defined or specially tagged symbols.

MAP SCALE

Scale is prescribed by the mapping media format and by its relation to the area being mapped. From a conceptual point of view, scale operates in subtle ways. The smaller the scale, the farther away the mapped area appears to be. This intuitive feeling of viewer distance should be matched with the character of the graphic design. This does not mean that symbols should merely shrink in size as map scale increases. Rather, the smaller the scale, the less feature detail there should be (see Chapter 24).

AUDIENCE

Both general reference and thematic maps are made for a variety of audiences. These range from young schoolchildren to senior citizens, from outdoor recreationalists to technical engineers. Each group may differ in terms of familiarity with graphic and symbolic conventions, geographical knowledge, and perceptual limitations.

For the geographically unsophisticated, important shapes, names, and the like should be emphasized in the graphic design. Novice map users also seem to prefer the more intuitive pictographic symbols. In contrast, skilled environmental professionals may get more information from a high-density representation using abstract symbols.

Perceptual considerations are also important. Older people have more difficulty seeing small symbols and type, especially if the map is crowded with symbols. Another example of this aspect of map design is provided by the problems of making maps for the partially sighted.

CONDITIONS OF USE

We use maps in many different situations. We use them in the field or laboratory, at normal reading distance or as wall displays, under sunlight or artificial light, and in stable or unstable viewing environments. Map designers must try to anticipate these conditions and create their maps accordingly. Rules for designing maps for one environment may be of little value under different viewing conditions. This factor explains the common failure of maps shown in newspapers or on television, which were "borrowed" from print media capable of higher spatial resolution.

TECHNICAL LIMITS

Technical limits refer to the way a map will be constructed or produced. In computer-assisted cartography, technical limits include the capabilities of the software and hardware. Available pen sizes, line widths, and value and hue variations all constitute design constraints. Various limits may be prescribed for a printed map: one color or four colors; hand-drafted or scribed; printed or plotted under computer control; with halftone or limited to line art. The list is almost endless. Before beginning a design project, you should know the full set of technical limits within which the map will be produced.

DESIGN PLANNING

Effective presentation of complex relationships demands careful planning. When creating a graphic design, you need to think it through as carefully as any other sort of communication. You need to evaluate each graphic ingredient in combination with the others in terms of its probable effect on the map reader. To do this requires a complete understanding of the objectives of the map to be made.

THE GRAPHIC OUTLINE

Before you write something, as a first step you commonly prepare an outline. It is also helpful to outline a graphic communication. We use the term "outline" in its broad sense, meaning a brief characterization of the essential features of a communication.

Although you can jot down notes to keep track of ideas, the outline for a graphic presentation should be graphic. You can do some of it by visualizing relationships. Often it requires preliminary sketches or experimentation on a computer monitor.

Your outline will differ depending on whether you are designing a general reference map or a thematic map. General reference maps tend to present a great variety of geographical data. Thus, your decisions will focus on how many items are in each class and how prominently they may be portrayed. All sorts of hierarchical problems arise involving road, stream, and boundary precedence and the visual sequence of symbols and type sizes.

An outline for a reference map or map series usually begins with a document listing all the categories of data to be included. The next stage is a preliminary design plan including a prototype of a map, or section of one, usually prepared in proof. This stage lasts through several trials until the final specifications are decided and production begins.

Thematic maps usually involve fewer categories of data, but they pose a different kind of graphic problem. The four panels in Figures 18.11 illustrate how you might prepare a graphic outline for a thematic map. Assume that the planned thematic map is to show two related (hypothetical) distributions in Europe. The fundamental organizational elements of the communication are as follows:

1. The place—Europe.
2. The features—the two distributions to be shown.
3. The position of the features with respect to Europe.
4. The relative position of the two distributions.

You can place any one of these four elements at the top of the graphic outline and vary the order of the others in any way you wish. In Figure 18.11A, the design places the organizational elements in the order 1-2-3-4; in B, 2-3-4-1; in C, 3-1-4-2; and in D, 4-2-3-1. Other combinations are, of course, possible.

After you structure the major elements, you can shift your attention to the second stage of graphic outlining—the position of detail within each major element. You will want an arrangement that achieves the maximum clarity, legibility, and relative contrast of the detail items.

COMPOSITION

Map symbolism can rarely stand alone. Explanatory aids such as titles, legends, insets, scales, and direction indicators are also standard components of map composition. These items give context to the mapped data. Their primary purpose is to identify the place, subject matter, symbolization, orientation, and so on.

On large-scale series maps, such as topographic quadrangles, these contextual items usually appear as marginal notations. Since a standard legend is common to all maps in the series, it often does not appear on the individual maps. Instead, the legend may be printed separately as a pamphlet, or it may be provided once at the beginning of a reference atlas.

But on smaller-scale maps, especially those of a thematic or statistical type, it is usually convenient to place explanatory items such as titles, legends, and insets within the map frame itself (Figure 18.12). When this is done, the explanatory items

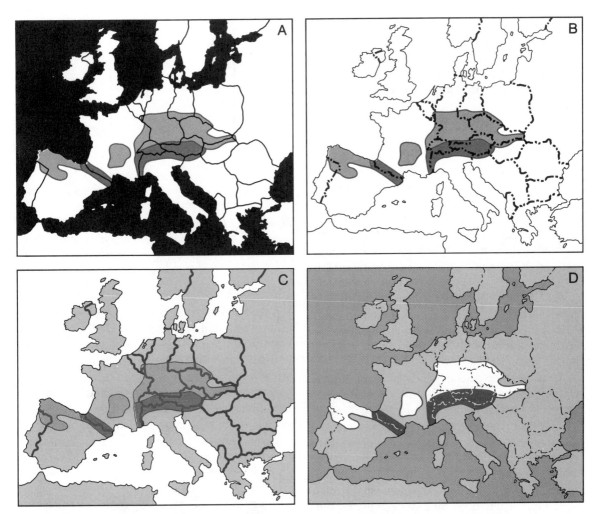

Figure 18.11   Examples of variations in the primary visual outline. See text for explanation.

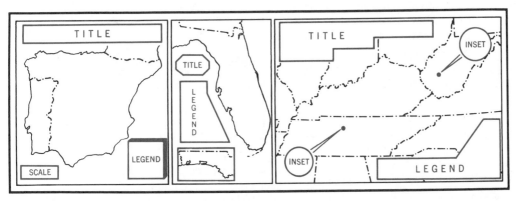

Figure 18.12   Titles, legends, scales, and insets may be arranged in various ways in the graphic organization of a map.

can form prominent visual masses, which must be integrated with map features such as land-water masses, color areas, and so on. In arranging these basic shapes, it is desirable to achieve visual balance within the overall graphic organization of the map.

Visual Balance

Balance in graphic design is the positioning of visual components so that their relationship appears logical. Nothing should seem out of place. **Layout** is the process of arriving at proper balance. In a well-balanced design, nothing is too light or too dark, too long or too short, too small or too large, in the wrong place, or too close to the edge.

Visual balance depends on the relative position and visual importance of the basic parts of a map. Thus, it is tied to the visual weight of each item. Size, color, pattern and contrast all contribute to a shape's visual weight.

Visual balance also depends on each item's relation to other items and to the map's **optical center**. The optical center is a point slightly (about five percent of the height) above the center of the bounding shape or the map border (Figure 18.13). Balancing items around the optical center is akin to balancing a lever on a fulcrum. Take a look at Figure 18.14. Note that a visually heavy shape near the fulcrum is balanced by a visually lighter but larger body farther from the balance point. You can probably think of many other combinations that will achieve visual balance.

The cartographer's job is to balance visual items so that they "look right" or appear natural in light of the map's purpose. The easiest way to accomplish this is to arrange the main shapes in various ways with respect to the map frame until you obtain a suitable combination. Figure 18.15 shows how you might do so. Here, the cartographer has tried out several ways of arranging the various shapes (land, water, title, legend, and shaded area).

The format—that is, the size and shape of the image area on which a small-scale map is to appear—is an important factor in balance and

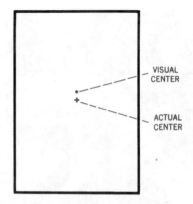

Figure 18.13 The visual as opposed to the actual center of a rectangle. Balancing is accomplished around the visual center.

layout. Shapes of land areas are prescribed, of course, and can be difficult to fit. They can also vary to a surprising degree on different projections. In many cases the desire for the greatest possible scale within a prescribed format may suggest a projection that produces an undesirable fit for the area involved. It can also be difficult to fit the necessary shapes (large legends, complex titles, captions, and so on) around the margins and

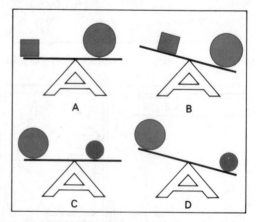

Figure 18.14 Visual balance. (A), (B), (C), and (D) show different degrees of balance. (A) and (B) are analogous to a child and an adult on a seesaw. (C) and (D) introduce relative density or visual weight, darker masses being heavier.

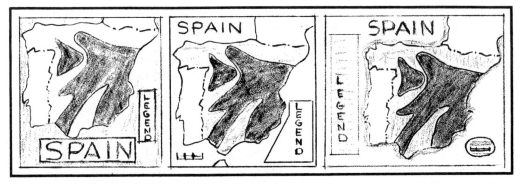

Figure 18.15   Preliminary sketches of a map made in order to arrive at a desirable layout and balance.

within the format. Generally, a rectangle with sides having a proportion of about three to five seems to be the most pleasing format (Figure 18.16).*

There are many ways to vary balance relationships to suit a map's objective. Try analyzing every visual presentation, from television displays to printed advertisements, in order to become more versatile and competent in working with this important factor of graphic design.

When the geographic configuration is itself unbalanced, leaving a large amount of open space within the map frame, the designer may welcome the chance to lay out contextual features. On the other hand, when geographic details fill the entire map space, you may have to try a number of tentative layouts before finding one that provides acceptable visual balance.

Importance of Contextual Items

After struggling with a difficult design situation, you may be inclined to solve the visual balance problem by leaving some standard contextual items off the map entirely. This is usually not a good idea, regardless of the frustration that layout problems pose. Before making such a decision, carefully consider the importance of each item you are

thinking of dropping. Since the purpose of the map is the most important design factor, let's begin by considering the importance of titles.

Titles   A title serves a variety of functions. Sometimes it reveals the map's subject or the area covered by the map. In such a case, it is as important as a label on a medicine bottle. But sometimes a map's subject or area is obvious, and no title is really needed. At other times, the title may be most useful as a shape to help balance the composition.

It is impossible to generalize about the form a title should take. It depends entirely on the map, its

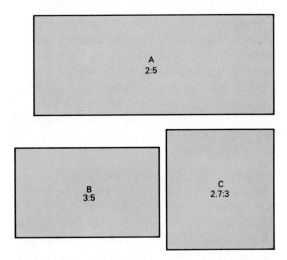

Figure 18.16   Various rectangles. (B), with the ratio of its sides 3:5, is generally considered to be the most stable and pleasing map format.

---

*For a large map to be viewed at normal reading distance, it is desirable to have the short dimension vertical. Not only does this reduce neck-stretching, it also allows easier use of bifocals.

subject, and its purpose. Suppose, for example, that a map shows the density of people per square kilometer in Canada according to the 1981 census. The following situations might apply.

1. If the map appeared in a textbook devoted to general worldwide conditions at that time with respect to the subject matter, then just

CANADA

would be appropriate because the time and subject would be known.

2. If the map appeared in a study of the current worldwide food situation, and if it were an important piece of evidence for some thesis, then

CANADA
POPULATION PER SQUARE KILOMETER

would be appropriate.

3. If the map appeared in a publication devoted to the changes in population in Canada, then

POPULATION
PER SQUARE KILOMETER 1981

would be appropriate, since the area would be known but the date would be significant.

Many other combinations might be appropriate, depending on the purpose. Similarly, the degree of prominence and visual interest given to the title, through the style, size, and boldness of the lettering, must be fitted into the map's whole design and objective. Novice cartographers usually err by making titles the most prominent item on their maps, a visual distraction that is rarely necessary or desirable.

Legends   Legends or keys are indispensable to most maps, since they explain the symbols, information sources, and data manipulation used in making the map. An explanation of data manipulation is particularly critical when a map depicts statistical data. For example, a frequency distribution of classed data provides the map user with valuable information. Similarly, when the map depicts the result of a modeling process, the model parameters (variables, weights, and links) should be given in the legend (see Chapter 28).

In theory, any map symbol that is not self-explanatory should be explained in a legend.* When symbols are included in the legend, they should appear exactly as they look on the map, drawn in precisely the same size and manner.

The arrangement of parts of a legend, such as the series of colors or patterns and point and line symbols used, is worthy of careful attention. As a rule, a range of values is arrayed vertically with the lowest values at the bottom.** The items making up a legend should also be positioned so as to achieve visual balance.

Map legends can be emphasized or subordinated by varying the shape, size, or value relationship. Figure 18.17 illustrates several variations. In the past it was the custom to enclose titles and legends in fancy, ornate outlines, called **cartouches**. With their intricate scrollwork, these outlines called attention to themselves. Today it is generally conceded that the legend's contents are more important than its outline. When an outline is used, therefore, it is usually kept simple.

Insets   Sometimes the general geographic context of a mapped region is not likely to be obvious to most map users. In such a case, it is good practice to include one or more inset maps of broader scope. This is a rather clever cartographic trick, since the effect is one of viewing

---

*In practice, we always take a great deal for granted when creating a map legend. At the least, we assume the map user will be familiar with the general concepts and practices of mapping. These assumptions of cartographic literacy permit us to limit legend items to those critical to reading the map in question. For an interesting discussion of this issue, see Wood and Fels (1986).

**Alternative suggestions have been made. See Alan A. DeLucia and Donald W. Hiller, "Natural Legend Design for Thematic Maps," *The Cartographic Journal*, 19, No. 1 (1982): 46–52

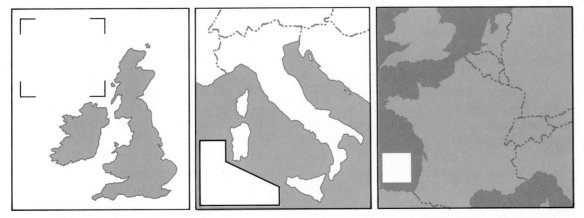

Figure 18.17 Examples of variations in the prominence of map legends. Note the operation of the principles of figure-ground relationships.

the environment from two or more vantage points simultaneously.*

When insets are used, it is important to make a clear graphic distinction between the several maps. An inset scale is not needed if a global perspective is provided. In other cases, however, inset maps should contain scale and orientation information (see next section).

ORIENTATION AND SCALE

The statement of map orientation and scale also varies in importance from map to map. As a rule, if the map has a conventional north-up orientation, a direction indicator is not necessary. A direction indicator also may not be needed on global maps where the orientation of familiar geographic features is obvious, or on maps where a terrestrial reference system

(such as latitude and longitude) is provided as base information.

A scale indicator is an important factor in making the map useful. This is particularly true on maps showing road or rail lines, air routes, or any other phenomenon or relationship that involves distance concepts. In such cases, the statement of scale should be displayed prominently and designed so that readers can use it easily.

The method of presenting the scale may vary. For many maps, especially those of larger scale, the representative fraction (RF) is useful because it tells the experienced map reader a great deal about how much simplification and selection went into preparing the map.* A graphic scale is much more common on small-scale maps, because it is easier to use and because an RF is not so meaningful in the smaller scales.

***

*This use of multiple vantage points in mapping predates cubism in the aesthetic arts. Although cubism is treated with considerable skepticism, the practice does take advantage of the liberating nature of representational technique. This freeing of expression from the constraints of our physical being has great creative power and probably deserves broader recognition in both art and cartography.

*It should be remembered that if a map to be printed will be changed in size by reduction, this will not change the printed numbers of the RF. On a map designed for reduction, the RF must be that of the final scale, not the construction scale.

DESIGN EXCELLENCE

Cartographers working on a reference atlas or government map series (topographic, soils, geology, etc.) may have little free choice in manipulating the design ingredients discussed in this chapter. They must map by the standards and conventions spelled out in the organization's official "recipe book." Since organizations naturally resist change, however, their standard design recipes quickly grow stale. We must usually look elsewhere for models of design achievement.

Cartographers not constrained by standard recipes have the greatest opportunity to be innovative and creative. But what ingredients make for excellence? It doesn't help much to turn to design experts for an answer. They will most likely tell you they just "do" design, just as golf experts may tell you they just "hit the ball." If you were persistent, you could get both design and golf experts to give you a list of rules for handling their tools. But knowing these means and processes would not in itself make you an expert at either profession. The ingredients do not make the whole; it also takes chemistry. You need to develop a basic sense of design, and you need a good deal of practice to polish your skills. These come with experience.

In the following chapters, we'll delve more deeply into the visual variables of color and pattern—powerful tools of graphic design. These will go a long way toward building that experience and background you need to become an expert designer.

SELECTED REFERENCES

Arnheim, R., "The Perception of Maps," *The American Cartographer*, 3 (1976): 5–10.

Arntson, A. E., *Graphic Design Basics*, 2nd ed., New York: Harcourt Brace Jovanovich College Publishers, 1993.

Arvetis, C., "The Cartographer-Designer Relationship, A Designer's View," *Surveying and Mapping*, 33 (1973): 193–95.

Bertin, J., *Graphics and Graphic Information Processing*, trans. W. J. Berg and P. Scott, New York: Walter de Gruyter, 1981.

———*The Semiology of Graphics*, trans. W. J. Berg, Madison, WI: University of Wisconsin Press, 1983.

Dent, B. D., "Visual Organization and Thematic Map Communication," *Annals of the Association of American Geographers,* 62 (1972): 25–38.

———*Cartography: Thematic Map Design*, 3rd ed., Dubuque: Wm. C. Brown Publishers, 1993.

Gilman, C. R., "Map Design at USGS: A Memoir," *The American Cartographer*, 10 (1983): 31–49.

Jones, J. C., *Design Methods*, 2nd ed., New York: Van Nostrand Reinhold, 1992.

Keates, J. S., *Cartographic Design and Production*, 2nd ed., London: Longman Group UK Limited, 1989.

Margolin, V., (ed.), *Design Discourse: History/Theory/Criticism*, Chicago: University of Chicago Press, 1989.

Muehrcke, P. C., "An Integrated Approach to Map Design and Production," *The American Cartographer*, 9 (1982): 109–22.

Petchenik, B. B., "A Map-Maker's Perspective on Map Design Research," in D. R. F Taylor, (ed.), *Progress in Contemporary Cartography*, Vol. 2., New York: John Wiley & Sons, 1983.

Porter, T., and S. Goodman, *Designer Primer for Architects, Graphic Designers, & Artists*, New York: Charles Scribner's Sons, 1988.

Taylor, D. R. F., (ed.), *Graphic Communication and Design in Contemporary Cartography*, Progress in Contemporary Cartography, Vol. 11, New York: John Wiley & Sons, 1983.

Tufte, E. R., *Envisioning Information*, Cheshire, CT: Graphics Press, 1990.

Wood, D., and J. Fels, "Designs on Signs: Myth and Meaning in Maps," *Cartographica*, 23 (1986): 54–103.

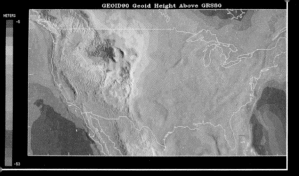

Color Figure 4.1 The GEOID90 geoid height model

Color Figure 8.4 High altitude infrared color photograph of Washington D.C. (NHAP 80 – Roll 581, Frame 27).

Color Figure 8.2 A 70-mm. infrared photograph of the Toronto, Ontario, shoreline. Images for Figures 8.2 and 8.3 are courtesy of D.J. King, P.W. Mausel, J.H. Everitt, and D.E. Escobar, and were published in *Photogrammetric Engineering and Remote Sensing*, August 1992, p. 1193.

Color Figure 8.3 A video (VHS) image of the same area shown in Color Figure 8.2.

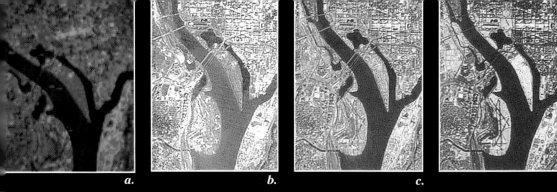

a.   b.   c.   d

Color Figure 8.5 Four Landsat scenes of Washington D.C.: **a.** Landsat 3 bands 4, 5, 7; **b.** Landsat 5 band 1, 2, 3; **c.** Landsat 5 bands 1, 3, 4; **d.** Landsat 5 bands 1, 5, 4;

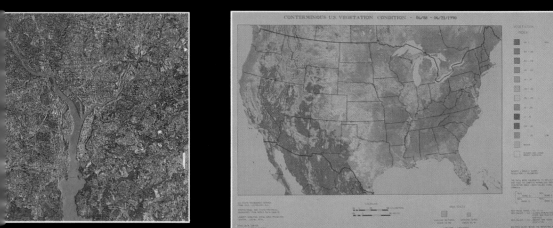

Color Figure 8.6 SPOT multispectral image of Washington D.C. (© 1994 CNES; Provided by SPOT Image Corporation).

Color Figure 8.7 Vegetation map of North America produced from AVHRR images.

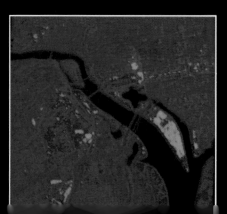

Color Figure 12.3 Contrast and terrain enhanced image made by combining Landsat TM multispectral data with a digitized SLAR image

Color Figure 12.4 Land use map of central Washington D.C. created by supervised image classification.

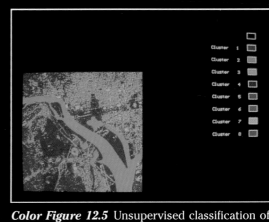

Color Figure 12.5 Unsupervised classification of Central Washington D.C. Land use categories corresponding to each color must now be determined from ancillary information.

Color Figure 12.6 Simplification of the land use classification of central Washington D.C. shown in Color Figure 12.4

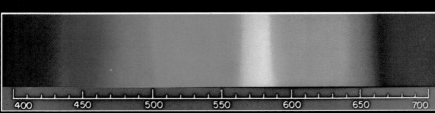

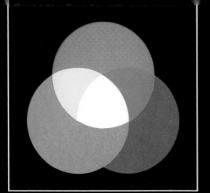

Color Figure 19.2 Mixing of light using additive color primaries as in electronic displays such as CRT monitors.

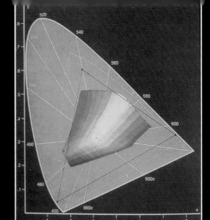

Color Figure 19. CIE chromacity diagram shown partially in color (Courte Eastman Kodak Co).

Color Figure 19.4 Munsell color circle showing the five principle hues — red, yellow, green, blue, and purple — and the five intermediates at mid-value and high chroma.

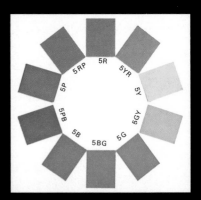

	Chroma				Value		Chroma			
	/8	/6	/4	/2		/2	/4	/6	/8	/10	/12

Hue : 5.0 B
8/
Hue : 5.0 YR

7/

6/

5/

↑— 5.0 YR 5/8

4/

3/

2/

Color Figure 19.5 Value and chroma variations of two complementary Munsell hues: 5.0 blue and 5.0 yellow red. Note

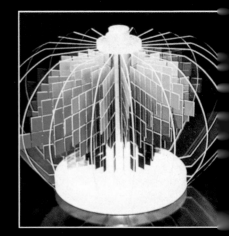

Color Figure 19.6 Munsell col chips, like those in Color Figure 19.
arranged as a three dimensional "

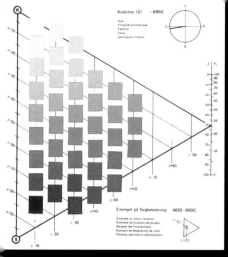

Color Figure 19.7 Reproduction from the green page of the Natural Color System (NCS). Separations made on a Crosfield Magnascan scanner. (Courtesy Crosfield Electronics Ltd.)

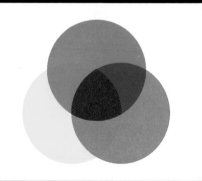

Color Figure 20.1 Mixing of subtractive primary pigments as in painting and printing.

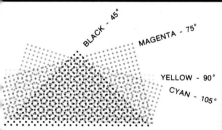

Color Figure 20.3 In four color process printing black (usually the most noticeable ink) is screened at a standard 45 [o], and cyan at 105[o].

Color Figure 19.8 Outer faces of the RGB color model.

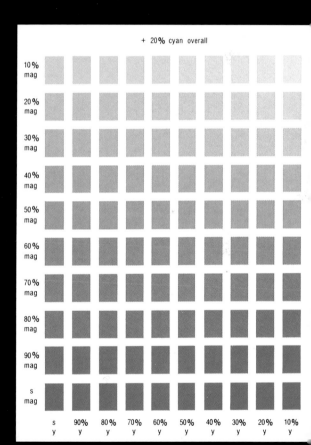

Color Figure 20.2 One block of a printer's color chart. Each row has the screen percentage of magenta ink indicated at the left, and each column has the percentage of yellow indicated at the bottom. S stands for solid (that is, no screening). A 20 percent cyan screen has been added overall. Other blocks on the complete chart have other percentages of cyan overall.

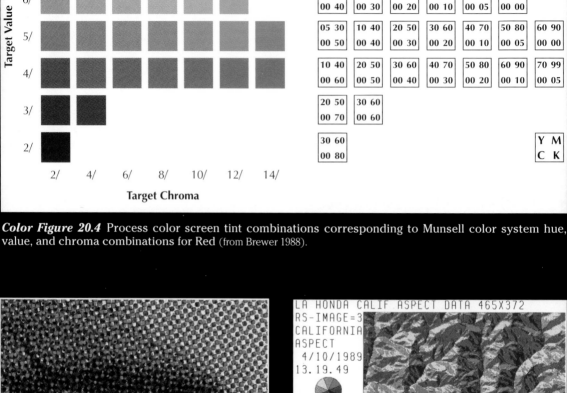

Target Value: 8/ 7/ 6/ 5/ 4/ 3/ 2/

Target Chroma: 2/ 4/ 6/ 8/ 10/ 12/ 14/

| 00 05 | 00 10 | | | |
| 00 20 | 00 10 | | | |

| 00 10 | 00 20 | 05 30 | 10 40 | 20 50 |
| 00 30 | 00 20 | 00 10 | 00 05 | 00 00 |

| 00 20 | 05 30 | 10 40 | 20 50 | 30 60 | 40 70 |
| 00 40 | 00 30 | 00 20 | 00 10 | 00 05 | 00 00 |

| 05 30 | 10 40 | 20 50 | 30 60 | 40 70 | 50 80 | 60 90 |
| 00 50 | 00 40 | 00 30 | 00 20 | 00 10 | 00 05 | 00 00 |

| 10 40 | 20 50 | 30 60 | 40 70 | 50 80 | 60 90 | 70 99 |
| 00 60 | 00 50 | 00 40 | 00 30 | 00 20 | 00 10 | 00 05 |

| 20 50 | 30 60 |
| 00 70 | 00 60 |

| 30 60 |
| 00 80 |

| Y | M |
| C | K |

Color Figure 20.4 Process color screen tint combinations corresponding to Munsell color system hue, value, and chroma combinations for Red (from Brewer 1988).

LA HONDA CALIF ASPECT DATA 465X372
RS-IMAGE=3
CALIFORNIA
ASPECT
 4/10/1989
13.19.49

LEVELS= 8
METHOD=KMA
L=   0.000
H= 359.000
3M=LOCAL
STROW=   10
STCOL= 120
NROWS= 240
NCOLS= 240

Color Figure 20.5 In process color printing, halftone separations are printed in transparent inks, one over the other, with a minimum of overlap between the cyan, magenta, yellow, and black dots. when the resulting dot pattern is viewed, the individual color dots combine and blend optically to produce the visual effect of full color. A magni-

Color Figure 20.6 Eight class slope-aspect map colored using a hue-value progression that results in a "relief shaded" appearance (from Moellering and Kimerling 1990).

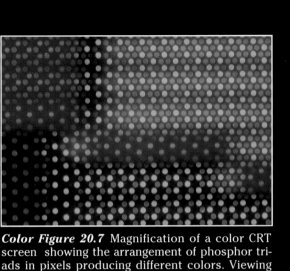

Color Figure 20.7 Magnification of a color CRT screen showing the arrangement of phosphor triads in pixels producing different colors. Viewing the illustration from a distance causes the pixels to visually blend into pixel colors, as they do on the CRT (from Thorell and Smith 1990, courtesy of Gunnar Tonnquist).

Color Figure 21.1 Simultaneous contrast, where background colors affect the appearance of central colors. The green discs are physically identical, but they do not appear so.

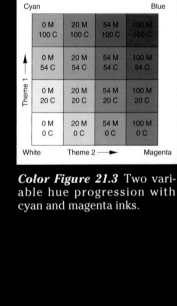

Color Figure 21.3 Two variable hue progression with cyan and magenta inks.

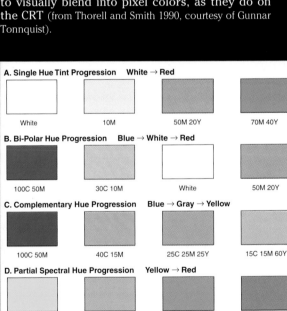

A. Single Hue Tint Progression   White → Red

| White | 10M | 50M 20Y | 70M 40Y | 100M 70Y |

B. Bi-Polar Hue Progression   Blue → White → Red

| 100C 50M | 30C 10M | White | 50M 20Y | 100M 70Y |

C. Complementary Hue Progression   Blue → Gray → Yellow

| 100C 50M | 40C 15M | 25C 25M 25Y | 15C 15M 60Y | 10M 100Y |

D. Partial Spectral Hue Progression   Yellow → Red

| 10M 100Y | 40M 90Y | 60M 80Y | 80M 70Y | 100M 70Y |

E. Blended Hue Progression   Yellow → Orange → Brown

| 10M 100Y | 30M 100Y | 50M 70Y 5K | 50M 70Y 20K | 50M 70Y 40K |

F. Value Progression

| 10K | 15K | 35K | 60K | 100K |

Color Figure 21.2 Five different color progressions for symbolizing quantitative data.

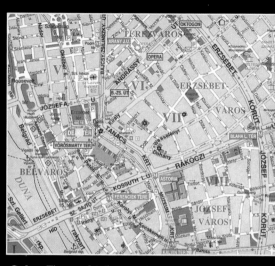

Color Figure 21.4 Visual acuity chart for screen tinted and solid yellow, magenta, and black ink. (Patterned after *Standard Printing Color Catalog*, U.S. Defense Mapping Agency, 1972.)

Color Figure 21.5 Complex multicolor map illustrating visual hierarchy development.

Color Figure 28.1 A triangular graph can be used to create a cross-variable map depicting the relationships between three phenomenon on a single map. (Courtesy D. White.)

Color Figure 28.2 This groundwater contamination susceptibility map of Wisconsin was created by weighting and linking five variables (surficial deposits, depth to groundwater, soil type, bedrock type, and depth to bedrock) to produce a single index value. (Courtesy Wisconsin DNR GeoServices Section.)

Wood, D., *The Power of Maps*, New York: Guilford Press, 1992.

Woodward, D., "Representations of the World," in R. F. Abler, M. G. Marcus, and J. M. Olson, (eds.), *Geography's Inner Worlds*, New Brunswick, NJ: Rutgers University Press, 1992: 50–73.

CHAPTER 19

COLOR THEORY AND MODELS

*U*sing color on maps is one of the most interesting and challenging aspects of cartography. Hand coloring of maps dates back at least to the early Egyptian cartographers. After the 15th century, when the printing of maps became common in Europe, most color still had to be applied by hand to each sheet, because color printing was prohibitively expensive and technically difficult. Map tinting with watercolors became a trade, and many people were so employed by mapmaking companies. Often templates and other elaborate ways of ensuring that colors were put in the right places were used. Such practices extended even to large-scale topographic maps.

With the development of lithography and then photography in the 19th century, techniques for printing color were developed. The quality of color map printing has increased steadily, so that today virtually all color maps published on paper are reproduced in this way.

In this century, the development of color television, coupled with the invention of digital computers, has made possible electronic color maps on cathode ray tube (CRT) displays. Faithfully transforming these ephemeral maps into tangible paper products challenges both cartographers and leaders of the computer graphics industry.

To fully understand the use of color in cartography, we must go beyond the techniques of applying color to maps. Our deeper understanding is based on knowledge of the physical, physiological, and perceptual nature of color. This knowledge leads to the study of color specification in cartography. Here we borrow heavily from systems developed by psychologists and colorimetrists (scientists who measure color).

Familiarity with color specification systems is a crucial part of color choice. Equally important are the historical, artistic, and logical guidelines and conventions for the selection of map colors. In this chapter, we will examine the nature of color and several systems for modeling color. Chapter 20 is devoted to the creation and specification of color for printed maps, CRT displays, and hardcopy output from these electronic devices. Chapter 21

will focus on the many aspects of color and pattern use in cartography.

THE NATURE OF COLOR

Color is a perceptual phenomenon, a product of our mental processing of electromagnetic radiation detected by our eyes. One important aspect of color is our response to **spectral colors** of the visible spectrum. Another is the fact that we see **reflected color**, resulting from selective absorption of visible radiation by different features. Let's look more closely at each of these facets of color.

SPECTRAL COLOR

The visual sensation of light occurs because receptors in our eyes are stimulated by electromagnetic radiation of certain wavelengths. The electromagnetic spectrum (see Chapter 8) ranges from short-wavelength gamma rays and X-rays to long-wavelength radio waves used in the broadcasting industry (Figure 19.1). Only wavelengths in a tiny portion of the spectrum—namely those with "visible" wavelengths ranging from approximately 400 to 700 nm*—stimulate the receptors in our eyes. This tiny visible portion of the spectrum corresponds with the sun's peak radiation emission, as our eyes are fine-tuned to our energy source.

When the sun or another source of illumination emits the full range of visible wavelengths in suitable portions, our brain interprets the electrical signals from the receptors in our eyes as white light. Since Isaac Newton's optics experiments in the 1670s, we have known that white light can be separated into a continuum of component wavelengths by refraction through a prism. When this continuum enters our eyes, the signals from our visual receptors are transformed by our mind into

---

*A nanometer (nm), previously called a millimicron (mμ) is one billionth of a meter (10^{-9}m).

the sensation of hue (blue, yellow, green, red, and so on).* The shorter wavelengths are seen as the violet-blues, near the 400 nm end of the spectrum; the longer wavelengths as the reds, near the 700 nm end. The order of these "spectral" hues is that of the rainbow, and is shown in Color Figure 19.1.

Pure spectral colors are not often seen except when white light is refracted, but they provide most of the basic names we use to identify hues (violet, blue, green, yellow, orange, and red). Their order is significant in cartography, and in Chapter 21 we will see that partial and full "spectral progressions" of hues are a popular way of mapping geographical data.

REFLECTED COLOR

The colors we see in nature and in all fabricated things, including maps, are rarely spectral hues but instead are almost always made by combinations of reflected wavelengths. This is because surfaces illuminated by white light absorb differing proportions of some of the wavelengths and reflect the remainder.

Graphs of **spectral reflectance curves** help us to understand reflected color (Figure 19.2). For example, curve 'A' could be from a white surface, such as typing or printing paper. From such a surface, sunlight reflectance is uniformly high throughout the visible spectrum. Such wavelength-by-wavelength surface reflectance data are collected using an instrument called a **spectrophotometer**. Curve 'D' might be measurements from a dark gray printed area, from which reflectance is uniformly low. Curve 'B' could be from a "greenish blue" inked area. From such an area, all wavelengths except those in the blue and green portions of the spectrum are heavily absorbed. Curve 'C' could be from a "red-purple" inked area, since green and yellow are heavily absorbed.

We have seen that ordinary surfaces, such as paper partially covered with ink, will reflect at least a small proportion of all visible wavelengths, even if the ink is black. The color we see comes from the wavelengths which are reflected the most.

---

*Defective color vision occurs in about 8 percent of men and 0.4 percent of women. The most common type is the inability to distinguish between red and green, which is also evidence for the opponent process theory of vision. Only a very small proportion are truly "color blind"—that is, see only in grays.

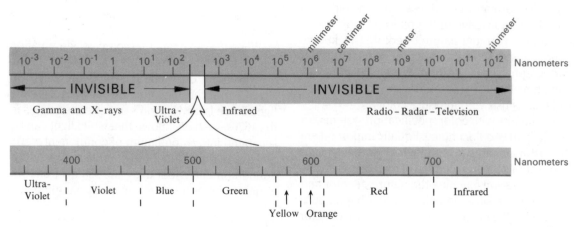

Figure 19.1 A portion of the electromagnetic spectrum with wavelengths shown in nanometers. Only a small portion of the entire spectrum is visible. See Color Figure 19.1.

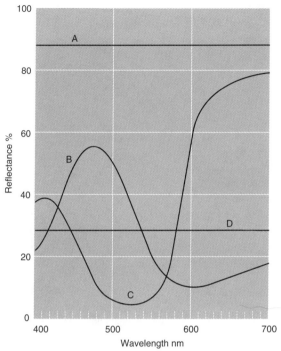

Figure 19.2   Four spectral reflectance curves: (A) a white surface; (B) a "greenish blue" ink; (C) a "red-purple" ink; and (D) a dark gray surface.

COLOR DIMENSIONS

Until now we have focused on hue, but color has additional dimensions. By "dimensions" we mean properties which can be varied systematically without changing other properties. Color is usually regarded as having three dimensions: hue, brightness or value, and saturation or chroma.

HUE

We have seen that **hue** is the color dimension associated with different dominant wavelengths. A nearly spectral hue of a single dominant wavelength will plot as a vertical line on a reflectance graph, such as Figure 19.3A. Non-spectral hues, such as printing inks, reflect light in a manner similar to curve 'B' in Figure 19.2. Most hues

have one definite dominant wavelength peak on the graph, although such hues as yellow, cyan, and magenta have two wavelength peaks of different heights.

BRIGHTNESS AND VALUE

All colors can be ranked in terms of lightness or darkness. **Brightness** is the general term for how light or dark a color appears. This depends on the amount of light reflected or emitted by a surface, and on the reflectance or emittance from neighboring areas.

Value is a parallel term referring to the sensation of lightness or darkness evoked by a color relative to standard black and white areas. A gray scale of value alone (Figure 19.4) illustrates the idea of an independent color dimension, since this scale could be superimposed on an area of constant hue and not change the dominant wavelength of any of the new colors. Reflectance curves 'A' and 'D' in Figure 19.2 might well be for steps at the left edge and middle right of the gray scale, since there is no dominant wavelength when the curve is a horizontal line. Two value levels of the same green hue are graphed in Figure 19.3B. We see that the shape of the reflectance curve has not changed, only its vertical position on the graph.

SATURATION AND CHROMA

The **saturation** of a color is the perceived amount of white in a hue relative to its brightness. **Chroma** is a closely related term meaning the perceived amount of white in a hue relative to a gray tone of the same value level. For a given hue at a certain value level, the chroma can vary from a hue that looks gray with a little hue added to a pure hue with no apparent gray. Three green hues at the same value level but differing in chroma are graphed in Figure 19.3C. Notice that the curves become flatter as the chroma decreases, so that a horizontal line would be reached at zero chroma. Here, of course, there also would be no hue, only value.

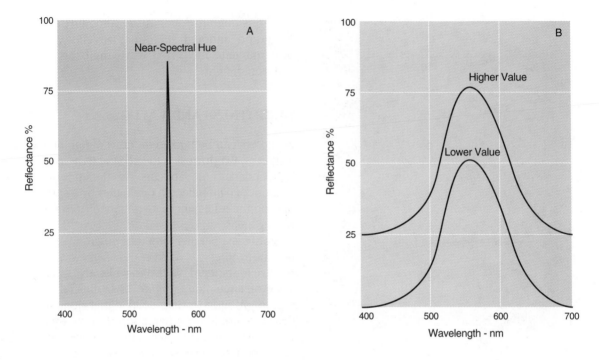

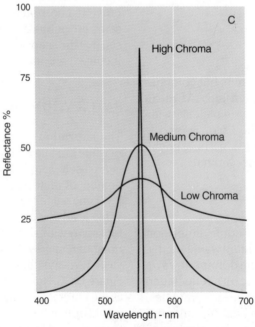

Figure 19.3   Reflectance curves for: (A) a near-spectral green hue; (B) two value levels of the same green hue at the same chroma; and (C) three chroma levels of the same green hue.

Figure 19.4 A photographic gray scale in which the steps are equal differences in surface reflectance. Note the apparent "waviness" caused by what is called simultaneous contrast. Note also that a middle gray does not appear to be halfway in reflectance between black and white. (From a Kodak Gray Scale, courtesy Eastman Kodak Co.)

Various terms, such as intensity and purity, are also used to describe this color dimension. These will appear in the different color specification systems we shall soon explore. Each term has a precise definition which sets it apart from the others.

THE NATURE OF COLOR VISION

Color not only is a three-dimensional phenomenon—it is something that cannot occur unless three elements are present: an illumination source, objects that selectively reflect light, and the human eye-brain visual processing system. Above all, color is a perceptual phenomenon, a mental response to the reception of light by our eyes. Theories as to the physiological basis for the creation of color through the eye-brain perceptual mechanism are very complex, and even today there is less than a full understanding of this mysterious phenomenon. In this section, we will explore two important (and for a long time competing) theories of how we mentally transform light from objects into the sensation of color.

TRICHROMATIC THEORY

The first plausible theory of human vision was an outgrowth of Scottish physicist James Clerk Maxwell's attempt in the 1860s to produce the first color photograph. His method was to take three black-and-white photographs of a colored object, with red, green, and blue filters placed in succession over the camera lens. This arrangement separated the light from the object into its red, green, and blue **additive primary** color components.

Maxwell then completed his "color photograph" by building what remote sensing specialists now call an additive color viewer. The three black-and-white photographic films were placed in projectors with red, green, and blue filters. This reversed the initial color separation so that a faithful color image was obtained when the three projected images were superimposed on a white screen (Color Figure 19.2).

The **trichromatic theory of color vision** which Maxwell helped develop, along with physicists Thomas Young and Hermann von Helmholtz, is tied closely to this color photography experiment. The basic premise is that our eyes act like the three filtered cameras. We must have visual receptors capable of separating light into its red, green, and blue components and judging the intensity of each. These additive primary color intensities are then transmitted directly to our brain, where the multicolor image of the object is recreated in a manner roughly analogous to image superimposition by three projectors.

We now know the trichromatic theory to be true only to the extent that the light-sensitive receptors in our eyes consist of around 115 million "rod" cells sensitive to all wavelengths from 400 to 600 nm, and of about 7 million "cone" cells of three types. These are called the α, β, and γ cones, with peak sensitivities of 440 nm (blue), 530 nm (green) and 570 nm (red). Cone cells are evenly intermixed and packed together closely in the fovea at the center of the retina, so that light from a small area on a colored object will be detected by numerous cells of each type. Rod cells are essentially absent from the fovea, but rapidly increase in density outward, where there is a corresponding rapid decrease in cone cells.

Electrical impulses proportional to the intensity of visible light detected by rod cells are indeed transmitted directly to the visual cortex region of our brain. Such signals are in large part responsible for "night vision," our ability to see the lightness and darkness of objects under very low illumination levels. However, physiologists have accumulated overwhelming evidence showing that a similar direct transmission from cone cells to the brain does not occur. The **opponent process theory** of color vision best fits this evidence.

OPPONENT PROCESS THEORY

The opponent process theory (OPT) of vision was first proposed by the physiologist Ewald Hering in 1877. It is based on the premise that electrical impulses from the rod and three types of cone cells do not travel directly through the optic nerve to the brain. Rather, they interact in such a way that three, not four, separate signals are transmitted. Figure 19.5 is a simplified diagram of how the transmission is now thought to occur. As it shows, rod and cone cells are connected to ganglion (nerve) cells that constantly send electrical signals through the optic nerve fibers to the visual cortex.

The transmission of color information to the brain hinges on the excitation and inhibition of the signals sent by three types of ganglion cells. These are labeled BY (Blue-Yellow), GR (Green-Red), and WBK (White-Black) in the diagram. A WBK ganglion cell is excited by impulses from the α, β and γ cone cells in a tiny spot on the fovea. Excitation is thought to be proportional to the output signal strength. The greater the excitation, the stronger the value (lightness) message sent to the brain.

GR ganglion cells work differently, being excited by impulses from the γ cones and inhibited by impulses from the β cones. Strong and weak signals from the GR ganglion cell are red and green messages to the brain. Red and green colors are

"opponent" in this way, since the message is either red or green, never both.

The BY ganglion cells work in a slightly different way. They are inhibited by impulses from the α cones and excited by impulses from both the β and γ cones. A strong signal from a BY ganglion cell is a yellow message; a weak signal indicates blue.

The exact hue is determined in the visual cortex by comparing the strengths of the signals from the GR and BY cells, once the green or red and blue or yellow message is received. For example, equally strong blue and green messages are interpreted as a cyan (blue-green) hue. Chroma is thought to be indicated by the strengths of the GR and BY signals. The stronger the signal, the more brilliant the color.

An everyday confirmation of the opponent process theory is the fact that we see no hues as mixtures of red and green, or of yellow and blue. This suggests our green-red and yellow-blue visual coding system is based on these four visually unique "primary" hues, with all other colors being mixtures of two non-opponent primary hues, as illustrated in Figure 19.6. We shall see that color modeling systems and map design guidelines have been developed around these opponent process primary hues and their mixtures.

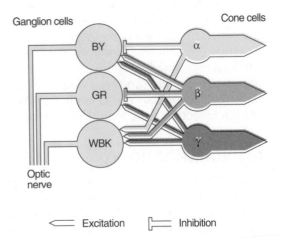

Figure 19.5 A simplified diagram of the opponent process theory of vision (after Eastman 1986).

COLOR CONSTANCY

Our ability to see an object as the same hue despite changing illumination conditions is known as **color constancy**. Color constancy greatly simplifies cartographers' use of color, since the white paper and map colors will look the same whether viewed in noon sunlight, shadowed areas, or under tungsten or fluorescent lamps.

Edwin Land, inventor of the Polaroid camera, worked 30 years to develop the **retinex theory of vision** to explain how we see the complicated multicolor scenes in everyday life. The basic premise is that we perceive an object's intrinsic hue by continuously adjusting for the illuminant color. Signals received by each of our three cone systems are thought to be scaled according to the reflectance across the entire visual field. This background color adjustment results in each object's hue remaining more or less constant despite large changes in illumination intensity and hue.

Color constancy does break down as the illumination source becomes more monochromatic. Selecting colors for maps to be viewed both under normal illumination conditions and under special lighting, such as in aircraft cockpits at night, is an interesting challenge that we will touch upon in Chapter 21.

COLOR MODELING SYSTEMS

Color modeling systems have been developed for several purposes by color scientists working in different professions. Cartographers have not been developers of systems, but heavy users of existing systems. In this section, we will first examine two approaches to color modeling used by researchers studying the phenomenon of color. These approaches are also widely used in a variety of artistic, industrial, and commercial activities, including cartography.

The "colorimetric" approach is based on spectrophotometer measurements of reflected light from maps and other surfaces. The best known colorimetric system is the CIE, often described as the most "objective" system. The second approach is grounded in human color perception and equal visual increments of color. We will examine two systems well suited to color map design: the Munsell and the Natural Color System. Throughout, we will show how the various systems are related to the physics of light and theories of color vision.

THE CIE SYSTEM

Beginning in 1931, the **Commission International de l'Éclairage (CIE)**, also known as the International Commission on Illumination, has developed a widely used "objective" system of colorimetry that allows the precise specification of any color in numerical terms. CIE color specifications are used not only to express results of cartographic research on color use, but also to produce maps. The printing profession now uses

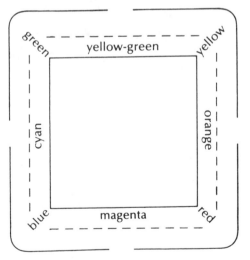

Figure 19.6 Opponent process primary (corner) and mixture (side) hues, forming a color circle (after Eastman 1986).

spectrophotometers that measure light reflected from map proof sheets and compute CIE color parameters. This means that CIE specifications for map colors can be written into printing contracts. Certain computer workstations also have color selection palettes defined in the CIE system.

There are three prerequisites for an "objective" scheme such as the CIE: (1) standard illuminants, (2) a standard observer, and (3) measurements of surface reflectance in small wavelength increments. Let's take a closer look at each of these essentials.

Standard Illuminants

A surface can reflect only the wavelengths contained in the illumination source. Reflectance is further limited by the amount of radiation at each wavelength. Several **standard illuminant** light sources are part of the CIE system. Three of these standard illuminants are relevant to cartography: (A) a tungsten incandescent lamp; (B) noon sunlight; and (C) average daylight from an overcast sky. Each standard illuminant is defined by a spectral irradiance curve giving the relative amount of energy at each wavelength (Figure 19.7).

Standard Observer

Since color is a product of the mind, any color specification system must be based in part on the characteristics of the human visual system. Thus, the CIE system incorporates the color-matching abilities of a **standard observer**, based on the response of an average, normal human eye to the spectral colors.

The standard observer is defined by the amounts of the additive primary colors—red, green, and blue—needed to match solar light at each wavelength in the visible spectrum. Hence, the standard observer "sees" according to the trichromatic theory of vision, which is appropriate for machine measurement of color. These varying amounts can be plotted as spectral irradiance curves, designated x, y, and z for the red, green, and blue primaries (Figure 19.8).

Chromaticity Coordinates

The CIE **chromaticity coordinates** are the heart of the system (see Box 19.A for a description of the computational method). A color is specified by its x and y chromaticity coordinates and the Y value, which is its **luminous reflectance** (lightness or darkness).

CIE colors can also be specified in terms of their three dimensions: dominant wavelength, purity, and luminosity. **Dominant wavelength** corresponds to hue and is defined as the wavelength of monochromatic light which, when mixed with light from the illumination source, matches the surface hue. **Purity** is the physical counterpart to chroma, defined on a scale from 0 to 100. Zero is

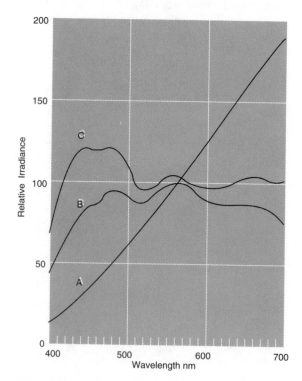

Figure 19.7 Relative spectral irradiance of CIE standard illumination sources: (A) incandescent lamp; (B) noon sunlight; and (C) average daylight from an overcast sky. (From Nimeroff).

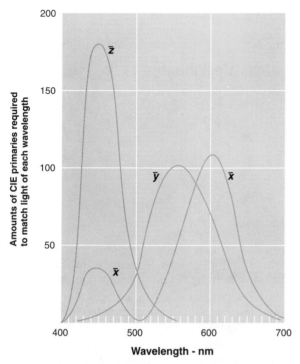

Figure 19.8 The CIE standard observer is defined by the amounts of the three selected primaries (lights) needed to be mixed to match the wavelengths of the visible spectrum.

the purity of the illumination source; 100 is the purity of the spectral hue of the same dominant wavelength. **Luminosity**, which corresponds to value, is the luminous reflectance of the color.

Dominant wavelength and purity are determined on the CIE **chromaticity diagram** (Figure 19.9 and Color Figure 19.3). This is simply a graph of *x* and *y* chromaticity coordinates. The horseshoe-shaped closed curve outlines the gamut of possible colors. Points along the curve are the coordinates of the spectral hues. The dominant wavelengths of these hues are written along the edge. Only a few are shown in Figure 19.9. The only exception is the straight line forming the bottom of the curve. Full purity magentas and purples, which fall between red and blue but are not spectral hues, fall along this line. You deter-

mine dominant wavelength and purity for a colored surface F in the following way:

1. Plot the chromaticity coordinates of the standard illuminant—for example, *C*. This is the achromatic (white) centerpoint of the diagram. You can visualize the diagram as showing all spectral hues of full purity along the edge of the horseshoe curve. All of these hues decrease in purity inward toward *C*.

2. Plot the chromaticity coordinates of the colored surface, such as $x = 0.212$, $y = 0.348$ for color F. Now draw a straight line from *C* through *F* to the edge of the closed curve. Read from the diagram the dominant wavelength at this intersection point *G*.

3. The ratio *CF* to *CG*, expressed as a percentage, defines the purity of the color. Determine this value simply by measuring the lengths of *CF* and *CG*.

The great advantage of the CIE system is that any color may be precisely specified by numerical values based on spectrophotometer measurements of light reflected from the surface. We must emphasize, however, that the system is based on the characteristics of additive primary color light mixtures, and not on the ink pigments placed on printed maps and our perception of them. There are also significant deviations between dominant wavelength, purity, and luminosity for two colors—and corresponding differences in hue, chroma, and value seen by people looking at the two colors.

Consequently, although the CIE system provides the most rigorous method for analyzing and specifying colors, which is extremely useful in cartographic production and research, it is of limited use in map design. The problem is that equal increments of dominant wavelength, purity, and luminosity do not represent equal visual differences.

It is possible to mathematically transform CIE coordinates into more visually uniform systems. Two such systems, the **CIELAB** and **CIELUV**, are widely used in the paint, dye, and electronic color industries and are of potential use in map design

Box 19.A

Computing CIE Chromaticity Coordinates

CIE chromaticity coordinates are calculated in a two-step process. The first step is a wavelength-by-wavelength summation of the product of: (1) the irradiance of the standard illuminant, (2) the irradiance of one of the three standard observer curves, and (3) the reflectance from the surface. Surface reflectance is given by a spectral reflectance curve such as for the green-blue ink in Figure 19.2B. The summation is repeated for each of the standard observer curves.

We can think of each summation as breaking the three curves into tiny wavelength increments (1 nm wide, for example), finding the height of each curve within the increment, multiplying these heights together, and summing all the products at the end. In mathematical terms, the three summations are:

$$X = \sum_{\lambda=400}^{700 \text{ nm}} R_\lambda \cdot e_\lambda \cdot x_\lambda$$

$$Y = \sum_{\lambda=400}^{700 \text{ nm}} R_\lambda \cdot e_\lambda \cdot y_\lambda$$

$$Z = \sum_{\lambda=400}^{700 \text{ nm}} R_\lambda \cdot e_\lambda \cdot z_\lambda$$

where R_l is the surface reflectance at each wavelength, e_l is the irradiance from the illumination source at each wavelength, and x_l, y_l, and z_l are the CIE spectral color matching functions for the standard observer.

The second step is to convert the X, Y, and Z summations (called the **tristimulus values** of the color) into CIE chromaticity coordinates. To make this conversion, calculate the ratios $x = X/(X + Y + Z)$ and $y = Y/(X + Y + Z)$.

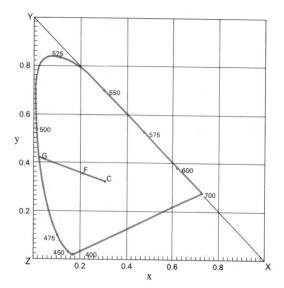

(see Box 19.B for a description of both systems).

CIELAB is designed for use with **subtractive primary** (cyan, magenta, yellow) color mixtures, such as printing inks and computer printer or plotter output. CIELUV is the companion model for emitted color such as on CRT color monitors. Quantitative measures of color difference can be computed, along with hue, lightness, and chroma values.

Figure 19.9   The CIE chromaticity diagram shows the gamut of the spectral hues by the smooth curve (locus). The sloping line between 400 and 700 nm shows the chromaticities of the purples. These do not occur in the spectrum. All physically possible chromaticities occur within these limits. See text for explanation of plotted points C, F, and G.

Box 19.B
CIELAB and CIELUV Color Models

The CIELAB and CIELUV color models are mathematical transformations of the 1931 CIE (x,y,Y) color space into a more perceptually uniform system. Both are three-dimensional color spaces derived from comparisons between colored samples of the same size and shape viewed on white and medium-gray backgrounds. The major difference between them is that straight lines between two colors on the CIE 1931 chromaticity diagram are also straight in the CIELUV color space, but not in the CIELAB.

CIELAB Equations

$$L^* = 116(Y/Y_n)^{1/3} - 16 \quad \text{for } Y/Y_n > 0.008856, \text{ or}$$

$$L^* = 903.3(Y/Y_n) \quad \text{for } Y/Y_n < 0.008856$$

$$a^* = 500[(X/X_n)^{1/3} - (Y/Y_n)^{1/3}]$$

$$b^* = 200[(Y/Y_n)^{1/3} - (Z/Z_n)^{1/3}]$$

L^* is the lightness coordinate associated with the CIE Y, and a^* and b^* are chromaticity coordinates. $X,Y,$ and Z are the CIE 1931 tristimulus values; X_n, Y_n, and Z_n are the tristimulus values of the white reference surface.

Perceptual color differences, the straight line distances between two colors in CIELAB space, are computed from the equation:

$$\Delta E^*_{lab} = [(\Delta L^*)^2 + (\Delta a^*)^2 + (\Delta b^*)^2]^{1/2}$$

The hue (h_{ab}) and chroma (L^*_{ab}) of a color can be found from its chromaticity coordinates through the equations:

$$h_{ab} = \tan^{-1}(b^*/a^*) \qquad C^*_{ab} = (a^{*2} + b^{*2})^{1/2}$$

Here h_{ab} is a hue angle beginning at red and progressing in reverse spectral order so that a hue circle results.

CIELUV Equations

$$L^* \text{ [computed exactly as in CIELAB]}$$

$$u^* = 13L^*[4x/(X+15Y+3Z) - 4x_n/(X_n+15Y_n+3Z_n)]$$

$$v^* = 13L^*[9Y/(X+15Y+3Z) - 9Y_n/(X_n+15Y_n+3Z_n)]$$

L^* is the lightness, and corresponds with the black-white opponent process in color vision. u^* is the red-green opponent chromaticity coordinate (-u^*: reddish; +u^*: greenish color). v^* is the yellow-blue opponent coordinate (-v^*: yellowish; +v^*: bluish color).

Like CIELAB, color differences are computed from the 3D Euclidian distance equation:

$$\Delta E^*_{uv} = [(\Delta L^*)^2 + (\Delta u^*)^2 + (\Delta v^*)^2]^{1/2}$$

Hue and chroma are computed and arranged as in the CIELAB system:

$$h_{uv} = \tan^{-1}(v^*/u^*) \qquad C^*_{uv} = (u^{*2} + v^{*2})^{1/2}$$

Nevertheless, at present "color appearance" systems based entirely on human judgments of hue, chroma, and value currently are the best map design tools. Several such systems have been developed and two, the **Munsell** and **Natural Color System**, are well suited to cartography.

THE MUNSELL SYSTEM

Originally developed in 1905 by A.H. Munsell, an American painter and student of color, the Munsell color system is used by industry and government for purposes ranging from specifying soil colors to checking color coding of wires. This system is popular with map designers because each of its three dimensions—hue, value, and chroma—are divided into a sequence of steps equally spaced from a perceptual point of view. These equal steps of hue, value, and chroma were determined through color difference experiments completed by thousands of people. Let's look at each of these color dimensions in the Munsell system.

Hue

In the Munsell color system, 100 equal visual increments of hue are arranged in a circle so that there is no starting or ending point (Figure 19.10). These increments are ordered according to the spectral hue progression, with the addition of purples to fill out the color palette and complete the hue circle. The clockwise progression of Munsell hues is similar to the counterclockwise progression of dominant wavelengths around the curved bounding edge of the CIE chromaticity diagram.

There are 10 major Munsell hues, arranged as shown in Figure 19.10 and Color Figure 19.4. There are five principal and five intermediate hues, referred to by their initials (for example, R for red or BG for blue-green). Artists call pairs of hues opposite each other on the circle the **complementary hues**. Complementary hues are two hues which produce gray when mixed.

The space between each major hue is divided into four equal perceptual increments. These are labeled in ascending order, such as 2.5G, 5G, 7.5G, and 10G. The principal hue is always prefaced by a 5 (for example, 5G). Today 10 increments between each principal hue form the circle, numbered from 1 to 100 and ending at 10RP.

Value

All Munsell colors are specified by value, the perceived lightness of a color relative to physically existing black and white endpoints. The Munsell **value scale** (Figure 19.11) is partitioned into 11 equal visual steps of grayness from 0/ (black) to 10/ (white). The reflectance of each step has been measured carefully, and an equation giving value as a function of reflectance now makes the scale

continuous. The scale is defined by the curve in Figure 19.12, which is the graph of a fifth-order polynomial equation giving percent reflectance as a function of value, or of a Chebyshev polynomial giving value level as a function of reflectance (see Box 19.C for equations).

Chroma

The dimension of chroma in the Munsell system is the degree to which a hue departs in "colorfulness" from a gray tone of the same value. Chroma is similar to purity in the CIE system, except that increments of chroma are visually equal. The chroma scale for each hue and value step extends from /0 for the gray value step upward in equal

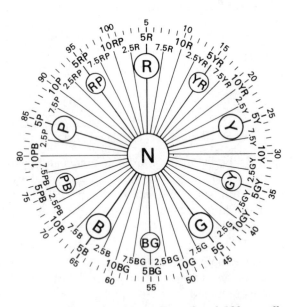

Figure 19.10 The Munsell circle of 100 equally spaced hues. The 10 major hues, 10 steps apart, are named R (red), YR (yellow-red), Y (yellow), GY (green-yellow), G (green), BG (blue-green), B (blue), PB (purple-blue), P (purple), and RP (red-purple). They increase in chroma outward from an achromatic neutral gray, N (neutral), at the center. Hues opposite one another on the circle are called complementary hues. (Courtesy Munsell Color, Macbeth Division of Kollmorgen Corporation).

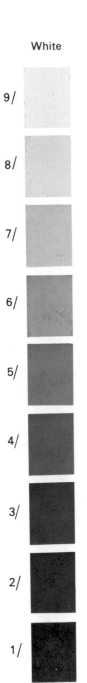

White

9/

8/

7/

6/

5/

4/

3/

2/

1/

Black

Figure 19.11 The Munsell value scale from 1 to 9 of equally spaced grays (0 is black and 10 is white). The range is continuous but is here shown in steps. (Courtesy Munsell Color, Macbeth Division of Kollmorgen Corporation).

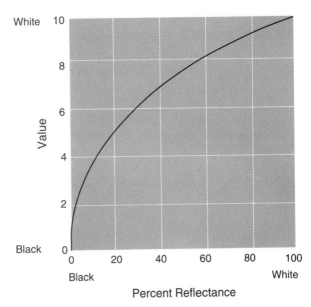

Figure 19.12 The relation between Munsell value scale and percent reflectance. The theoretical limits are pure black (0 percent reflectance from a surface and pure white (100 percent surface reflectance).

increments to the maximum for the hue and value level. Each hue and value combination has a different maximum chroma level, as Color Figure 19.5 shows. The maximum chroma range tends to be in the middle value steps.

Munsell Color Solid

The relationship among the three dimensions of color in the Munsell system can be visualized as forming a three-dimensional color solid. This solid is best likened to a circular index file, as Color Figure 19.6 illustrates. Each card in the circular file corresponds to one of the 100 Munsell hues, arranged and labeled as described above.

Value and chroma steps are arranged on each card. Two of these steps appear in Color Figure 19.5. The steps are arranged so that equal increments of value form the rows and equal steps of chroma form the columns on the card. Each card is irregular in shape due to inherent differences in

Box 19.C
Munsell Value-Reflectance Equations

The reflectance equation is $R = 1.2219V - 0.2311V^2 + 0.23951V^3 - 0.021009V^4 + 0.0008404V^5$. The inverse equation is a Chebyshev polynomial of the form:

$$V = \sum_{i=0}^{9} W_i \cdot \left(R^{\frac{1}{2}} \right)^i$$

W_i is the weighting factor given below:

i	W_i
0	-1.5323707
1	30.6459571
2	-97.1723668
3	292.9487603
4	-608.8463883
5	870.8293729
6	-840.7901111
7	521.7276980
8	-187.6371487
9	29.7284594

value among hues (yellow is intrinsically higher in value than blue, for example) and in the number of possible chroma steps.

In Munsell notation, each color chip is designated by H V/C. H is the hue designation, such as 5R; V/ is the value step; and /C is the chroma level. A strong red might be 5R 5/14, a pale pink 5R 8/6, and a deep red 5R 3/10.

Links with Other Systems

CIE chromaticity coordinates have been measured and published for the 1,500 color chips that physically make up the Munsell system. The two systems may be compared by viewing a plot of the 10 major Munsell hues on the CIE chromaticity diagram (Figure 19.13). You can see that the circular arrangement and order of the primary hues is preserved, but the spacing on the graph between hues is not uniform. Furthermore, all primary hues were at the same chroma level, but in the CIE system they vary markedly in purity. Finally, note that these chromaticity coordinates outline only a small part of the total color gamut defined by the spectral hues. This is typical of reflecting surfaces, including maps.

We shall soon examine a computer workstation color specification system based on the Munsell system. In Chapter 20 we will see how the Munsell system has been linked to color printing using screen tints.

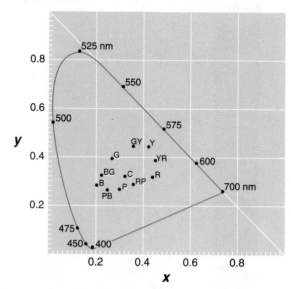

Figure 19.13 The chromaticity coordinates (x and y) of the 10 major Munsell hues (at value 5/ and chroma /6) plotted on the CIE chromaticity diagram.

THE NATURAL COLOR SYSTEM

The Natural Color System (NCS), developed by Swedish color scientists beginning in the mid-1960s, is the best known practical application of the opponent process theory of vision. The NCS is structured around the three pairs of opponent process primary colors, so that the dimensions of color are greenness, redness, blueness, yellowness, blackness, and whiteness. We can simplify these to hue, blackness, and whiteness.

Hue

NCS hues are arranged into a circle with the yellow, red, blue, and green opponent process primaries spaced 90° apart in clockwise order, as shown in Figure 19.14. All other hues are mixtures of two neighboring primaries—that is, yellow-red, red-blue, blue-green, or green-yellow mixtures. A medium gray is in the circle center, meaning that mixing equal proportions of any two opposite "complementary" hues on the circle produces gray.

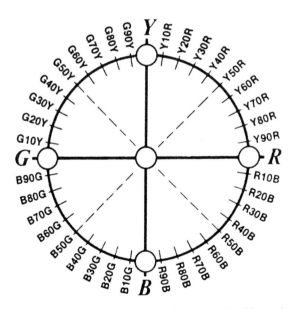

Figure 19.14 Arrangement of hues in the Natural Color System (from Hunt).

Any hue along the circle edge is specified by the beginning letter of the first primary, the percentage of the second primary, and the first letter of the second primary hue. The first primary hue is yellow, red, blue, and green (Y,R,B, and G) in quadrants one through four of the circle. For instance, a mixture hue composed of 70% yellow and 30% red is specified by Y30R. Color samples have been produced for 10% increments between each pair of primary hues, so that 40 hues form the basic system. Finer increments are, of course, possible.

Whiteness and Blackness

In addition to hue, NCS colors are defined by their whiteness and blackness. The percentages of the "pure" hue, white, and black that comprise the color are plotted on a triangular graph (Figure 19.15). White, black, and the pure hue are at the three corners of the triangle, as seen in Color Figure 19.7. Colors seen as combinations of white and the pure hue are found at the edge of the triangle along line W-Hue. The distance along the line indicates the percentage of white and pure hue in the color. For example, a color one quarter the distance from white to pure blue is 75% white and 25% blue. Artists would call combinations along this line the **tints** of the hue.

Combinations of black and the pure hue fall along the Blk-Hue line. These are the **shades** of the hue. Similarly, combinations of white and black along the line W-Blk are the achromatic gray **tones**. The sequence of tones is similar, but not identical, to the Munsell value scale.

NCS Color Solid

Inside the triangular graph are percentage combinations of the pure hue, white, and black. Each triangle is like a complete color definition card in a circular index file of hues. The NCS can be visualized as a 40-card circular file that forms a geometrical solid called a double cone (two cones glued together).

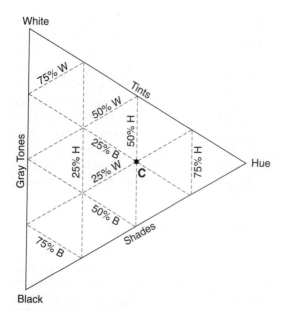

Figure 19.15 Triangular graph of the Natural Color System showing locations of shades, tints, and gray tones.

Each card is read like any other triangular graph. Color C, for example, is made of 50% pure hue, 25% white, and 25% black. The three percentages always sum to 100%, so that only two need to be quoted. The convention is to omit the whiteness percentage. Hence, colors are specified by percent blackness, percent of pure hue, and their hue label. For example, a color 25% black and 50% of a pure hue labeled B50G is specified by 25,50,B50G.

Links With Other Systems

We have seen the similarity between the NCS triangle edges and the artist's idea of tints, shades, and tones. The NCS circular file structure with a gray tone scale at its center is roughly similar to the Munsell structure. However, basing the hue circle on opponent process primaries and using whiteness and blackness as color dimensions instead of chroma makes the NCS quite different in concept from the Munsell. CIE chromaticity coordinates have been calculated for all NCS color samples, and dominant wavelengths, purities, and lumi-

nosities are known for each. In the next chapter we shall see that NCS whiteness, blackness, and hue combinations can be converted into screen tint specifications for lithographic printing.

COMPUTER ELECTRONIC DISPLAY COLOR MODELS

Today cartographers commonly design color maps on computer workstations. The **modeling of color** on electronic displays is a frequent and important part of their work.

A basic requirement for any electronic display color modeling system is that colors have unique positions in a three-dimensional "color space" large enough to contain all possible combinations of the additive primaries. Let's look at three such color modeling systems: (1) the Red, Green, Blue (RGB); (2) the Hue, Lightness, Saturation (HLS); and (3) the Hue, Value, Chroma (HVC).

RGB COLOR MODEL

The **Red, Green, Blue (RGB)** system may be visualized as a cube with positions specified by x, y, and z integer coordinates. These coordinates control the intensities of the red, green, and blue light in the color (Figure 19.16 and Color Figure 19.8). The range of integers defines the number of colors in the cube. The maximum normally is 256 increments of red, green, and blue, ranging from 0 to 255. This gives 256^3 or 16,777,216 possible color combinations, far more than the eye can distinguish, let alone display simultaneously on a CRT monitor screen. Most microcomputer mapping systems allow you to select only a small subset, usually 16 or 256 colors, whereas more expensive workstations normally have larger palettes.

In the RGB cube, the (0,0,0) corner position designates black (without emitted light), while the diagonally opposite corner (255,255,255 in a system with the maximum increments) denotes white. The RGB gray scale is along the diagonal line. The

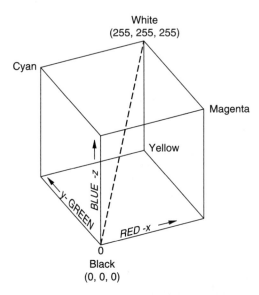

White
(255, 255, 255)

Cyan

Magenta

Yellow

BLUE -z

Y- GREEN

RED -x

0

Black
(0, 0, 0)

Figure 19.16 Diagram illustrating the RGB color model. The colored diagonal line is the locus of equal amounts of the three additive primaries and is thus the progression of the achromatic values (neutral grays) from black to white.

additive primaries are found at the three corners of the cube adjacent to black. Mixtures of pairs of additive primaries—cyan = blue + green, magenta = blue + red, and yellow = red + green—are positioned at the other three corners.

HLS COLOR MODEL

The **Hue, Lightness, Saturation (HLS)** color system was originally developed in the 1970s by *Tektronix, Inc.*, formerly a large producer of CRT monitors. The HLS color solid is a double cone, as in the Natural Color System, but the use of hue, lightness, and saturation (the latter two similar to value and chroma) makes it more akin to the Munsell system (Figure 19.17). However, increments of hue, lightness, and saturation are not perceptually equal.

The double cone is composed of hue circles at different lightnesses. In all circles, hues are arranged

identically in a spectral sequence, with additive and subtractive primary hues alternating around the circle. Each hue is specified by its angular distance counterclockwise from blue, the Tektronix starting hue. The gray tone scale is along the vertical axis, with lightness ranging from 0 (black) at the bottom to 1.0 (white) at the top of the double cone. Within each hue circle, saturation ranges from 0 at the vertical axis to 1.0 at the circle edge.

Hue, lightness, and saturation numbers are related to RGB coordinates through a set of simple mathematical equations. Thus, you can make conversions between HLS and RGB quickly and precisely.

HVC COLOR MODEL

The **Hue, Value, Chroma (HVC)** color modeling system was recently developed by Tektronix Laboratories as an electronic version of the Munsell color system for high-performance computer workstations. The three-dimensional color space is identical in appearance and structure to the Munsell color

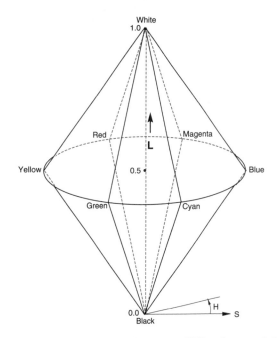

White
1.0

Red

Magenta

L

Yellow

0.5

Blue

Green

Cyan

0.0

Black

H

S

Figure 19.17 Double hexcone HLS color model (adapted from Foley and Van Dam).

solid (Figure 19.18). The major difference lies in the color notation. Each hue is specified by its angular distance counterclockwise from red. Value is on a scale from 0 (black) to 100 (white). Chroma varies from 0 at the central axis to somewhere between 1 and 100 at the outer edge, depending on the color's hue and value. Of course, increments of hue, value, and chroma are perceptually equidistant.

The HVC system was created by determining the CIE coordinates for many thousands of screen colors, and relating these to similar coordinates for Munsell colors. Consequently, there is no simple relationship between HVC and RGB or HLS color specifications.

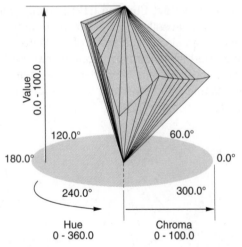

Figure 19.18 HVC color model components (reproduced with permission of Tektronix, Inc.)

SELECTED REFERENCES

Eastman, J. R., "Opponent Process Theory and Syntax for Qualitative Relationships in Quantitative Series," *The American Cartographer*, 13 (1986): 324–33.

Evans, R. M., *An Introduction to Color*, New York: John Wiley & Sons, 1948.

Foley, J. D. and A. Van Dam, *Fundamentals of Interactive Computer Graphics*, Reading, MA: Addison-Wesley, 1982.

Hard, A. and L. Sivik, "NCS—Natural Color System: A Swedish Standard for Color Notation," *Color Research and Application*, 6 (1981): 129–38.

Hunt, R. W. G., *Measuring Colour*, Chichester, UK: Ellis Horwood Ltd., 1987.

Hurvich, L. M. and D. Jameson, "An Opponent-Process Theory of Color Vision," *Psychological Review*, 64 (1957): 384–404.

Munsell, A. H., *A Color Notation*, 12th ed., Baltimore: Munsell Color Co., 1975.

Nimeroff, I., *Colorimetry, National Bureau of Standards Monograph 104*, Washington, D.C.: U.S. Government Printing Office, 1968.

Shellswell, M. A., "Towards Objectivity in the Use of Colour," *The Cartographic Journal*, 13 (1976): 72–84.

Taylor, J. M., G. M. Murch, and P. A. McManus, "Tektronix HVC: A Uniform Perceptual Color System for Display Users," *SID Digest* (1988): 77–80.

CHAPTER 20

COLOR AND PATTERN CREATION AND SPECIFICATION

*C*artographers frequently design and construct multicolor maps on printed surfaces or electronic displays. In both cases, they must be able to specify colors and patterns in a way that is clear, logical, and allows the maps to be reproduced faithfully.

Color and pattern specification is closely tied to the way in which color is physically created on the map. Three major color creation technologies are in common use: lithographic printing, electronic color display, and plotted or printed hardcopying of electronic display images. We will examine each of these in this chapter, showing the links among color creation, color specification, and color models where possible. In the concluding section, we will delve into the problems associated with accurate reproduction of CRT monitor colors on computer printers.

COLOR AND PATTERN FOR LITHOGRAPHIC PRINTING

In lithography, the printing surface either accepts or does not accept ink. Consequently, there are two ways to print colors: (1) completely cover the area with an ink of the desired color, or (2) print arrays of tiny, closely spaced dots of one or more basic colors which, when overlaid and seen in combination, form the desired colors. In either case, the ink colors must be specified prior to printing.

PRINTING INK SPECIFICATION

Printers specify ink colors by using a standardized color matching system, and the *Pantone Matching System (PMS)** has become the North American standard. The basic PMS ink colors are the four "process" inks—cyan, magenta, yellow, and black—plus white, warm red, rubine red, rhodamine red, purple, process blue, reflex blue, and green. Nearly

---

*The *Pantone Matching System (PMS)* is a registered trademark of *PANTONE, Inc.*, Moonachie, N.J. 07074.

600 different PMS ink colors can be made by mixing two or three basic inks in suitable proportions. Ink colors are selected from a numerically coded **swatch book**. Proportions of basic inks needed to make the color are printed below each swatch.

PREPRINTED TINTS AND PATTERNS

Several decades ago, large numbers of maps were constructed using preprinted tints and patterns. These were cut out of preprinted sheets, positioned over a map area, trimmed along the area boundary with an x-acto knife or razor blade, and permanently burnished onto the map sheet. Different map sheets were required for each ink color if the map was to be lithographically printed. An ever diminishing number of maps are still constructed in this manner, using commercially available materials. Sheets of commonly used patterns may be purchased, some with different orientations and textures (Figure 20.1).

Tints are available in different percentages and rulings (see the next section, *Screen Tints*, for definitions of percentages and rulings). Most commercially marketed tints range from 10 to 70 percent black in rulings from 27.5 to 85 lines per inch.

When all tints and patterns have been burnished onto the map sheet, a lithographic negative of the sheet is made for platemaking. Problems with inconsistent symbol blackness across the preprinted sheet often become visible on the negative. Tints and patterns can also lighten or

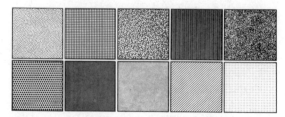

Figure 20.1 Examples of preprinted symbols, in this case *Zip-A-Tone*. A few of the maps and diagrams in this book have been prepared with the aid of these kinds of symbols.

darken by several percent due to film exposure and processing inconsistencies. These problems, along with the lack of finely ruled tints needed for multicolor printing, make preprinted tints and patterns less popular for printed maps than lithographic film screen tints and pattern negatives or positives. However, digital screening and patterning is rapidly replacing both of these manual methods.

SCREEN TINTS

Although simple color maps may be printed with a different printing ink for each map color, this procedure becomes very expensive when more than a few colors are needed. A better method is to create a large number of colors from a small number of printing inks by breaking each colored area into an array of tiny dots of a given size. This is done using **screen tints**, which are film sheets filled with clear or black dots of the same size arranged in a rectangular pattern (Figure 20.2). Screen tints are used with open window negatives (see Chapter 31), which mask out all light coming through the screen except within the desired areas on the map. Consequently, screen tints can be re-used, and no manual cutting and trimming is necessary.

Each screen tint is specified by its ruling and its percentage. The **ruling** is the number of **lines of dots per inch (lines/in.** or **lpi.)** or **lines per centimeter (l/cm.)** Screen rulings of 65, 85, 100, 120, 133, and 150 lpi. are produced in the United States. A set of 65 lpi. screens is considered coarse, because individual dots can be seen, while a 150 lpi. set is considered fine, since no pattern is visible and only a smooth tonal effect results. Cartographers usually use the 133 or 150 lpi. rulings for hue and value progressions, and the 65 lpi. ruling for pattern-value color series.

In the United States, the screen tint **percentage** is the percent of incident light transmitted through the screen to a printing plate. This means that screen tints are film "negatives"; therefore, an 80 percent screen is composed of small opaque dots, whereas a 10 percent screen is made of small

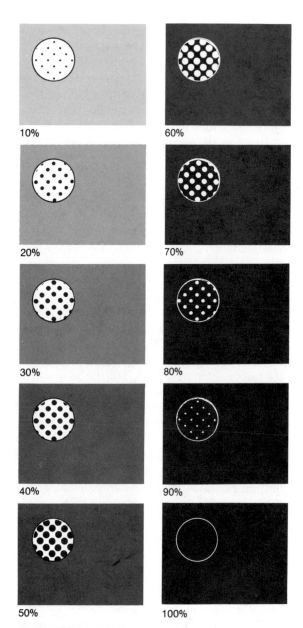

Figure 20.2 A series of dot screen tints with enlargements showing the shapes of the dots. In the United States, screens are negative. Thus, a screen producing a 30 percent tint will actually look like a 70 percent screen.

transparent dots. Map printing in Europe usually employs "positive" screen tints, so that what you see on the screen tint is what will be printed.

The screen tint percentage ideally will equal the percent of the printed area on a map covered by dots of ink. Unfortunately, printing ink tends to spread slightly when applied to paper, so the screen tint percentage is only approximately correct.

Each set of commercially available percentage calibrated screen tints includes 10, 20, 30, 40, 50, 60, 70, 80, and 90 percent. Some series also include 5, 7.5, 15, and 25 percent.

Two-Ink Combinations

When one printing ink is used, the selection of screen tints is usually limited to around 10 available gradations, although computer-driven laser screening techniques now allow us to produce dots of any size and ruling. In either case, we are still limited to tints of a single hue.

Two inks provide a vastly greater number of color combinations. For instance, 10 gradations of two printing inks give 100 different color combinations. Many of these will be virtually indistinguishable from their neighbors, but the increase in usable colors is still dramatic.

Since ink spread makes faithful duplication of screen tint dots highly unlikely, it is best to specify colors by viewing a printed two-color chart, such as Figure 20.3. The color combinations on the chart will be specified by the component screen tint percentages, and the small errors in printed dot sizes are of no consequence.

Process Color Combinations

The two-color combination method is easily extended to three printing inks, again greatly increasing the number of color combinations. However, the fullest color palette is obtained by going one step further and using **four-color process** printing.

The four process ink colors are the three subtractive primaries: cyan, magenta, and yellow (Color Figure 20.1), plus black. When the many combinations of screen tint percentages are overprinted, the process inks produce the greatest range of hues possible through lithographic printing.

As with two-color combinations, the best procedure is to specify process colors from a three-color or four-color chart, such as the three-color chart section in Color Figure 20.2. After you select the colors, you can read their cyan, magenta, and yellow percentages from the chart edge. Similar charts can be created for three- and four-color combinations of non-process inks, but the color range will not be as great.

Monochrome Screen Tint Combinations

In monochrome printing, up to three screen tints can be overprinted. You can often use this fact to advantage. Sometimes, for example, screen tints called for on the basis of good design principles may not match the limited set on hand. Combining screens to form composite tints may be the only way to create the desired screen percentages.

A possible drawback of this method of creating custom screen tints is the fact that the resulting blackness (percent area inked) will be less than the sum of the component screen percentages. The reason for this discrepancy is that a certain percentage of the dots will overlap completely or partially when two or more tints are overprinted. Fortunately, the resulting percent area inked can be predicted by the following equations for two (P_{1+2}) and three (P_{1+2+3}) screen combinations:

$$P_{1+2} = P_1 + P_2 - (P_1 \times P_2)/100;$$

$$P_{1+2+3} = P_1 + P_2 + P_3 - (P_1 \times P_2 + P_2 \times P_3 + P_1 \times P_3)/100 + (P_1 \times P_2 \times P_3)/10,000.$$

Table 20.1 lists predicted percentages for various two-screen combinations.

Special-Purpose Screen Tints

A number of screen tints have been developed in addition to those composed of regular dot arrays. Cartographers periodically use the special-purpose screen tints described in the following sections.

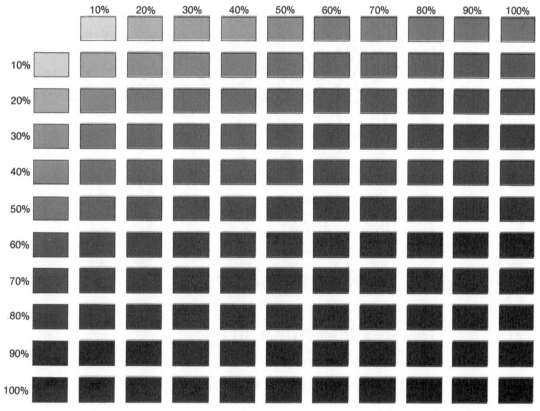

Figure 20.3 A two color screen tint chart showing the possible combinations of the two inks used in printing this book. It is apparent that relatively few of the combinations are usable, especially when both ink colors are dark.

Table 20.1

Percent Area Inked for Various Combinations of Two Dot Screen Tints

Screen 1 (%)									
10	19								
20	28	36							
30	37	44	51						
40	46	52	58	64					
50	55	60	65	70	75				
60	64	68	72	76	80	84			
70	73	76	79	82	85	88	91		
80	82	84	86	88	90	92	94	96	
90	91	92	93	94	95	96	97	98	99
	10	20	30	40	50	60	70	80	90

Screen 2 (%)

Bi-Angle Screen Tints Fine lettering and linework is sometimes tinted using **bi-angle screen tints** (Figure 20.4). These are created by photographically superimposing two screen tints rotated 30° to each other to eliminate any moiré patterns (see "Moiré Avoidance" later in this chapter). The component screen tints often are of a fine ruling, such as 200 lpi. (80 l/cm.) At this ruling, the composite screen tint increases the readability of finely detailed lettering, thin linework, and other symbols that appear to have slightly "jagged" edges when tinted using standard screen tints.

Parallel Line Screens Parallel line screen tints are rarely used in the United States, but are popular among European cartographers. These are avail-

BI-ANGLE TINT SCREENS ENLARGED (x10)

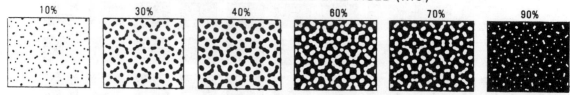

30% BI-ANGLE TINT SCREEN

30% BI-ANGLE TINT SCREEN

40% BI-ANGLE TINT SCREEN

40% BI-ANGLE TINT SCREEN

50% BI-ANGLE TINT SCREEN

50% BI-ANGLE TINT SCREEN

60% BI-ANGLE TINT SCREEN

60% BI-ANGLE TINT SCREEN

Figure 20.4   Bi-angle screen tints.

able in the same range of rulings and percentages as dot screen tints. Line tints may be used in place of dot screen tints, and also can be combined in a colored area, often allowing more than four inks to be overprinted.

Random Dot Screens   Uniform colors and gray tones can be produced using screen tints created by randomly placing equal sized dots on the screen (Figure 20.5). Different screen percentages are made by increasing or decreasing the number of dots on the screen. The result is a screened area with a "pointillist" texture. The advantage of such **random dot screens** is that moiré patterns never occur when tints are overprinted.

Pattern Screens

Map pattern master sheets can be photographed onto large sheets of lithographic film to create pattern screens that are used like screen tints. Pattern negatives are commonly used, but pattern positives can also be employed to create reverse pattern screens such as those in the bottom row of Figure 21.2. The pattern master can also be photographically reduced or enlarged to create a set of fine to coarse pattern negatives and positives.

Pattern screens are specified by name or number code, and there is no standard pattern specification system. Fortunately, a number of pattern names, such as marsh, have become conventional terms that define a specific arrangement of graphic marks.

MOIRÉ AVOIDANCE

Screen tints must be used in a particular manner to avoid introducing a usually undesirable "**moiré**" pattern into a colored area (Figure 20.6). Over-

Figure 20.5   Random dot screen tints.

printing screen tints of different rulings will always create a moiré pattern, as will incorrect screen tint orientation.

Moiré patterns are minimized by orienting screen tints 30° apart, and 15° apart when a 30° angular separation is not possible. With process inks, the normal practice (Color Figure 20.3) is to orient all black ink screen tints at 45° from horizontal, magenta screen tints at 75°, cyan screens at 105°, and yellow tints at 90°. This insures that any possible moiré introduced by the 90° tint is in the lightest, least visible ink. With other ink combinations, the rule is to orient the darkest ink at 45°, and the lightest at 90° if four inks are used.

You can use a mechanical screen angle gauge or a graphic template to orient screens at the proper angle (Figure 20.7). Or you can purchase pre-angled screens or make a set ahead of time.

LINKS WITH COLOR MODELS

The color models introduced in Chapter 19 serve as a way to control color effects in map design and production. To take advantage of the insight each model provides, we must link model attributes to tint screen percentages and printing ink characteristics. We will do so here with the three color models we discussed in Chapter 19, beginning with the CIE system.

Figure 20.6   Screen tints should be aligned with an angular separation of 30 degrees. Other angles can produce various moiré patterns such as the one shown.

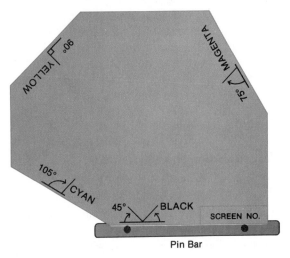

Figure 20.7   A graphic template can be used to trim standard screen tints so that they can be quickly and accurately positioned to achieve proper angular separation between successive screen exposures.

CIE

Color scientists have determined the chromaticity coordinates for the process color and other basic inks when printed on various types of white paper. Today printers routinely use spectrophotometers that compute the chromaticity coordinates of printed color bars on the edge of the map sheet. This not only is a quality control measure, but also is proof that inks specified by CIE coordinates in map printing contracts are being used.

Chromaticity coordinates can also be determined for the colors selected from process ink or other color charts. This is done from spectrophotometer measurements made on the chart for each color selected, or by using equations to predict the chromaticity coordinates based on the reflectance curves and screen tint percentages of the component printing inks. The former method is more reliable at present, but this type of overall quality control is rarely used at present.

Munsell

Recently, cartographers have determined the process color screen tint combinations needed to match the value and chroma steps associated with each of the 10 major Munsell hues. One page from the first process ink color chart for the Munsell color system is presented in Color Figure 20.4. Note the yellow, magenta, cyan, and black screen tint percentages needed to match each Munsell color sample. As you can see, no more than three of the four inks need to be overprinted. This holds true for all 10 major hues.

Natural Color System (NCS)

NCS hue, whiteness, and blackness combinations can be approximated by screen tint combinations. A sequence of low to high percentage screen tints for a printing ink matching a NCS hue inherently produces the tints along the W-Hue line on the hue's triangular graph. This same screen tint sequence used with black ink gives the progression of gray tones along the W-Blk line. The shades along the Blk-Hue line are formed by printing the

black ink screen tint sequence over areas of solid hue. Screen tint combinations of the hue and black ink create the colors in the triangle interior.

The approximation will be close if the printing ink is mixed to match the chromaticity coordinates of the NCS hue. A far less exact but still useful approximation can be made using cyan, magenta, and yellow process inks. To begin, we find the process ink combinations for red, blue, and green (Figure 20.8). From the theory of subtractive primary colors, we know that 100% red = 0% cyan + 100% magenta + 100% yellow; 100% blue = 100% cyan + 100% magenta + 0% yellow; and 100% green = 100% cyan + 0% magenta + 100% yellow. Knowing this, we can specify the other colors in the NCS hue circle by simple linear interpolation. For example, the Y50R hue is approximated by 0% cyan + 50% magenta + 100% yellow (0C 50M 100Y).

If we know the screen tint percentages for an NCS hue, we can specify the percentages needed to approximate any hue, whiteness, blackness combination in the triangular graph (Figure 20.9). Again, we do so by linear interpolation. Hence, the NCS tint 0,50,R—halfway between white (0C 0M 0Y) and red (0C 100M 100Y)—is approximated by (0C 50M 50Y).

The cyan, magenta, yellow screen tint specifications will only roughly approximate the actual NCS colors. This is because NCS hues, whitenesses, and blacknesses are based on visual appearance ex-

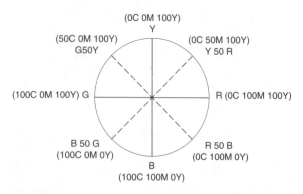

Figure 20.8 Process color screen tint combinations (in parentheses) approximating major NCS hues.

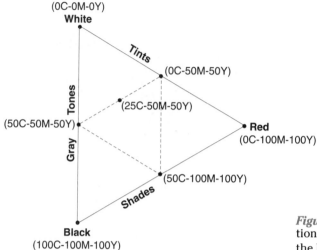

Figure 20.9 Process color screen tint combinations that approximate shades, tints, and tones for the NCS red hue.

periments with painted color chips. A Y50R hue will appear to be halfway between yellow and red, but it is unlikely that the corresponding (0C 50M 100Y) process ink color will appear to be exactly halfway between printed red and yellow. The same holds for the shades, tints, and tones in each NCS triangular graph. Nevertheless, this simple approximation method captures the essence of the NCS system, and provides many color sequences useful in map design.

CONTINUOUS TONE CREATION BY HALFTONING

Until now we have focused on the creation and specification of printed uniform color using screen tints. Many maps, of course, contain continuous tone features such as relief shading or satellite images. Printing these continuous tones requires transforming the colors or gray tones into varying sized dots on the negatives used in platemaking. The goal is to produce a negative on which the size of the clear dots relative to the opaque areas between them is in direct proportion to the variations in the lightness and darkness of tones on the continuous tone original (Figure 20.10).

To understand how this is done, it is convenient to think of the continuous tone original as consisting of three main types of tones: highlights, midtones, and shadows. The immediate aim is to produce a negative on which the lightest highlights are represented by small clear dots, while the

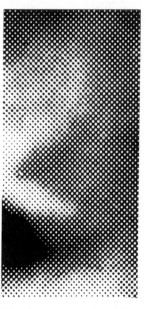

Figure 20.10 How a halftone looks under magnification. The illustration shows clearly that the number of dots per unit area remains uniform; only their sizes vary. Compare with any photograph in this book under a magnifying glass.

darkest shadows are represented by tiny black dots (Figure 20.11). Dots that just begin to touch or connect, forming a checkerboard pattern, should occur in the midtones.

This conversion of a continuous tone image to a reproducible set of continuous varying dot sizes is accomplished using a **contact halftone screen** (Figure 20.12). This screen is made on flexible plastic film and is used in contact with the emulsion of the film negative. The varying sized dots* are created by adjusting light intensity through the optical action of the vignetted dot pattern on the screen. The vignetted dots are produced by a dye that is thicker near the centers of the dots. Therefore, the size of each halftone dot depends on the amount of light that is able to pass through the dye.

Contact halftone screens are available in typical rulings of 100, 133, 150, 200, or 300 lines/in. The finer the ruling, the smoother and more natural the result will be. To eliminate moiré patterns, however, we must match the ruling to any screen tints that might be overprinted. Thus, 133 or 150 lpi. contact screens are most often specified.

Care should be taken when specifying the orientation of contact screens. With monochrome reproduction of continuous tone originals, the most visually pleasing copy results when the screen is oriented 45° to the horizontal. It is, of course, also possible to overprint solid or screen tinted symbols using up to three other ink colors, as long as one is a high reflectance ink.

Multicolor continuous artwork, such as "natural" landform coloring, is reproduced by color separating the image into its blue, green, red, and lightness components using a graphic arts camera with blue, green, red, (and no) filters placed over the lens. The contact screen between the lens and negative film must be correctly oriented for each of the four successive exposures, with 90° used for blue, 75° for green, 105° for red, and 45° for the lightness separation. These negatives are then used to expose the

---

*The "dots" on these halftone screens may be round, square, or elliptical. Each of these dot structures produces slightly different effects in the final product.

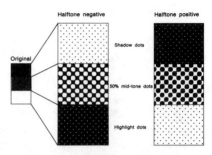

Figure 20.11 A good halftone negative will exhibit a range of density from very small clear dots in the highlight areas to tiny black dots in the shadow areas. All other tones will be represented by dot sizes falling between these extremes.

yellow, magenta, cyan, and black ink plates used to print the map (Color Figure 20.5).

Since the four screen angles are used in the color separation process, only solid ink symbols of the process or other ink colors can be overprinted. Solid black symbols are most commonly used, since their color is not changed by the continuous tone or color background.

DIGITAL SCREEN TINTS AND PATTERNS

The latest and most dramatic development in screen tint, halftone, and pattern production is electronic image-setting. Conventional mechanical screens and pattern negatives are replaced by a laser scanner controlled by a computer program.

The computer-driven scanner arranges clusters of microscopic dots—up to 2,540 per inch—to form screen dots, patterns, and all other map symbols. As the digital file representing the map is exposed on lithographic film or directly onto the printing plate, each digitally specified tint or pattern is translated into an appropriate dot size and shape, according to the ruling selected. The translation process is called **dithering**. Dithering refers to any mathematical procedure that creates the illusion of uniform or continuous tone images by the judicious arrangement of binary picture ele-

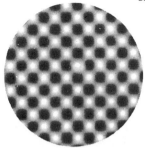

Figure 20.12 The vignetted dots on a contact half-tone screen are produced by a dye that is thicker near the centers of the dots. The size of the halftone dot depends on the amount of light that is able to pass through the dye. (Enlarged and reproduced in halftone).

			143	127	119	98	115	123	139					
		130	110	95	79	70	75	91	111	131				
	134	102	86	67	55	38	51	63	87	103	135			
138	106	82	59	43	31	22	32	44	60	83	107	144		
122	90	62	42	26	15	10	16	27	45	68	96	128		
114	74	50	30	14	9	5	6	17	33	56	80	120		
101	73	41	25	13	4	1	2	11	23	39	71	99		
118	78	54	37	21	8	3	7	18	34	52	76	116		
126	94	66	49	29	20	12	19	28	46	64	92	124		
142	109	85	58	48	36	24	35	47	61	84	108	140		
	137	105	89	65	53	40	57	69	88	104	136			
		133	113	93	77	72	81	97	112	132				
			141	125	117	100	121	129	145					

Figure 20.13 Spiral dither pattern used to create digital screen tints. Gray area is a 10 percent dot.

ments (pixels). Dithering is also the basis of gray tone and color production on computer output devices such as dot matrix and laser printers.

Dithered Screen Tints

Dithering is based upon dividing an area into cells, which are further subdivided into a large number of pixels. Cells may be squares or hexagons that completely fill the area. These can be rotated to any desired screen angle. Square cells oriented horizontally will be examined here.

Each pixel is made black or clear, depending on the screen percentage. The number of pixels in each cell is determined by the resolution of the image-setting device and the screen ruling specified. For example, a 2,540 lpi. image-setter creating 150 lpi. screen tints will need 2,540/150 or 17 pixels (to the nearest integer) per line of dots. This means that there will be 289 pixels per cell, or a maximum of 289 different screen tint percentages (Figure 20.13). As an example, a 10 percent screen tint requires blackening pixels 1-29. The numbers in this figure indicate the order in which pixels will be turned black as the screen percentage increases. The pixel arrangement is called a **spiral dither**, because the pixels spiral clockwise outward from the cell center, with each additional pixel placed so as to keep the dot as symmetrical and circular as possible.

Only half the pixels need to be addressed. This is because for dots greater than 50 percent the entire cell is first turned black; then the required number of pixels are turned white. A 60 percent screen tint, for instance, will have 40 percent white pixels. Pixels 1-116 are selected so that the white dot is centered in the cell. This pixel arrangement and procedure will produce a series of screen tints virtually identical to traditional film screen tints (Figure 20.14) and has the advantage that screens of any percentage can be created. For multicolor printing, cells are mathematically rotated to the correct screen angle and the pixels falling within each cell are mathematically computed.

Dithered Continuous Tone

Continuous tone features can be dithered so as to closely resemble halftone screening. We can imagine a grid of cells laid over the continuous tone original, so that the reflectance within each cell can be measured. An image-scanning device does just this. A mathematical algorithm

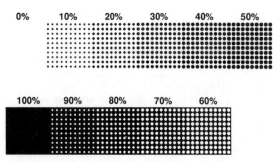

Figure 20.14 Digital screen tints from spiral dithering (enlarged so that individual dots can be seen).

next relates reflectance to the number of pixels that should be blackened in each cell.

The dithering procedure for digital halftones is different than for screen tints, since a 50 percent toned area should look like a checkerboard. The arrangement of pixels in Figure 20.15, called a **classical dither**, produces this pattern, plus the smaller black or white dots such as the two 10-percent black dots outlined in the 16 × 16

124	110	94	74	76	96	112	126	134	148	164	184	182	162	146	132
108	72	64	46	48	50	66	114	150	186	194	212	210	208	192	144
92	62	30	16	14	16	52	98	166	196	228	242	244	240	206	160
88	44	22	4	2	12	34	78	170	214	236	254	256	246	224	180
86	42	20	6	8	10	36	80	172	216	238	252	250	248	222	178
90	60	32	26	24	28	54	100	168	198	226	232	234	230	204	158
106	70	58	40	38	56	68	116	152	188	200	218	220	202	190	142
122	120	104	84	82	102	118	128	136	138	154	174	176	156	140	130
129	139	155	175	173	153	137	135	127	117	101	81	83	103	119	121
141	189	201	219	217	199	187	151	115	67	55	37	39	57	69	105
157	203	229	233	231	225	197	167	99	53	27	23	25	31	59	89
177	221	247	249	251	237	215	171	79	35	9	7	5	19	41	85
179	223	245	255	253	235	213	169	77	33	11	1	3	21	43	87
159	205	239	243	241	227	195	165	97	51	17	13	15	29	61	91
143	191	207	209	211	193	185	149	113	65	49	47	45	63	71	107
131	145	161	181	183	163	147	133	125	111	95	75	73	93	109	123

Figure 20.15 Classical dither pattern used in digital halftoning. The two gray dots in total cover 10 percent of the dither cell.

pixel cell. The resulting image is strikingly similar to a traditional halftone when viewed under magnification (Figure 20.16).

Other Types of Dithering

Different pixel blackening sequences produce different forms of screen or pattern tints. One example is a parallel line tint or pattern. The pixel arrangement within each 64-pixel cell shown in Figure 20.17 will produce the line patterns shown in Figure 20.18. A 17 × 17 cell could be used to create 150 lpi. parallel line screen tints.

Randomly blackening the pixels in each cell, called **random dithering**, produces random screen tints identical to Figure 20.5, since the master random screen tints for the illustration were produced in this manner. This method suffers from the ever increasing number of random number pairs that must be generated to created darker tints. This problem occurs even when white and black backgrounds are reversed, as in the spiral dither, so that no more than a 50 percent screen need be computed. The number of duplicate random numbers rises steadily as the screen percentage increases. Fortunately, the total random number pairs (N) required for any random screen tint can be accurately predicted by the equation:

$$N = \ln(1-T) / \ln(1-P),$$

where T is the proportion of each cell to be blackened and P is the proportion of the total cell occupied by each pixel.

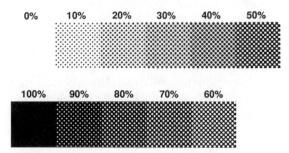

Figure 20.16 Classical dithered gray scale (enlarged so that individual dots are visible).

64	62	60	58	57	59	61	63
48	46	44	42	41	43	45	47
32	30	28	26	25	27	29	31
16	14	12	10	9	11	13	15
8	6	4	2	1	3	5	7
24	22	20	18	17	19	21	23
40	38	36	34	33	35	37	39
56	54	52	50	49	51	53	55

Figure 20.17 An 8 × 8 dither cell for parallel line screens.

COLOR AND PATTERN FOR COMPUTER PLOTTERS AND PRINTERS

The forms of dithering examined above are well suited to high-resolution laser image-setting devices creating film negatives or positives for lithographic printing. However, these are not appropriate for lower-resolution computer output devices such as line plotters or dot matrix, electrostatic, ink jet, or thermal printers. We shall see in this section how colored and patterned areas are created with these devices.

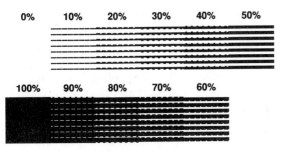

Figure 20.18 Parallel line screen gray scale.

LINE PLOTTER COLOR AND PATTERN CREATION

Pen plotters are computer-controlled devices. They use ink pens to draw line segments, defined by (x,y) coordinate vectors, on plotting paper or plastic film sheets. Pen plotted vector symbols generally have sharper edges than the slightly jagged edged raster symbols. Plotter inks tend to give highly saturated colors, particularly if plotted on glossy rather than matte finish paper. A limited number of ink colors is available, normally including black, brown, and the major spectral hues.

The major disadvantage of plotted color is the long time and expense involved in filling areas. Solid colors are produced by drawing hundreds to thousands of closely spaced parallel lines. Any gaps or line overlaps will cause streaks in the colored area. Desaturated inks are not available, and plotting arrays of tiny dots to simulate printed screen tints is impractical.

Coarse-textured tint and shade progressions, as well as mixture hues, can be created by ink overplotting. Plotting black parallel lines, cross-hatched lines, or dots over a solid inked area gives color shades. Mixture hues result when parallel lines of two or more ink colors are overplotted at different angles. Tint progressions are often made by a sequence of dotted, dashed, continuous, and cross-hatched lines of increasing percent area inked (Figure 20.19). In all cases the symbol pattern is clearly visible, but if the colors are viewed from a distance the eye blends the marks into an impression of shade, tint, or a new color.

Area patterns are plotted from mathematical templates called **pattern primitives**. These are nothing more than strings of coordinates defining the exterior boundary and interior fill for all symbols that make up one cycle (repetition) of the pattern. These coordinates can be scaled or

Figure 20.19 A tint progression created by plotting different patterns at increasingly finer textures.

rotated, and the symbol ink color or colors can be changed within the mapping program. Their disadvantage again is the long time required to fill an area with plotted vector symbols.

COMPUTER PRINTER COLORS AND PATTERNS

In contrast to vector plotters, computer printers produce colors and patterns in a raster cell format. As a result, the complexity of the color or pattern has no effect on the speed of production. Each cell is processed regardless of whether it is empty or fully colored. Some printers are mechanical in design, with colors or patterns created by striking one or more inked ribbons against the paper in typewriter fashion. Other printers are electronic in design, producing graphic images by electrostatic, thermographic, ink jet, or laser technologies (see Chapter 30 for detailed descriptions of each). Some of these devices print one whole character per cell, just as the standard typewriter does. Others create the image in each cell by printing columns or arrays of tiny pixels. The latter "dot" printers are well suited to cartographic output because of their higher resolution of up to 400 lpi. and the ability to control which combination of pixels will be printed in each cell.

We can better understand how color and pattern are produced with dot printers by studying how one such device, the dot matrix printer, operates. The dot matrix printhead normally consists of a column of eight tiny pins that strike the printer ribbon (Figure 20.20) thousands of times along each raster line. Each pin creates a pixel on the printed surface. The combination of eight pins striking the ribbon is determined by the binary number (8-bit byte) controlling the pins at any instant. A value of 7 causes the lowest three pins to strike the paper, for example. Cells are made by grouping the columns of pixels. Eight consecutive columns form an 8×8 or 64-pixel cell; two adjacent rows of pixels 16 columns wide form a 16×16 or 256-pixel square; and so on. At 300 lpi., each 16×16 cell will be a small 0.05" $\times$ 0.05" (0.135cm $\times$

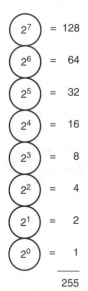

$$2^7 = 128$$
$$2^6 = 64$$
$$2^5 = 32$$
$$2^4 = 16$$
$$2^3 = 8$$
$$2^2 = 4$$
$$2^1 = 2$$
$$2^0 = 1$$
$$\overline{255}$$

Figure 20.20 How printer pin firings are stored as a binary number (from Plumb and Slocum).

0.135cm) square. Colors and patterns are formed by dithering and bit-mapping (see *Bit-Mapped Patterns* later in this chapter) within cells.

Dispersed-Dot Ordered Dithering

Spiral and classical dithering are well suited to high-resolution laser image-setting devices, but are inappropriate for lower- resolution devices, such as 300 lpi. dot matrix printers. Here colors must be created by dithered patterns that disperse the blackened pixels within each cell. **Dispersed-dot ordered dithering** is used with many printers, either with square or hexagonal cells. This method is designed to give the illusion of a constant gray tone or color within each cell, at the expense of sometimes introducing noticeable patterns into the cell.

A pixel blackening arrangement for a 256-pixel cell is shown in Figure 20.21, along with the pixels blackened for a 10% screen tint. Notice that the pixel placement algorithm spaces each pixel as far as possible from previous assignments and from pixels in neighboring cells. Patterns formed in each tone due to the assignment procedure are visible in a gray scale such as Figure 20.22.

66	188	124	144	80	184	120	131	67	185	121	141	77	181	117	130
194	18	252	48	208	32	248	35	195	19	249	45	205	29	245	34
98	146	82	176	112	160	96	163	99	147	83	173	109	157	93	162
226	50	210	6	240	64	224	11	227	51	211	7	237	61	221	10
74	178	114	134	70	192	128	139	75	179	115	135	71	189	125	138
202	26	242	38	198	22	256	43	203	27	243	39	199	23	253	42
106	154	90	166	102	150	86	171	107	155	91	167	103	151	87	170
234	58	218	14	230	54	214	1	235	59	219	15	231	55	215	4
68	186	122	142	78	182	118	129	65	187	123	143	79	183	119	132
196	20	250	46	206	30	246	33	193	17	251	47	207	31	247	36
100	148	84	174	110	158	94	161	97	145	81	175	111	159	95	164
228	52	212	8	238	62	222	9	225	49	209	5	239	63	223	12
76	180	116	136	72	190	126	137	73	177	113	133	69	191	127	140
204	28	244	40	200	24	254	41	201	25	241	37	197	21	255	44
108	156	92	168	104	152	88	169	105	153	89	165	101	149	85	172
236	60	220	16	232	56	216	3	233	57	217	13	229	53	213	2

Figure 20.21 Dispersed-dot ordered dither within a 16 × 16 dither cell. Ten percent of the pixels are darkened to show the regularity in the dither pattern.

Multicolor Dithering

We have seen that classical, spiral, and other types of dithering were developed for monochrome dot matrix printers, or for laser image-setters used to produce lithographic negatives or positives for platemaking. Multicolor dithering is also widely used in thermal, ink jet, and electrostatic color printers. Thermal printers, for example, operate by placing electrically heated pins against colored

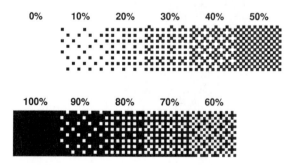

Figure 20.22 Dispersed-dot order dither gray scale (enlarged so that pixels are visible).

ribbons containing waxy ink (see Chapter 30). Heat liquifies a tiny spot of ink, allowing it to be transferred to the paper as a colored pixel. Cyan, magenta, yellow, and often black ribbons are used in succession, placing the subtractive primary colors in adjacent pixels as required to create each color. As with screen tints, we mentally integrate the individual pixels into the perception of a single color.

Dither patterns for red-orange, orange, and orange-yellow hues are illustrated in Figure 20.23. Such 4 × 4 (16-pixel) cells can be seen under magnification in the red to yellow section of the legend circle in the slope aspect coloring example (Color Figure 20.6). Larger cells allow a greater color range, as we can imagine filling the pixels in the 16 × 16 (256-pixel) dispersed-dot ordered dither with different proportions of subtractive primary inks. Many thousands of combinations are possible.

Bit-Mapped Patterns

The term **bit** is an acronym for **binary digit**, which can only take a value of 0 or 1. A **bit-map pattern** is a two-dimensional binary (0 or 1) array containing one or more repetitions of the pattern.

Computer printer patterned areas are normally created using bit-mapped pattern cells of the same size as (or larger than) dither cells. When the pixel has a value of 1, it is blackened. Otherwise, it is left blank.

A simple example (Figure 20.24) is a 12 × 12 (144-pixel) cell defining six full repetitions of the tailing pond pattern used on U.S. Geological Survey topographic maps. Notice that the pattern will

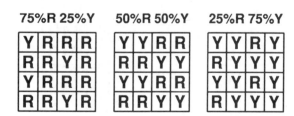

Figure 20.23 Multicolor dither patterns for yellowish red, orange, and reddish yellow areas.

repeat perfectly across two or more adjacent cells.

Bit-mapped patterns are available from a variety of sources in a form called **clip-art**. The term clip-art has come to mean any raster image, either scanned or mathematically specified, that can be used as a cell or entire image in a computer graphics program. Both monochrome and multicolor clip-art patterns can be created, but more than a single level or "plane" bit-map is required to define multicolor patterns. This, we shall now see, is closely related to how colors and patterns are electronically displayed on color CRT monitors.

ELECTRONIC COLOR AND PATTERN CREATION

As on color TVs, color on a cathode ray tube (CRT) monitor is created by a rapidly sweeping electron beam. Depending on its intensity at any instant, this beam controls the intensity of light emitted by tiny triads of blue, green, and red phosphor dots. A simplified diagram of a color monitor (Figure 20.25) shows that electrons from three separate beams (called Red, Green, and Blue) sweep the phosphors on the screen. A metal **shadow mask** filter insures that electrons from the Red, Green, and Blue beams strike only the correct red, green, and blue phosphors on the screen. The mask "shadows" neighboring phosphors from errant electrons.

This method of color creation has its roots in the trichromatic theory of color vision. A large color palette is created by combining different proportions of these nearly spectral additive primary colors. Indeed, if we plot the maximum extent of a typical CRT color gamut on the CIE chromaticity diagram (Figure 20.26), we see how much larger the range of color is than that for process color printing inks.

Several triads of red, green, and blue light-emitting phosphors form one square-shaped pixel on the monitor screen (Color Figure 20.7). Color monitor screens used in cartography vary in resolution from 350×640 to $1,024 \times 1,024$ rows and columns of pixels. A medium resolution 512×512 screen will be assumed throughout the following discussion.

Computers control pixel colors through a frame

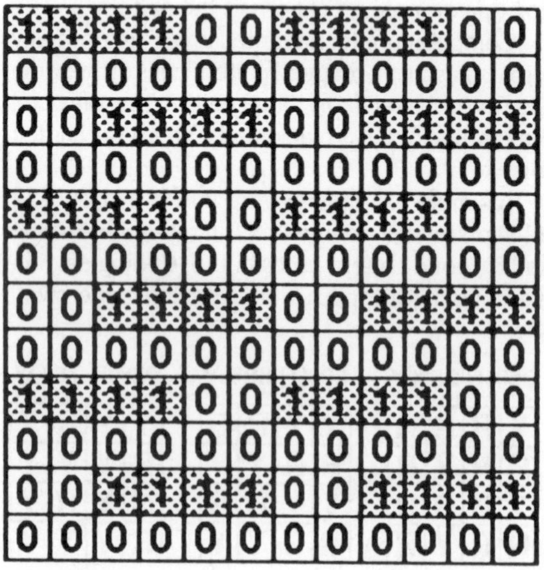

Figure 20.24 Bit-map for U.S. Geological Survey tailing pond pattern.

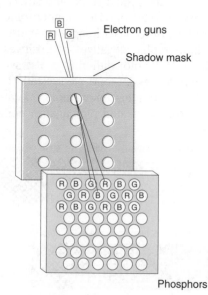

Figure 20.25 Diagram of a CRT shadow mask, showing a "delta" electron gun arrangement.

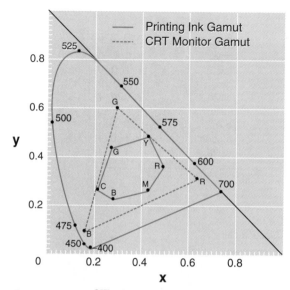

Figure 20.26 CIE chromaticity diagram showing the gamut of process color printing (solid outline) and CRT monitor (dashed outline) color. Note the greater gamut with the additive primary phosphors in the CRT (process color data from Kimerling 1980, CRT phosphor data from Hunt 1987).

buffer and display controller. A **frame buffer** is a section of computer memory containing a binary representation of the array of pixels comprising the CRT screen. We can think of the frame buffer as a matrix of numbers, in which each (row,column) value specifies the color of the pixel at the corresponding location on the screen.

Frame buffers for color monitors are more than one bit deep. Early color-monitor frame buffers were three bits deep, so that a 512×512 pixel screen would have a $512 \times 512 \times 3$ bit buffer (Figure 20.27).

We can think of this three-dimensional buffer as being divided into one-bit-deep red, green, and blue arrays, called **bit planes**. Each bit plane specifies the intensities of the red, green, and blue electron guns for each pixel. A value of 0 means that no electrons will strike the phosphors within the pixel; a value of 1 indicates a full-intensity electron beam.

If all three bit planes are set to 0, the corresponding pixel will be black on the screen. Setting all

three bit planes to 1 will produce a white pixel; setting only the first bit plane to 1 gives a red pixel; and so on. The binary specifications for these and the five other additive and subtractive primary colors that complete the color palette are listed in the table in Figure 20.27. These eight sets of binary numbers are the (x, y, z) coordinates of the corners of the RGB color cube model described in the previous chapter.

The number of colors that can be specified increases according to 2^n with each additional bit plane (n) in the frame buffer. The practical limit is usually 24 bit planes, giving 2^{24} or 16,777,216 possible color specifications. This, of course, is far greater than the number of pixels on the screen. The 24 bit planes are divided into thirds so that eight bit planes are used to control each electron gun (Figure 20.28). This means that 2^8, or 256, electron gun voltages can be specified by binary numbers ranging from 00000000 (0) to 11111111 (255). Hence, 256^3 RGB color combinations, including 256 gray tones, are possible. These form the full RGB color model.

A **display controller** must be used with frame buffers containing more than three bit planes. This part of the computer reads the red, green, and blue binary values for each pixel from the three sets of bit planes. These values serve as index numbers for red, green, and blue look-up tables* that contain integer numbers proportional to gun voltages. These numbers are then read into a digital-to-analog converter. This converter adjusts the gun voltages that control the electron beam intensities.

We would like the binary numbers for each electron gun to specify voltages that will produce equal perceived changes in phosphor luminance

---

*A look-up table is a one-dimensional array in computer memory that stores a sequence of values controlling the intensity of an electron gun. Separate look-up tables are needed for the red, green, and blue electron guns. In a look-up table system, the RGB values for each screen pixel do not directly control the electron guns, but rather serve as index numbers into the look-up table for each gun. A red pixel value of 20, for example, indexes the 20th entry in the red look-up table, which might contain a gun intensity value of 90.

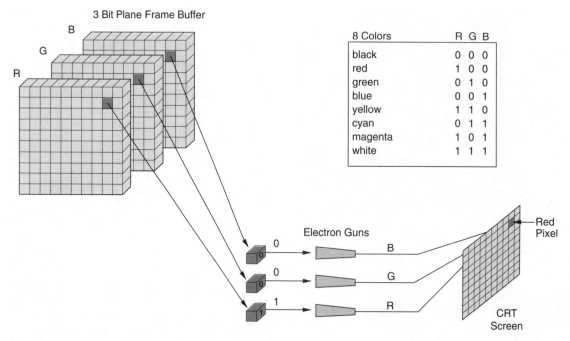

3 Bit Plane Frame Buffer

8 Colors	R	G	B
black	0	0	0
red	1	0	0
green	0	1	0
blue	0	0	1
yellow	1	1	0
cyan	0	1	1
magenta	1	0	1
white	1	1	1

Figure 20.27 A frame buffer system for a simple CRT display. Eight different colors can be created from the three electron guns controlled by one-bit frame buffers.

on the screen. Unfortunately, voltage and luminance are not proportional, and a non-linear **gamma correction** must be applied to each electron gun to have binary numbers linearly proportional to luminance. The gamma correction procedure described in Box 20.A should be applied whenever equal luminance intervals are required. However, luminance and perceived lightness are not necessarily proportional. Thus, care must still be taken in selecting colors for quantitative progressions and in other cases which require equal value increments.

You can also use display controller look-up tables to increase the range of color possible with smaller frame buffers. In this case, a $512 \times 512 \times 3$ bit buffer will hold color numbers between 000 (0) and 111 (7) for each pixel (Figure 20.30). The color number gives the row in 8×8 look-up tables for the red, green, and blue electron guns. Values from 0 to 255 can be loaded in each row, allowing you to select eight colors from the full 16.8 million color palette.

These look-up table numbers can be changed easily. Thus, you can instantly change screen colors without having to rewrite the frame buffer.

Dithered Screen Colors

The multicolor dithering technique for color printers can also be used on the CRT monitor screen. Here each dither cell consists of a matrix of pixels, with each pixel binary number giving a location in the display controller look-up table. Different dither cell color combinations can be specified in the computer mapping program, or the program may automatically determine the pattern needed to approximate a given color.

Dithered cells are usually kept small (2×2 or 4×4 pixels) because the texture of the dithered color is easily seen in larger cells. Using large cells to color very small areas can also produce unintended color due to only part of the cell being used to fill in the area.

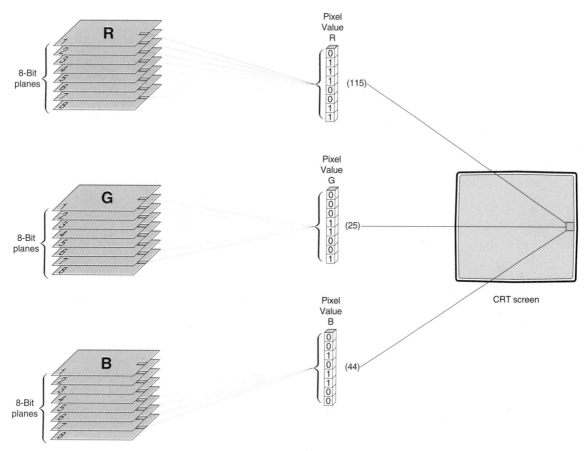

Figure 20.28 A 24-bit-plane color CRT monitor. Eight bit planes control each electron gun, giving 2^8, or 256 possible intensities of red, green, and blue, and 256^3 (16,777,216) possible color combinations (adapted from Jensen).

LINKS BETWEEN ELECTRONIC AND HARDCOPY COLOR

Exactly matching the gamut of screen colors with plotted or printed colors is still a dream. This is because the additive primary colors on CRT screens and their mixtures inherently cover more of the chromaticity diagram than do the subtractive primary pigment combinations used in lithographic printing and color plotters and printers (see Figure 20.26). A large part of this difference lies in the greater brightness of screen colors, which causes colors to look slightly different on the screen and in print. This occurs even if they are measured as having the same dominant wavelength and are seen as the same hue. Differences in lightness and chroma will be readily apparent.

Nevertheless, we are rapidly approaching the day when colors on the CRT screen can be approximated, within these inherent limits, on hardcopy output, either directly on paper or indirectly on lithographic film separations for printing. One approach is to define the color gamut of our CRT monitor and hardcopy device using the same

Box 20.A
Gamma Correction

In general, CRT color monitor phosphor luminance is a power function of electron gun voltage. Since this voltage is directly proportional to the R, G, or B value specifying the color, three equations relate RGB color specifications to phosphor luminance (L):

$$L_R = R^{\gamma}_R \qquad L_G = R^{\gamma}_G \qquad L_B = R^{\gamma}_{B}.$$

The R, G, and B values for a particular luminance are obtained by inverting each equation to give:

$$R = L_R^{1/\gamma}{}_R \qquad G = L_G^{1/\gamma}{}_G \qquad B = L_B^{1/\gamma}{}_{B}.$$

The γ value for each phosphor is determined by measuring the screen luminance of each red, green, and blue digital value with a photometer. Plotting luminance against digital number on log-log graph paper (Figure 20.29) should give a straight line for each phosphor. The slope of each line is the corresponding gamma value. Using these measured gamma values in the R, G, and B equations allows a look-up table to be created in the display controller so that the R, G, and B values will give equal luminance steps.

---

perceptual color space.* For example, we could determine the CIELAB color coordinates for RGB combinations throughout the color cube and for all overprintings of process ink screen tints in 10% increments. We would find the printed color gamut to be smaller than the CRT monitor color range, so that matches based on identical CIELAB coordinates are not possible for many of the colors on the screen. However, we can compute the hue angle and chroma of all measured colors. This allows us to select printed colors of the same hue and "relative" lightness and chroma. For instance, we could match a green hue of 50% chroma and 80% lightness relative to black and pure green on a white CRT screen with a printed green of the same 50% chroma and 80% lightness relative to black and green ink on white paper. The colors will match relative to the color gamut possible with the hardcopy device, but will still appear to differ in lightness and chroma.

To do this matching for thousands of screen and printed colors requires either complex mathematical algorithms or look-up tables so that we can convert all additive color screen specifications into percentages of subtractive primary color inks. The look-up table approach appears to offer the greatest promise, but very large tables will be required to accommodate all possible conversions.

Since **automatic color matching** is still in its infancy, it is best to be able to predict what a screen color will look like in printed form. To do this requires creating or using an available color chart or swatch book showing the appearance of printed colors throughout the CRT monitor gamut. The screen colors thus only serve as symbols for the actual printed colors created by the mapping software and printer hardware.

---

*Several companies have developed **color management** systems for color matching between monitors and hardcopy devices. Most systems consist of look-up tables (called characterization tables) containing the RGB and CIE specifications for the colors that can be produced on a number of color printers and monitors. This allows an optimal color match to be made between electronic and hardcopy colors. Current color management systems include *Tektronix Tekcolor, Adobe Photoshop 2.0, Pantone Professional Color Toolkit*, and *Letraset Color Studio 1.5*.

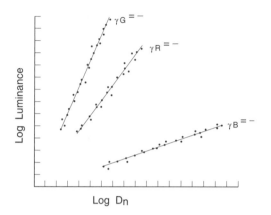

Figure 20.29 Determining γ for red, green and blue phosphors.

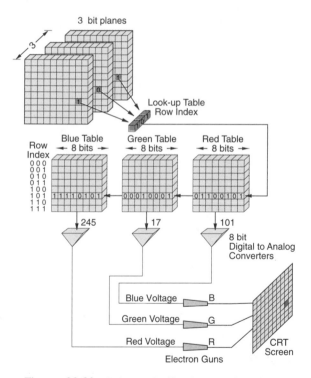

Figure 20.30 A frame buffer for a color display using look-up tables.

SELECTED REFERENCES

Billmeyer, F. W., Jr. and A. K. Bencuya, "Interrelation of the Natural Color System and the Munsell Color Order System," *Color Research and Application*, 12, No. 5 (1987): 243–55.

Brewer, C. A., "The Development of Process-Printed Munsell Charts for Selecting Map Colors," *The American Cartographer*, 16 (1989): 269–78.

Kimerling, A. J., "Visual Value as a Function of Percent Area Inked for the Cross-Screening Technique," *The American Cartographer*, 6, No. 2 (1979): 141–48.

Kimerling, A. J., "Color Specification in Cartography," *The American Cartographer*, 7, No. 2 (1980): 139–53.

Kimerling, A. J., "Predicting Dot Coverage for Colored Areas Produced by Single and Multiple Pixel Random Dots," *The American Cartographer*, 16, No. 1 (1989): 17–27.

Plumb, G. A. and T. A. Slocum, "Alternative Designs for Dot-Matrix Printer Maps," *The American Cartographer*, 13, No. 2 (1986): 121–33.

Pocket Pal, 13th ed., New York: International Paper Company, 1983.

Roth S., "All About Color," *Macworld*, Jan. 1992, 141–45.

Stefanovic, P., "Digital Screening Techniques," *ITC Journal* (1982-2): 139–44.

Stoessel, O. C., "Standard Printing Color and Screen Tint Systems for Department of Defense Mapping, Charting and Geodesy Services," *Proceedings*, ACSM Technical Sessions, ACSM-ASP Fall Convention, 1972.

Thorell, L. G. and W. J. Smith, *Using Computer Color Effectively*, Englewood Cliffs, NJ: Prentice Hall, 1990.

Travis, D., *Effective Color Displays*, London: Academic Press, 1991.

Ulichney, R., *Digital Halftoning*, Cambridge, MA: The MIT Press, 1987.

CHAPTER 21

COLOR AND PATTERN USE

Colors are used on maps in a great variety of ways. On some maps, large areas, such as oceans and continents, are all colored alike. On others, only a few line symbols, such as roads, streams, or contours, are colored. The sequence of numerical categories on quantitative maps is often symbolized with a series of different colors, as on maps showing statistics by enumeration districts or relief maps showing elevation ranges. In all cases, the choice of colors is vital to the map's success in portraying the geographic information effectively and, equally important, in not creating bothersome combinations and contrasts that are garish or draw unwanted attention.

Even a small amount of color can make an enormous difference in a map's appearance. A map composed of numerous black lines, representing boundaries, coasts, rivers, roads, and lakes—to say nothing of such features as canals, railroads, contours, and transmission lines—may convey little but confusion to the map user. Yet, if the various features are carefully color coded, visual order replaces chaos and the map becomes much easier to read.

Pattern is the systematic repetition of dots, lines, or other graphic "marks" on the map. Patterns may vary in the shape of marks used, as well as in texture (spacing), orientation and arrangement of marks (see Chapter 18). Pattern is not as versatile an element as color because its use is generally restricted to area symbols. Nevertheless, pattern has always been a major element on many types of maps. For some 400 years before lithography and color printing were developed, the production of patterned area symbols on various kinds of printing plates called for ingenuity and great manual dexterity. Today pattern, like the color attributes of hue, value, and chroma, is used alone on single color maps or, very often, as a component of multicolor map symbols. Color and pattern are often used in combination. Patterns often are colored, and colors are often patterned.

In this chapter, we will focus on the major functions of color and pattern in map design. Color and pattern selection guidelines arising from each function will be presented where appropriate, along with color specifications in one or more of the systems described in the last two chapters.

FUNCTIONS OF COLOR AND PATTERN IN MAP DESIGN

From earliest times, color has been applied to maps, both for decoration and to improve the communication of mapped information. The brightly colored flags, emblems, tents, camels, and other artistic ornaments filling empty spaces on medieval sea charts are but one example of color used purely for decorative purposes. On these same maps, ports and famous cities were often colored red, with the rest in black. Here color served a function: to differentiate classes of cities and to show their relative importance.

Creating an aesthetically pleasing map and distinguishing classes of features are only two of color's many roles in cartography. Colored symbols also allow a greater number of features to be on the map, since color can greatly increase symbol legibility. Certain color sequences are inherently seen as being ordered from less to more, and are used to portray variations in magnitude within a set of mapped numerical data. Color also is a way to place certain features in the visual foreground, while making other information less visually prominent. In this way, the desired figure-ground relationship (see Chapter 18) among symbols is developed.

SYMBOLIZING QUALITATIVE FEATURES

Both pattern and the color dimension of hue are appropriate primary graphic elements to use when portraying different types or classes of qualitative features (see Chapter 18). This is because colors varying only in hue and patterns of the same texture do not carry a "magnitude message"; they are only seen as different. Many factors enter into the selection of hues and patterns for different features. Let's look at some of these factors, beginning with hue conventions that have been followed over the centuries.

Hue Conventions

The one universal color convention is blue for water bodies and rivers. Our earliest surviving maps, such as a military map from the Han Dynasty in China and an early medieval copy of a Roman surveying manual, show water in blue. Virtually all medieval European maps do the same, the only exception being a red colored Red Sea.

Other conventions are less strictly adhered to and are more likely to vary from culture to culture. In our Western culture the following hues are normally used for land-related features:

(a) green for vegetated areas

(b) brown for land surface symbols such as contours

(c) yellow or tan for dry, sparsely vegetated areas.

Not using a hue convention, particularly blue for water, can easily confuse the map user, sometimes bringing angry responses to the "wrong" color on the map. Conventions, of course, restrict

map design options, but the list is short and places few real limits on our freedom of choice.

Color Standards

Government agencies, mapping companies, and even professional societies have adopted color standards for the types of maps they produce. Hue conventions are usually a part of each standard, along with colors unique to each type of map. These standards are most often specified by ink color and screen tint percentage. As examples, hue standards for two major types of maps produced by U.S. government agencies are listed in Table 21.1.

Notice that the U.S. Geological Survey almost exclusively uses solid (100%) inks of six different colors, whereas many Defense Mapping Agency hues are made by single or combined process ink screen tints (with a red-brown ink substituted for magenta). A fifth, dark purple ink for aeronautical features highlights critical flight information.

Standard colors are used for map series, such as USGS topographic quadrangles, which require all adjoining maps to be colored identically.

Table 21.1

Ink Color and Screen Tint Standards for Features on U.S. Government Maps

USGS 7½-minute Quadrangles (Topographic Maps)		U.S. Defense Mapping Agency JOG (Aeronautical Charts)	
Lettering	100% Black	Lettering	100% Black
Boundaries	...	Hard Surface Roads	...
Railroads	...	Other Roads	61 Blk+79 Br+100 Yel
Buildings	...	Rivers	100 Cyan
Light Duty Roads	...	UTM Grid	...
SPC Grid Ticks	...	Inland Open Water	54 Cyan
Rivers	100 Blue	Open Water	21 Cyan
UTM Grid Ticks	...	Woods	21 Cyan+42 Yellow
Water Bodies	30 Blue	Contour lines	61 Brown
Major Roads	100 Red	City Tint	67 Brown
USPLS Grid	...	Continuous Habitation	21 Brown
Urban areas	30 Red	Low Elevation Tint	42 Yellow
Forested areas	100 Green	High Elevation Tint	42 Yel+21 Br
Contour lines	100 Brown	Aeronautical Info.	100 Purple
Sand/Gravel areas	...	Aerodrome Tint	54 Purple

Standardized hues are also needed for maps of the same area which are updated periodically, since long-time users come to expect and rely on the same color scheme. Many companies also establish standards to give their maps a certain identity which sets them apart from their competition. The observation in the computer industry that "the nice thing about standards is that there are so many from which to choose" certainly holds true in color map design.

Color standards have been established in many ways. Some are based on the properties and theories of color vision, some on design principles borrowed from other fields, and others on special map user requirements. We should always look into the rationale behind any existing standard before adopting it ourselves, for as the Swiss cartographer Eduard Imhof notes:

Tradition, partiality and whim, preconceived opinions, aesthetic sensitivity or barbarity of taste often play leading roles in the selection of colors. There are "brown supporters," "green fans," "blue enthusiasts," "yellow admirers," and "red worshippers." Many map makers and map users do not like to change and stick by their first loves.

Pattern Conventions and Standards

Probably the most common use of pattern is as a qualitative area symbol for depicting such features as bedrock type, climatic regime, or administrative jurisdiction. An associated use is to add graphic distinctiveness to uniformly colored areas, especially on maps with a large number of classes (Figure 21.1).

Figure 21.1   An example of a two-color map in which pattern has been used to make areas appear more distinctive. Note that numerals in the classes provide an additional aid to recognition. (From *Fundamentals of Physical Geography*, 3rd ed., by Trewartha, Robinson, Hammond, and Horn, Copyright 1977, McGraw-Hill Book Co., Inc.)

Hundreds of distinctive patterns are possible. Although there is no systematic classification of patterns, it is convenient to recognize four general categories (Figure 21.2):

1. **Coarse line patterns** composed of usually straight but sometimes wavy parallel lines. When two sets of lines are crossed, the result is called **cross-hatching**, and the sets of lines may intersect perpendicularly or at another angle.

2. **Dot patterns** composed of round dots in rectangular, triangular, or irregular arrays. A rather evenly spaced, irregular distribution giving a more or less smooth appearance is called **stippling**.

3. **Pictographic patterns** that in some way resemble the feature being portrayed. These range from the tufted grass symbols commonly used to show swamps and marshes to arrays of crosses representing cemeteries.

4. **Reversed patterns** of the first three categories. Here the pattern appears white on a uniformly colored background. It is often easier to read text and other map symbols placed over reversed patterns.

Many of these patterns have become conventions or standards in specific subject matter fields, such as geology or soil science. U.S. Geological Survey topographic map standard patterns, for example, include Pictographic Pattern B in Figure 21.2, printed in solid blue ink for marshes, overprinted in blue on a green background for wooded marshes, and overprinted on a lighter blue for submerged marshes. Solid brown random dot patterns such as Dot Patterns C and D indicate sand areas, washes, and gravel beaches, whereas Pattern C in solid green indicates scrub lands. Solid green rectangular dot patterns such as B are used in fine and coarse rulings to represent vineyards and orchards. Map users are accustomed to these symbol associations, and they should be used for the same or similar features on other maps if at all possible.

It is important not to choose line or dot patterns that vary systematically in texture from coarse to fine. Such texture progressions will appear to represent magnitude differences and detract from the perception of different classes. We shall soon see, however, that texture progressions are well suited for portraying quantitative features.

Unique and Mixture Hues

Cartographers often select hues based on their uniqueness. As we discussed in Chapter 19, the opponent process theory of vision is based on our perception of blue, green, red, and yellow as unique hues, with other hues appearing as mixtures of these. All mixtures of unique hue pairs are possible except blue-yellow and red-green. The opponent blue-yellow and red-green combinations are seen as bluish, yellowish, reddish, and greenish grays. Some cartographers also include brown as a unique hue, while others see brown as a mixture of red, yellow, and black.

Unique and mixture hues are used in two ways to display qualitative features. The unique hues should be used to symbolize distinctly different phenomena. Mixtures of two non-opponent unique hues can be used to portray phenomena that share some of the characteristics of the features symbolized by the two unique hues.

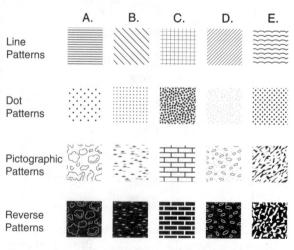

Figure 21.2 An assortment of common line, dot, pictographic, and reverse patterns.

For example, areas entirely populated by ethnic group A might be yellow, areas composed of ethic group B might be red, and mixture areas might be an orange color made of equal proportions of yellow and red. Munsell and NCS color specifications for such a progression are given in Table 21.2, along with approximate screen tint percentages and RGB coordinates for the Munsell specifications. Screen tint percentages can only be approximate due to variations in paper color and ink absorption, platemaking procedures, and printing presses. RGB or other computer color specifications are also general approximations due to different assignments of phosphor intensity to each red, green, and blue integer value.

Hue-Value Combinations

As we saw in Chapter 19, value is the perceived lightness or darkness of a feature relative to a standard white surface. In the previous example, it was impossible to keep the value level constant for all hues. This is because a "pure" yellow hue is inherently lighter than a "pure" red. The relationship between "pure" hue and value can be illustrated on a graph of printed surface reflectance or CRT screen luminance for a systematic hue progression across the spectrum, such as the HLS hue circle described in Chapter 19 (Figure 21.3). Here pure yellow has the highest luminance and value level, followed by cyan, green, magenta, red, and blue.

These inherent value differences among pure hues complicate color selection, but they also can be used to advantage for certain phenomena. The

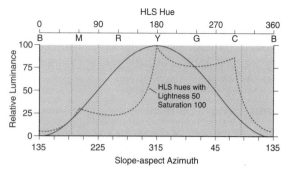

Figure 21.3 Relative luminances (dashed line) of HLS hues, plotted on the cosine shading equation curve. (From Moellering and Kimerling 1990).

best example is slope aspect, the compass direction of maximum downhill slope at a point. Aspect is a circular qualitative phenomenon since a compass azimuth of 0° is not physically greater than a 359° azimuth; it is only closer to 359° than to other azimuths such as 180°.

The maximum discrimination among aspect classes should occur when the opponent process unique hues are 90° apart on a hue circle, as in the HLS and NCS systems. Either can be used as the basis for aspect class coloring, particularly if the hue circle is rotated so that yellow corresponds to the 315° (NW) aspect angle.

Slope aspect is displayed most effectively if the value level or luminance of hues in the color circle follows the basic rule of relief shading (see Chapter 25).* This rule states that gray tones for shading should be the lightest at the 315° aspect direction and darkest at the 135° (SE) aspect. Intermediate aspect angles should be shaded according to a value or luminance function, similar to the solid line in Figure 21.3. With a circular color progression, the curve is fit best by lowering the value level of higher luminance hues, such as green, and eliminating small portions of the hue circle, such as those around magenta and cyan. The resulting

Table 21.2
Munsell, Natural Color System, and Approximate Screen Tint and RGB Specifications for Red, Orange, and Yellow.

Munsell	NCS	Screen Tints	RGB		
5R 5/14	0,100, R	60Y 90M	255	0	0
5YR 6/12	0,100, Y	50R 80Y 60M	255	127	0
5Y 8/12	0,100, Y	100Y 10M	255	255	0
Y=yellow	M=magenta	R=red			

---

*The slope aspect display system described here is patented (U.S. Patent No. 5067098) and more fully described in Moellering and Kimerling, 1990.

color progression (Color Figure 20.6) portrays slope aspect in a relief shaded manner. This color progression further enhances our ability to determine the aspect azimuth, particularly when a large number of classes are used.

Four-Color Theorem

We sometimes must apply colors to homogeneous areas in such a way that no adjoining areas will be the same hue. Coloring the nations of Europe so that no two neighboring countries are the same hue is a prime example. To do this requires a minimum of four different hues. Using the four opponent process unique hues provides the maximum contrast among colors, although these are often reduced in chroma to serve as background coloring. Five or more hues will, of course, work as well, but only four are necessary.

Symbolic Connotations of Hue

We commonly associate hues with different physical phenomena, sensations, and emotions. These associations evoked by different hues can be used when portraying certain features. Important symbolic connotations include blue for wetness or coldness, tan and yellow for dryness, green for lushness or vegetation, red and orange for warmth, and brown for soil or bare ground. Notice that several of these also are color conventions.

Pattern is often used along with hue to create more complex symbolic connotations. The blue marsh pattern over green or blue areas that represents wooded or submerged marshland on USGS topographic maps is a classic example. Such combinations are often used to establish the hierarchy among map classes. In this example, the blue pattern represents the major class of marsh, with the blue and green background hues denoting the subdivision into wooded and submerged marshland.

One of the more interesting uses of symbolic connotations is in the design of "natural color" landscape maps. One of the earliest examples is William Folkingham's instructions from 1610 for the coloring of English estate plans. Arable land was to be painted a tan, meadows a light green, pasture land a darker green, and heather a still darker green. Similar color schemes are used today.

Physiological Limitations

Physiological reactions to different hues are essentially the same from person to person, but not perfectly identical. Hue sensitivity and simultaneous contrast are two universal physiological phenomena that constrain the use of hue for symbolizing both qualitative and quantitative features. Let's look at each of these factors in more detail.

Hue Sensitivity   Most people are quite sensitive to very slight differences in hue when colors are placed side by side against a medium-gray background. We are able to distinguish hundreds of different hues on a color chart, particularly when they also vary in value and chroma. On the other hand, our ability to match hues on the map with those in the legend is severely restricted, especially when different colors surround each hue on the map.

Cartographers disagree on the maximum number of hues that can be used on a map before legibility is severely diminished. Depending on the size of colored areas and map complexity, the limit is reached somewhere between eight and 15 hues. Also, when hues are used to distinguish one class from another, they should be made as different as possible. This suggests using the Munsell or NCS hue circle as the basis for color selection.

Simultaneous Contrast   Whenever two colors are adjacent, they modify each other's appearance. Not only does hue appear to change, but also chroma and value. When a hue is surrounded by another color, it shifts in appearance toward the complementary hue of the surrounding color. For example, a green hue surrounded by yellow will appear slightly bluer than the same green on a blue background (Color Figure 21.1).

Because of this **simultaneous contrast** phenomenon, it's best not to use similar hues on the same map. Similar hues may be difficult to match with the legend and to identify accurately on the multicolored map. If you hold to this rule, of course, you will have fewer choices of hues to use on the map. Thus, if a large number of classes must be mapped, you may have to add patterns or category codes to the hues to distinguish one from another.

SYMBOLIZING QUANTITATIVE FEATURES

Cartographers often must select a color progression to represent a set of quantitative data at the ordinal, interval, or ratio level of measurement (see Chapter 12). Whether the quantitative data describe point or line features, homogeneous areas, or continuous surfaces, the color progression should be designed to:

1. allow the user to correctly match the color on the map with the corresponding legend color
2. convey a correct impression of how the feature changes in magnitude across the map area
3. allow the comparison of quantitative data for features on two or more maps of the same area.

Recognizing that these three functions are difficult to incorporate into one color progression, you can take four general approaches: (1) Use a series of value and chroma combinations for one hue. (2) Use a sequence of colors grading between two endpoint hues. (3) Use a progression of several hues. (4) Use a white to black value progression. Examples of each approach are illustrated in Color Figure 21.2, and color specifications are provided in Table 21.3.

Single-Hue Progressions

A progression in which value decreases and chroma increases systematically as the colors grade from white to the pure hue is called a **single-hue progression**. This type of progression is one of the

most effective ways to show the spatial pattern of magnitude change on a map. This is because value and chroma are both primary graphic elements that impart a spontaneous visual association with magnitude—what we call a **magnitude message**—when used in a color progression.

The white to red single-hue progression seen in Color Figure 21.2A and detailed in Table 21.3A is the easiest to produce on printed maps, since only screen tints of a single ink are required. Color specification in RGB color coordinates is also straightforward. Such progressions are easiest to define in the Natural Color System, since you merely need to specify the tints along the W-Hue edge of each triangular graph. In Munsell terms, value decreases and chroma increases from white to red. Such simultaneous changes in value and chroma invariably occur in color progressions and can be put to good use in map design.

Bi-Polar Progressions

We are sometimes called upon to symbolize quantitative data describing features that range from negative to positive, below to above average, or bad to good. A **bi-polar color progression** is often used to portray the geographic pattern of such data, while focusing on the positive-negative character of the phenomenon.

A bi-polar progression from dark blue to red is illustrated in Color Figure 21.2B and specified in Table 21.3B. The pure blue and red ending classes emphasize the data extremes, with intermediate blues and reds increasing in value and decreasing in chroma toward the center white, which is always the zero or average class. With this design, the map user will instantly know whether the colored feature is negative or positive, and to what extent it deviates from the zero dividing point.

Complementary Hue Progressions

Bi-polar progressions can also be created by an ordered mixing of complementary hues. In Chapter 19, we saw that blue and yellow are complementary hues (Color Figure 21.2C and Table 21.3C),

Table 21.3
Munsell, Natural Color System, with Approximate Process Ink Screen Tint and RGB Specifications for Examples of Six Different Color Progressions Used to Symbolize Five Classes of Quantitative Data

A. Single-Hue Tint Progression (white — red)

Munsell	NCS	Screen Tints			RGB		
— 10/0	0,0,—	—	—	—	255	255	255
5R 8/4	0,25,R	10M			255	191	191
5R 7/8	0,50,R	50M	20Y		255	127	127
5R 6/12	0,75,R	70M	40Y		255	63	63
5R 4/14	0,100,R	100M	70Y		255	0	0

B. Bi-Polar Hue Progression (blue — white — red)

Munsell	NCS	Screen Tints			RGB		
5PB 5/12	0,100,B	100C	50M		127	0	255
5PB 7/6	0,50,B	30C	10M		191	127	255
— 10/0	0,0,—	—	—	—	255	255	255
5R 7/8	0,50,R	50M	20Y		255	127	127
5R 4/14	0,100,R	100M	70Y		255	0	0

C. Complementary Hue Progression (blue — gray — yellow)

Munsell	NCS	Screen Tints			RGB		
5PB 5/12	0,100,B	100C	50M		127	0	255
5PB 7/6	25,50,B	40C	15M		127	63	191
— 8/0	25,0,—	25C	25M	25Y	127	127	127
5Y 8/6	12,50,Y	15C	15M	60Y	191	191	63
5Y 8/12	0,100,Y		10M	100Y	255	255	0

D. Partial Spectral Hue Progression (yellow — red)

Munsell	NCS	Screen Tints		RGB		
5Y 8/12	0,100,Y	10M	100Y	255	255	0
10YR 7/12	0,100,Y25R	40M	90Y	255	191	0
5YR 6/12	0,100,Y50R	60M	80Y	255	127	0
10R 5/12	0,100,Y75R	80M	70Y	255	63	0
5R 4/14	0,100,R	100M	70Y	255	0	0

E. Blended Hue Progression (yellow — orange — brown)

Munsell	NCS	Screen Tints			RGB		
5Y 8/12	0,100,Y	10M	100Y		255	255	0
10YR 7/12	0,100,Y25R	30M	100Y		255	200	80
5YR 6/10	0,100,Y50R	50M	70Y	5Bk	230	120	120
5YR 5/8	0,80,Y50R	50M	70Y	20Bk	190	115	75
5YR 4/6	40,60,Y50R	50M	70Y	40Bk	180	100	60

F. Value Progression (white — black)

Munsell	NCS	Screen Tints			RGB		
— 9/0	0,0,—	—	—	—	255	255	255
— 7.5/0	0,25,—			15Bk	191	191	191
— 6/0	0,50,—			35Bk	127	127	127
— 4.5/0	0,75,—			60Bk	63	63	63
— 3/0	0,100,—			100Bk	0	0	0

B=blue Bk=black C=cyan M=magenta PB=purple-blue R=red Y=yellow as are any other pair of hues situated opposite each other on the Munsell, NCS or HLS hue circles. Colors will grade from both complementary hues toward gray in the center of the progression.

Since complementary hues are likely to be inherently different in value, we must often arbitrarily select the value level of the center gray. Choosing a lighter central gray usually gives a greater range of value and chroma combinations for the two hues.

Partial Spectral Hue Progressions

Cartographers regularly design color progressions for data describing mixtures of two unique classes, such as the percentage of men and women in each county. Colors grading between two adjacent opponent process hues, such as yellow and red, will depict the pattern of magnitude variation while emphasizing the degree of mixture. When the endpoint and mixture hues match part of the solar spectrum, they are called a **partial spectral progression**.

Color Figure 21.2D shows a classic partial spectral progression from yellow through orange to red. For a five-class map, the ideal colors would be: full yellow, ¾ yellow–¼ red, ½ yellow–½ red, ¼ yellow–¾ red, and full red. Munsell and NCS hues that, in theory, will appear as these mixtures are listed in Table 21.3D.

Blended Hue Progressions

The spatial pattern of magnitude variation can also be symbolized by a progression of related hues, one or more of which is not part of the spectrum. These hues must blend smoothly between the two endpoint hues, changing uniformly in hue, value, and chroma. We call this type of progression a **blended hue progression**.

A very popular blended hue progression used to show elevation differences is from yellow through orange to brown (Color Figure 21.2E). As Table 21.3E indicates, this is a difficult progression to define and specify, particularly when brown is an endpoint hue. Notice that brown is created by overprinting yellow, magenta, and black, rather

than cyan, magenta, and yellow. Specifying RGB coordinates for a progression of brown colors is also a challenge, since these lie in the interior of the RGB cube.

Value Progression

On monochrome maps, a **value progression** from white to black, or from light to dark gray, is the most effective way to depict the geographical pattern of magnitude variations across the map. This is because value is the primary graphic element that best carries a magnitude message. Like single-hue progressions, the white to black value progression is easy to produce on printed maps, since only screen tints of a single ink (in this case black) are needed.

Notice the white to black value progression shown in Color Figure 21.2F and specified in Table 21.3F. Screen tints in the table give approximately equal Munsell value increments under typical map printing conditions. The RGB specifications are only rough estimates of equal value increments, since the value-RGB relationship varies considerably among monitors and monitor calibrations.

Full-Spectral Progression

Using the **full-spectral progression** of hues from blue through red to symbolize quantitative data has been shown to help the user match map colors with legend classes. Spectral progressions have been used for over a century to portray successive elevations on relief maps so that water is blue, lowlands are green, intermediate elevations are yellow, and high areas are reddish. Many modern weather maps also use a spectral progression for mapping the temperature surface, presumably since people often associate blue with coldness and red with warmth.

Except for its familiarity and map-legend matching advantage, there is little reason to use spectral progressions for quantitative data. Communicating the geographic pattern of magnitude variation generally is a more important objective, and a spectral hue progression does not inherently carry a magnitude message. In addition, the symbolic connotations of some spectral hues detract from the map's effectiveness. Large value changes associated with printed spectral hues also result in marked variations in the perceptibility of other data being shown.

Two-Variable Color Progressions

Color progressions have been developed for choropleth maps (see Chapter 27) that simultaneously depict magnitude variations within homogeneous areas for two map themes, called data variables. **Two-variable** (or **bi-variate**) choropleth mapping began in earnest in the early 1970s when the U.S. Census Bureau published maps relating such topics as education level and per capita income by county. The Census Bureau created a two-dimensional color scheme based on a full-spectral progression running around the outer edge of a box subdivided into 16 legend classes arranged in a 4 × 4 grid. This hue-based scheme, like spectral progressions for single-variable quantitative data, was roundly criticized on perceptual grounds. A simpler, more visually logical color progression is described here.

A simple yet effective two-variable color scheme is that resulting from combining two single-hue tint progressions. The cyan and magenta color scheme seen in Color Figure 21.3 is a typical example of this approach. Screen tints used for each class are specified in Figure 21.4. A white to cyan tint sequence forms the four rows in the legend, and a white to magenta progression fills the four columns. When these are overprinted, the legend box corner classes are seen as white, cyan, magenta, and dark blue.

Blue and adjacent low value hues indicate high data values for both variables; magenta and surrounding reddish hues depict high values for theme 2 and low for theme 1; and so on. Map users often see the legend as divided into cyan, red, light color, and dark blue-purple quadrants. These help us to correctly interpret the quantitative relationships between the two themes.

User testing of Census Bureau two-variable maps indicates that they are more easily understood if three things are included on the map:

1. a very prominent and clear legend
2. small black-and-white versions of the two single-theme maps that were composited
3. notes explaining the color scheme and drawing attention to the types of information that can be obtained from the map.

Hue Guidelines for Quantitative Progressions

The hues used in the color figures to illustrate different types of color progressions are good map design examples, but we have provided little rationale for their selection and arrangement. As with qualitative features, hue selection guidelines are based on convention, symbolic connotation, and the perceptual characteristics of vision.

Perhaps the most important symbolic connotation is the fact that most people see the darker, lower-value colors in a progression as indicating *more* of whatever is being symbolized. The perception of *darker as more* has been shown to hold true for both printed maps and electronic displays. It makes no difference which hue is used; we naturally associate darker color with deeper water, greater per capita income, or more densely populated areas. For bi-polar data, "more" means the positive and negative extremes.

Symbolic connotations play an important role in hue selection. Using a blue ink tint progression to show the percent of the labor force in blue-collar occupations is a classic example. Another is a bi-polar color series with progressively bluer tints for sub-zero centigrade temperatures and progressively redder tints for above freezing to hot temperatures. Solar radiation data have been mapped with a blended progression from a cloudy gray to a sunny yellow, or with a partial spectral progression from a warm yellow to a hot red. The list of creative applications of color connotations is endless.

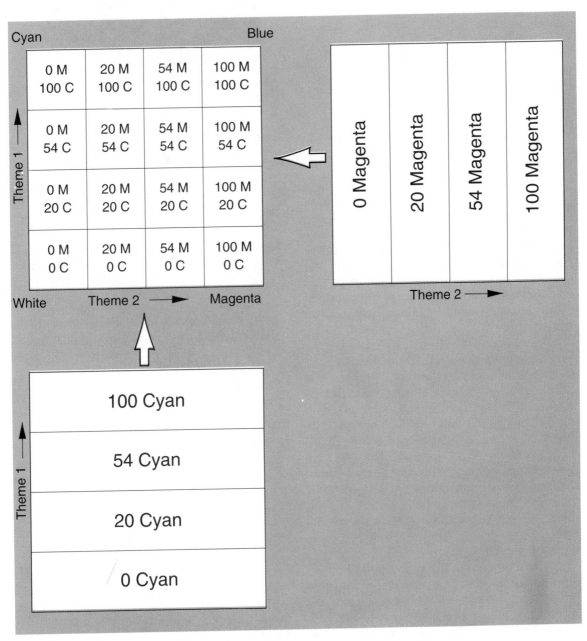

Figure 21.4   Two-variable map color scheme based on combining cyan and magenta screen tint progressions depicting themes 1 and 2.

The maximum number of steps in a progression is a matter of color perception and the nature of the phenomenon being mapped. We can have many more color increments in a progression of layer tints between contour lines than on a choropleth map of homogeneous areas. There are several reasons for this difference. For one thing, on a continuous surface we always know which layer tints are next to each other. Therefore, we can plan a smoothly blended progression with many steps. This is not possible on choropleth maps, on which we usually can't predict colors of neighboring areas.

Also, simultaneous contrast of hue, value, and chroma is not consistent from area to area on choropleth maps. Thus, in order to be sure map users see classes correctly and don't confuse similar hues, we must use fewer color increments on choropleth than on layer-tint maps. Indeed, one researcher found that for choropleth maps with green or magenta tint progressions, the upper limit is five color increments for consistently accurate map reading.

The maximum number of increments in a single-hue tint progression also depends on the difference in surface reflectance between the pure hue and its background color. A graph of the ideal relationship (no ink spread) between screen tint percentage and surface reflectance on white paper shows three groupings of ink colors (Figure 21.5). Black, dark brown, and blue allow the greatest number of inherently distinguishable tints. Red, green, cyan, and magenta have fewer easily discerned increments, and yellow has the fewest by far. Hence, it's a good idea to use partial spectral or blended progressions when yellow is an endpoint hue.

Mapped data do not always have to be grouped into classes. Modern computer map production systems can produce screen-tinted areas of any percentage, allowing each feature to be precisely colored according to its exact magnitude. In this case, the map legend shows the continuous progression of tints or blended hues. Such progressions are created by interpolating between the

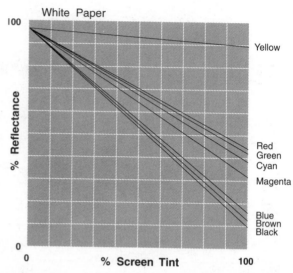

Figure 21.5 The theoretical relationship between screen tint percentage and surface reflectance.

screen tint percentages or RGB coordinates of the endpoint hues, or endpoint hues and white or gray for bi-polar schemes.

Value Guidelines for Quantitative Progressions

Value is an indispensable graphic variable in monochrome map design. As with hue, the effective use of value is based on a number of perceptual aspects. Some of these are physiologically based; others are founded in our subjective reactions to value differences.

A basic design rule is: The greater the value contrast among symbols, the greater the clarity and legibility. This rule reflects the fact that our visual system is not very sensitive to differences in value. Our ability to recall or recognize a particular value level is limited. Simultaneous contrast also comes into play, as illustrated in Figure 21.6. A gray area will appear slightly darker when on a white background, and slightly lighter when surrounded by a darker gray or black.

When quantitative geographical data are represented by value differences alone on a black-and-white map, it is wise to limit the symbolization to four or five value steps, not including solid black and white. For maximum readability, equal contrast value steps should be used. We must again remember that a value progression from light to dark inherently indicates a magnitude progression from less to more.

Screen Tint Selection   Printed solid black ink on white paper reflects about eight percent of incident light, and white paper reflects only 75-85 percent. This means that the ordinary range of perceived black to white extends only from about 3.3 to between 8.7 and 9.3 on the Munsell scale.

For a given paper and printing system, we can determine the relationship between surface reflectance and percent area inked. Knowing this relationship, we can create a 0 to 100 equal-contrast gray scale in terms of printed tint percentages to provide guidance in determining tonal steps for map production. Such an equal-contrast gray scale relative to black ink and white paper is graphed in Figure 21.7 for a 75 percent reflectance white printing paper.*

The graph may be used to derive black ink dot screen tint percentages needed to produce desired visual differences in a series of tones as follows:

1. Determine the number of classes to be symbolized—for example, five in an equal-interval series.

2. Choose the tones to be used for the first and last classes. Example: The first will be white

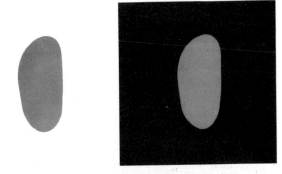

Figure 21.6   Value is a perception that cannot be measured instrumentally. The two gray spots have the same reflectance, but they are clearly different in value due to simultaneous contrast.

and the last black. Thus, the range of perceived value contrast is 100.

3. Divide the range by one less than the number of classes: $100/4 = 25$. The number 25 is the wanted difference in value contrast between each of the five tones.

4. Find the printed screen tint percentages that correspond to 0, 25, 50, 75, and 100 perceived blackness percentages on the graph. In our example $0 = 0, 25 = 13, 50 = 35, 75 = 62,$ and 100 perceived blackness = 100 percent screen tint.

Printed screen tint percentages needed for equal contrast gray scales having from four to eight steps are listed in Table 21.4.

Suppose that we do not wish to use either solid black or white as part of a four-step equal-contrast gray scale. Instead, we propose to use a 20 percent screen for the first class and a 70 percent screen for the last. By reversing the above use of the graph, a 20 percent screen is found to have a perceived blackness of 33; and a 70 percent screen, a blackness of 81. The blackness range is then 48, giving an increment of 48/3, or 16. This gives a relative blackness series of 33, 49, 65, and 81. Proceeding as in the previous example, the corresponding printed screen tints needed are 20, 34, 50, and 70 percent.

---

*Based upon (1) the Munsell value scale, and (2) measurements of reflectance of known percent area inked samples of Defense Mapping Agency process black ink, several 120 line/in. screen tinted areas, and white paper (Kimerling 1985). The graph is unique to this printing situation (paper, ink, press, and so on). Consequently, for another set of circumstances, the curve might be somewhat different, but probably not much. See Monmonier 1980 for a detailed discussion of these printing inconsistencies.

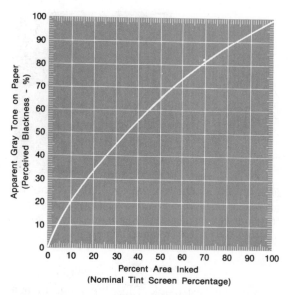

Figure 21.7 Graph showing the relation between apparent gray tone on paper (percentages of perceived blackness) and nominal screen tint percentages. See text for explanation and use.

Pattern-Value Guidelines for Quantitative Progressions

Value and pattern are related graphic elements. For dots or lines, the difference between value and pattern lies only in the size and texture of these graphic marks. Value alone is seen when the marks are too small and closely spaced to be individually discerned. We know, for example, that dot screen tints with rulings higher than 75 lines/in. (30 lines/cm.) will usually be seen as having only value, whereas tints with rulings from 40-75 lines/in. are normally perceived as having both pattern and value (Figure 21.8). Textures coarser than 40 lines/in. (16 lines/cm.) will be seen primarily as pattern without value.

Value progressions that inherently carry a magnitude message can be designed using regularly spaced dots, parallel lines, or cross-hatchings that are seen as pattern and value. If the pattern texture is fine enough that the individual marks are barely

Table 21.4
Printed Screen Tint Percentages for Five Different Equal Value Scales (from Kimerling 1985).

Four	Five	Six	Seven	Eight
0	0	0	0	0
20	13	10	8	6
54	35	26	20	16
100	62	45	35	28
	100	68	54	42
		100	73	58
			100	76
				100

noticeable, the Munsell value scale can be used to specify the percent area inked for each tint in the progression. However, for coarser texture patterns, in which lines or dots stand out clearly, the Williams gray scale for pattern-value is more appropriate. The Williams scale (Figure 21.9) was experimentally determined using coarse dot and parallel line patterns. Like the Munsell value scale, the relationship between the percent area inked of the pattern and its perceived value was found to be non-linear.

The Williams scale is used in exactly the same manner as the Munsell. Thus, the five-step pattern-value progression between white and black listed in Table 21.5 is read from the graph. Comparing percent area inked numbers with the Munsell scale, we see that the two scales are virtually identical at the light end of the gray tone scale. They differ in that higher percent area inked tones are required to produce lower value tones on the Williams scale. This is probably due to the fact that patterns with the same percent area inked have lower surface reflectances as their ruling increases, so that coarser rulings will be perceived as lighter.

Dot and line patterns have an orientation in the eyes of viewers. Map users tend to move their eyes in the direction of this perceived orientation. If irregular areas are symbolized by line patterns, as in Figure 21.10A, readers' eyes will

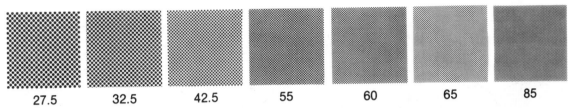

| 27.5 | 32.5 | 42.5 | 55 | 60 | 65 | 85 |

Figure 21.8 A series of dot pattern symbols in which the orientation and percent area inked are held essentially constant but the texture is varied. The numbers below the blocks show the numbers of lines of dots per inch (lpi) in each symbol. Note that textures coarser than about 40 lpi are primarily seen as pattern without value, whereas a texture as fine as 85 lpi appears primarily as value without pattern.

be forced to change direction frequently. Consequently, readers will have difficulty focusing on the area boundaries.

If these line patterns are replaced with dot patterns, as in Figure 21.10B, the map becomes more stable. The eye no longer has a tendency to jump, and boundaries are easier to distinguish. Names are also easier to read against a dot background.

Placing all parallel line patterns at the same orientation is irritating to the eye. This is particu-

larly true of vertically aligned patterns, such as Figure 21.11.

Printers have long known that images "look better" if the dot or line pattern is oriented at a 45 degree angle. Physiologists now know that our sensitivity to screen rulings is highest at the 45 degree orientation.

On the other hand, our sensitivity to screen rulings is poorest at the 0 or 90 degree orientations. At these orientations, we are often unable to focus clearly on one vertical line. The effect is somewhat reduced if the lines, regardless of their width, are separated by white spaces greater than the thickness of the lines. Generally, we should be wary of using anything but finer ruled line patterns to show classes of quantitative data.

Pattern-Texture Progressions A progression of coarse to fine texture dot, parallel line, or crosshatch patterns may also be used to symbolize quantitative data (see Figure 21.12). Texture

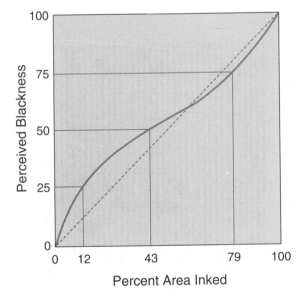

Figure 21.9 The Williams pattern-value gray scale.

Table 21.5
Percent Area Inked Needed for Five-Step Williams Pattern-Value and Munsell Value Scales from White to Black

Williams Pattern-Value	Munsell Value
0	0
12	13
43	35
79	62
100	100

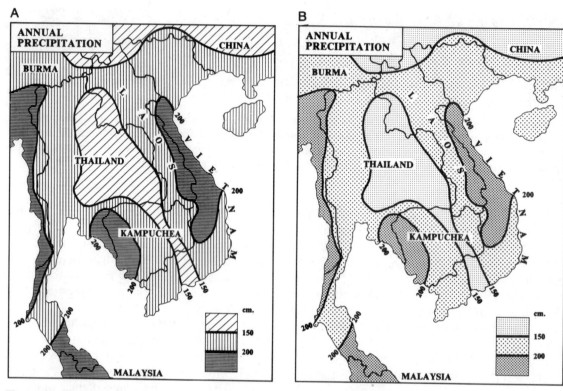

Figure 21.10   A simple monochrome map contrasting the use of parallel line and dot patterns. Line patterns are perceptually unstable, and all but the finest textures should be used with caution.

progressions inherently carry a less-to-more magnitude message (see Chapter 18). This magnitude message results from two factors. First, as marks of the same width are spaced more closely, the percent area inked increases, causing a perceived change in value. Second, we perceive a change in value as patterns of the same percent area inked increase in texture (see Figure 21.8).

Parallel line, dot, and cross-hatch patterns are commonly used in pattern-texture progressions. Of these, parallel line progressions are the least desirable, due to the visual irritation of viewing closely spaced parallel lines such as those in Figure 21.11.

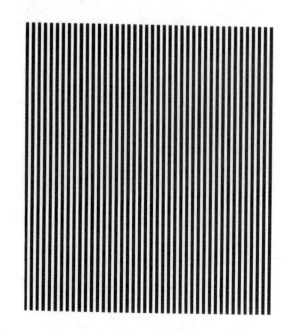

Figure 21.11   Parallel line patterns, particularly in a vertical orientation, can be irritating to the eyes.

ENHANCING DESIGN EFFECTIVENESS

Color and pattern are often used to enhance the graphic effectiveness of monochrome and multicolor map designs. Perceptual phenomena may be heightened by the selection of appropriate contrasting colors and patterns. In the following sections, we will look at some of the factors which contribute to design effectiveness.

INCREASING VISUAL ACUITY

Many maps show a wealth of detail by fine lines and point symbols. **Visual acuity** refers to the ability to detect and differentiate between these small symbols. Color is often applied to increase the symbol legibility. On the other hand, the more monochromatic the background, the easier it is for the eye to resolve fine detail. The visual efficiency of the symbol and its background color is a good estimate of legibility.

Visual Efficiency

The difference in percent reflectance or luminosity between the symbol and its background color is called **visual efficiency**. This is a measure of value contrast between colors, as can be seen in the visual efficiency table for white paper and eight ink colors (Figure 21.13).

When colors are arranged from high to low reflectance, the visual efficiencies of the possible symbol-background color combinations can be grouped into classes of excellent to poor legibility. Notice that excellent and good legibility for all inks except yellow is only obtained with white and yellow backgrounds. The most legible combinations are black, dark brown, and blue on white. The reverse is also true, as can be seen on the many electronic displays that use white symbols on black or blue backgrounds.

Combinations of pure hues such as yellow on white, green on red or cyan, cyan on magenta, and blue or dark brown on black should not be used. The visual efficiency of these combinations can, of course, be raised by using a low percentage screen tint of the background color.

The visual efficiency of symbol-background combinations of the same hue is also important, since symbols often must be placed over screen tints of the same ink. A visual efficiency table for printed screen tints of various inks, such as the eight-ink table in Figure 21.14, shows us two things. On the left edge of the table is a scale of screen tint percentages for the background color. Here the visual efficiencies in the table are for each background screen tint relative to the pure (100%) hue. If a 20% visual efficiency is used as a cut-off value for minimal readability, the table shows that all symbol hues except yellow are legible on background tints up to 60% of the same hue. Blue, dark brown, and black symbols are still discernible on a 70% background tint. Naturally, it is wise to select background tints in the "good" and "excellent" sections of the graph.

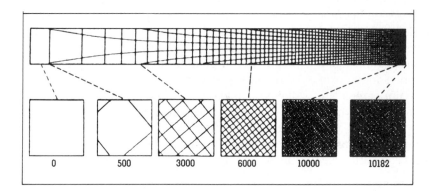

Figure 21.12 A progression of coarse- to fine-textured patterns inherently carries a less-to-more magnitude message (from Peterson 1979).

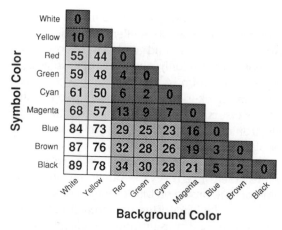

Symbol Color	White	Yellow	Red	Green	Cyan	Magenta	Blue	Brown	Black
White	0								
Yellow	10	0							
Red	55	44	0						
Green	59	48	4	0					
Cyan	61	50	6	2	0				
Magenta	68	57	13	9	7	0			
Blue	84	73	29	25	23	16	0		
Brown	87	76	32	28	26	19	3	0	
Black	89	78	34	30	28	21	5	2	0

Background Color

Figure 21.13 Visual efficiencies for a range of symbol and background color combinations. Zero is the lowest efficiency.

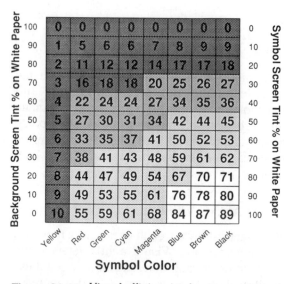

Background Screen Tint % on White Paper	Yellow	Red	Green	Cyan	Magenta	Blue	Brown	Black	Symbol Screen Tint % on White Paper
100	0	0	0	0	0	0	0	0	0
90	1	5	6	6	7	8	9	9	10
80	2	11	12	12	14	17	17	18	20
70	3	16	18	18	20	25	26	27	30
60	4	22	24	24	27	34	35	36	40
50	5	27	30	31	34	42	44	45	50
40	6	33	35	37	41	50	52	53	60
30	7	38	41	43	48	59	61	62	70
20	8	44	47	49	54	67	70	71	80
10	9	49	53	55	61	76	78	80	90
0	10	55	59	61	68	84	87	89	100

Symbol Color

Figure 21.14 Visual efficiencies for screen tints of various symbol colors. The scale on the right is for screen tinted symbols on white paper. The left scale is for solid colors on screen tinted backgrounds of the same color.

Along the right edge of the graph is a scale of screen tints for each symbol hue. The visual efficiency values now refer to the difference in reflectance between the screen tint used for the symbol and white paper. Again taking a 20% visual efficiency as the legibility cut-off, we see that symbols screened 40% or higher will be readable for all hues but yellow, with 30% screen tints acceptable for the inherently darker hues. This again assumes ideal viewing conditions, so it is prudent to use only "good" and "excellent" efficiency tints.

Acuity and Symbol Size

Visual acuity is also a function of symbol size, line width, or letter height and weight. A general design rule is to use color to increase the visual efficiency as the symbol size is decreased. As seen in Color Figure 21.4, thin lines and small lettering become more legible as their visual efficiency is increased by using higher screen tint percentages and finally solid ink. Conversely, larger lettering and thicker lines are required as their screen tint percentage lowers, particularly for higher reflectance hues such as yellow and orange.

A related perceptual phenomenon is **irradiation**, the apparent spread of light at a dark edge.

Irradiation occurs at the highest visual efficiencies, especially black on white and the reverse. A line of constant width will appear slightly wider when white is used on a black background, and slightly narrower when black is used on white paper or a white CRT screen.

PROMOTING FIGURE-GROUND

In Chapter 18 we noted that when viewing anything, we spontaneously organize the entire image into a "figure" which stands out from a more or less formless "ground." In cartography, we often describe this **figure-ground phenomenon** as **levels of visual prominence**. Well-designed maps should have at least two visual levels: The main theme should be at a higher, more visually prominent level than the locational base information in the visual background. Often one or more visual levels are, in turn, developed within the main theme, creating a multi-layer visual hierarchy on the map. We can use color in several ways to help establish this hierarchy, as we shall see in the following sections.

Advancing and Retreating Colors

Pure higher-value colors such as yellow, orange, and red are more likely to be seen as figure than lower-value greens, blues, browns, and grays. This inherent visual prominence may be due in part to the physiological phenomenon that light rays entering the eye are refracted in inverse proportion to their wavelength. This means that blue features are focused slightly in front of the retina, and red slightly behind. The eye must accommodate for this fact, with the result that a red object should appear slightly closer than a blue feature. This effect can be quite dramatic when red and blue features are juxtaposed on a CRT monitor screen. The effect is usually slight on printed maps, however.

Contrasting Colors

Color contrast is the basis of a good figure-ground relationship. Value is the primary color dimension cartographers use to systematically vary the contrast among map features to promote figure-ground. The interplay of light and dark, as for example in the type on this page, is essential to map clarity, legibility, and the recognition of the type as figure. Without value contrast, it would be difficult to see differences among map symbols. What this means is that symbol-background color combinations exhibiting high visual efficiency enhance the visual prominence of symbols.

Another important principle is that high-chroma colors should be used for the map theme, while lower-chroma colors should fill the background areas. The reason for this rule is that we tend to group symbols of similar chroma as either figure or ground. In other words, when viewing the map, the smaller areas high in chroma will usually be seen as figure.

The tourist map of central Budapest (**Color Figure 21.5**) illustrates the use of advancing and retreating colors, visual efficiency, and chroma contrast to establish a multi-level visual hierarchy. Notice that the white and yellow roads appear to stand above the low chroma brown and blue background. The numerous point symbols, in turn, are at the highest visual level.

Pattern and Figure-Ground

Patterned areas tend to be seen as figure. There are several reasons for this fact. For one thing, the similar shape, size, and arrangement of marks in the pattern lead us to visually group the marks into a single entity. If similar symbols are arranged close together, they are often seen as figure, too. The array of marks in a pattern also provides the greater **articulation** (density of symbols) that promotes the emergence of the area as figure.

Figure-ground development often requires that we reduce the visual prominence of eye-catching map symbols, such as coarse parallel line patterns, so that they become part of the visual background. Screen tinting the pattern to a light gray or desaturated hue is one way to reduce prominence, since visual efficiency is decreased. Reverse (white) patterns on gray or screened hue backgrounds is an equally effective alternative. Using a pattern composed of small black marks overprinted on a low-value color such as dark blue is another option.

A related problem is designing a pattern-background color combination so that both pattern and background appear to be one entity on the same visual level. One possibility here is to use related hue patterns and backgrounds of nearly the same reflectance, such as red or orange on yellow, to link the pattern with its background color.

PROMOTING MAP AESTHETICS

Well designed maps are easily interpreted and aesthetically pleasing. Selection of pleasing colors, particularly when used in combination, is an important part of **map aesthetics**.

Color Preferences

Researchers in several fields have studied human color preferences. Some colors are definitely liked more than others, and preferences vary from culture to culture. Color preferences also change with age. For example, young children tend to like pure warm colors, such as high-chroma red, followed by blue, green, and the other spectral hues. These bright colors are often used in children's atlases and similar products.

Studies of adult color preferences in the United States—which probably reflect Western culture—suggest that blue, green, and red are generally considered "pleasant." Blue is liked most, while orange and yellow are rated significantly lower. We appear to like greenish-yellows the least. These preferences also depend on the value and chroma level, with lower-chroma "pastel" hues preferred over the full-chroma color in a few instances.

Color Combinations

Psychologists have studied the color combination preferences of adults in our culture. For cartography, the most important result of these studies is that pleasant colors are those which stand out from their background by being significantly lighter or darker. Pleasantness, then, appears largely a matter of high visual efficiency. Given a constant value level, high-chroma colors on gray or other low-chroma backgrounds are also pleasing combinations. Hence, it appears that visual acuity, figure-ground development, and map aesthetics are all tied to good visual efficiency and chroma contrast.

SPECIAL COLOR DESIGN PROBLEMS

Not all maps will be read under normal indoor or outdoor illumination by people with normal color vision. Designing maps for special lighting conditions, for people who are partially sighted, or for people with color-deficient vision (color blindness) is an interesting cartographic challenge cen-

tered around judicious color and pattern selection. We shall look at two of these special color design problems.

SPECIAL ILLUMINATION CONDITIONS

Ships, aircraft, and many military vehicles are navigated at night by people reading maps under low-level filtered light. Red, blue, and blue-green filtered lights are used most often. Such illuminants can significantly alter the appearance of map colors.

One test of line symbol appearance under these lighting conditions found the color changes listed in Table 21.6. When viewed under red light, only black and purple were correctly perceived by the majority of people tested. This is in contrast to the generally correct perception of most colors under blue and blue-green lighting. The implication is that the maps best suited to red illumination may be monochrome maps with symbols varying only in value.

DESIGNING FOR COLOR-DEFICIENT VISION

About four percent of the public (8% of men and 0.4% of women) are congenitally color deficient. About one-fifth of these individuals are **dichromats**, who are very poor at distinguishing between certain color combinations. There are two types of dichromats: **protanopes**, who appear to be missing β cone cells; and **deuteranopes**, whose γ cone cells are absent.

The colors confused by protanopes and deuteranopes fall along "confusion lines" that are plotted on the CIE chromaticity diagram (Fig. 21.15). Protanopes confuse such colors as yellow and orange, green and reddish orange, or blue and reddish purple. Deuteranopes often confuse yellowish green and reddish orange, or greenish blue and purple.

Colors selected for maps designed for protanopes and deuteranopes should fall along curved lines on the chromaticity diagram that are every-

Table 21.6
Appearance of Colored Lines Under Different Illumination Sources (percentages refer to the percent of test subjects). Data from Erikson (1991).

White Light	Red Light	Blue Light	Blue-green Light
Black	Black 90% Brown 20%	Black 80%	Black 90%
Gray	Gray 55% Purple 35%	Gray 85%	Gray 95%
Purple	Purple 40% Gray 25%	Purple 60% Blue 40%	Purple 60% Blue 40%
Blue	Black 75% Blue 25%	Blue 75% Purple 25%	Blue 75% Purple 25%
Cyan	Black 80% Purple 10%	Blue 95%	Blue 100%
Green	Black 85%	Green 95%	Green 100%
Yellow	White 50% Yellow 40%	Yellow 100%	Yellow 100%
Red	Gray 30% Red 20%	Brown 60% Red 20%	Brown 50% Red 35%

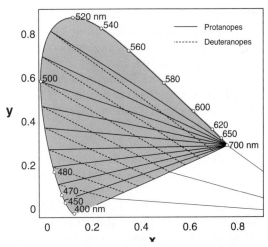

Figure 21.15   Confusion lines for protanopes (solid lines) and deuteranopes (dashed lines). All colors with CIE chromaticity coordinates lying along a given line radiating from the lower right portion of each diagram will appear similar.

where perpendicular to the confusion lines. It is also better to select colors from the left edge of the diagram, where the spacing between confusion lines is greater. This means that we should use colors grading from yellow through blue. It is also important to select colors that vary significantly in lightness, since all color-deficient individuals can see lightness differences.

SELECTED REFERENCES

Cuff, D. J. and M. T. Mattson, *Thematic Maps: Their Design and Production*, New York: Methuen, 1982.

Dent, B. D., *Principles of Thematic Map Design*, Reading, MA: Addison-Wesley, 1985.

Erikson, C., "The Perception of Colored Lines Under White, Red, Blue, and Bluegreen Lighting Conditions," Unpublished Master of Science Thesis, Department of Geography, University of Wisconsin-Madison, 1991.

Gilmartin, P. and E. Shelton, "Choropleth Maps on High Resolution CRTs/The Effects of Number of Classes and Hue on Communication," *Cartographica*, 26, No. 2 (1989): 40–52.

Helson, H. and T. Lansford, "The Role of Spectral Energy of Source and Background Color in the Pleasantness of Object Colors," *Applied Optics*, 9 (1970): 1513–62.

Kimerling, A. J., "The Comparison of Equal-Value Gray Scales," *The American Cartographer*, 12, No. 2 (1985): 132–42.

Kumler, M. P. and R. E. Groop, "Continuous-Tone Mapping of Smooth Surfaces," *Cartography and Geographic Information Systems*, 17, No. 4 (1990): 279–89.

McGranaghan, M., "Ordering Choropleth Map Symbols: The Effect of Background," *The American Cartographer*, 16, No. 4 (1989): 279–85.

Mersey, J. E., "Colour and Thematic Map Design: The Role of Colour Scheme and Map Complexity in Choropleth Map Communication," *Cartographica*, 27, No. 3 (1990): 1–157.

Meyer, M.A., F.R. Broome, and R.H. Schweitzer, Jr., "Color Statistical Mapping by the U.S. Bureau of the Census," *The American Cartographer*, 2 (1975): 100–17.

Moellering, H. and A.J. Kimerling, "A New Digital Slope-Aspect Display Process," *Cartography and Geographic Information Systems*, 17, No. 2 (1990): 151–59.

Monmonier, M.S., "The Hopeless Pursuit of Purification in Cartographic Communication: A Comparison of Graphic-Arts and Perceptual Distortions of Graytone Symbols," *Cartographica*, 17, No. 1 (1980): 24–39.

Olson, J.M., "The Organization of Color on Two-Variable Maps," *Proceedings*, International Symposium on Computer-Assisted Cartography, Falls Church, VA: American Congress on Surveying and Mapping, 1977, pp. 289–94 and color insert ff. p. 250.

Peterson, M.P., "An Evaluation of Unclassed Cross-Line Choropleth Mapping," *The American Cartographer*, 6 (1979): 21–37.

Robinson, A.H., "Psychological Aspects of Color in Cartography," *International Yearbook of Cartography*, 7 (1967), 50–59.

Sargent, W., *The Enjoyment and Use of Color*, New York: Dover Publications, Inc., 1964.

Sharpe, D., *The Psychology and Use of Color*, Chicago: Nelson-Hall, 1974.

Sibert, J.L., "Continuous-Color Choropleth Maps," *Geo-Processing*, 1 (1980): 207–16.

Sorrell, P., "Map Design—With the Young in Mind," *The Cartographic Journal*, 11 (1974): 82–91.

Standard Printing Color Catalog for Mapping, *Charting*, Geodetic Data and Related Products, Defense Mapping Agency, 1972.

Williams, R.L., "Map Symbols: Equal Appearing Intervals for Printed Screens," *Annals*, Association of American Geographers, 50 (1958): 132–39.

CHAPTER 22

TYPOGRAPHY AND LETTERING THE MAP

Some cartographers have claimed that the names on maps are a "necessary evil" because they crowd and complicate the representation. They argue that a view of the earth from above is unencumbered by names. On the other hand, maps are made to show where things are, and to do this it is important to be able to tell what's what. Only a person familiar with an area would not need names to identify the mapped features.

When names are put on a map, they become an important component of the visual display. They catch our attention, and obscure other symbols. As a matter of fact, the graphic quality of a map depends heavily on the design and placement of the names.

The process of selecting a typeface design, preparing the names, and placing them in position is collectively called "lettering the map." When there are a considerable number or variety of names, it is among the more complex and time-consuming parts of the cartographic process. Computers have brought new efficiency to preparing and placing type. But these computer methods still require some intervention by the mapmaker.

There is a close association between the process of lettering and the techniques for map construction and production (see Chapters 30 and 31). Computers treat type no differently than they treat other map symbols. To put cartographic typography in perspective, we will first survey the major changes which have taken place in this essential aspect of mapmaking.

HISTORY OF MAP LETTERING

In the manuscript period, a map had to be laboriously hand lettered by the calligrapher. The styles were varied and ranged from cramped and severe to free flowing (Figure 22.1). **Freehand lettering** dominated mapmaking through the 15th century.

Toward the end of the 15th century, when maps began to be duplicated by woodcut printing and copper engraving, lettering became the chore of craftspeople. Since small lettering in woodcuts was difficult, they devised ways of incorporating movable type, such as inserting type in holes cut in the blocks. The copper engraver cut letters with a

Figure 22.1 A portion of one of the maps in Sir Robert Dudley's atlas, *Dellá Arcano del Mare* (1646–1647). The engraver, A. F. Lucini, was fond of ornate calligraphy.

burin or graver in reverse on the plate. The map lettering during this period was generally well planned in the classic style and well executed.

As might be expected, when hand lettering was done by those more interested in its execution than in its function, it became excessively ornate. The trend toward poor lettering design accelerated after the development of lithography in the early 19th century and continued well into the Victorian era. Lettering and type styles became so bad by the close of the 19th century that there was a general revolt against them, which caused a return to classic styles and greater simplicity.

Figure 22.2 is an example of ornate lettering in the title of a 19th-century geological map. The fancy lettering demonstrates manual dexterity, since it is intricate and difficult to execute. But it is also an example of poor communication design, because it is hard to read and calls undue attention to itself.

Until the mid-19th century, most names on maps were created by craftspeople in one of three ways: drawn freehand with a pen or brush on a manuscript, carved with a knife in a woodblock, or incised with a graver in a copperplate. Metal type was used in mapmaking only sparingly, such as for titles or blocks of lettering.

In the 1840s, a process called cerography (wax engraving) allowed the mapmaker to press type into a wax surface to make a mold from which a printing plate could be made (Woodward, 1977). This direct use of type made it much easier to letter maps and led to some overcrowding.

Maps of the highest quality were still lettered by hand, either by engraving or pen, well into the first half of the present century. But freehand lettering is slow and therefore costly. If it is to be done well, it requires much more calligraphic skill than most cartographers possess.

A desire for lower costs, speed, and standardization in lettering maps led to two developments in the first half of the 20th century that made freehand lettering of maps all but a thing of the past.* The first blow came with the introduction

---

*Freehand lettering is now largely restricted to the compilation phase, interim or provisional map products, and to the occasional special map produced by freelance cartographers and illustrators who desire special effects. Freehand lettering is also practiced in the most demanding type-placement situations, such as when placing names in small regions of odd shapes and orientations. For example, parcel labels on cadastral maps are still primarily hand-lettered.

Figure 22.2 Ornate lettering in the title of a lithographed map of 1875. The name "Illinois" is more than one foot long on the original.

of the **stick-up** or **press-on type** process in the 1920s and 1930s. In this procedure, names set in type were printed on thin, clear plastic, which was then wax-backed for adhesion. The pre-printed type was then affixed to the map at the proper location (see *Stick-Up Placement* later in this chapter).

The second blow to freehand lettering came with development of a variety of **mechanical lettering** jigs and guides. These tools could be used with drafting pens to letter directly on the map sheet. Both stick-up and mechanical procedures largely circumvented the skilled eye-hand coordination needed for freehand lettering.

The development of photographic typesetting devices made use of metal type obsolete and further expanded the role of typography in mapmaking. **Phototypesetters** work in two ways. Either a template is stored for each type size and style within a typeface family, or a range of type sizes is produced through enlargement or reduction of a primary template for each style within a font family. In either case, the result tends to be rather stiff and mechanical, since there is limited opportunity to adjust type design axes to fit a particular mapping situation.

The electronic form of phototypesetting now dominates the typographic industry and brings new flexibility to lettering. You can replace the stiff mechanical type associated with photo-mechanical typesetting with lettering that is adjusted simultaneously along several design axes (spacing, stroke width, etc.) to fit the mapping situation at hand.* Since each character is treated as a graphic object, it can be stretched and distorted into various shapes (see *Outline Type* later in this chapter). In a sense, computer typesetting takes cartographers back to the era when freehand calligraphy was dominant. This is calligraphy in its modern electronic reincarnation, of course.

---

*Adobe Systems' (1585 Charleston Road, Mountain View, CA 94043, 415-961-4400) new *Multiple Master* typeface technology enables you to create thousands of instances of font characteristics by selectively varying the width and weight of a single set of primary font outlines.

What all this means is that today's mapmaker is faced with scores of type designs in a full range of sizes. With these many options, you must plan the typographic component of a map very carefully. Your goal is to get the lettering to meld into the overall graphic design so that it will function effectively.

FUNCTIONS OF LETTERING

The utility of a general reference map depends greatly on the characteristics of the type and its positioning. Recognizing the feature to which a name applies, finding the name, and reading it are all important to the functionality of a general reference map.

The type on a thematic map does not run the gamut of functions that it does on general reference maps. Still, the type must be made to fit into the display so as to enhance communication without drawing undue attention to itself.

Like all other marks on a map, the type is a symbol, but its function is more complicated than most symbols. Its most straightforward service is as a **literal symbol**: The individual letters of the alphabet, when arrayed, encode sounds that are the names of the features shown on the map. This is the most important role of the lettering in the communication system that is the map.

In several ways, the type on a map also functions as a **locative symbol**. First, by its position within the structural framework, it helps to indicate the location of points (such as cities). Second, its spacing may show linear or areal extent (as of mountain ranges and national areas). And by its arrangement with respect to the graticule, it can clearly indicate orientation.

Type can also be a **nominal symbol**. By systematically using design attributes such as upright/slant, standard/italic, serif/sans serif, and color (hue), cartographers can arrange type to show nominal classes to which features belong (see *Elements of Typographic Design* later in this chapter). For example, they might show all hydrographic features in blue type. Within that general class, they might identify open water with standard (upright) letters and running water with italic type.

By variations of size, upper/lowercase, tone, and boldness, type can also serve as an **ordinal symbol**, showing hierarchy among geographical phenomena. Thus, cartographers may use type to rank features with respect to size, importance, and so on. For example, they might use all uppercase letters for cities and lowercase with initial capital letters for towns.

In view of these many functions of type in cartography, the issue deserves serious consideration by mapmakers. We must approach the problem of choosing type for a map as we do other aspects of cartographic design. We can do this by settling clearly on the objectives of the map and then fitting the selection of type to those objectives. To pursue this matter, we will begin by discussing the character of type itself and go on to discuss how we can integrate type into the map design.

NATURE OF TYPOGRAPHY

Typography has many applications beyond its use on maps. With respect to each application, two questions arise. How are we to use the design characteristic of type effectively to serve different needs? And, what is the most effective way to render the chosen type? We will consider both issues.

ELEMENTS OF TYPOGRAPHIC DESIGN

Regardless of the kind of map, the lettering is there to be seen and read. Consequently, the elements of visibility and recognition are the major yardsticks against which we must measure our choices of type. Visibility and recognition are determined by the type's style, form, size, and color. Let's look at each in turn.

Type Style

The term **type style** refers to the design character of the type. It includes elements such as **serifs**—smaller lines used to finish off a main stroke of a letter, as at the top and bottom of M—and line thickness. You are faced with an imposing array of choices when planning the lettering for a map. Not only are there many different alphabet designs, but you must also settle on desired combinations of capital letters, lowercase letters, small capitals, roman, italic, slant, and upright forms. There is no other aspect of cartographic design that provides such opportunity for individualistic treatment. Once you become acquainted with type styles and their uses, you will find that every map presents an interesting challenge.

Letter style has undergone a complex evolution since Roman times.* The immediate ancestors of our present-day alphabets include the capital letters the Romans carved in stone and the manuscript writing of the long period before printing was invented. After the development of printing, the type styles were copies of manuscript writing, but it was not long until designers went to work to improve them. Using the classic Roman capitals and the manuscript writing as models for the small letters, there evolved the alphabets of uppercase and lowercase letters that we use today. There are three basic classes.

One class of designers kept much of the freeflowing, graceful appearance of some freehand calligraphy, so that their letters carry the impression of having been formed with a brush. The distinction between thick and thin lines is not great, and the serifs are smooth and easily attached. Such letters are known as **classic** or **oldstyle**. They appear "dignified" and have an air of quality. This style has a neat appearance, but at the same time it lacks any pretense of the geometric (Figure 22.3).

*A well-illustrated, interesting treatment of the development of lettering and type styles is Alexander Nesbitt, *The History and Technique of Lettering* (New York: Dover Publications, 1957).

CHELTENHAM WIDE

Cheltenham Wide

CHELTENHAM WIDE ITALIC

Cheltenham Wide Italic

GOUDY BOLD

Goudy Bold

GOUDY BOLD ITALIC

Goudy Bold Italic

CASLON OPEN

Figure 22.3   Some classic or old-style typefaces. (Courtesy Monsen Typographers, Inc.)

A radically different kind of face was devised later; for that reason it (unfortunately) is called **modern**. Actually, the modern faces were tried out more than two centuries ago and came into frequent use around 1800. These typefaces look precise and geometric, as if they had been drawn with a straightedge and a compass—and so they were. The difference between thick and thin lines is great, sometimes excessive, in modern styles (Figure 22.4). Both old-style and modern types have serifs. All upright forms of these type styles are sometimes loosely classed together as **Roman**.

A third style class includes some varieties that are modern in time but not in name, as well as some of older origin. This class, which is more and

BODONI BOLD

Bodoni Bold

BODONI BOLD ITALIC

Bodoni Bold Italic

Figure 22.4   Some modern style letter forms. (Courtesy Monsen Typographers. Inc.)

MONSEN MEDIUM GOTHIC

Monsen Medium Gothic Italic

COPPERPLATE GOTHIC ITALIC

FUTURA MEDIUM

LYDIAN BOLD

DRAFTSMANS ITALIC

Figure 22.5   Some sans serif letter forms. (Courtesy Monsen Typographers, Inc.)

more important in modern cartography, is called **sans serif** (without serifs) and has an up-to-date, clean-cut appearance. There is nothing subtle about most sans serif forms. There are many variants in this class, some of which include variations in the thickness of the strokes (Figure 22.5). Sans serif forms are sometimes called **Gothic**.

Several less common styles, such as **text** and **square serif** (Figure 22.6) are occasionally used on maps. Text, or black letter, is dark, heavy, and difficult to read.

This listing by no means exhausts the type style possibilities. There are hundreds of variations and modifications, such as light or heavy face, expanded or condensed, and so on.

In selecting type, you can rely on principles derived from a considerable amount of research. For instance, researchers have found that recognition depends on the occurrence of familiar forms

CLOISTER BLACK

Cloister Black

STYMIE MEDIUM

Stymie Medium

Figure 22.6   Examples of text and square serif letter forms. (Courtesy Monsen Typographers, Inc.)

and on the distinctiveness of those forms from one another. For this reason, "fancy" lettering or ornate letters are hard to read, and text lettering is particularly difficult. Conversely, well-designed classic, modern, and sans serif forms are all easy to read.

Ease of recognition also depends on the thickness of the lines forming the lettering. The thinner the lines relative to the size of the lettering, the harder it is to read. The cartographer thus faces a design conflict. Although bold lettering may be more easily seen, the thicker lines may overshadow or mask other equally important data.

The type on a map may not be the most important element in the visual outline. When type is a secondary design matter, you may want the type to recede into the background. If so, light-line type may be an effective choice.

The position of the lettering in the visual outline is significant. For example, the title may be of great importance, while the rest of the type may be of value only as a secondary reference. Size is usually much more significant than style in determining relative prominence, but the general design of the type may also play an important part. For example, rounded lettering may be lost along a rounded, complex coastline. In the same situation, angular lettering of the same size may be sufficiently prominent.

Cartography conventionally uses different styles of lettering for different nominal classes of features, but this may be easily overdone. The average map reader probably does not detect type differences nearly as well as cartographers have assumed. Using many subtle distinctions in type is probably a waste of effort, and should be avoided.

As a rule, the fewer the styles, the better harmony there will be. Most common typefaces are available in several variants. It's better to use variants of a single type style rather than many different styles (Figure 22.7). If styles must be combined for emphasis or other reasons, it is good practice to use sans serif with either classic or modern. It is not advisable to combine classic and modern. Differences in style are best used to show nominal differences. Size and boldness are more appropriate for ordinal, interval, and ratio distinctions.

Sans serif is becoming more common in cartography, while old-style and modern types are used more often in literary composition. You will often find a sans serif type used in the body of a map with a contrasting style, usually old-style, used for the map's title and legend.

There is no question that readers respond subjectively to type styles. They may, for example, see certain styles as "authoritative," "delicate," "strong," "arty," or "clean." But cartographers are usually less concerned with this "congeniality" of type style than they are with the readability of the map.

Type Form

The term **typographic form** refers to whether type is uppercase or lowercase,* whether its stance is upright or slanted, whether it is roman or italic, and combinations of these and similar elements.

Individual alphabets of any one style consist of two quite different letter forms, capitals and lowercase letters. These two forms are used together in a systematic fashion in writing, but conventions as to their use are not well established in cartography. Past practice has been to put more important names and titles in capitals and less important names and places in capitals and lowercase. Falling between capitals and lowercase in visual weight are capitals with the initial letter larger than the rest. Names requiring considerable separation of the letters are commonly limited to capitals.

The tendency in cartography is for hydrography, landform, and other natural features to be labeled in slant or italic, and for cultural features (features created by humans) to be identified in upright or Roman forms. This can hardly be called a tradition, since departure from it is frequent. But

---

*In common typographic terminology, capital letters (for example, A) are called "uppercase" and small letters (for example, a) are called "lowercase." The terms are a holdover from the time when all type was set, one letter at a time, by a compositor standing before a case holding the type. The capitals were in the upper part of the case and the small letters in the lower.

Futura Light

Futura Light Italic

Futura Medium

Futura Medium Condensed

Futura Medium Italic

Futura Demibold

Futura Demibold Italic

Futura Bold

Futura Bold Italic

Futura Bold Condensed

Futura Bold Condensed Italic

Figure 22.7 Variants of a single type style. The Futura style has an especially large number of variants. The list is representative except that expanded (opposite of condensed) is missing. (Courtesy Monsen Typographers, Inc.)

Kennerley—an upright Classic

Kennerley in the Italic form

Monsen Medium Gothic—upright

Monsen Medium Gothic Italic

Figure 22.8 Differences between italic and slant forms. (Courtesy Monsen Typographers, Inc.)

in the case of water features it is clearly a strong convention. The slant or italic form seems to suggest the fluidity of water.

There is a fundamental difference between slant and italic, although the terms are sometimes used synonymously (Figure 22.8). True italic in the classic or modern faces is a cursive form similar to script or handwriting. Gothic or sans serif italic and slant are simply like the upright letters tilted forward. Classic and modern italic forms, being more cursive, are harder to read than their upright counterparts. It is doubtful that there is much difference in ease of recognition between the upright and slant letters of sans serif, however.

Type Size

When we are concerned with legibility, the subject of type size is quite complex, because different type styles at one size may actually appear to be different sizes. The light, medium, bold, and extra bold variations of each type style further complicate the matter. To appreciate why the apparent size of type can vary so greatly requires a brief description of how the letters of our alphabet are formed.

As already noted, there are two basically different kinds of letter forms, lowercase and capitals. The majority of lowercase letters are usually the same vertical height (called the **x-height**), namely a, c, e, m, n, o, r, s, u, v, w, x, and z. Forms of this same size make up the **body** of other lowercase letters. These lowercase letters may have an element called an ascender that extends upward from the body, as in b, d, f, h, k, l, and t. Or they may have a descender below the body, as in g, j, p, q, and y. The dot in an i and a j may or may not be placed as high as an ascender. The ratio between the x-height of the body and the length of the ascenders and descenders is entirely arbitrary; the body may be small and the ascenders and descenders long or the body may be large and the ascenders and descenders short.

In contrast, capital letters (of a given size and style) are all the same height, usually about as tall as the distance from the base of the body of lowercase letters to the top of the ascenders in lowercase letters of the "same" size. These relationships are illustrated in Figure 22.9.

The size of type, which refers to its height as on a printed page, is commonly designated by **points**. One point is about 0.35 mm. (1/72 in.) The system of designating by points originated when all type was metal, and point size refers to the distance

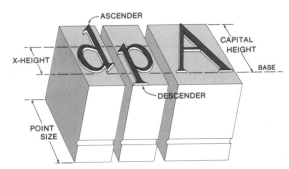

Figure 22.9 The point size of a type is a specific dimension of the block on which the typeface is located. Among different type styles, there is no consistent relation between lengths of ascenders/descenders and body height. But capitals usually are as tall as from the base of the body to the top of the ascender.

between the upper and lower edges of the cast metal block on which the typeface appears. In most styles, the point size of type comes close to being the distance between the upper limit of an ascender and the lower limit of a descender (see Figure 22.9). Thus in 18-point type, that total distance will be approximately 6 mm. It is important to note, however, that the capital letters of 18-point type are likely to be only about four millimeters high, while capital letters that are actually six millimeters high (approximately .25 inch) will be those of a 24-point size type.

The lowercase letters of dissimilar styles of type, all of a given point size, may differ markedly in the sizes of the x-height or body, whereas there will be little variation in the heights of the capitals. Although the vast majority of composition (the production of type for printing) is now done photoelectronically, the system of designating sizes by points is still standard procedure.

Reading type in continuous, spaced lines, as on this page, is very different from reading isolated names on maps. Through long experience in reading books, we have learned to recognize the shapes of thousands of words, based mostly on the patterns of ascenders and descenders. On the other hand, type on maps is mainly used for unfamiliar

names which must fit in given spaces. Furthermore, the type on maps may be angled, curved, or widely letterspaced—that is, there may be some map distance between each letter of a name. Except in occasional descriptive legends and symbol annotations, type on maps is not set in lines or blocks. Consequently, since there is great variation among type styles in the apparent sizes of lowercase letters (x-height with ascenders and descenders), a more useful index of type size in cartography is the height of the capital letters. Therefore, Figure 22.10, which equates point size with letter height, is based upon the normal height of the capital letters in various point sizes, not the distance between the top of an ascender and the bottom of a descender.

Perhaps the most common typographic decision facing the cartographer is what size to use for names. Traditionally, this decision is based on the size of the object being named or space to be filled,

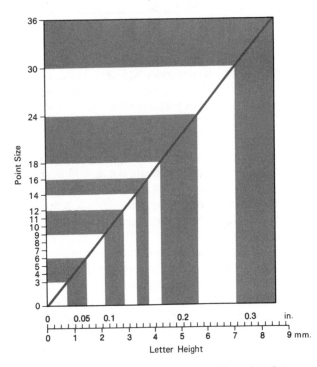

Figure 22.10 The relation between letter height and the corresponding approximate point sizes of capital letters.

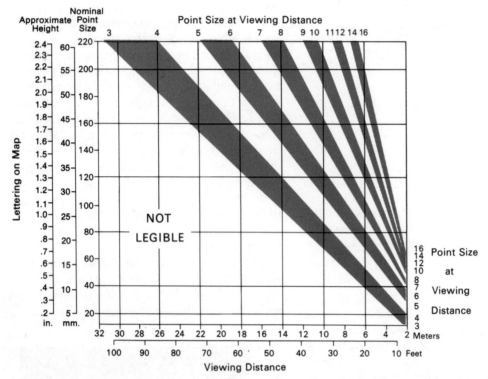

Figure 22.11 A nomograph showing the approximate apparent point sizes of lettering (capital letters) when viewed from various distances. You may extend a viewing distance vertically to a desired point-size band (in color) and then read horizontally to the left to the equivalent size. Or you may start at the left with a size on a map and reverse the process. Note that equivalent sizes less than three points are not legible.

but the lettering must also be graded with respect to the map's total design and content. Much of the criticism of map type is aimed at size. Many cartographers overestimate the eye's ability to read small type against a background cluttered with graphic symbols.

In Chapter 18, we noted that the eye reacts to size in terms of the angle the object subtends at the eye. With normal sight, an object that subtends an angle of one minute can just be recognized. Letter forms are complex, however, and it has been determined that three-point type is the smallest that people can recognize at usual reading distance. Also, "normal" sight is not the same as average sight. Thus, it is safer to assume that five-point type is the lower limit of visibility for the average person. For the growing

proportion of people with "aging eyes," the practical lower limit is much larger, especially under poor lighting conditions.

Cartographers commonly vary type size to show differences in population. But studies have shown that most map users aren't sensitive to small type size variations unless the different type sizes are next to one another, which is usually not the case on a map. One study, which tested discrimination among sizes of Futura Medium type (see Figure 22.7) with names set in capitals and lowercase letters, obtained the following results (Shortridge, 1979):

1. Size differences of less than 15 percent (for example, between 72 and 82 points) are consistently unrecognizable.

2. Differences of more than approximately 25 percent in letter height are highly desirable.

3. Within the range of 5 ½ to 15 points, any paired combination differing by 2 to 2 ½ points can be used safely—for example, 5 ½ and 7 ½ in the smaller sizes and 8 ½ and 11 points in the larger.

These results suggest that when we wish to use one style and form of type in several sizes to differentiate categories of data (for example, sizes of cities), we should limit the number to three. People seem to be able to tell "small," "medium," and "large" apart, but have difficulty with four or more. For more than three categories, we probably should introduce significant differences in form. In the interest of promoting ease of recognition ("findability" when looking for a name), the cartographer should use the largest type sizes possible, consistent with good design.

Type on wall maps and on maps projected on a screen before a large audience is commonly too small to be readable. Often the reason for this failure is that the projected slide was made from a map designed to be read at normal viewing distance. With growing frequency, however, such slides are printed to an imagesetter from a screen image that is not properly designed for projection. How can we remedy this situation?

If you are called on to prepare map presentations for groups, you must ask yourself a question such as this: "What is the minimum type size that may be read, under normal conditions, on a wall map or chart from the back or middle of a 10-m. (33-ft.) room?" The nomograph seen in Figure 22.11 is designed to answer your question. This diagram is constructed on the assumption that type readability is a function of the viewing angle it subtends. Regardless of distance, type that subtends the same angle at the eye is, for practical purposes, the same size. Thus, 144-point type at about 10 m. (33 ft.) from the observer is the same as approximately 8-point type at normal reading distance, since each circumstance results in nearly the same angle subtended at the eye. You can see from the graph that legibility diminishes very rap-

idly with distance. For example, any type of 16-point size or smaller cannot be read even at 3 m. (10 ft.) from the chart or map.

The guidelines for type size discussed here are generally applicable regardless of the mapping medium. But you should take extra care when preparing maps and diagrams to be projected by means of 35-mm. slides or overhead transparencies. These media involve several complicating factors that should be investigated with care.*

Type Color

In typography, "color" of type refers to the relative overall tone of pages set with different faces. Here it is restricted to the actual hue (such as black, blue, gray, or white) of the letters and the relation between the hue and value of the type and that of the background on which it appears. Commonly, lettering of equal intrinsic importance does not appear equal in visual weight from one part of the map to another because of background differences. By being aware of these possible effects, you may be able to correct or at least alleviate a graphically inequitable situation.

The legibility of lettering on a map (other effects being equal) depends on the amount of visual contrast between the type and its ground (Figure 22.12). Black type on a white ground is the most readable, since this combination has the highest visual efficiency (see Chapter 21). As the tonal value of the lettering approaches the tonal value of the ground, visibility diminishes. Visibility is of special concern when large regional names are "spread" over a considerable area composed of units that are colored or shaded differently. It also is a concern when names of equal rank must be placed on areas of different values.

Commonly, the lettering on maps is either dark (such as black) on a light ground or light

---

*An excellent treatment of these problems with specific directions is given in H. W. Brockemuehl and Paul B. Wilson, "Minimum Letter Size for Visual Aids," *The Professional Geographer*, 28, No. 2 (May 1976): 185–89.

CARTOGRAPHY

Figure 22.12 Perceptibility and legibility depend on lettering-background contrast (visual efficiency).

(such as white) on a dark ground. The latter is called reverse (or drop-out) lettering. Type is often added to one of the color flaps of a multi-colored map (such as blue for hydrography). Regardless of the color of the print and of the ground, if the value contrast is great, the lettering will be visible.

FORMING THE LETTERS

We now turn from the design aspects of type to methods of creating the characters. The technology of lettering has always influenced the use of type on maps (see "History of Map Lettering" earlier in this chapter). Modern electronic technology is no exception. The raster and vector approaches that now dominate the graphic arts and printing industries define today's typographic industry. Since the approach taken in forming characters can greatly influence your typographic options, let's take a closer look at the process of forming characters electronically.

Bit-Mapped Type

In the raster approach to typography, we define a character by a special arrangement of pixels (or dots). Since it takes one bit of data to turn each pixel in the character pattern on or off, the type is often said to be **bit-mapped** (Figure 22.13). Since a bit-map defines the actual image of a character, each type design variation requires a new bit-mapped description. Thus, a collection of type bit-maps suitable for mapping must be quite large.

The close relation between the spatial resolution of a character's bit-map and the graphic quality of typography is clearly seen in Figure 22.13. Coarse bit-maps save computer storage and processing, but they yield ragged characters. Similarly, enlarging a bit-mapped character significantly will produce jagged-edged type. The graphic quality of bit-mapped type depends on the original raster definition. Bit-mapped type tends to appear somewhat rigid and mechanical, lacking details that require subtle variations in line widths and curvatures.

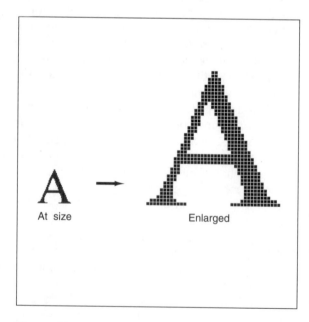

At size   Enlarged

Figure 22.13 The graphic quality of bit-mapped type depends on the spatial resolution of the pixel array within which characters are defined.

In view of these drawbacks, it is natural to wonder why we use bit-mapped type at all. There are several reasons. The technology of bit-mapping is relatively simple, inexpensive, and fast to implement in the native raster environment of computer screens and printers. Indeed, until recently, bit-mapping was the only way to display type on raster-based computer monitors. You can also shift between typefaces exhibiting different characteristics without slowing the printing process significantly.

Outline Type

The alternative to bit-mapped type is to treat characters as graphic objects. The object outline is then defined mathematically and procedurally, and consists of vertexes (often called "handles"), straight lines, and curves (Figure 22.14). This outline description permits a **master character** to be scaled to a range of sizes with no loss of graphic quality. It also allows you to stretch, bend, shear, and otherwise modify the shape of type to create special design effects. As you might expect, **scalable outline type** exhibits excellent graphic characteristics.

Since computer screens and printers are raster devices, however, outline type must be converted to bit-mapped format for purposes of display and

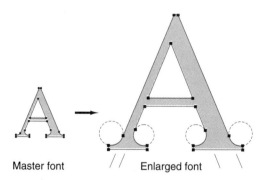

Master font   Enlarged font

Figure 22.14 Outline type is treated as a graphic object defined by points (vertexes), straight lines, and curves. This means a master character can be scaled to different sizes and transformed in shape without loss of graphic quality.

printing. This conversion process is called rasterization, or raster image processing (RIP). The rasterization may take place within the computer, or it may be handled by the peripheral device (monitor or printer). In either case, it takes the added expense of both memory and microprocessor time to carry out the scaling and rasterization. Since it takes time to scale a character template to the desired shape and size and then convert its outline description to bit-mapped form, you pay a price for flexibility gained with outline type. The cost is a reduction in operating speed. This slowdown can be overcome by using more powerful microprocessors. But even then, extensive shifting between outline typefaces of different characteristics may slow the character-forming process significantly.

From Print to Display

Scalable typefaces were first introduced for printing devices. The breakthrough came in the mid-1980s with the introduction of *PostScript Type 1* fonts by Adobe Systems.* You now have access to large Type 1 font libraries from a variety of vendors. Many vendors also market their own versions of scalable fonts.**

Printers that can handle outline fonts are of two forms. The printer may be a dumb slave to a host computer, in which case the computer must carry out the scaling and rasterization, and send the bit-mapped data to the printer. Although this makes for a relatively inexpensive printer, it places extra processing demands on the host computer.

The alternative is to download outline descriptions to a printer that contains an on-board microprocessor and memory. The smart printer can then do its own RIP. This unburdens the host computer, but increases the cost of the printer, since it must be a computer in its own right.

---

*Adobe Systems Inc., 1585 Charleston Road, Mountain View, CA 94043 (415-961-4400).

**Speedo* is the proprietary format of Bitstream Inc., 215 First Street, Cambridge, MA 02142 (617-497-6222); *Intellifont* is marketed by Agfa-Compugraphic, 90 Industrial Way, Wilmington, MA 01887 (800-424-8973).

Having outline fonts for printers radically changed the printing industry, but caused a problem for graphic designers who worked with computer monitors. The typographic effects you could produce with a printer could not be seen on the computer display where designing was done. What was needed was a true **"what you see is what you get"** (abbreviated **WYSIWYG**) working screen environment. This arrived with the introduction of "on-the-fly" type managers/rasterizers in the early 1990s.* When using software that takes advantage of these type managers, you now have typographic capability on screen and in print that is faithful to the original typeface design. This greatly enhances your ability to incorporate type effectively into a map design.

LETTERING THE MAP

The phrase "lettering a map" means deciding where type should go and placing it on the map sheet. Let's look at the details involved in this phase of mapmaking.

POSITIONING GUIDELINES

Map reading is greatly affected by the positioning of names. When properly placed, the lettering clearly identifies the phenomenon to which it refers, without ambiguity. The positioning of type has as much effect on a map's graphic quality as does the selection of type styles, forms, and sizes. Incongruous, sloppy positioning of lettering is just as apparent to the reader as are garish colors or poor line contrast.

Large map-making establishments often develop policies regarding positioning of type. They do so partly to obtain uniformity, partly because it is

---

*Adobe Type Manager (ATM) renders Type 1 PostScript fonts on screen. Apple Computer's System 7 and Microsoft Windows (Version 3.1) operating systems implement *TrueType* technology, which renders TrueType fonts on screen. Products from other vendors clone or emulate these formats.

cheaper, and partly because they can assign parts of such activity to a machine. The excessive result—all names parallel, for example—sets an unfortunate standard that should not be blindly followed. The cartographer should be guided by principles based on the function of the map as a medium of communication. The object to which a name applies should be easily recognized, the type should conflict with the other map material as little as possible, and the overall appearance should not be stiff and mechanical.

As we saw earlier, one important function of type is to serve as a locative device. The lettering can do this in three ways: (1) by referring to point locations, such as cities; (2) by indicating orientation and length of linear phenomena, such as mountain ranges; and (3) by designating the form and extent of areas, such as regions or states. The first rule of positioning type is to arrange lettering so that it enhances the locative function as much as possible. Here are some guidelines:

1. Names should be either entirely on land or on water.
2. Lettering should be oriented to match the orientation structure of the map. On large-scale maps, this means that type should be parallel with the upper and lower edges of the map; in small-scale maps, type should be parallel with the parallels.
3. Type should not be curved (that is, different from rule 2 above) unless it is necessary to do so.
4. Disoriented lettering (rule 2 above) should never be set in a straight line but should always have a slight curve.
5. Names should be letterspaced as little as possible. That is, there should not be wide spaces between letters in the name.
6. Where the continuity of names and other map data, such as lines and tones, conflicts with the lettering, the data, not the names, should be interrupted.
7. Lettering should never be upside down.

Conflicts among these guidelines frequently occur because of opposing requirements. There is no general rule for deciding such issues; the cartographer must make a decision in light of all the special factors. Figures 22.15 and 22.16 illustrate the foregoing principles of type positioning as well as the precepts that follow.*

*For many illustrations of good and bad practice in positioning lettering, see Eduard Imhof, "Positioning Names on Maps," *The American Cartographer*, 2, No. 2 (1975): 128–44.

Let's look now at some other rules of type positioning. In the following sections, we will organize these positioning guidelines in terms of place, linear, and areal features.

Place Features

When positioning lettering with respect to place locations, you have a number of options, which can be ranked in order of decreasing preference

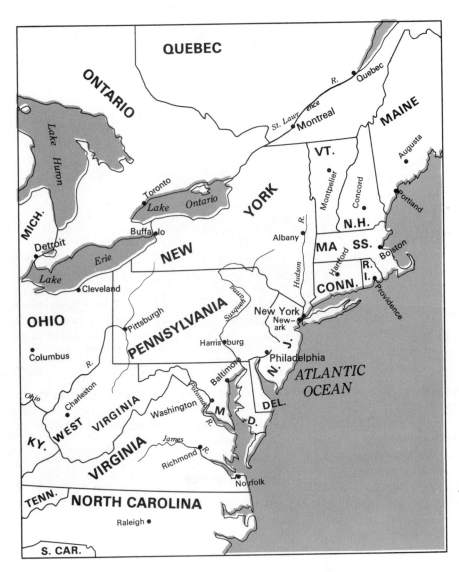

Figure 22.15 Most of the rules about positioning type have been violated on this map. Compare Figure 22.16.

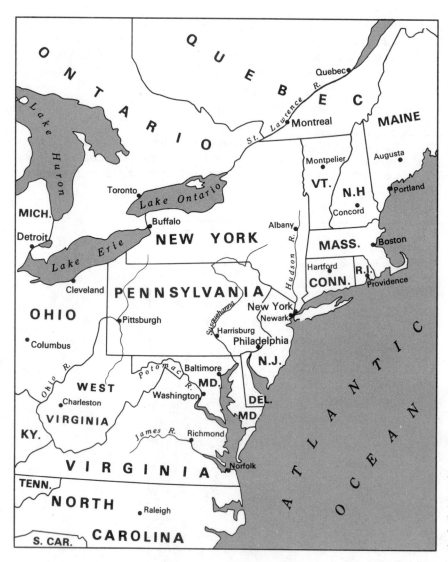

Figure 22.16 The same lettering as in Figure 22.16 has been positioned in accordance with good design practice on this map.

(Figure 22.17). The most preferred position for lettering is above and to the right of the place in question. When a conflict occurs with other map symbolism in this preferred position, you can look to the next most desirable positioning option. If a conflict again occurs, go to the next option, and so on, until you find a lettering position.

The names of places located on one side of a river or boundary should be placed on that same side. Places on the shoreline of oceans and other large bodies of water should have their names

entirely on the water. Where alternative names are shown (Köln, Cologne), one should be symmetrically and indisputably arranged with the other so that no confusion can occur.

Linear Features

Lettering should always be placed alongside and parallel to the river, boundary, road, or other linear feature to which it refers. Names are better placed above than below linear features because there

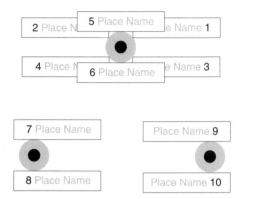

Figure 22.17 We have considerable flexibility when positioning type next to place features, although some positions are preferred over others.

are fewer descenders than ascenders in lowercase lettering.

The lettering should never be separated from the linear feature by another symbol. Names along rivers need to be repeated occasionally.

If the linear feature is curved, the lettering should correspond. Ideally, we would place the type along an uncrowded stretch where the lettering could be read horizontally, but often this is not possible. If we must position lettering nearly vertically along a linear feature, it is easier to read it upward on the left side of the map and downward on the right.

Complicated curvatures should be avoided, and names should be somewhat letterspaced but not much. Where possible, it is good practice to curve lettering so that the upper portions of the lowercase letters are closer together because there are more clues to letter form in the upper parts.

Areal Features

Lettering to identify areal phenomena should be placed within the boundaries of the region to which it refers. The name should be letterspaced to extend across the area, but the letters should not be so far apart that they do not look like part of the name. Letters should not crowd against boundaries. If tilting is necessary, use clearly noticeable curvature so that the name will not look like a printed label simply cut out and pasted on the map. Curvatures should be simple and constant.

METHODS OF LETTERING

Today few mapmakers rely on freehand or mechanical lettering systems for finished map construction. Instead, they letter their maps with some form of stick-up type or with the aid of a computer. Let's look at both these lettering methods.

Stick-Up Placement

With stick-up lettering, you simply affix preprinted type to the map. This method is fast and requires few skills beyond a good design sense and normal hand-eye coordination. If the position selected for a name turns out to be unsuitable, you can relocate the type without difficulty. You may use the letters as composed in a straight unit, or you may cut them apart and apply them separately to fit curves. Sans serif typefaces are easier to work with in this respect, since there are no serifs to interfere with cutting and curving.

The type used for stick-up lettering is commonly printed on opaque white paper with a standard laser printer. The type is then photographically reduced to a special **stripping film**, which consists of a thin emulsion layer laminated to a thicker base material. Increasingly, the type may be printed directly to film using an imagesetter and then contacted to stripping film using a platemaker (see Appendix D). In either case, the back of the emulsion layer is then given a light coating of colorless wax.

The words, numbers, or letters found on the waxed emulsion layer may then be cut and stripped from the base sheet, and burnished onto the artwork by rubbing. Rubbing is usually done using a smooth tool, with a protective sheet laid over the lettering strip. Light rubbing is sufficient to hold the lettering in place and will allow it to be later repositioned if necessary without damage to the type. Final burnishing is best postponed until all positioning decisions have been made.

Prior to desktop computers, a cartography laboratory either had a dedicated typesetting machine in house, or sent lists of type out to special facilities for contract typesetting. In most modern mapping establishments, however, the type list is merely keyed into word processing or text editing software installed on a desktop computer. Then the digital file is sent to a laser printer or imagesetter for output. Laser printer output is usually reduced photographically to improve graphic quality of the type.

The cartographer must make a complete type list prior to typesetting, regardless of the method used. This list consists of all names and other alphanumeric characters to appear on the map or series of maps for which stick-up is to be used. All lettering should be grouped in the listing according to type style, size, and capital and lowercase requirements. Spellings must, of course, be reviewed carefully. Unfortunately, commercial "spelling checkers" usually lack the necessary place name vocabulary, so your checking must be done manually.

It is usually desirable to obtain several copies of the composition. Then, it is only necessary to repeat names, such as "river" or "lake," as many times as the total number of map occurrences of that size divided by the number of copies of the composition to be obtained. The names will be set as listed and, since there will probably be some inadvertent omissions, it is good practice to obtain extra characters of the styles and sizes being used.*

Computer-Assisted Placement

Although stick-up type is still widely used, cartographers are increasingly lettering their maps directly with the aid of computers. At this time semi-automated methods are the rule. But progress is being made toward developing fully automated procedures.

Semi-Automated Methods Computer software packages currently used in map production (see Chapter 31) commonly make provision for handling scalable type (see "Outline Type" earlier in this chapter). When using this software, you have the option of entering type from a keyboard, importing digital text lists, or importing a digital cartographic file with the location of type defined. You then can select from a variety of typefaces options, and can specify type design parameters. With these decisions made, you can place the type at the desired map location. Tools are commonly available to help you letterspace and curve the type to fill a region, or to follow a curved feature such as a river or boundary. Finally, with the type properly positioned, the map layer holding the type can be output to a printer or imagesetter.

If software used in map production does not permit output of outline type, you may want to export the digital files to a software package that does have this capability. This permits you to use the second software package to create the desired typographic effects, and then send the enhanced type files to a printer that can handle scalable type.

Automated Methods Our ultimate goal, of course, is fully automated type placement.** Yet, in spite of advances in computer-assisted cartography, the placement of names on maps has yet to be automated in a way that meets accepted carto-

---

*Press-on lettering consisting of alphabets (and symbols) attached to the "underside" of a plastic carrier sheet are available in stores that sell graphic arts supplies. When burnished, the letters will come off the plastic base material and adhere to the map. Unfortunately, it is difficult to get an acceptable match between these press-on alphabet characters and those printed by other means. Thus, these press-on characters are primarily useful for a large name (title) or two, and when you need a few letters or numbers to use as special symbols.

**It may turn out that the whole concept of map lettering will change in adjustment to computer capabilities. Imagine, for instance, using an interactive display screen, on which a map appears without any type at all. You could simply touch a feature on the map, and its name would be verbally announced or displayed on the side of the map. Alternatively, you could call out a place name, and the computer would show its location with a flashing symbol on the map. Fully developed systems of this sort would make far more information available to the map user than simple place-name data. If this were to happen, the nature of maps would change dramatically from what we have come to expect.

graphic standards. Effective integration of lettering with the other elements of the map image by strictly automated means has proven to be extremely difficult. The reason is that good placement requires making numerous decisions that involve spatial synthesis and judgment. People speed through tasks of this type. But the current generation of mapping software lacks these human abilities.*

To date, most automated positioning has been restricted to standard lettering set in a straight line, usually horizontally, or letterspaced type set along simple curves. This may be sufficient for constructing titles and legends but, within the map image, only crudely approximates results achieved by hand-eye placement of type.

The most effective type placement algorithms take a hierarchical approach in following "accepted" type placement guidelines (see "Positioning Guidelines" earlier in this chapter). Names are placed in increasing orders of freedom. Since there is least flexibility in placing the names of areal features, the algorithm deals with them first. Linear features provide the next least place-name flexibility, so the algorithm places these names second. Since point features provide the most type placement options, they are dealt with last.

The results produced by current type placement algorithms can be quite impressive for graphically simple maps calling for moderate type density (Figure 22.18). However, for complex maps, or maps requiring high type density, the current generation of type placement algorithms usually run into some conflicts that cannot be resolved automatically. If these problems are few, something of benefit still has been achieved. Some estimates show that lettering accounts for almost 40 percent of total map production costs. A type placement

system that calls for human intervention with only a small minority of names could still result in considerable savings.

In fact, research shows that semi-automated type placement using human-machine interaction may be the most practical approach at the present time. With this procedure, you select type from an existing database of geographical names. The computer system automatically places the type on the map according to a set of good positioning principles. Using this "first approximation," you can shuffle names around and change type parameters to avoid overlap and to make all names legible. Once you achieve an acceptable arrangement, you can send the positioning data to a suitable printing device to create the finished lettering flap. The result thus demonstrates an efficient merging of the advantages of the computer and the cartographer.

GEOGRAPHICAL NAMES

In addition to selecting and positioning type on maps, we need to understand the nature of geographic names. Furthermore, in view of the high

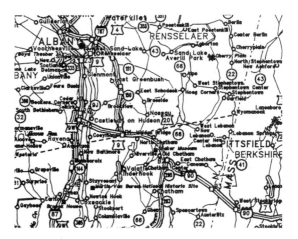

Figure 22.18 Example of map produced using automated name placement techniques (courtesy Doerschler and Freeman).

---

*Ongoing research in the area of artificial intelligence is leading to the development of expert systems in a variety of fields. This work suggests that future generations of mapping software may be better able to mimic the human thought processes needed to place type effectively.

cost associated with building computerized lists of geographic names, mapping efficiency may hinge on access to existing databases of place names. Let's briefly look at both topics.

NAMING CONVENTIONS

Often the most difficult question associated with place names on maps concerns their appropriate spelling. For example, do we call an important river in Europe Donau (German and Austrian), Duna (Hungarian), Dunav (Yugoslavian and Bulgarian), Dunarea (Romanian), or do we spell it Danube—a form not used by any country through which it flows! Is it Florence or Firenze, Rome or Roma, Wien or Vienna, Thessalonkie, Thessoloniki, Salonika, or Saloniki, or any of a number of other variants? The problem is made even more difficult by the fact that names change because official languages are altered or because internal administrative changes occur.

Indeed, governments that produce many maps have established agencies just to specify spellings to be used. Examples are the *U.S. Board on Geographical Names (BGN)*, an interdepartmental agency, and the *British Permanent Committee on Geographical Names (PCGN)*. Most such governmental agencies concern themselves only with problems associated with domestic names, but these two oversee the spelling of all geographical names.

One of the major concerns of such an agency (and of every cartographer) is how a name in a non-Latin alphabet will be rendered in the Latin alphabet. Various systems of transliteration from one alphabet to another have been devised, and the agencies have published the approved systems. The Board on Geographic Names publishes bulletins of place-name decisions. It also furnishes guides recommending spellings for many foreign places. These are available upon application.*

The practice in the United States is to use the conventional English form whenever one exists. Thus, Finland (instead of Suomi) and Danube River would be preferred. Names of places and features in countries using the Latin alphabet may, of course, be used in their local official form if that

is desirable. The problem is complex, and it is difficult to be consistent. It is easy for English speaking readers to accept Napoli and Roma for Naples and Rome, but it is harder for them to accept Dilli, Mumbai, and Kalikata for Delhi, Bombay, and Calcutta.

Cartographers should always be alert to the meanings of place names. They must be careful not to fall into toponymic blunders by using such names as Rio Grande River, Lake Windermere, or Sierra Nevada Mountains.

DATABASES

To take advantage of computer assistance in typesetting and type placement, a list of place names and other typography must be available in computer compatible (digital) form. If this is done on a project-by-project basis, the cartographer will need to prepare the usual type list and then enter this information into a computer file. The computer can then use this digital record to drive the selected lettering device. Alternatively, the cartographer may use an existing database, created by a central mapping authority, which contains all place names for a region. By working from this computer-generated list, the cartographer can select only those names that are needed for a given map.

In the interest of standardizing geographical names usage and automating the use of place names, the U.S. Geological Survey (USGS) has developed its *Geographical Names Information System (GNIS)*. As the core of the system, a machine-readable national geographical names database is being constructed. The primary source for the place names in these files is the standard 7.5-minute quadrangle topographic map series published by the USGS. Names information de-

---

*To inquire about domestic names, contact: U.S. Board on Geographic Names, National Center, Stop 523, Reston, VA 22092. For foreign names, contact: U.S. Board on Geographic Names, Defense Mapping Agency, Building 56, U.S. Naval Observatory, Washington, DC 20305.

rived from the records of the U.S. Board on Geographical Names, the U.S. National Ocean Survey charts, and other federal agency sources are also being integrated into the files. To date, millions of names have been incorporated.

The comprehensive files include all named natural features (about 80 percent of the database) and most major and minor civil divisions, dams, reservoirs, airports, and national and state parks. Named streets, roads, and highways are not included at this time. A number of data elements are coded with each feature included in the database. In addition to the official name of the feature, these include one of 72 feature classes (school, cemetery, stream, summit, populated place, and so on), location (state, county, and geographic coordinates), elevation, variant names (aliases), and the map sheet on which the feature is found.

Information from the GNIS is presently available in several forms. You can use an outside computer terminal for direct on-line access to the database. Alternatively, you can purchase USGS computer products such as magnetic tapes and disks, computer printouts, and microfiche through the USGS Earth Science Information Center (ESIC).* You can also purchase printed and bound products, such as alphabetical and topical lists (Figure 22.19).

Regardless of the form of output, the GNIS offers accurate, ready reference information for automated name placement and map production. Because the database is in machine-readable form, the information it contains can be correlated easily with the information in other computerized files.

SELECTED REFERENCES

Ahn, J., and H. Freeman, "A Program for Automated Name Placement," *Proceedings Auto Carto 6*, 2 (1983): 444–453.

Bartz, B. S., "An Analysis of the Typographic Legibility Literature," *The Cartographic Journal,* 7 (1970): 10–16.

Blok, D. P., "Terms used in the Standardization of Geographical Names," *World Cartography*, 18 (1986).

---

*ESIC, U.S. Geological Survey, 507 National Center, Reston, VA 22092 (703/648-6045).

GEOGRAPHIC NAMES INFORMATION SYSTEM (GNIS)**
STATE OF WISCONSIN
ALPHABETICAL FINDING LIST
18 SEPTEMBER 1981

NAME	FEATURE CLASS	STATE COUNTY		COORDINATE	BGN	ELEV FT	SOURCE	MAP			
Herbster	ppl	55007		464957N0911530W		614		0015			
Herbster Cemetery	cem	55007		464933N0911304W		858		0016			
Herby Lake	lake	55095		453521N0923817W	1969	951		0277			
Herde Lake	lake	55017		451452N0912326W		1074		0417	0371		
Herman Center	ppl	55027		432450N0882736W		1118		0968			
Herman Coulee	valley	55063		435943N0910119W			435940N0905920W	0829	0830		
Herman Creek	stream	55115	55087	443425N0882834W			443823N0882245W	0670	0671	0628	0627
Herman Lake	lake	55135		442919N0890538W				0705			
Herman School	school	55027		432530N0882704W				0968			
Hermans Landing	locale	55113		455603N0911115W				0166			
Hermansfort School	school	55115		444739N0884632W				0578			
Hermit Island	island	55003		465312N0904106W				0010			

Figure 22.19 A convenient product of the Geographical Names Information System is the Geographical Names Alphabetical Finding List, which consists of a set of spiral-bound volumes for each state. Information is presented as it is coded in the database.

424 PART V/PERCEPTION AND DESIGN

Brockemuehl, H. W., and P. B. Wilson, "Minimum Letter Size for Visual Aids," *The Professional Geographer,* 28 (1976): 185–89.

Doerschler, J. S., and H. Freeman, "A Rule-Based System for Dense-Map Name Placement," *Communications of the ACM*, 35 (1992): 68–79.

Ebinger, L. R., and A. M. Goulette, "Noniterative Automated Names Placement for the 1990 Decennial Census," *Cartography and Geographic Information Systems*, 17 (1990): 69–78.

Gottschall, E. M., *Typographic Communications Today*, Cambridge, MA: MIT Press, 1989.

Hirsch, S. A., "An Algorithm for Automatic Name Placement Around "Point Data," *The American Cartographer*, 9 (1982): 5–17.

Imhof, E., "Positioning Names on Maps," *The American Cartographer*, 2 (1975): 128–44.

Johnson, D. S., and U. Basoglu, "The Use of Artificial Intelligence in the Automated Placement of Cartographic Names," *Proceedings Auto Carto 9* (1989): 225–230.

Labuz, R., *Typography & Typesetting: Type Design and Manipulation Using Today's Technology*, New York: Van Nostrand Reinhold, 1988.

Nesbitt, A., *The History and Technique of Lettering*, New York: Dover Publications, 1957.

Rogondino, M., *Computer Type: A Designer's Guide to Computer-Generated Type*, San Francisco, CA: Chronicle Books, 1991.

Shortridge, B. G., "Map Reader Discrimination of Lettering Size," *The American Cartographer*, 6 (1979): 13–20.

Updike, D. B., *Printing Types, Their History, Forms and Use: A Study in Survivals*, 3rd. ed. Cambridge, MA: Harvard University Press, 1966.

van Roessel, J. W., "An Algorithm for Locating Candidate Labeling Boxes within a Polygon," *The American Cartographer*, 16 (1989): 201–209.

Woodward, D., *The All-American Map, Wax Engraving and Its Influence on Cartography*, Chicago and London: The University of Chicago Press, 1977.

Yoeli, P., "The Logic of Automated Map Lettering," *The Cartographic Journal*, 9 (1972): 99–108.

MAP COMPILATION

*I*n cartography the term **compilation** refers to assembling and fitting together the geographical data that you will include in your map. "Fitting together" means locating data in their proper horizontal positions according to the map projection system and scale being used.

In compiling a map, your objective is to prepare a composite containing all the base reference data, lettering, geographical distributions, and everything else that will appear. This composite becomes the guide for constructing the map, either manually by scribing or ink drawing or automatically by printing or plotting. It's important to compile the map in a way that will make the ultimate map construction as easy as possible. (For more on map construction, see Chapter 31).

THE COMPILATION PROCESS

When you compile data, you may need to use other maps, text and tabular sources, and digital records to find the information you need. The maps (or the digital records that represent the maps) may be on different projections; they may differ markedly in level of accuracy; the dates of publication may vary; their scales will probably be different; and they are likely to have different forms (photo versus line imagery). Text and tabular data may exhibit an equal variety of characteristics. You will have to pick and choose, discard this, and modify that, all the while placing the selected data on the new map, locating each item precisely.

An important **rule of compilation** is to work from larger to smaller scales, because even the largest-scale maps show data that have been generalized. These data may be "accurate" for the scale at which they are presented but not for a larger scale. If you compile from smaller to larger scale, you may be building error into the compilation. You may also give a false impression of the environment, because there may appear to be a lack of feature detail when that is not the case.

THE WORKSHEET

The composite that results from the compilation process is called the **compilation worksheet**. The worksheet may be constructed by hand or machine on a sheet of drafting material, or it may consist of nothing more than an electronic image on a display screen. For a simple map, the worksheet contains everything. But complex maps may need more than one worksheet, each carefully registered (accurately fitted and positioned) with its companions. When completed, a worksheet is comparable to a corrected, "rough draft" manuscript. All that is then necessary is to prepare the final artwork by drafting, scribing, printing, or plotting.

You may be able to use many of the elements of one map when compiling another. General features such as boundaries, hydrography, and even lettering may not vary much, if at all, from one to another in a series of maps. Consequently, you can save effort by anticipating future use of your compilation worksheet. You should prepare the worksheet with future possibilities in mind. Modern reproduction methods make it relatively easy to combine different separation drawings or images, even when printing in one color (see Chapter 30).

SOURCE DATA AND MAP SCALE

As suggested in Chapter 3 (History of Cartography), an important distinction in mapmaking is between large-scale topographic and special reference maps and small-scale general reference and thematic maps. One of the major differences lies in the methods used to acquire and compile data for these two types of maps. Let's take a look at the techniques used to compile large-scale, medium-scale, and small-scale maps.

Large-Scale Mapping

Large-scale maps (maps with scales larger than about 1:75,000) are usually compiled by photogrammetric methods (Chapter 12) or field survey (Chapter 7). The planimetric accuracy of such maps is controlled as carefully as possible.

Two types of survey—plane survey or geodetic survey—may be used in compiling large-scale maps. There are important differences between the two. Plane (or cadastral) survey is commonly performed for a limited area. Thus, earth curvature may not be taken into account, because it is relatively insignificant over a small area. The lines of a plane survey are determined from ground observations and are mapped as observed, without being referred to as a spheroid. Topographic maps, on the other hand, are based on a framework developed by geodetic survey, and the ground observations are referred to as a spheroid. Consequently, the two kinds of surveys usually do not match.

Most large-scale plane survey and cadastral maps do not show much physical environmental data. If you wish to combine physical and cultural data on one map, you may have trouble. For example, suppose you wish to make a map showing up-to-date information concerning (1) streams, lakes, and swamps, and (2) land ownership of a region. You are likely to find the first, but not the second, category of information on topographic maps. Ownership data will probably be available from county surveyors' maps, but these may not show the drainage details. The two sources will be essentially "accurate" according to the definitions used for their mapping, but they will not match each another. The smaller the scale of the map you are compiling, the fewer such difficulties you are likely to have.

Medium-Scale Mapping

You can compile medium-scale maps (those with scales between 1:75,000 and 1:1,000,000) in several ways. You can trace selected data from larger-scale maps and reduce the result manually, mechanically, photomechanically, or electronically. The degree to which interpretation and generalization take place in this process will vary with the technique. With manual methods, "eyeball" generalization has long been practiced. Strictly mechanical and photomechanical processes involve relatively little interpretation and generalization. Electronic methods, on the other hand, involve complex mathematical generalization procedures.

Medium-scale maps are being compiled more and more frequently from remotely sensed imagery taken from high-altitude aircraft and space vehicles. Sometimes such maps may be compiled by creating a graphic image from digital sensor data. At other times (as in land cover mapping from satellite imagery), maps may be compiled by obtaining a line image from a photographic record.

Small-Scale Mapping

Small-scale general reference and thematic maps (those with scales smaller than 1:1,000,000) are compiled quite differently from large-scale or medium-scale maps. On small-scale thematic maps, the main subject matter is usually presented against a background of locational information, which is called the base data. This base material is ordinarily compiled first. Then it is used as a skeleton on which to hang the subject-matter data. The subject-matter information, usually consisting of coasts, rivers, lakes, and political boundaries, is available from large-scale general reference maps.

More and more base data useful for small-scale mapping are being converted to digital records and stored in cartographic data banks. Statistical information is also being stored with increasing frequency in digital data files. If you have access to these records and to appropriate computer hardware and software, you can compile small-scale maps in electronic form. If you lack access to suitable base data files, it doesn't mean you can't compile small-scale maps electronically. You may, however, have to create your own digital records (see Chapter 11).

THEMATIC DATA

We make thematic maps to show the distribution of something. There is no limit to the kinds of information that can be called "thematic." The information may relate to environmental attributes, such as land cover, soils, vegetation, or climate. Or it may be statistical, showing population density, incidence of disease, tax rates, and so forth. Due to this diversity, sources of thematic data are practically unlimited.

National and state governments have many bureaus which compile and publish thematic data on an enormous array of subjects. Many libraries maintain separate collections of government publications, and the catalogs (and the librarians) are rich sources of information. Professional societies, of which there are hundreds, will usually respond to requests for information by suggesting sources or people who might help. There are hundreds of national, state, and regional atlases, which contain diverse kinds of data and which usually list sources. There are also scores of specialized "thematic" atlases, which deal with almost any subject imaginable. Information is not in short supply, and a careful search is usually successful.

BASE DATA ON THEMATIC MAPS

The importance of including an adequate amount of base data on a thematic map cannot be overemphasized. It is disconcerting for map readers to see a large amount of thematic detail with no "frame" of basic geographic information to which they can relate the distributions. The most important objective of thematic mapping is to communicate geographical relationships. Since few people are able to conjure up an acceptable "mental map" and project it on the thematic map, it is up to you, the cartographer, to provide it.

Feature Types

The appropriate variety of base data will vary from map to map. But almost every thematic map should include such features as major highways, coastlines, major rivers and lakes, political boundaries, and latitude-longitude lines. Let's look more closely at how to compile such base data.

Highways Highways and major road networks provide a useful structure to anchor thematic subject matter. Due to their hierarchical character, they are relatively easy to generalize to a level of

detail appropriate to a wide range of map scales. Care must be taken that up-to-date source material is used.

Coastlines Your main problem in compiling coastline information will be finding the right source material. One difficulty is that coasts may look quite different on hydrographic charts than on topographic maps. Hydrographic charts are made with a datum (or plane of reference) of mean low water, while topographic maps are made with a datum of mean sea level. The two are not the same elevation. Thus, there will be a difference in the resulting outline of the land. In places that experience high tidal ranges or that use special planes of reference, the differences will be even greater.

Another problem is that coloring may be inconsistent on charts and maps of the same area. On a chart, marshland is likely to be colored as land, because it is not navigable. By its appearance, you would assume it to be land. On the other hand, low-lying swamp on a topographic map is likely to be colored blue, as water, and only a small area may be shown as land. For many low-lying coasts, you may have to decide what is land and what is water; the charts and maps do not tell you.

An even greater problem is that coastlines may change their outline through the years. Such changes may be significant even on small-scale and medium-scale maps. This problem is particularly evident on coastal areas of rapid silting. Note, for example, the changes in the north coast of the Persian Gulf shown in Figure 23.1.

In many regions, such as the poles, the coastlines are not well known, and they vary surprisingly from one source to another (Figure 23.2). On some maps, a region may appear as an island; on others, as a series of islands; and on still others, as a peninsula. On simple line maps, you can use a broken or dashed line to show unknown coastlines. But if you color the water, the color change clearly outlines the coast, making it difficult to show the map reader that the coastline is unknown.

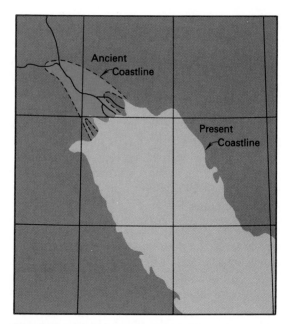

Figure 23.1 Major changes in coastlines occur over long periods of time, and these changes may be significant even on small-scale maps. A portion of the Persian Gulf is shown.

Administrative Boundaries Compiling political boundaries can be a remarkably complex problem. Almost all humanly created boundaries change from time to time, and it is surprising how difficult it is to search out the minor changes.

The problem is compounded on some kinds of thematic maps. Say, for example, that you are compiling a map of central European population prior to World War II, showing present boundaries. You first must find maps showing the original census districts as well as enough latitude, longitude, and other base data so that you can transfer the boundaries to the worksheet in proper planimetric position. Your second problem is to place today's international boundaries, which are greatly changed from their pre-World War II positions, in correct relation to the earlier census boundaries. Some present-day international boundaries actually may cut through the original census districts.

It is not uncommon for official civil division boundary maps to lack any other base data, even

projection lines. After using a few such maps yourself, you will become a more sympathetic cartographer, understanding the need to provide base data on the maps you make.

Boundary compilation entails more than thoroughness and hard work, however. It also has a political dimension. There are constant disputes between adjoining countries concerning jurisdiction. Most disputes involve territory too small to be visible on medium-scale to small-scale maps. Some claims, however, involve considerable areas and must be addressed even on small-scale maps. This is the case, for example, with respect to the boundary between Ecuador and Peru in the Amazon

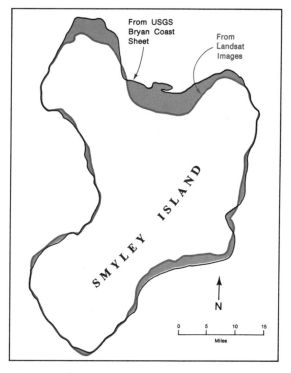

Figure 23.2 There are marked differences in the planimetry of features shown on the best medium-scale maps of the Antarctic coastline and modern remote sensing imagery. Here the Smyley Island portion of the USGS Bryan Coast map (1:500,000) is compared with Landsat images E-1536-12563 and -12570. (Courtesy James W. Schoonmaker, Jr.)

basin, and the boundary between India and Pakistan in the Kashmir region.

To further complicate your task as a compiler, there are usually several wars being waged over territories someplace in the world. Two current examples are the portions of neighboring states under Israeli military occupation, and the country of Yugoslavia. In such cases, you need to decide whether to show the *de facto* (in fact) situation or the *de jure* (by law) jurisdiction.

As a cartographer, then, you are forced to make political decisions. If your map indicates that a territory is in dispute, some—perhaps all—parties involved may be upset. If you show a boundary but give no indication of a dispute, at least one party will be angry. Whatever option you choose will represent a political statement.

Hydrography Rivers, lakes, and other hydrographic features are an important part of the base data you are compiling. On many maps, these are the only relatively permanent interior geographical features shown. They provide helpful anchor points for compiling other data and for helping map readers understand place relationships.

How many rivers and lakes you include depends on your map's objective. On maps that contain well-known state or other boundaries, you may need to include only the larger rivers. Maps of less well-known areas require more hydrography, since the drainage lines are sometimes better known than the internal boundaries.

When a stream has several branches, you must be sure to show the main one. This "main stream" may not be the one with the greatest width, depth, or volume. Rather, it may be better known than the other branches because of some economic, historical, or other significant element.

Just as coastlines have characteristic shapes, so do rivers. On larger-scale maps, these shapes are very helpful in identifying types of streams. For instance, you can use shape to identify the braided streams of dry lands, intermittent streams, or meandering streams on floodplains. On small-scale thematic maps, it may not be possible to include enough detail to differentiate among stream types.

Still, the larger sweeps, angles, and curves of the stream's course should be faithfully delineated.

Likewise, the way a stream enters the sea is significant. Some enter at a particular angle, some enter bays, and some break into a characteristic set of distributaries.

Swamps, marshes, and mud flats are also important locational elements on the map base. These features are particularly helpful in explaining land use patterns. The absence of towns or roads, for example, may be a direct result of these hydrographic features.

Landform Historically, thematic maps and small-scale reference maps have been designed to show environmental data planimetrically. This focus on horizontal rather than vertical position greatly simplified mapping. For one thing, it is difficult to render the landform effectively, especially at smaller scales. Furthermore, including landform in the background can detract from map symbols in the foreground. Thus, maps tend to be more readable when they don't show the landform.

On the other hand, there is a fundamental relation between many environmental features and the terrain surface. Therefore, abstracting away the landform can diminish people's ability to interpret a map. Landform depiction will generally make a map more meaningful, especially if visual clutter can be minimized.

Before deciding not to represent the landform, you should make every effort to incorporate this important aspect of the environment into a map's design. The development of computer compatible databases, called digital elevation models (DEMs), makes this task much easier (see Chapter 13). Of course, these terrain data are of little value without effective software for representing the landform. Fortunately, such software is now widely available in the mapping community (see Chapter 27).

Level of Generalization

As you compile your data, you should try to depict all features with approximately the same degree of generalization. Violating this important compilation principle can result in a deceptive map. The problem occurs when a person using the map is misled into believing that all the mapped data are of equal quality. This is particularly troublesome if base data are depicted in far greater detail than thematic data, since the person using such a map may unconsciously give greater credibility to the theme data representation than is warranted. For instance, a map based on a crude global weather prediction model does not justify precise weather determinations at specific sites, such as at a small jog in a continent's coastline. To avoid the possibility that a map may be wrongly used in this way, it is better not to encourage misinterpretation through unwarranted use of base data details.

The issue of deceptive map design falls in the realm of **cartographic ethics**. Although distortion is fundamental to all forms of representation, the cartographer has a professional responsibility to keep distortion to a minimum or, at least, to provide map users with tools to assess the distortion inherent in a map's design. It takes more mapping effort to be sure that thematic data and base data correspond with respect to level of generalization. But doing so is an effective safeguard against subsequent map abuse.

ANALOG COMPILATION WORKSHEET

When you plan to produce a map by traditional manual or photomechanical methods, you should prepare a worksheet of the compilation before you begin constructing the final artwork.* You should view this analog compilation worksheet as an essential planning document. If it is well thought out and prepared, it will guide you smoothly through subsequent production and reproduction stages. But if it is poorly conceived and executed, you may have no end of frustration in trying to carry the artwork through subsequent production and reproduction steps.

Ideally, the worksheet should be a clear, accurate representation of everything to be shown on the finished map. It should be laid out, designed, and formatted to make subsequent production and reproduction procedures as easy as possible. Let's look at the factors you need to consider in preparing your worksheet.

BASE MATERIALS

Your first step in preparing a compilation worksheet is to select a base material. In making this decision, you need to consider two things. The first is the degree of **dimensional stability** needed to attain the desired fidelity in artwork construction. As a rule, drafting papers are less stable than drafting films (polyesters). Consequently, papers contract and expand more with changes in temperature and humidity, thus altering the dimensions of the artwork. If you are constructing all artwork for a small map on a single sheet of material, and it will not have to be precisely registered with any other image (see "Registry" later in this section), then paper may suffice, and it is cheaper. But if a very large, accurate map is needed, or if precise registry (that is, the matching of several compilation or final artwork sheets) is a requirement, you should use a dimensionally stable material such as plastic. In fact, the added cost of drafting film over paper is such a small part of overall mapping costs that it is good practice to use plastic base materials for all compilation chores. In the long run, the benefits will be substantial.

Your second consideration in choosing a base material is its level of transparency. Nothing helps the compiling procedure so much as having a

---

*An analog compilation worksheet can also be useful when using digital production methods. It is common, especially with small-format maps, to first create a worksheet in traditional analog form, and then enter this document into a computer through scan-digitizing. Then, with the digital version of the worksheet displayed on the computer monitor, you can proceed with electronic production of the final artwork (see *Digital Compilation Worksheet* later in the chapter).

translucent material with which to work. When you use translucent material, tracing is easier. You can use the worksheet to make contact negatives or to contact images to other sensitized materials. You can lay out lettering and move it around under the compilation worksheet. To draw a series of parallel lines, letter at an angle, or place dots regularly, you need only place graph paper under the material for a guide.

FORMING THE GUIDE IMAGE

When you are preparing for positive artwork, you will usually compile your worksheet on one sheet of material. You may mark the worksheet with pencil, ballpoint pen, or any satisfactory instrument. It's a good idea to use different colors to indicate different categories of features that will be shown on the final map. It also helps to lay out lettering with approximate size and spacing as it will be on the final map. If the first try does not work, you can erase it and do it over. Borders and obvious linework need only be suggested by ticks.

SEPARATIONS

Sometimes you will need to prepare the worksheet on more than one sheet of base material. You will need to do so, for example, when the artwork is to be reproduced in multiple colors. You will also need more than one sheet when the artwork is to undergo complex photomechanical manipulations to obtain special effects (see Chapter 31). Finally, separation artwork may be necessary when images are to be contacted to sensitized materials, say for subsequent scribing.

The artwork can be separated in two ways. One approach is to place different categories of information, such as roads, rivers, political boundaries, and so forth, on separate sheets, called **feature separations**. Or, features to be reproduced eventually in the same color may be placed together on a separate sheet. In this second case the results are called **color separations**.

Just as with a single worksheet, it may be helpful

to use different colored pencils to distinguish between categories of symbols drawn on each separation. The exception is a separation which is to be contacted to sensitized materials, in which case black linework provides the best image transfer. In that case, instead of using different colors, you can use varied line forms, such as dots or dashes, to show different categories of features.

REGISTRY

Multiple worksheets or artwork flaps require some means of registration to ensure that the final map composite is accurate. Two methods of registration are commonly used: pins and graphic marks. Each technique serves a special purpose in subsequent map production and reproduction.

The most precise method of registering the various flaps is to fasten them together with **registry pins** (Figure 23.3). These pins are available in two forms. The most common type consists of a flat

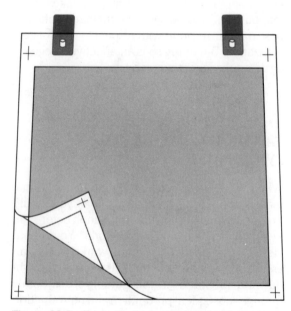

Figure 23.3 Each sheet of material used to prepare map artwork is referred to as a flap. The ¼-inch pins hold the flaps in register. The small crosses position the printing plates on the press for successive colors.

piece of material (metal or plastic) to which one or more machined ¼-inch round studs have been attached. Studs of various heights are marketed. The shorter pins will accept several sheets of material in overlay and can be used in vacuum frames and camera copy boards without danger of breaking the glass. The longer pins cannot be safely used in camera or contact work, but they may be handy during compilation or artwork construction when it is necessary to overlay many flaps.

To use registry pins, you need to punch precisely matched holes in the worksheet and each proposed flap before you begin artwork construction. Special **precision punches** are manufactured in a variety of sizes especially for this purpose (Figure 23.4). Large, precision punches are quite expensive, however, and their cost may be hard to justify. Fortunately, for maps of relatively small size, you can usually get by with an ordinary two- or three-hole adjustable office punch.

Another way to produce properly spaced holes is to use **prepunched tabs**. These are available commercially in gummed and ungummed form. (Be careful if you use gummed tabs, because they tend to slip under changing temperature and humidity conditions). Stick or tape two tabs along the top of the worksheet, and push the pins through

Figure 23.4 Pin registry of graphic arts materials requires that precisely punched holes be made in each flap. Specially designed punch machines can save time and simplify operations, as well as improve registry precision. (Courtesy ARKAY/HULEN).

the holes. Then place each flap successively onto the worksheet. Carefully position the holes in another set of tabs over the pins, and fix the tabs rigidly onto the material.

Two pins, however, may not provide the rigidity needed for large sheets of material. In that case, you may place an extra pin at the bottom of the sheet, or use a pin at the midpoint on each of the four sides.

When all flaps are properly punched or tabbed before artwork construction begins, each sheet will fit precisely on the pins throughout the entire production and reproduction process (see Film *Registration* in Appendix D). You can then interchange flaps quickly and accurately.

Using three slots instead of ¼-inch circular holes results in more precise registry and is a more satisfactory system for large maps. This system consists of a flat table surface with three movable punch assemblies that can be locked into position with the axes perpendicular. The rectangular pins used in this system are the same width as the slots but slightly shorter. This lengthwise clearance lets you reset the punch to conform with previously punched sheets. The arrangement of slots and rectangular pins keeps the map center in register. Any misregister caused by contraction or expansion is allocated proportionately on both sides of the center axis.

Pin registration may not be sufficient when multiple-flap artwork is to be reproduced by plate-based methods. Although pins usually provide the printer with an adequate means of registry, it may be necessary to make small adjustments in the paper feed to the press to achieve the highest quality printing. To serve this purpose, a second method of registering the negatives and the printing plates is needed; it can be used in conjunction with register pins.

Graphic registry marks are commonly used for this purpose. Marks such as small crosses (crosshairs) are placed in each of the four margins outside the map border (see Figure 23.3). During map preparation, you can use these registry marks to line up each flap in the same position on the worksheet. When the negative is prepared, the marks are retained and transferred to the printing plate along with the map image. When the printer is satisfied that the plate is positioned properly on the press, the registry marks are removed with an abrasive. You should be sure to make the crosses with very fine lines and place them far enough outside the map border so that the printer can remove them without damaging the map. On the other hand, do not place them too far outside the map border, or much more film will be necessary to record their images.

IMAGE GEOMETRY

The process of compilation may require a change in the geometry of source materials. If so, you will probably not be able to use mechanical means of reduction. Instead, you may need to transfer much of the data to the worksheet by eye. You can use the transformation geometry established by the graticules on the source map and worksheet as guidelines to help you estimate the positions of features. Do not be too concerned about the accuracy of this process. The eye is remarkably discriminating and, with practice, can position data with the necessary precision. Before the advent of computer-assisted cartography, 90 percent of all small-scale maps were compiled in this manner, and many still are, because the desired data are not available in a form that can be handled mechanically.

Sometimes your sources will have projections different from the map you are compiling. You must therefore become adept at imagining the shearing and twisting of the graticule from one projection to another so that you can position features accordingly. The problems caused by map projection differences between sources and the compilation can be largely eliminated by making the graticules comparable. Having the same interval on each will greatly facilitate the work (Figure 23.5).

You will find compilation easiest if you first outline on the new projection the areas covered by the source maps. This outlining is similar to the index map of a map series. You may draw the sheet outlines and lightly indicate the special spacing of the graticule (5°, 2°, and so on) on each source.

SCALE

Much of your compiling work is associated with map scale. First, you must determine the scale of the finished map on the basis of map size and area to be covered. So let's look more closely at how to determine the map scale.

Determining the Map Scale

All maps are constructed to a scale. Many thematic maps are prepared to fit the size and shape of the sheet on which the map will appear. The format may be a whole page in a book or atlas, part of a page, a separate map requiring a fold, a wall map, or a map of almost any conceivable shape and size. Whatever the format may be, the map must fit within it.

The shapes of earth areas vary greatly when mapped, depending on the projections on which they are plotted. When you compile a map, therefore, it's important that you choose a projection carefully. To do so, you need to match shape variations of the mapped area on different projections against your format to see which will provide the best fit.

The easiest way to do this is to establish the vertical and horizontal relationship of the format shape on a proportion basis. Then compare the proportion with different projections to see which is the closest match. The actual length and breadth dimensions are not critical. It is the proportion between them in relation to the format that is important.

When you have selected the projection that best fits the purpose and format of the proposed map, you may then determine the scale of the finished map as explained in Chapter 6. It is good practice to use either a round number RF or a simple stated scale (map distance to ground distance).

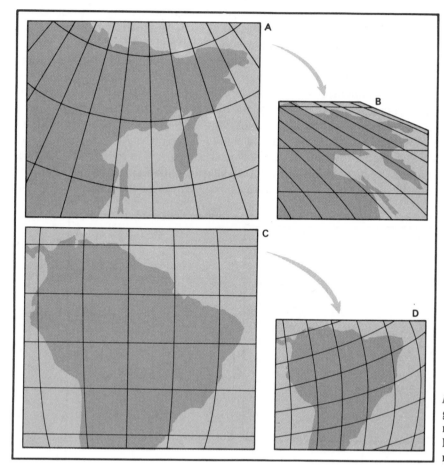

Figure 23.5 Shape and geometry changes can be made during compilation. Maps (B) and (D) are derived from (A) and (C).

Compilation Scale in Relation to Finished Scale

Once you have selected a scale for the finished map, you can then determine the worksheet scale. You may want to construct your worksheet at the scale of the final map. This is common, for example, if a map is to be scribed. If done in sharp linework on a translucent drafting medium, the compiled information can be photomechanically transferred directly to the scribe coat. You will also want to compile your map at the finished scale if it is to be reproduced from an original drawing on plastic or paper by a process that does not allow changing scale.

On the other hand, if reproduction will be by a process that allows reduction, you may want to compile and draft your map at a larger scale and then reduce it to the finished scale. Sometimes the complexity of the map or other characteristics make it desirable to compile at a larger scale even for a map to be scribed. In this case, you would use reduction to place the image on the scribe coat.

You can usually reduce an ink drawing to advantage. It is often impossible to draft with the detail and precision desired at the map reproduction scale. At the larger scale, drafting is easier, and reduction "sharpens up" the lines. The amount of reduction will depend on the reproduction process and the map's complexity. It will also depend on whether definite specifications have been determined for the reproduction, as may be the case with maps of a series.

Ink-drafted maps are usually made to be reduced by 20 to 30 percent in the linear dimensions. It is unwise to make a reduction greater than 50 percent. At greater reductions, it becomes difficult to make the design decisions required to make the final map look right. Drafting a map in ink for reduction does not mean merely drawing a map that is well designed at the drafting scale. On the contrary, it requires anticipating the finished map and designing each item so that when it is reduced and reproduced, it will be "right" for that scale.

The greatest problem in designing for severe reduction involves line widths. Lines that appear correct at the drafting scale usually appear a little "light" when the map is reduced. Consequently, you must make the map overly "heavy" so that it won't look too faint after reduction. You also need to "overdo" type sizes somewhat, just as you must make lines and symbols a little too large on the drawing. Figure 23.6 illustrates some of these relationships.

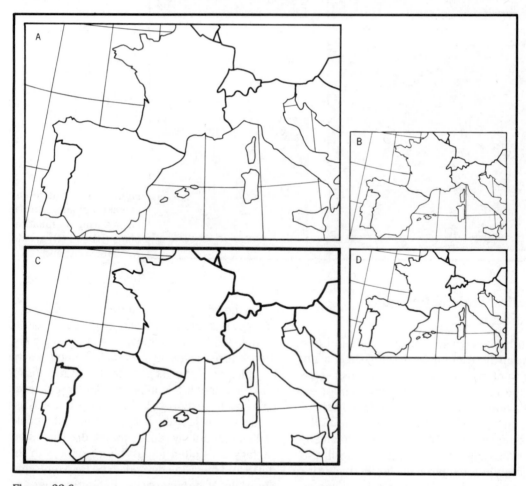

Figure 23.6 Effects of reduction. In (A) when manual artwork is designed for appropriate line contrasts at drawing scale, then reduction (B) will decrease the contrasts too much. In (C) when artwork is designed for reduction, then reduction (D) produces appropriate line contrast relationships.

If you plan to reduce maps of a series, be sure to draw each map for the same amount of reduction. Then, when they are reduced, similar lettering, lines, and the like will appear comparable on all maps in the series. This may necessitate changing the scale of base maps, which is troublesome, but it will insure that the line treatment and lettering in the final maps will be uniform.

It is common practice in photo-processing labs and printing plants to photograph several illustrations at once. This practice saves both labor and materials. For economy, therefore, it is desirable to make series drawings for a common reduction or, at most, for a limited number of reduction sizes.

Specifications for drafting and for reduction are given in terms of linear change, not areal relationships. It is common to speak of a drawing as being "50 percent up," meaning that it is half again wider and longer than it will be when reproduced. The same map may be referred to as being drafted for one-third reduction; that is, one-third of the linear dimensions of the original will be lost in reduction. Figure 23.7 illustrates these relationships.

Changing the Map Scale

After you have determined the worksheet scale, you will often find that it differs from the scale of available source maps. If so, you will have to change the scale. You can change the scale of a base map or a source map in one of three ways—by optical projection, photography, or similar squares.* Let's look more closely at each of these procedures.

Optical Projection Most cartographic establishments have a projection device. This device may operate by projection from overhead onto an opaque drawing surface or from underneath onto a translucent tracing surface (Figure 23.8).

Projectors are handy if you want to change the scale of relatively simple map artwork and transfer

---

*The pantograph, a mechanical device for changing scale used in past centuries, is rarely used today because it is cumbersome and slow.

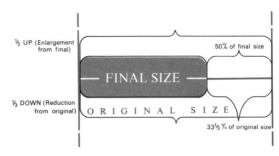

Figure 23.7   Relation of enlargement to reduction.

details from one map to another when their scales differ. To do so, place your source map in the projector. Adjust the enlargement or reduction to change the scale of the source map to that of the map you are compiling. You may then trace the desired information.

This task is simple if you need to transfer only a few image details, but it can be a chore if a complex map image is involved. Furthermore, if the

Figure 23.8   Optical projection devices are handy for changing the scale of relatively simple map artwork and for transferring details from one map to another when their scales differ. (Courtesy Artograph, Inc.)

transformation geometry of the source map and the map being compiled are very different, a projector will not be much help.

Occasionally, you may be called on to produce an oversized chart or map that involves great enlargement. In most cases, extreme accuracy is not required. If you cannot sketch the outlines satisfactorily because of their intricacies, the projector provides a solution. You can project a slide or film transparency to paper affixed to a wall. Then you can trace the image onto the paper.

Photography Photography can produce high-quality copies at desired sizes on dimensionally stable materials, making it ideal for compiling maps. You can use the film process to enlarge or reduce existing maps. Then you can use either negatives or positives to trace detail onto a new map.

You need to specify the reduction or enlargement by a percentage ratio, such as 25 percent reduction or 150 percent enlargement. Take care in specifying the percentage change. To obtain a 50 percent enlargement in the linear dimension of the original, you would set the camera to 150 percent. Conversely, you would obtain a 25 percent reduction in the linear dimension by setting the camera to 75 percent. To achieve a 75 percent reduction in the linear dimension of the original, you would set the camera to 25 percent (see Figure 23.7).

Photographic enlargement or reduction has several advantages over projection techniques that require hand tracing of the image. For one thing, the photographic process is not affected by the level of image detail. Photography is especially useful when working with complex maps, since it takes no more time to process detailed than simple artwork.

Moreover, the photographic process is more precise than hand copying. The image may be either projected or viewed in the camera, making it possible to scale the dimensions of the image more exactly. All you need to do is specify a line on the piece to be photographed and then request that it be reduced to a specific length. The ratio can be worked out exactly, and the photographer needs only a scale to check the setting. Any clearly

defined line or border will serve as the guide. If none is available, you may place one on the drawing with light blue pencil so that it will not be apparent after photographic processing.

You can use photography or a projector to transform a scale difference of as much as four or five times, but much larger changes are difficult. For any larger reduction or enlargement, you may have to repeat the process (that is, make one or more intermediate copies). If the image gets too large, you may have to work with it in sections and piece it together later.

Another problem is found in the relation between linear and areal scales. The size (area) of maps of the same region will vary as the square of the ratio of their linear scales (see Figure 23.7). For example, a region shown in 1 cm^2 at a scale of 1:50,000 will occupy only 1 mm^2 (1/100 cm^2) at a scale of 1:500,000. Therefore, although you can determine positions of individual points or simple lines, anything that involves complications in areal spread will be reduced to an almost indecipherable complexity.

You can sometimes combine photographic and xerographic processes to advantage. For example, if you find a source map in a book, you can copy it at scale on a copy machine. Then you can place the resulting image in the camera to be reduced or enlarged. Be careful, however, because some xerographic copiers change the scale of the image slightly, while others stretch the image somewhat in one dimension. Although some copiers can reduce or enlarge, most machines offer only a limited number of preset reduction or enlargement options.

Similar Figures When appropriate projection or photographic equipment is not available, you may be forced to change scale by a method called "similar squares." To do so, you draw a grid of squares on the original and then draw the "same" squares, only larger or smaller, on your compilation. You can then transfer the lines and positions by eye from one grid to the other (Figure 23.9). If you perform this process carefully, it is quite accurate. But it is the most tedious form of compilation and can usually be avoided.

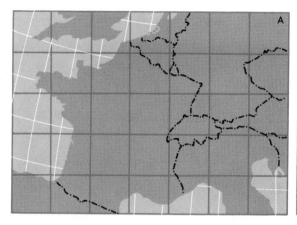

Figure 23.9 Changing scale by similar squares. (B) has been compiled from (A).

USE OF THE ANALOG WORKSHEET

When you have completed the worksheet and compilation, you may then transfer the image to a scribe sheet for processing. Or you may draft in ink on translucent material directly over the worksheet. If the map is simple and you are preparing the artwork yourself, you will have planned such factors as the character of the lines. But if the artwork is to be constructed by someone else, you will need to prepare a sample sheet of specifications as a guide in map production. Even for a map you are producing yourself, it is wise to prepare such a sample sheet. The task is simple if each category is in a different color or otherwise clearly distinguished. You can easily prepare separation drawings for small maps from a single worksheet, and they will register.

DIGITAL COMPILATION WORKSHEET

Electronic technology provides another way to prepare a compilation worksheet. Although an electronic worksheet serves the same purpose as an analog worksheet, there are major differences between the two. Many of the factors that were critical in manual and photomechanical compilation are not relevant in digital compilation. Conversely, digital compilation introduces factors not found in analog compilation. Still other factors are relevant in both methods.

Digital methods generally provide you with greater compiling flexibility than analog methods. You have much more freedom to try alternative designs, to merge data from diverse sources, to change scale and geometry of the worksheet at will, and to skip costly steps that manual and photomechanical techniques require. But these advantages are gained only at a price. Obviously, digital records must be available, along with the software to manipulate them and the computer hardware to carry out data entry, processing, and display functions. While cost of these items is high, it is rapidly declining. Digital compilation now lies within the reach of many cartographic compilers.

Despite its promise, however, digital compilation is still a long way from replacing analog methods completely. You have been provided with a second option, and must carefully weigh the costs and benefits of each before making a commitment to one method or the other. Among the key determining factors, access to suitable digital data ranks at the top.

SOURCES OF DIGITAL DATA

Electronic compilation begins with digital data. You must decide whether the desired data are available in the form of existing databases and whether you can easily gain access to those databases. If the needed digital data are not available, you must decide if it makes sense to create the data yourself or contract to have it done.

Existing Databases

As an early step in digital compilation, it's a good idea to check to see if the information you need already exists in digital form. A diversity of digital databases is now available, and the pace of new database development is increasing (see Chapters 11, 13, and 15). For medium to large mapping projects, it is usually more economical to buy existing databases, or to pay a user fee, than to try to create the digital files yourself (see next section).

The more existing databases you use, however, the greater the likelihood that they will be coded with respect to different spatial reference systems. Thus, you may need to convert them to a common base (see *Establishing a Common Base* in Chapter 10). Additionally, different databases are not usually coded according to the same quality standards. As a rule, you should try to discover the origin of each database used, and then convey that information to the map user through a legend notation.

When compiling from existing databases, it pays to remember that the data may have been collected from maps having different scales. Since we compile from larger to smaller scales, but not the reverse, the scale of the database is a critical limiting factor in the scale of digital map compilation. In practice, existing databases were developed from maps representing a hierarchy of scales. You should not use these databases to compile a map much larger in scale than the original map that was digitized. Mapping at smaller scales than the original is justified, but you may find it necessary or desirable first to remove excess detail from the database through generalization (see Chapter 24).

Digitized/Scanned Input

For small mapping projects, or when you don't have access to suitable databases, you can create your own digital data. This usually entails using a vector digitizer or raster scanner to convert an existing printed map or analog compilation worksheet to digital form (see Chapter 11). As noted in Chapter 11, several editing and file manipulation steps are usually required before the digital data are usable.

The issues we discussed with respect to using existing databases also apply when you create your own digital data. Namely, you must pay attention to scale, resolution, and the quality of source maps. You have a responsibility to clearly convey critical information to the map user through legend notations.

PROJECTIONS AND COORDINATE SYSTEMS

One of your first choices when compiling a base map is the projection or coordinate system on which you will compile features (see Chapters 5 and 6). The impact of computer assistance has been particularly great in this area of the mapping process. Computers give you great flexibility in using different projections and coordinate systems with unconventional orientations and origins.

Most common projections are now available in software packages. When used with cartographic databases, this software can produce plots of the graticule, coastlines, boundaries, and so on at almost any desired scale (Figure 23.10). It is no longer necessary, except in unusual circumstances, for a cartographer to perform the calculations, plot the graticule, and compile general base data by hand.

Free of such manual tasks, cartographers can now concentrate on the design aspects and fundamental purposes for which the map is being prepared. For example, it is literally possible to make the map fit the format that is available. For special effects, cartographers can easily produce elongated, sheared, or unconventionally rotated projections and overlay them with coordinate systems. They must be thoroughly schooled in the use of projections and able to select the one that best fits the required purpose and format.

In order for the computer to plot geographic information into a chosen projection format, the information must be available in suitable digital

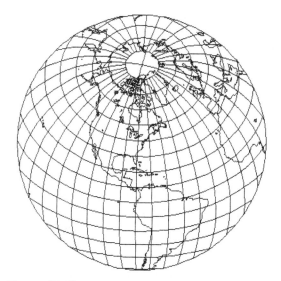

Figure 23.10 Even relatively small desktop computers can be used for map compilation work. This oblique Lambert Azimuthal Equal Area projection was programmed for an Atari 800 microcomputer and prepared on a dot matrix printer by John P. Snyder. Shorelines were taken from a 9,000-point data file in the Atari "Mapware" Software Program (c), written by Koons and Prag.

form. Unfortunately, there is a great deal of incompatibility between the format and structure of digital databases. As we saw in Chapters 10 and 11, digital data are gathered in both vector and raster form using arbitrary machine coordinates. These coordinates may then be converted to a variety of terrestrial coordinate systems, such as latitude/longitude, State Plane Coordinates, and Universal Transverse Mercator Coordinates. If you are compiling data based on more than one spatial reference system, you will have to convert those data to a common system. Only then can you plot them into the chosen projection format (see *Establishing a Common Base* in Chapter 10).

PREPARING THE WORKSHEET

When you work with computer assistance, the preparation of the compilation worksheet is under software control. Someone has already taken responsibility for the program algorithm and for providing options and defaults* for its use. Of course, you must still make most of the same decisions as when you compile a map manually. In this case, however, you implement these decisions through software instructions to the computer and computer-driven peripheral devices. If all needed digital records are compatible with respect to their data structure format and their locational reference base, worksheet preparation can proceed with great speed and efficiency.

Your role as map compiler is to choose the elements associated with the cartographic processes of selection, classification, simplification, and symbolization. These choices include: type of projection, area of coverage, scale, features to be mapped, mapping technique, and mapping parameters (class intervals, number of classes, and so on). In other words, although you prepare the worksheet with the aid of the computer, you bear much the same responsibility as with manual compilation.

Once you have given the computer all the information necessary to construct the worksheet, you can call for a plot, print, or display of the selected data. You may receive this portrayal of the compilation material in the form of a composite image or as a set of registered separations.

USING THE DIGITAL WORKSHEET

You can use digital compilation worksheets in many ways. Cartographers have always used worksheets to test layout and design concepts, but time and expense kept this function to a minimum. The great speed and relatively low cost of electronic compilation have encouraged cartographers to expand their use of worksheets to preview design alternatives. Indeed, you might

---

*Most mapping programs have built-in specifications for such elements as number of classes, class interval scheme, shading patterns, and so on. Unless you instruct the program to do otherwise, these standard or default specifications will be used in map production.

make a number of rough worksheets, especially if you are working interactively with a video display screen image, before you settle on the one that you will carry through into production.

You can also use digital worksheets as a basis for subsequent production by nonautomated methods.* It is common, in fact, to start with a digital worksheet and finish the map by hand. If your computer graphic devices are limited, you may be able to achieve better quality artwork by hand than by machine.

A third use of digital worksheets is to provide a preproduction test of data files, mapping program controls, and graphics equipment. In this case, the worksheet serves as an intermediate step prior to going ahead with electronic map reproduction. Commonly, for example, a map that is to be produced finally in liquid ink on drafting film is first plotted in preliminary form in ballpoint pen on paper.

Finally, the digitally produced worksheet may constitute the one and only map that is constructed. In other words, the mapping process may actually end with the compilation step. This may occur, for instance, when a researcher or planner is using cartographic output as an aid in tackling environmental problems. A quick look at the rough map image may be all that is needed. The essence of the cartographic portrayal, not a finished product, is what is important in such a case.

ACCURACY AND RELIABILITY

When you are compiling a map, you must be constantly alert to the quality of source data. Errors can easily be propagated and exaggerated through the compilation process (see Chapter 16). Unfortu-

---

*It is becoming common to compile map artwork using a mix of manual and electronic methods. For example, you might compile a city street map by hand and then use it as a base, overlaying it with digitally compiled thematic information, such as traffic accident sites, schools, fire stations, and other data.

nately, the quality of source data is not always apparent. For this reason, you need to be sensitive to potential sources of error.

One thing to watch closely is the age of source data. Be sure you note the date that source material was collected. If possible, pass that date on to the map user. When you choose source data, it's a good idea to pick newer sources over older ones. Since environmental knowledge and methods for acquiring data are continually improving, newer sources tend to be more reliable.

Also pay close attention to the scale of source material. Progressive generalization with smaller scales is an inevitable part of mapping. For this reason, you should always compile from larger-scale sources rather than smaller. The temptation to enlarge a smaller-scale source map is bad enough. But it would be even worse to blow up a smaller-scale map of one feature (such as soils categories) to be overlaid on a compilation worksheet containing other features (such as land-ownership parcels) that were compiled from larger scales. The different degree of cross-variable generalization on the resulting map could quite easily invalidate it with respect to its intended use.

You should also ask yourself how careful the data collection or mapping agency was in gathering the source data. For instance, land-ownership boundaries in County Plat Books produced by many commercial vendors are highly generalized and therefore notoriously "inaccurate" in terms of any coordinate reference system (Figure 23.11). These cadastral maps are designed more for topological than planimetric accuracy. If map users don't realize that fact, they could easily misinterpret the map. On the other hand, the standard topographic quadrangles published by the United States Geological Survey are produced to strict map accuracy standards and are relatively reliable within the limits of the map scale, date, and so forth.

Many relatively large-scale compiled maps have been put together from a variety of sources in a kind of patchwork fashion. One part of the map may be derived from modern topographic maps, another part from older compiled maps, and so on.

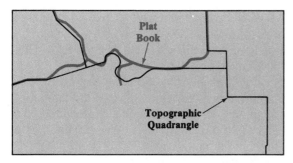

Figure 23.11 Map accuracy standards are not widely defined or adopted by the various private and government mapping establishments. As a consequence, there may be marked variations in the planimetry of features from one map to the next, as illustrated by this stream segment traced from a county plat book (1:53,350) and a USGS topographic quadrangle (1:24,000).

If there is any significant difference in the quality of the sources, then you should prepare a coverage diagram for the map. An example of such a diagram is shown in Figure 23.12. You can use a coverage diagram to show many kinds of information, such as the quality of source maps, dates of censuses, or names of compilers or investigators.

RIGHTS AND RESPONSIBILITIES

Map compilation is creative work, but it also builds upon the work of others. You must be sure to respect the legal rights of these other people. Once you move beyond compilation to production, you will enjoy similar rights with respect to your maps.

In addition to your legal rights as a mapmaker, you also have certain duties. Let's take a look at these legal rights and responsibilities, beginning with your obligation to cite your sources.

SOURCES AND CREDITS

Much of the geographic information compiled as base data—such as coastlines, rivers, and bound-

aries—may be classed as general knowledge. This information has been mapped many times, and you don't need to cite a source when you use it. But what if your portrayal differs significantly from the way the phenomenon has usually been mapped? Say, for example, that you include recent boundary changes or use some very new source, such as a new survey or remotely sensed imagery (such as Landsat; see Chapter 8). In such cases, it is helpful to the map user if you indicate your sources.

Whenever you use thematic maps and large-scale specialized reference maps, you can assume that the data being mapped will *not* be common

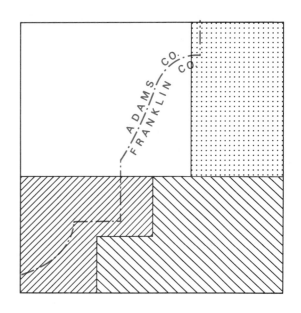

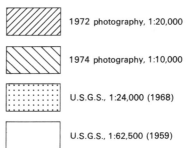

1972 photography, 1:20,000

1974 photography, 1:10,000

U.S.G.S., 1:24,000 (1968)

U.S.G.S., 1:62,500 (1959)

Figure 23.12 Coverage diagram showing the kind and date of sources used in compilation.

knowledge. Consequently, you should give the source and, if important, the date of the special information. You might say, for example, "U.S. Bureau of the Census, 1980," or "Land use data from provisional 1982 map by Special Project Committee," or "Zoning boundaries as approved by the County Board in 1976." Such notations will give map users the necessary information.

Whenever you obtain specialized information from a nonpublic source, such as a research foundation paper, a scholarly publication, or an industrial report, it is common courtesy to acknowledge the source. Judgment is always involved. The information may be widely known and may appear in substantially the same form in several sources. In that case, citation may be unnecessary. Whenever you have any doubt, however, you should acknowledge the source.

COPYRIGHT

A large portion of published material is protected by **copyright**, the name given to the legal right of creators to prevent others copying their original work. Copyright law is aimed primarily at the protection of literary works. But copyright also encompasses such materials as musical compositions, photographs, and maps.

In the past, the labor used to compile facts was sometimes rewarded in court by extending copyright protection to the facts themselves. This "sweat of the brow" protection came under the **merger doctrine**, which states that expression and idea sometimes cannot be separated. The courts have now rejected the merger doctrine, however. In its place we have the **originality doctrine**. This doctrine requires that there be some degree of "intellectual invention" before a work can be protected by copyright.

Over the years, the law has been quite restrictive with regard to maps' originality. Judges have tended to see maps only as collections of facts and not as very original. But recent rulings recognize the abstract nature of maps, and require only that a **minimal degree of creativity** be involved in a map's design.* Copyright protection depends on demonstrating a minimal degree of creativity in either "the selection, coordination, and arrangement" or the "pictorial/graphic expression" of the facts.

Under this interpretation, the vast majority of maps qualify as copyrightable works. If a map is copyrightable, it is automatically protected. It is legal to "build upon" the facts contained in a copyrighted map to create a similar product, as long as you do not copy in a technical, mechanical way the protected "pictorial/graphic expression" or "selection, coordination or arrangement" of these facts. Finally, similar products developed through "independent creation" do not violate the law.

You have a responsibility not to copy another map without obtaining permission in writing from the copyright holder. The term "copy" is used here in its literal sense, meaning "reproduce exactly." You must not, therefore, either through photography or tracing, copy exactly the way some other mapmaker has represented geographical data.

Your request in writing for copyright permission should specify (1) the work in which the requested material is to appear, (2) the publisher, (3) the exact reference to the material (author, book or periodical, pages, figures, tables, and so on), and (4) how the material is to be used (as an illustration, a quotation, or whatever). It is important to direct the request to the copyright holder; in many instances this is the publisher, not the author. Copyright holders almost always will grant you permission, although sometimes they will ask you to pay a fee for using their work. They will tell you how they want the credit line to read, and you must be sure to follow their directions exactly.

Obtaining permissions can be a tedious and lengthy operation, involving much letter writing, so it is wise to start early. The extra effort is well worth

---

*For a good summary of recent rulings concerning maps, see Wolf, D.B., "Is There Copyright Protection for Maps After Feist?," *Journal of the Copyright Society of the USA*, Vol. 39, No.3 (1992), 224–242.

it if it prevents a copyright lawsuit. No matter who wins, a lawsuit is bound to be upsetting and expensive, and it's best to do all you can to avoid one.

Most publications, including maps, that are produced by agencies of the U.S. government are in the **public domain**. This means they are not copyrighted and can be used freely. You ought always to give credit where it is due, however (see previous section). In the case of specialized publications containing judgments and opinions of named authors, it is not only courteous but wise to request permission from the issuing office. The material may have been copyrighted by the authors separately, or some of it may have come from copyrighted sources.

Although topographic maps, census information, and similar materials produced by the U.S. government may be used freely as sources of data, this is not true of such materials produced by foreign governments. In many instances reproduction of such materials without permission is strictly prohibited. The United States is a party to the international Universal Copyright Convention, which requires citizens to comply with copyright laws of the agreeing nations. Thus, you must be very careful whenever you use foreign sources.

LIABILITY ISSUES

The map-using community has a right to expect cartographers to conduct their business in a professional manner. At best, mapping errors inconvenience map users. In extreme cases, they can contribute to bodily injury, property damage, and financial loss. To what extent is the cartographer liable for pain and suffering caused by inaccurate maps?

The answer depends in part of the origin of the inaccuracy. Court cases strongly suggest that if the information and technology is available to do a better job, and your work does not meet community standards, you are liable. If the error can be traced to professional negligence, you risk being sued for malpractice.

You risk a lawsuit even if your maps do not contain outright errors, but do exhibit deceptive design work or misleading artifacts of the mapping process. A "user beware" defense is not sufficient. If you hold yourself an expert—which you do when making a map—and the person using your product is not an expert, the courts have ruled that you are liable for misinterpretation by "trusting" laypeople.

Commercial mapping firms historically have been especially vulnerable to lawsuits. As a result, their maps often contain a prominent disclaimer, such as:

"This map should not be used for navigation. It is intended for reference use only."

This may seem a strange note to find in a nautical atlas, but it does warn the user that the map is just one of several tools available to a navigator. A prudent person will double-check the accuracy of each tool if grave danger is a consequence of making an error.

Government cartographers have long enjoyed relative immunity from lawsuits. But the trend clearly is toward greater accountability on their part. The "untouchable sovereign" defense no longer appears to be sufficient. Citizens increasingly are asserting that they have legal rights to quality information from their government servants. There has now been successful litigation against a variety of government agencies, and the list is rapidly growing. Although lawsuits regarding government maps were rarely successful in the past, this may change dramatically in the next few years.

How do you minimize your liability risk? In the context of compilation activities, you must be able to document the fact that you made a reasonable effort to use available information and technology to minimize mapping errors. You also must be able to show that you warned map users in an unambiguous way about potential problems with your maps. Proper citation of data sources, a statement of data manipulation processes, and notations concerning known cartographic artifacts are essential.

SELECTED REFERENCES

Bond, B. A., "Cartographic Source Material and its Evaluation," *The Cartographic Journal*, 10 (1973): 54–58.

Cerny, James W., "Awareness of Maps as Objects for Copyright," *The American Cartographer*, 5, No. 1 (1978): 45–56.

Dando, L. P., "Open Records Law, GIS, and Copyright Protection: Life After Feist," *URISA Proceedings* (1991), pp. 1–17.

Davies, J., "Copyright and the Electronic Map," *The Cartographic Journal*, 19 (1982): 135–36.

Rhind, D., "Data Access, Charging and Copyright and their Implications for GIS," *International Journal of Geographical Information Systems*, Vol. 6, No. 1 (1992): 13–30.

Snyder, J. P., "Efficient Transfer of Data Between Maps of Different Projections," *Technical Papers*, ACSM Annual Meeting, Washington, D.C. (1983), pp. 332–40.

Tosta, N., "Copyrights, Servicerights?," *Geo Info Systems*, Vol. 2, No. 2 (1992), pp. 22, 24, 31.

Wolf, D. B., "Is There Any Copyright Protection For Maps After Feist?" *Journal of the Copyright Society of the USA*, Vol. 39, No. 3 (1992): 224–242.

PART VI

CARTOGRAPHIC ABSTRACTION

SELECTION AND GENERALIZATION PRINCIPLES

We are surrounded by detail and complexity. To avoid being mired in confusion, we often focus on universal characteristics of things rather than on their individual, unique qualities. For example, we speak of average annual rainfall or median family income. We categorize features and eliminate visual complexity by simplifying outlines. These are everyday actions that help us comprehend our environment. Collectively they are called generalization.

Many methods of generalization are applied in cartography. Reducing reality to map scale makes generalization necessary. Also, the act of generalization gives the map its *raison d'etre*. Only in recent years has it been possible for a person to see much of the earth at once. Although the view from space high above the planet may arouse awe, the scene itself is not likely to be very meaningful. The earth is simply too large and its phenomena too complex for anyone to grasp much by direct observation from afar.

The solution to deciphering this complexity is to reduce scale systematically and eliminate or de-emphasize unwanted detail. At the same time, we must emphasize certain features, which form the map's subject. This quality of maps to typify and simplify is one of the fundamental reasons maps are needed by society. Otherwise we could operate with photographs alone.

Unmodified reduction, when applied to earth phenomena, is accompanied by unavoidable changes. Distances separating features and widths and lengths of features are shortened. Adjacent discrete items become more and more crowded. Intricate configurations may appear chaotic.

Such consequences of reduction are a mixed blessing. On one hand, space shrinkage allows us to see the geographical arrangement of phenomena. On the other hand, the increased complexity and crowding promote visual confusion. To counteract these undesirable effects, we must perform two key operations on the wealth of geographic features. One is to limit our concern to those classes of information that will serve the map's purpose. We call this operation **selection**. The other is to fit portrayal of selected features to the map scale and to the requirements of effective communication. This we call **generalization**.

SELECTION

As the term is used here, selection is the intellectual process of deciding which classes of features will be necessary to serve the map's purpose. No modification takes place; the choice is either to portray roads or not to portray roads, to include or not include major hydrographic features, or to name or not name all cities with populations over 150,000.

To perform selection properly, cartographers must have a clear idea of the information to be presented via the map. The overall conception of the map—that is, its purpose and preliminary design—should govern the features and attributes to be selected for portrayal.

In analog cartography, selection has received too little attention. Once a cartographer manually began rendering a map, it became very costly to change the features selected. In contrast, digital cartography allows experimentation in the selection process. Provided that data are available in digital form, it is easy to view those data on a monitor and to overlay them with other feature data sets. Cartographers can thus pick and choose during the selection process without greatly increasing the time and cost of mapping. In fact, spending sufficient time on the selection process can significantly reduce overall mapping costs.

Once cartographers have selected features and attributes for mapping, they are ready to move to the next step—generalization. In the remainder of this chapter, we will be discussing generalization in detail. Let's begin with an overview of important generalization concepts.

GENERALIZATION CONCEPTS

In discussing cartographic generalization, it is important to define five terms: classification, simplification, exaggeration, symbolization, and induction. These five operations are controlled by cartographers to varying degrees.

When cartographers use **classification,** they order, scale, and group features by their attributes and attribute values. When they use **simplification**, they determine important characteristics of feature attributes and eliminate unwanted detail. When they use **exaggeration**, they enhance or emphasize important characteristics of the attributes. These three operations require cartographers to make a series of operational decisions.

After making these decisions and applying any necessary algorithms to the selected data, cartographers use graphic marks to encode the information for visualization. This process of graphically coding information and placing it into a map context is called **symbolization**. Before assigning marks to represent features and their attributes, cartographers must make at least two conceptual decisions. These decisions will be discussed below (and in Chapters 25 and 26).

Finally, **induction** occurs when we make inferences from interrelationships among features on the map. Cartographers have little control over induction. When we apply induction or inductive generalization, we extend the map's information content beyond its features. We do so by making logical geographic inferences.

For example, suppose you have average January temperatures for a series of weather stations. You can, by suitable "logical contouring," construct a set of isotherms. The isotherms allow inferences about probable January temperatures in areas *between* weather station locations. Thus, they convey far more information than the temperatures recorded at the weather stations themselves. Any such logical extension of data, founded on accepted associations, is inductive.

Induction may extend beyond what was consciously added by the cartographer. A map user, with personal knowledge about a feature, may further amplify information on the map. Through induction, map users often uncover hypotheses that cannot be generated in any other way. This is one of the primary utilities of a map. It is also one reason cartographers can't completely plan inductive generalization. Induction can best be encouraged by good clean cartographic design after careful classification, simplification, exaggeration, and symbolization.

Each map provides a different set of requirements. The "mix" of generalization processes varies from map to map. So does the inductive potential of the resulting map.

Now that we have looked at the basic concepts of generalization, let's explore the important generalization elements. Before we begin, it is only fair to observe that separating cartographic generalization into elements is itself a generalization about generalization. The aim, like that of all generalizing, is to simplify a complex process to manageable proportions.

THE ELEMENTS OF GENERALIZATION

In many cartographic situations, it's impossible to separate the four elements of generalization (classification, simplification, exaggeration, and symbolization). This is especially true in analog cartography. In digital cartography, it is easier to distinguish among the different generalization processes. This is because an explicit program or set of commands must be written to initiate each operation.

The overall intent of generalization is to enhance map communication. Each selected feature and its attributes should contribute to the effective communication of information. And each feature should appear in its rightful place in the visual hierarchy of the design. For instance, suppose cartographers include a country boundary merely as a frame of reference. In that case, they can greatly simplify it—perhaps slightly exaggerating distinctive parts of the boundary. They can then emphasize the subject of the map. (See Figure 24.1). On the other hand, what if the purpose of the map is to show border crossings? In that case, the cartographers might perform less generalization to the boundary and perhaps place more names near the border than elsewhere on the map. (See Figure 24.2).

Now let's look individually at the four elements of generalization. Keep in mind that in this discussion we are treating each element without regard for the others. In reality, it may be difficult to control one element without affecting the others.

Classification

The goal of classification is to express the salient character of a distribution. As noted earlier, classification is the ordering, scaling, and grouping of features by their attributes and attribute values. Classification modifies feature attributes in an attempt to typify.

Figure 24.2   A map of the international boundary of Poland and its neighbors. Note the detail with which the boundary of Poland is delineated, reflecting its importance with respect to international matters.

Figure 24.1 A map of the major highways in Poland. Note the simplified country boundary.

Classification is an intellectual process that groups similar phenomena in order to gain relative simplicity. It is hard to imagine any intellectual understanding that does not involve classification. We classify without thinking about it: We calculate attribute value averages, modes, or extremes, and we sort all the varieties of features into simple classes such as chairs, books, roads, rivers, or coastlines.

There are two common ways we perform classification on maps:

· We allocate similar qualitative attributes, such as land use or vegetation, into categories (cropland, forest) or quantitative attribute values into numerically defined groups.
· We modify the attribute value at a selected location to create a "typical" feature for portrayal on the map.

The manipulations cartographers perform in the process of classification include class interval selection and various agglomeration routines. One manipulation, commonly called **clustering**, is illustrated here since it makes a clear distinction between simplification and classification.

Clustering is necessary when numerous discrete features characterize a distribution and, at the reduced map scale, it would be undesirable or impossible to portray every individual feature. Two options are available to the cartographer: One is a method of simplification. The other is a method of classification. Both seek to generalize the distribution. In Figure 24.3, features have been grouped, and then a typical location is designated to represent each group. This is classification. In Figure 24.4, a predetermined number of features are portrayed on the reduced-scale map, and the others are eliminated. This is simplification.

Simplification

When we practice cartographic simplification, we determine important characteristics of feature attributes and eliminate unwanted detail. In doing so, we have two main objectives:

· We must reduce the amount of information to the map's ability to portray it legibly at the chosen scale. That is, we must decide how much information to portray.
· We must maintain as far as possible the essential geographical characteristics of the mapped phenomena.

Since symbols take up room on a map, it's clear that as map scales get smaller, fewer features can be represented. Part of the solution is to select only necessary features to be portrayed. But that is usually not enough. In most cases, information to be mapped must also be simplified. We can simplify information by eliminating some of it and by smoothing (reducing the details of) the remaining features.

The map space available to portray selected features is a function of scale. Available map space

classification

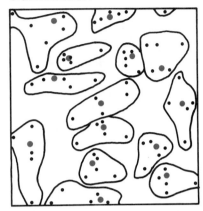

Figure 24.3 Classification of a point pattern. After clustering the points, the cartographer selects a position within each cluster and places a dot (shown in red) to "typify" the cluster. The "typical" position need not coincide with the position of any of the original data points.

simplification

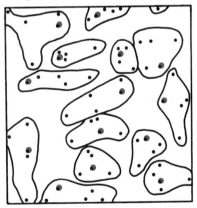

Figure 24.4 Simplification by point elimination. In the illustrated clusters of points, one original point (designated by red overprint) is selected to represent each cluster of original points on the generalized map.

is reduced by the square of the difference in linear scales. Thus, reducing the scale by one-half reduces the map area to one-quarter. For example, a region mapped at 1:25,000 will only occupy one-fourth as much map space when mapped at 1:50,000.

Clearly, the compression that results from scale reduction allows only limited portrayal. Figure 24.5 shows two examples of the pressure exerted by reduction and the consequent simplifications. Whether a feature is discarded or retained depends on:

- the feature's relative importance in the visual hierarchy
- the relation of that class of feature to the map's purpose
- the graphic consequences of retaining the feature.

By analyzing what cartographers have done to reduce items from one compilation scale to another, F. Topfer developed what he called the **radical law**. See Box 24.A.

When we apply the radical law, we obtain a statement of the number of items we can expect on a newly compiled map. Although the law's primary value is theoretical rather than practical, it is useful in several important ways. We can apply it to: (1) point feature sets (for example, towns on a road map), (2) linear feature sets, such as roads or streams, and (3) areal feature sets that consist of numerous small similar items within a region, such as lakes or islands (see Figure 24.5). The radical law can also be used in computer algorithms to evaluate computer generalization schemes.

The radical law specifies, with high probability, how many features can be retained when working from larger to smaller scale. But, since it's a statistical law, it can't tell us which items to include and which to discard. How, then, do we make this choice? We can't base our decision on just one attribute of the feature, such as size. Suppose, for instance, that we mapped only cities of more than 100,000 inhabitants. We would thus eliminate many key cities in the western United States. At the same time, we would retain many cities in the more populous eastern United States that are not as "important" to their regions.

We must base our choice of features on a variety of factors. In doing so, we must consider the map's purpose. We must also note how each feature type fits into the visual hierarchy planned for the map.

Some problems are unavoidable when we simplify a region. For instance, the lengths of irregular lines (rivers, coasts) become shorter, and areas bounded by irregular lines (lakes, countries) become smaller as the lines themselves become simpler. Despite these problems, we should strive to maintain the distinctive geographical character of a region. We should make sure, for example, that winding roads appear winding and that highly serrated coasts retain their irregular character. To do so, we will need to move from simplification to another generalization element—exaggeration.

Exaggeration

Only at very large scales (such as 1:2,500) can we show such features as roads, buildings, and small streams without greatly enlarging them. Imagine, for instance, that we are making a map with a 1:25,000 scale. We want to include a street which is 20 meters (66 feet) wide. If we showed the street true to scale, it would be symbolized by a line 0.8 millimeters (about 0.03 inches) wide. If reduced photographically to 1:100,000, the street symbol would be only 0.2 millimeters wide. If reduced to 1:500,000, the lines probably would disappear to the unassisted eye.

When we use exaggeration, we deliberately enlarge or alter a feature in order to capture its real-world essence. Suppose, for example, we wish to map a highly meandering stream, such as the lower Mississippi River. At small scale, using only simplification, the river may lose its meandering character. The portrayal should have some

Figure 24.5 Reasonable delineations of two regions at two scales. Keep in mind that the larger scales are in themselves relatively small-scale, so that a great amount of simplification had already been introduced into the maps.

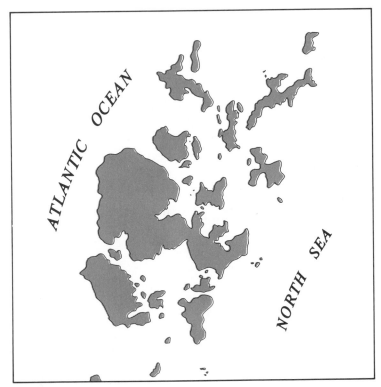

1:1,000,000

THE
ORKNEY ISLANDS

1:10,000,000

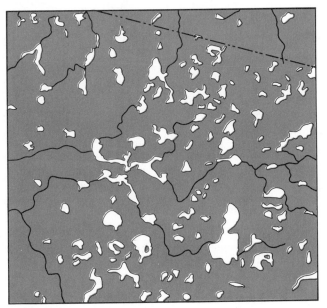

1:500,000

PART OF THE LAKE REGION
OF WISCONSIN AND MICHIGAN

1:2,500,000

Box 24.A

Topfer's Radical Law

We can express the relationship between map scale and number of items on the map by the equation:

$$n_c = n_s \sqrt{\frac{S_c}{S_s}}$$

where

n_c = the number of items on a compiled map with a scale fraction of S_c

n_s = the number of items on a source map with a scale fraction of S_s.

The problems of simplification are compounded by the nature of the data. For example, such features as islands or lakes are generally represented without exaggeration, and a relatively large number can be retained. In contrast, items such as settlement symbols, with their associated names, take up a much larger proportion of space, and fewer can be retained. For these reasons, the author of the radical law found it necessary to introduce two constants, C_e and C_f, to the right side of the basic equation. Thus:

$$n_c = n_s C_e C_f \sqrt{\frac{S_c}{S_s}}$$

C_e is called the constant of symbolic exaggeration and takes three forms:

C_{e1} = 1.0 for normal symbolization—that is, for elements appearing without exaggeration.

$$C_{e2} = \sqrt{\frac{S_s}{S_c}} \text{ for features of areal extent}$$

shown in outline, without exaggeration, such as lakes and islands.

$$C_{e3} = \sqrt{\frac{S_c}{S_s}} \text{ for symbolization involving great}$$

exaggeration of the area required on a compiled map, such as a settlement symbol with its associated name.

Constant C_f is called the constant of symbolic form and also takes three forms:

C_{f1} = 1.0 for symbols compiled without essential change.

$$C_{f2} = \left(\frac{w_s}{w_c}\right)\sqrt{\frac{S_s}{S_c}} \text{ for linear symbols in}$$

which line widths on the source map (w_s) and the newly compiled map (w_c) are the important items in generalization.

$$C_{f3} = \left(\frac{a_s}{a_c}\right)\sqrt{\frac{S_c}{S_s}} \text{ for area symbols in}$$

which the areas of the symbols on the source map (a_s) and the newly compiled map (a_c) are the important items in the generalization.

meanders even though the mapped loops may be far larger than reality at that scale.

Or, imagine a region studded with hundreds of small lakes. If we merely simplify the region, all the lakes will disappear at the smaller scale. But to capture the identity of the area, we need to show that it abounds with lakes.

As these examples show, a map's purpose and scale often require **diagrammatic generalization**. This process has some characteristics of a caricature. It involves analyzing the shape of outlines (continents, countries, lakes, islands, and so on) and retaining those distinctive shapes on the map (Figure 24.6). It is often better to use clean, firm, smooth lines to emphasize, or exaggerate, the main characteristics of an area's form than to try to be highly "accurate." Rudolph Arnheim put it well:

> ...an image of reduced size is not obtained simply by leaving out details. Artists as well as cartographers realize that they face the more positive task of creating a new pattern, which serves as an equivalent of the natural shape represented. Such a newly created pattern is by no means the same thing as a sandpapered copy of the original. The reduction of size provides the mapmaker with a degree of freedom, which he can use to make his visual images more readable and more pertinent.

Symbolization

After cartographers apply classification, simplification, and exaggeration to features selected for mapping, they are ready to translate these features to graphic marks on the map. We call this process symbolization.

All graphic marks on a map are symbols. Cartographers can use these marks to symbolize a concept, a series of facts, or the character of a geographical distribution.

Symbolization may also require intellectual generalization. That is, cartographers may change a feature's dimensionality. Or they may change the measurement scale of a feature's attribute values. (We will discuss these procedures in detail in Chapter 25).

The degree of generalization by symbolization can vary widely within a map. On one map, for instance, you may find symbols, such as latitude-longitude lines, which involve almost no generalization. On the same map, you may find a large, smooth arrow symbolizing population migration from one region to another. Such an arrow is highly generalized, since it takes no note of the many different routes traveled by individuals.

Figure 24.6   Two representations of Great Britain and Ireland at the scale of about 1:15,000,000. The map on the left (A) is simplified to fit the scale and is suitable for a reference map intended to give the impression of detailed precision. The map on the right (B) is a diagrammatic generalization suitable as a base on which to display thematic data. Note that (B) captures basic shapes, which tend to be masked by the detail in (A).

THE CONTROLS OF GENERALIZATION

As we noted earlier, cartographers don't have complete control over the processes of generalization. Generalization is also guided by a number of external forces. These controls are similar to the influences on map design which we explored in Chapter 18. As we explained there, the following factors affect the design process: reality, map purpose, quality and quantity of available data, map scale, audience, conditions of use, and technical limits. These same forces affect cartographic generalization.

In the following discussion, we will treat reality, map purpose, and conditions of use as one set of controls, which we call **map purpose and conditions of use**. **Map scale** will be treated independently because of its importance in selecting generalization algorithms, as will **quality and quantity of data**. A final set of generalization controls is provided by **graphic limits**. Graphic limits include the technical limits of production and the perceptual limits of the audience.

Map Purpose and Conditions of Use

The basic reason for making a map greatly influences the generalization process, as does the conditions under which the map will be used. Will it be studied over a long period, as with a topographic map or a general atlas map? Or will it be shown briefly on a screen during a slide presentation? Is it designed to provide a great deal of general geographic information or to show the structure of a particular distribution? Will the display be manipulated by the user? Before beginning to make a map, the cartographer must answer these questions.

Map Scale

The scale of the finished map also has a major impact on the amount of generalization that will be used. The smaller the scale, the more generalization will usually be required. At large scales, most of the generalization is classification and symbolization. At smaller scales and in thematic mapping, the situation is quite different. Simplification and exaggeration become the most impor-

tant elements of generalization, although classification is important as well.

There is a range of generalization that will "fit" each scale. That is, the treatment will be neither too detailed nor too general for that scale. The cartographer's problem is to maintain a balance among and consistency within feature categories.

At very small scales, as on global maps, there may be a great scale range within the map. On Mercator projections, for example, there is four times more map space at 60° latitude than at the equator. The natural tendency is to use all the space available. Consequently, maps using Mercator projections usually contain far more simplification and classification in the equatorial regions than at higher latitudes. Whether cartographers should use the same degree of generalization throughout the map depends on the map's purpose. It is not always easy to be consistent. Algorithms which enable digital generalization rarely compensate for areal distortion or angular compression resulting from the map projection.

Quality and Quantity of Data

The quality and quantity of data available to cartographers also greatly affect the generalization processes. The more reliable and precise the data, the more detail is available for presentation.

One of the most difficult tasks for cartographers is to indicate to map readers the quality of the data used. When writing or speaking, we use words such as "almost," "nearly," and "approximately." These words let people know how precise we mean to be. It is not easy to do the same thing on a map.

Cartographers meet this challenge in several ways. On large-scale maps, they often include a reliability diagram, which shows the relative accuracy of various parts of the map. They may also insert a statement in the legend, explaining how accurate the information is. Here they can use such terms as "position approximate," "generalized roads," or "selected railroads." Such phrases give map readers an idea of how complete and precise features are. (See Chapter 23 for more on accuracy and reliability).

Intellectual honesty is vital in cartography, because maps have an authoritative appearance of truth and exactness. Therefore, cartographers must take special pains to ensure that information is correct. They must also be sure that their maps don't convey a greater impression of completeness and reliability than is warranted.

Digital cartography compounds the problem. As information in cartographic databases increases, the need for more metadata (which includes data accuracy information) explodes (see Chapter 15). The chances for misuse also increase.

When using existing map sources for analog compilation, cartographers abide by the rule "Always compile from larger scale to smaller scale." But they have no idea whether this rule has been followed in a database unless information regarding input scale has been included. Thus, it's imperative that every database indicate the quality of data to help cartographers avoid misuse.

In addition to the **quality** of data, the **quantity** of data available has a great impact on the generalization process. If *not enough* information is available, cartographers should either make the map at a smaller, more generalized scale or not make it at all. The definition of "not enough" is subjective, however. Thus, many poor maps have been created with too little data. When map readers draw inductions from such maps, the results can be tragic.

The other extreme occurs when *too many* data exist. This situation is becoming more of a problem in digital cartography as our databases become more detailed. At this extreme, cartographers are inundated with data and can't separate the important or "typical" features from all the clutter. In this case, it may be necessary to use generalization algorithms such as those discussed below under *Elimination Routines.*

Graphic Limits

Other factors which affect generalization are called graphic limits. We can break these factors into two groups: (1) technical limits set by the cartographer's tools and (2) perceptual limits of the human eye.

We create map symbols by combining the basic graphic elements: point, line, and area marks. (See Chapter 18). Our ability to form symbols from these elements is subject to three types of limitations: physical, physiological, and psychological. **Physical limits** are imposed on the graphic elements by the equipment, materials, and skills available to the mapmaker. **Physiological and psychological limits** are imposed by the map user's perceptions and reactions to the primary visual variables. (As we saw in Chapter 18, the visual variables are shape, size, orientation, hue, value, chroma, arrangement, and texture). All three types of limitations place important controls on the amount of generalization cartographers can successfully use.

Limitations on the visual variables include maximum format size, available line width sizes, lettering styles and sizes, color screens, preprinted symbols, dimensionally stable film or plastics, specialized symbol templates, and the cartographer's or machine's abilities to work within these limits to create a symbolization.

In digital cartography, the graphic limits are more restrictive only in the size of the area (on the color monitor) we can use to create a visualization at any one time. In most other respects, we gain flexibility by moving to digital cartography. In fact, computers have extended the capabilities of the physical limits beyond what is physiologically and psychologically possible for humans to perceive. Also, the consistency obtained by computer assistance exceeds that of manual rendering.

The audience for whom the map is aimed (geographically sophisticated or ignorant, children or adults) is also an important factor. Map readers' perceptions vary consistently from one symbol type to another. For example, a line twice as wide as another will usually look that way, but a circle with twice the area of another circle will look significantly less than twice as large.

Physiological limits refer to our ability to distinguish among hues, sizes of type, shades of gray, and so on. Our ability to judge differences in values is much less than our ability to produce them on maps. Thus, it's important to keep these limits in mind when creating map symbols. These limits also control the number of classes we should use in classification. In addition, physiological limits affect the minimum distance between points that a computer-assisted simplification algorithm can use.

CLASSIFICATION, SIMPLIFICATION, AND EXAGGERATION MANIPULATIONS

We have now looked at the elements of generalization and the controls that influence them. Next, we will explore the manipulations cartographers perform to achieve each type of generalization.

Cartographers working manually must complete each generalization independently. No two generalizations done by humans are identical. Also, it's difficult for cartographers working manually to provide detailed descriptions of how they accomplish each type of generalization.

Computerized cartography has changed all this. Computers demand unambiguous instructions. Thus, cartographers have been forced to rethink and to define more explicitly the processes used to generalize a map. Computers have opened the door to a vast array of sophisticated, often statistical, computation processes which cartographers can use on either vector or raster data to aid generalization.

The manipulations cartographers use will vary depending on the form of the data about features and their attributes and values. These data may exist in the following forms: (1) tabular values, (2) maps or photographically recorded images, (3) text (verbal) accounts, (4) computer-readable strings of coordinates (vectors), or (5) machine-stored arrays of picture elements (rasters).

Let's now see how cartographers manipulate these data to achieve generalization. In the following discussion, we will again differentiate among the generalization elements of classification, simplification, and exaggeration. (We will discuss the fourth element, symbolization, in Chapters 25 and 26). In practice, of course, it's not always easy to distinguish one element from another. Most generalization routines contain a mixture of all three.

CLASSIFICATION MANIPULATIONS

As we discussed earlier, the goal in classification is to typify the data set. In the process of typifying, none of the original features may actually be retained. Rather, a more general feature replaces the true feature. (See Figure 24.7). Because of this emphasis on typifying, the most convenient way to discuss classification manipulations is by using the categories of point, line, area, and volumetric features. Let's begin with point features.

Point Feature Methods

To classify point features, we use aggregation techniques. There are two types of aggregation:

1. **Collapsing.** This process refers to the loss of dimension by the feature being mapped (see Figure 24.8). We use this process when we must group individual elements and specify a typical location for the group. We might use this technique, for example, in constructing a dot map. We might use one dot (point) on the map to represent 10 houses (areas).

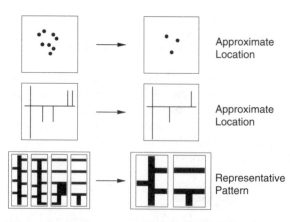

Figure 24.7 Representations illustrating typification by classification of point, line, and area features. (Courtesy of D. Lee.)

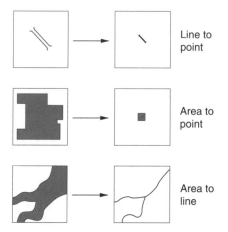

Figure 24.8 Illustrations of the collapsing process in cartographic generalization. Each feature represented in the left diagrams has lost at least one dimension in its portrayal in the right diagram. (Courtesy of D. Lee).

2. **Typification.** We use this process for a complex pattern consisting of individual features. This process allows us to retain the essence of the pattern on the reduced-scale map.

These manipulations remain rather poorly defined and subject to the whim of cartographers performing them. The cartographers must subjectively specify criteria that allow the aggregation procedure to be applied. For example, they might specify a starting feature for the aggregation. Or they might specify the direction of movement to be followed during the aggregating procedure. Different results are obtained by starting with different features or by specifying different directions of movement. Although we know that the choice of these parameters significantly affects the outcome, we don't know the advantages of one set of parameters over another.

Even less is known about capturing the essence of a pattern of point features. No computer algorithm generalizes a pattern while retaining its "essence." Certain patterns can be characterized by a statistical index referred to as the **nearest neigh-**

bor statistic. (See also *Method-Produced Error* in Chapter 26). Computers can generate generic point patterns with a given statistical index, thereby typifying the nature of a distribution of point features. But we are unaware of any cartographic applications of these capabilities.

Line Feature Typification

While it is uncommon for cartographers to aggregate lines, they do so occasionally. They might, for example, combine lines representing airline passengers traveling between pairs of cities. The flow of passengers from Chicago to New York can be individually shown by airline X, airline Y, and airline Z, or these may be aggregated into one flow line. Likewise, distributary streams may be aggregated into a few lines to convey the essence of the distribution.

Eventually geographers will develop a model for the outline of a fjorded coast or a meandering stream. When they do, these models will allow cartographers to typify meandering streams or fjorded coasts for generalization on maps.

Aggregation of Areas

The aggregation of areas is a very important manipulation in cartography. With nominal data, this process depends on the size of the area units to be aggregated. For example, there might be two areas of cropland separated by a tree-lined stream, as shown in Figure 24.9. Suppose the area is broken into 15 small unit areas. At a comparatively large scale, we could map each unit separately, depending on the percentage of cropland in each small unit (See Figure 24.9A). At a small scale, the area might be mapped as a single unit consisting entirely of cropland (See Figure 24.9B).

Now suppose we want to aggregate ordinal, interval, or ratio data. The method is similar to that for aggregating nominal data. The only difference is that we must consider quantitative differences between the regions. Such aggregation of areas, based on quantitative differences, is a form of dasymetric mapping (see Chapter 26).

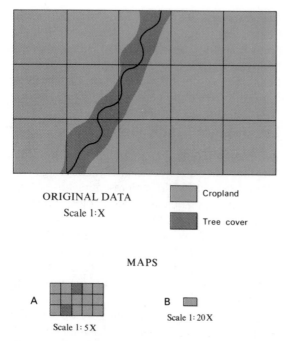

ORIGINAL DATA
Scale 1:X

Cropland

Tree cover

MAPS

A

Scale 1: 5 X

B

Scale 1: 20 X

Figure 24.9 The original data are mapped at a scale of 1:X. (A) represents a smaller-scale agglomeration of the original data. (B) represents the further aggregation of areas for an even smaller-scale representation.

Aggregation of Volumes

When we aggregate volumes at points, lines, or over areas, we group similar attribute values into classes. The resulting categories are bounded by class limits, such as 0-2, 2-4, 4-8, and 8-16. This operation is called **range-grading**. On choropleth maps, for example, cartographers range-grade attribute values associated with census areas.

When classifying volume data, we may want to pay special attention to the range of the data being portrayed. For example, we may be more concerned with relative changes than with absolute values. Thus, we would use a smaller interval in the lower portion of the range. We would do so because a change from 2 to 4 is the same relative change (100 percent) as a change from 50 to 100. On the other hand, we may be more interested in absolute values of the higher part of the range. We would then make an opposite choice of intervals.

Before they specify class limits, cartographers must decide on the number of classes for the map. This choice is also a generalizing process. Obviously, fewer classes means greater generalization. Likewise, minimal generalization implies more classes. (See *Classless Choropleth Mapping* in Chapter 26). The section on *Digital Image Classification* in Chapter 12 gives examples of these aggregation algorithms applied to pixels.

The controls on generalization limit the number of classes cartographers can show. For example, the graphic limits determine the maximum number of classes, because the human eye has limited capabilities. On the other hand, data quality, along with the desired simplification, establishes a minimum number of classes that can be used effectively.

Because of these controls on generalization, determining number of classes is not as much a problem for cartographers as determining class limits. Class limit determination is treated more fully in Chapter 26.

SIMPLIFICATION MANIPULATIONS

Using manual methods, cartographers can simplify map data rather easily, if somewhat subjectively. Such simplification mainly takes the form of omissions or deletions.

With digital cartography, cartographers have more sophisticated simplification methods available. Information exists in machine-usable form in either a vector or a raster format. Statistical techniques include multiple regression and correlation (see Chapter 16), factor analysis, discriminant analysis, curvilinear interpolations, and surface-fitting processes (see Chapter 26). In dealing with raster data, cartographers perform routines such as ratioing, principal component analyses, density slicing, and low pass filtering (see Chapter 12).

These computer manipulations are more objective than manual ones in the sense that the outcome is repeatable. Subjectivity remains, however, in that cartographers must select an algorithm and its input parameters. In this sense generalization is like map projections. Cartographers must be trained to select and use correctly the generalization or projection most appropriate for a given purpose. Cartographers no longer must perform the actual manipulations; the machine does that with a degree of precision cartographers could never match.

The amount of information available, in every aspect of our lives, is constantly increasing. This is useful if we need finer and finer detail about a topic. But too much information can be as much of a problem for cartographers as too little. When mapping a region, they must avoid excess detail, or "noise," that will detract attention from overall trends in the distribution.

Too often, cartographers do not use simplification processes rigorously enough—perhaps because they are reluctant to eliminate what they consider "good" data.

In remotely sensed data, the mechanics of the sensor system often cause small variations from true pixel values. These aberrations are also termed "noise." Cartographers are usually eager to remove this "bad" noise. But whether noise is considered bad or good, cartographers need to remove it to clarify distributions for map readers.

Cartographers also use simplification when they convert features to digital records. The accuracy of digital records depends on the scale of the source. It also depends on the resolution of the computer hardware used for the conversion. Often these digital records are then used at output scales smaller than that of the source. In addition, the output device may have a coarser resolution than the capture device. In such cases, more information exists in the digital record than can be plotted by the output device. Cartographers use simplification to eliminate this unneeded and unusable information.

Let's look now at how cartographers apply simplification techniques. For discussion purposes, we will class simplification processes into two categories: elimination and smoothing.

Elimination Routines

There are two reasons to use elimination routines:

- To accompany a scale reduction of the map being made. (See Figure 24.10).
- To de-emphasize a class of minor features at a constant scale. (See Figure 24.11).

Simplification accompanied by scale reduction is a straightforward concept. As the map's scale is reduced, the physical space available for detail is reduced. Therefore, simplification of details is mandatory.

Simplification applied at a constant scale is more subtle and perhaps more important in good map design. For example, if the outline of a feature (such as a state boundary or coastline) is only intended to give general reference information, it can be greatly simplified, as in Figure 24.11D. On such a map, the thematic information is the primary message. On the other hand, suppose the map's purpose is to show ports or coastal shipping. On such a map, coastline detail is important, and an outline like that in Figure 24.11A is more appropriate. In both cases, cartographers used elimination routines to change the degree of simplification, without changing the map scale.

When cartographers perform the process of elimination, they can use algorithms that eliminate either point or areal features. Let's look at both these types of elimination, beginning with point features.

Point Elimination   When cartographers want to simplify a line, they can use a process called point elimination. The map in Figure 24.12 illustrates point elimination used to simplify the outline of Portugal. The points that are emphasized by black dots on the left map have been retained and connected by straight-line segments to make the map at the right.

Cartographers use two basic computer-assisted simplification algorithms to eliminate points. The first method is to systematically retain every nth point in the data file. When cartographers use this process, they create a new data file by taking the

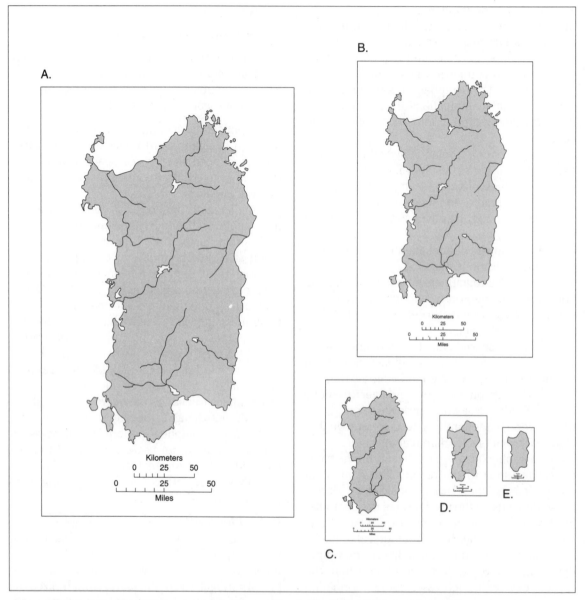

Figure 24.10 Simplification accompanied by scale reduction. Since the scale is successively reduced from (A) to (E), an increasing number of points in the outline of Sardinia must be eliminated.

first coordinate point in a data string and every *n*th point thereafter. The value of *n* may be determined by the radical law (see Box 24.A). The larger the value of *n*, the greater will be the simplification. If possible, cartographers usually modify this basic algorithm by assigning importance values to the points. When they do so, perhaps by some classification algorithm, the systematic nature of the elimination depends on the importance values assigned to each point.

A.   B.   C.   D.

Figure 24.11   Simplification applied at a constant scale. The four maps of Sardinia (A through D) represent increasing simplifications of the coastline and hydrography.

The second procedure is for cartographers to create a new file by randomly selecting l/nth of the points. Different values of n may be applied to different sets of points or segments of lines. Again, the larger the value of n, the greater will be the simplification.

Both these simplification processes are applied to the *xy* location attributes of a feature. These processes often form the basis for more complex algorithms that simplify coastlines, political boundaries, roads, networks, and so on.

However, neither of these routines considers the geographic characteristics of the lines being simplified. Therefore, these routines are most often used as an initial pass to reduce large digital files to a size in keeping with intended output. They thus cut down subsequent processing time of more sophisticated generalization algorithms.

Imagine, for example, that data are digitized on equipment with a resolution of .0001 inch and will be plotted by equipment with a resolution of .001 inch. The amount of data potentially available in the digitized file is 10 times what can be physically plotted. (Note: Forget for the moment the fact that map readers cannot perceive .001 inch, let alone .0001 inch). In this case, one of

Figure 24.12   Simplification of the outline of Portugal by point elimination. The points indicated on the map to the left were retained on the map to the right where they were connected with straight-line segments. All points not selected on the map to the left were eliminated in producing the map on the right. (Courtesy of the American Congress on Surveying and Mapping).

these basic elimination schemes is appropriate as an *initial* simplification procedure. Cartographers may still have to apply an elimination routine that retains the basic character of the line after using one of these procedures.

Cartographers sometimes manually eliminate points in a line, based on the location attribute defining a feature. This manipulation consists of deleting points that are visually unimportant. Cartographers gain a "feel" for this elimination process only after considerable experience, and the radical law is only marginally useful for these manual applications.

Other criteria can be established to retain or eliminate points in a file. One of the most common algorithms was developed by Douglas and Peucker. The Douglas-Peucker algorithm retains points at which a line abruptly changes direction. This algorithm allows cartographers to specify a threshold that controls the amount of simplification.

Figure 24.13 shows how the Douglas-Peucker algorithm works. For a specified line segment, the two end points are connected by a straight line.

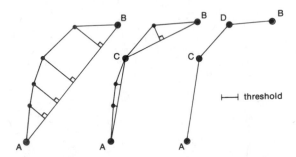

Figure 24.13 Successive stages in applying the Douglas-Peucker algorithm for line simplification: (1) The initial line, whose endpoints are connected by a straight line, AB. (2) Point C, having the greatest perpendicular distance to line AB in (1) is selected for retention. Lines AC and CB are drawn. (3) The elimination of points between points A and C, because no perpendicular exceeds the threshold, and the retention of point D, because its perpendicular distance to line CB does exceed the threshold. (Courtesy of M.S. Monmonier).

The perpendicular distances from all intervening points to that line are calculated. If a perpendicular distance exceeds the threshold, the point with the greatest perpendicular distance is used as the new end point for subdividing the original line. The perpendicular distances from all intervening points to the two new lines are calculated and compared to the threshold. If none of the perpendiculars exceeds the threshold, all the intervening points are eliminated. The routine continues until all possible points within the threshold have been eliminated.

The amount of data in machine-readable form is increasing rapidly. Thus, cartographers must pay close attention to the simplification manipulations that can be applied to machine-readable files. At the same time, they must remember that human subjectivity and hardware limitations have determined the contents of those files.

Area Elimination Cartographers can use elimination routines to simplify area features as well as point features. Simplification by area elimination is illustrated in Figure 24.14. This figure shows the areas of forest cover in Sheboygan County, Wisconsin.

In area elimination, a feature is either shown in its entirety or omitted. In Figure 24.14, the smaller areas on the left map have been eliminated from the map on the right. Obviously, it is possible to combine point and area elimination simplification by eliminating areas and point-simplifying the outlines of the retained features.

Area elimination, if done manually, will usually be inconsistent. In machine processing, the criteria may be minimum feature size or proximity to neighboring features. The simplification routine often requires cartographers to specify either the minimum size of features to be retained or minimum distance between features based on output scale and line width. (See Figure 24.15).

First, cartographers need to obtain relative importance rankings for each feature. They can do so by applying a classification algorithm. They then include these importance rankings as attributes of the features. Next, they specify output

Figure 24.14 Simplification by feature elimination. Areas on the left map are either shown in their entirety or completely eliminated in the feature-simplified map on the right. (Courtesy of American Congress on Surveying and Mapping).

containing all lakes of rank 5 or above, or all roads of rank 2 or above. All other lakes and roads will be eliminated.

Cartographers can also apply area elimination to raster data. They use a process called **stenciling** to change values of pixels in the image. In Figure 24.16, for example, cartographers have removed land areas from the image on the left. The result is the simplified image on the right. To perform this operation, cartographers only needed to know the different reflective values of land and water. They then blackened all pixels with reflective values indicating land. This emphasizes the remaining pixels.

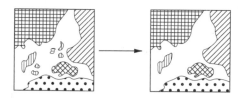

Figure 24.15 Simplification by area elimination. Example from Intergraph Corporation algorithm using size and proximity to determine which features to eliminate. (Courtesy of D. Lee).

Smoothing Routines

Instead of *eliminating* features, cartographers can simplify a map by *smoothing* features. Simplification by smoothing includes: (1) smoothing operators, and (2) surface-fitting models. Cartographers have begun to use these routines only recently. Prior to computers, these processes were computationally too complex to apply.

Both these processes modify feature attribute values. After the generalization has been applied, a "simplified" attribute value is substituted for the original attribute value.

Smoothing Operators The first class of simplification manipulations is referred to as smoothing operators. These processes are appropriate for vector data representing linear features or areal boundaries, for regular matrices of attribute values, or for raster data stored as pixels.

One example of smoothing operators is provided by **moving averages**. Moving from one picture element to the next, each pixel in turn is averaged with its neighbors. After averaging, pixels will be more like their neighbors, and spatial variation will be reduced. It is important that this modification not destroy the individual

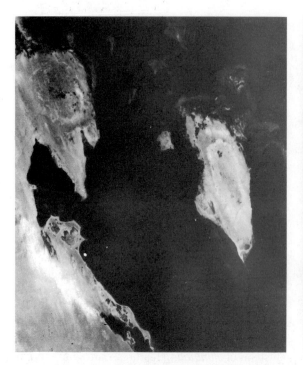
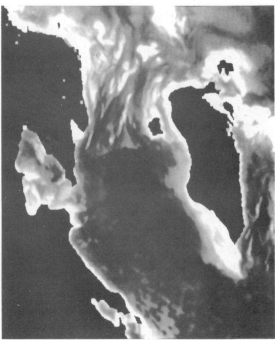

Figure 24.16 An example of stenciling. The image on the right shows the results of stenciling the satellite image on the left. On the right image, all nonwater pixels have been printed in black. This enables the user to concentrate only on the water areas. (Courtesy of P. Chavez, U.S. Geological Survey, Flagstaff, AZ).

pixel or combine it with another pixel; it must merely modify its value. (See *Digital Image Simplification* in Chapter 12). Figure 24.17 illustrates the simplification of area/volume (raster) data using an ordinally-scaled smoothing operator.

Figure 24.18 represents a commonly used moving average applied to ratio-scaled point feature attribute values. The assigned weights and the

included neighboring points determine the generalizing effect of applying a moving average. For example, as the number of neighboring values included in the moving average increases, the amount of simplification is increased. (Compare line C to line B in Figure 24.18). Furthermore, the more evenly apportioned the weights are, the greater will be the simplification. (Compare line D to line B in Figure 24.18). If large weights are

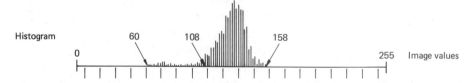

Figure 24.17 A smoothing operator has been applied to the previously classed picture elements in (A) to obtain (B). (Courtesy of S. Friedman).

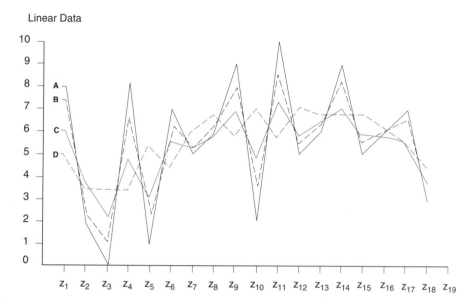

Linear Data

Smoothing operator:

for B

$$\hat{Z}_1 = .9z_1 + .1z_2$$

$$\hat{Z}_i = .1z_{i-1} + .8z_i + .1z_{i+1}$$

$$\hat{Z}_n = .1z_{n-1} + .9z_n$$

for C

$$\hat{Z}_1 = .75z_1 + .15z_2 + .1z_3$$

$$\hat{Z}_2 = .25z_1 + .5z_2 + .15z_3 + .1z_{i+2}$$

$$\hat{Z}_i = .1z_{i-2} + .15z_{i-1} + .5z_i + .15z_{i+1} + .1z_{i+2}$$

$$\hat{Z}_{n-1} = .1z_{n-3} + .15z_{n-2} + .5z_{n-1} + .25z_n$$

$$\hat{Z}_n = .1z_{n-2} + .15z_{n-1} + .75z_n$$

for D

$$Z_1 = \frac{2z_1 + z_2}{3}$$

$$Z_i = \frac{z_{i-1} + z_i + z_{i+1}}{3}$$

$$Z_n = \frac{z_{n-1} + 2z_n}{3}$$

Figure 24.18 Examples of smoothing operators applied to linear data. Line *A* represents the original data. Below the graph are given the smoothing operators used to produce line *B* (a three-term moving average with unequal weights); line *C* (a five-term moving average with uneven weights); and line *D* (a three-term, equally weighted moving average)

assigned only to close neighbors, simplification is minimized.

Figure 24.19 illustrates the effects of two-dimensional smoothing operators applied to a matrix of elevation data. Consider, for instance, the moving average with equal weights demonstrated by matrix *C* in Figure 24.19. Or consider the line-*D* moving average in Figure 24.18. Both are low-pass filters, as described in Chapter 12. In contrast, note the moving average with unequal weights demonstrated by line B in Figure 24.18. Also observe the

matrix-*D* moving average application in Figure 24.19. Both are high-pass filters, as described in Chapter 12.

Theoretically, we should select weights based on the amount of autocorrelation in the distribution being mapped (see Chapter 16). If autocorrelation is high, large weights assigned to nearby points result in little simplification (as illustrated by line *B* in Figure 24.18). We can obtain greater simplification only by assigning even weights to neighboring feature attribute values (as illustrated

Area Data

A - Original data array

0	6	7	0	9	5	3
0	0	10	2	9	1	6
6	3	6	5	1	10	3
8	1	10	7	0	10	2
9	3	4	4	6	1	10
8	6	0	2	7	1	7
7	6	5	5	4	1	1

Smoothing operator: for B

for $Z_{i\,j}$

$\hat{Z}_{i\,j} = .03z_{i-1,j-1}+.03z_{i,j-1}+.03z_{i+1,j-1}+.03z_{i-1,j}+.76z_{ij}+.03z_{i+1,j}+.03z_{i-1,j+1}+.03z_{i,j+1}+.03z_{i+1,j+1}$

for corners (appropriately rotated)

$\hat{Z}_{11} = .85z_{11}+.06z_{21}+.06z_{12}+.03z_{22}$

for edges (appropriately rotated)

$\hat{Z}_{12} = .06z_{11}+.79z_{12}+.06z_{13}+.03z_{21}+.03z_{22}+.03z_{23}$

B - Smoothed data array

0.36	5.46	6.25	1.59	7.77	5.15	3.24
0.63	1.14	8.47	2.93	7.83	2.14	5.58
5.34	3.51	5.70	5.15	2.08	8.56	3.48
7.43	2.23	8.59	6.40	1.32	8.59	2.99
8.37	3.66	4.03	4.12	5.52	2.05	8.80
7.73	5.82	1.05	2.57	6.04	1.87	6.28
6.97	5.88	4.85	4.76	3.82	1.54	1.36

Smoothing operator: for C

Matrix of weights

.01	.015	.02	.015	.01
.015	.025	.04	.025	.015
.02	.04	.5	.04	.02
.015	.025	.04	.025	.015
.01	.015	.02	.015	.01

Boundary conditions are met by summing appropriate weights e.g. corner conditions are:

.685	.08	.045
.08	.025	.015
.045	.015	.01

C - Smoothed data array

1.32	4.55	5.70	2.68	6.90	4.77	3.795
1.735	2.105	7.09	3.665	6.855	3.055	5.125
5.215	3.805	7.67	4.925	3.25	7.15	3.975
6.77	3.255	7.175	5.71	2.685	7.175	3.715
7.675	4.355	4.41	4.175	5.11	3.025	7.255
7.29	6.1	2.595	3.125	5.34	2.62	5.52
6.78	5.835	4.935	4.465	3.785	2.155	2.16

Smoothing operator: for D

Matrix of weights

.11	.11	.11
.11	.12	.11
.11	.11	.11

Boundary conditions are met by summing appropriate weights e.g. corner conditions are:

.45	.22
.22	.11

D - Smoothed data array

2.67	4.00	4.22	5.89	4.44	5.56	3.89
3.33	4.22	4.33	5.44	4.67	5.22	4.44
3.56	4.89	4.89	5.67	5.00	4.67	4.78
5.89	5.78	4.78	4.78	4.89	4.78	5.67

Figure 24.19 Examples of smoothing operators applied to area data. Matrix A represents the original data. The smoothing operators and boundary conditions used to create Matrix B, a nine-term, two-dimensional, unequally weighted moving average; Matrix C, a twenty-five-term, two-dimensional, unequally weighted moving average; and Matrix D, a nine-term, two-dimensional, equally weighted moving average; are given in the illustration.

by line D in Figure 24.18) or by increasing the number of neighbors considered (as illustrated by line C in Figure 24.18). The effect of such a selection is to maximize the smoothing of the surface. Be careful, however, because greatly smoothing a distribution with low autocorrelation results in a meaningless simplification.

In contrast, what if we want to capture the major features of a distribution having low auto-correlation? In that case, we should assign high weights only to a few nearby attribute values.

Surface-Fitting Models Another group of smoothing routines is referred to as surface-fitting techniques. In this category are several procedures that can be used to approximate lines and surfaces. These techniques are also used to estimate feature attribute values for logical contouring (leading to induction). Examples of line-type fitting techniques are regression line analyses (mentioned in Chapter 16). These constitute a kind of linear or profile "surface-fitting" in two dimensions only. For area or volume data, the procedure can be likened to fitting a regression equation in three dimensions.

The simplest surface-fitting technique is the "plane of best fit" (see Figure 24.20). To fit such a plane, we need at least three **noncollinear points** (points not lying in a straight line), located in a three-dimensional space. Preferably, more than three such points are used. One coordinate of each point (say the z value) is selected as the dependent variable, as in regression. The other two coordinates are independent variables. An equation $z = ax + by + c$ is fitted to the points by the following matrix operation:

$$
\begin{bmatrix} N & \sum x & \sum y \\ \sum x & \sum x^2 & \sum xy \\ \sum y & \sum xy & \sum y^2 \end{bmatrix}
\begin{bmatrix} c \\ a \\ b \end{bmatrix} =
\begin{bmatrix} \sum z \\ \sum xz \\ \sum yz \end{bmatrix}
$$

Most computers have software routines to compute the values for "a," "b," and "c." The resulting

Figure 24.20 The shaded plane, with respect to the surface shown, is situated so as to minimize the sum of the squares of the deviations between points on the surface and corresponding points on the plane.

equation represents a simplification of the original values. This is because modified values for each location can be obtained from the equation. The process is one-to-one mathematically.

We can easily extend this procedure to higher-order polynomial equations by using machine computation. These simplification techniques are also widely used in subsurface mapping. Each fit gives a percentage explanation, in the regression sense, that indicates the reliability of the fit.

A second group of similar techniques uses trigonometric series expansions. These methods approximate a linear profile in two dimensions or a surface in three dimensions.

Often we are interested in showing trends through time or over areas. Simplification by surface-fitting techniques allows us to estimate these trends by computing equations for them. The equations in turn allow the induction or interpolation of predicted values for points or areas for which no data are available. The calculations also give an indication of the trend's reliability. It is often more visually meaningful to map the trend than to map the complexities of the actual distribution. This is because the intricacy of the actual data may be visually confusing to the map reader. Thus, mapping the results of surface-fitting can "enhance" the readability of a map. Surface-fitting techniques are not usually thought of as enhancement routines, however.

EXAGGERATION MANIPULATIONS

Exaggeration is rather ill defined by cartographers. Increasingly, however, it is being recognized as a valid concern. Perhaps this is because of increasing computer use. Computers apply exaggeration algorithms more consistently than is possible with analog production. Thus, computers highlight the importance of exaggeration.

Take another look at Figure 24.11. It is the result of successive simplification routines. Clearly, the simplified shape looks weak and lacks character. We can enhance it by using exaggeration.

With raster data, contrast enhancements and image combination are examples of using exaggeration to enhance images. (See Figures 12.14 and 12.15 and Color Figure 12.4). With vector data, Figure 24.6 illustrates diagrammatic generalization, which is another example of exaggeration.

Cartographers have only begun to explicitly define exaggeration. They are now researching algorithms to determine critical points in lines or area outlines. But it is still easier to perform some exaggeration routines (for example, identifying points that can summarize a pattern's character) visually than automatically.

We can anticipate that exaggeration will receive increasing attention in the cartographic literature. Meanwhile, we can expect that critical points will be identified in digital databases. For example, a point's level of importance may be recorded as one of a feature's attributes.

SELECTED REFERENCES

Arnheim, R., "The Perception of Maps," *The American Cartographer*, 3 (1976): 5–10.

Bertin, J., *Semiology of Graphics: Diagrams, Networks, Maps*, W.J. Berg, translator; Madison WI: University of Wisconsin Press, 1983.

Buttenfield, B.P., and R.B. McMaster, *Map Generalization: Making Rules for Knowledge Representation*, Essex, U.K.: Longman Scientific and Technical, 1991.

Court, A., "The Inter-Neighbor Interval," *Yearbook of the Association of Pacific Coast Geographers*, 28 (1966): 180–182.

Cromley, R.G., and G.M. Campbell, "Integrating Quantitative and Qualitative Aspects of Digital Line Simplification," *The Cartographic Journal*, 29, 1992: 25–30.

Dent, B.D., *Cartography: Thematic Map Design*, 2nd ed., Dubuque, IA: William C. Brown Publishers, 1990.

Douglas, D.H., and T.K. Peucker, "Algorithms for the Reduction of the Number of Points Required to Represent a Digitized Line or Its Caricature," *The Canadian Cartographer*, 10 (1973): 112–22.

International Cartographic Association, *Basic Cartography: For Students and Technicians*, Vol. 2, R.W. Anson, (ed.); El Sevier Applied Science Publishers, 1988.

Lam, N.S., and L. DeCola, *Fractals in Geography*, Englewood Cliffs, NJ: Prentice-Hall, 1993.

McMaster, R.B., "The Integration of Simplification and Smoothing Algorithms in Line Generalization," *Cartographica*, 26, 1989: 101–121.

McMaster, R.B. (ed.), "Numerical Generalization in Cartography," *Cartographica*, 26, 1989. (Note that the entire issue is devoted to generalization).

Monmonier, M.S., *Computer Assisted Cartography: Principles and Prospects*, Englewood Cliffs, NJ: Prentice-Hall, 1982.

Robinson, A.H., "Psychological Aspects of Color in Cartography," *International Yearbook of Cartography*, 7 (1967): 50–59.

Salichtchev, K.A., "History and Contemporary Development of Cartographic Generalization," *International Yearbook of Cartography*, 16 (1976): 158–72.

Tobler, W.R., "Numerical Map Generalization," *Cartographica*, 26, 1989: 9–25.

Topfer, F., and W. Pilliwizer, "The Principles of Selection," *The Cartographic Journal*, 3 (1966): 10–16.

Tufte, E.R., *The Visual Display of Quantitative Information*, Cheshire, CT: Graphic Press, 1983.

Wang, Z., and J.C. Muller, "Complex Coastline Generalization," *Cartography and Geographic Information Systems*, 20, 1993: 96–106.

CHAPTER 25

SYMBOLIZATION: FEATURE ATTRIBUTES AT POINTS, LINES, AND AREAS

We are all aware that column after column of numbers relating to points, lines, or areas on the earth look forbidding. A map can present important geographic characteristics in a far more understandable, interesting, and efficient way. The process of creating graphic symbols to represent feature attribute values is part of what we call **symbolization**.

Since tabular data exist in many forms and are available from a wide variety of sources, it isn't surprising that cartographers spend great effort deciding how to symbolize feature attributes. To complicate things further, many feature attribute values are **derived parameters**—that is, generalizations of numerical arrays. When such values are being symbolized, cartographers must proceed warily. Figures can easily *lie*—not only statistically, but also cartographically (Monmonier, 1991). When cartographers symbolize attributes which are derived parameters, two different generalizations have taken place. The first is the parameterization of the attribute value, and the second is the symbolization.

As we saw in Chapter 24, symbolization is a generalization process. We will explore symbolization further in this chapter and in Chapter 26. Let's begin with a short review of what cartographers are trying to accomplish through symbolization.

Features exist in the real world, and the cartographer's task is to portray those features relevant to the map being created. In the digital technological world that is increasingly the standard cartographic environment, the features existing in reality have been digitized. Cartographers must take these digital records and symbolize them so that they communicate reality to map users. In one sense, digital technology has made the cartographers' job harder. No longer can they simply symbolize reality; using digital technology, they must symbolize the digital representation of reality. Thus, cartographers must balance their knowledge of reality with the digital attribute data available for any feature to be shown on the map.

As we learned in Chapter 24, cartographers turn to the symbolization process after they have ap-plied classification, simplification, and exaggeration routines to features selected for mapping. Symbolization is the use of visual variables (see Chapter 18) to represent the data summarizations resulting from classification, simplification, and exaggeration. This graphic coding makes the generalization visible.

Clearly, symbolization is critical to any map's success. Good simplification and classification procedures can be nullified by poor symbolization. On the other hand, good symbolization can enhance the effectiveness of simplification and classification. Unfortunately, good symbolization can also impart an unwarranted impression of accuracy to poorly simplified or classified data.

THE SYMBOLIZATION PROBLEM

Symbolization begins with stylizing, or typifying, attribute values. The symbolization process includes selecting the **level of measurement** (nominal, ordinal, interval, or ratio) to use in the feature's visualization. It also includes choosing the **dimensionality** of the feature (point, line, area, or volume). All features *exist* in a combination of these two aspects. And all features can be *mapped* in a combination of these two aspects. The way features exist and the way they are mapped need not be the same.

Cartographers, therefore, must perform two important symbolization tasks *before* they choose the symbols. They must (1) select (and possibly change) the measurement level of the original data values, and (2) conceptualize the dimension for each feature they wish to portray. Making these two choices is the essence of the cartographic symbolization problem. Let's look more closely at each of these tasks.

MEASUREMENT LEVEL

All attribute values of features existing in reality can be measured on one of the four measurement scales defined in Chapter 16. As we have learned, the four scales of measurement are nested: nominal<ordinal<interval<ratio. In other words, when we symbolize data on a map, we can generalize "down" the measurement scale from ratio to interval, ordinal, or nominal; from interval to ordinal or nominal; or from ordinal to nominal. But we can't generalize "up" the scale. We can't change from nominal to ratio, for example. And we must map nominal data as nominal data; we have no choice.

Most symbols connote only qualitative (nominal) or quantitative (ordinal, interval, or ratio) data. While in measurement theory the distinction between interval and ratio data is important, this distinction makes little or no difference in symbolizing features. The only way we can portray interval or ratio scales is to use textual or numerical annotations to enhance the symbols.

FEATURE DIMENSIONALITY

We think of features as having dimensionality ranging from zero to three dimensions. A point feature is dimensionless (zero dimensions). A line feature has one dimension. An area feature has two dimensions; and a volume feature has three dimensions.

We use point, line, or area marks to represent data associated with point, line, area, and volume features. It would be nice if there were unique symbols for each class of data. Since there aren't, the problem is more interesting. Cartographers have the choice of symbolizing a feature attribute as it exists or by generalizing it.

If they decide to generalize a feature's dimension, cartographers have a number of options. Some of the possibilities actually displace data from their geographic location. For example, cartographers may generalize an area to a point or line when they construct a network (Figure 25.1).

Curiously, cartographers have traditionally not considered losing one or both of an area's two dimensions during symbolization as a cartographic problem. They have, however, considered reducing the number of dimensions from three (globe)

to two (flat surface) as a major cartographic problem. (Indeed, that is the "projection problem" and is the subject of Chapter 5).

In this chapter, we will discuss symbolizing point, line, and area data. Symbolizing volume data is more complex and is the subject of Chapter 26.

VISUAL VARIABLES IN SYMBOLIZATION

As we saw in Chapter 18, there are six primary visual variables—size, orientation, shape, and three aspects of color (value, chroma, and hue). (See Figure 18.2). Each of these primary visual variables inherently connotes a scale of measurement. Likewise, each of these variables has greater usefulness for portraying some feature dimensions than others. Cartographers must learn the level of measurement connoted by each visual variable and the efficiency with which each can symbolize a conceived feature dimension. They can then link their desired conception of the feature to a visual variable and create an acceptable symbol.

Ordering Visual Variables

When cartographers use the primary visual variables of **value (color)**, **size**, and **chroma (color)** in symbols, they create a visual ranking or "ordering." Thus, these three primary variables are called the **ordering variables**.

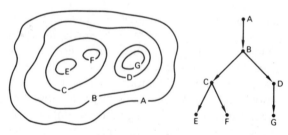

Figure 25.1 An example of generalizing a contour plot to a network. (From *Computer Assisted Cartography: Principles and Prospects*, by M.S. Monmonier, 1982, Prentice-Hall, Englewood Cliffs, NJ. Reprinted by permission of Prentice-Hall, Inc.)

The ordering variables are useful for point, line, or area referenced feature attributes. They can be used to portray feature attributes at the ordinal, interval, or ratio measurement scales. But because of their inherent ordered nature, they should not be used for any nominally scaled attributes.

Perhaps **value (color)** is the most basic of these ordering variables, because it is impossible to have a visualization without variations in value. As illustrated in Figure 19.11, value can be used to display ordinally-scaled feature attributes. By adding numbers to the legend, cartographers can also use value to portray interval or ratio scaled feature attributes.

Size is one of the most useful variables in cartographic symbolization. Size inherently connotes differences in quantities.

Chroma (color) is the most subtle of these three ordering visual variables. It is less visually effective than either value or size. At times it is useful to combine chroma and value and to vary these two visual variables simultaneously to reinforce the magnitudes being symbolized.

Differentiating Visual Variables

In contrast to the above "ordering" visual variables, the remaining three primary visual variables—**hue (color)**, **orientation**, and **shape**—are **differentiating variables**. They can be used to differentiate among feature attributes referencing points, lines, or areas. They should only be used to connote a nominal scale of measurement. They should not be used for higher levels of measurement.

Admittedly, **hue** is sometimes used in the spectral progression to represent heights of land and depths of seas—an example of using hue for a nonnominally scaled attribute. This use of hue is a poor symbolization choice. It is accepted only because it is such a widely learned cartographic convention. Whenever hue is used in this way, the map must be carefully designed and clearly supplemented with numbers and text.

Like hue, **orientation** is best used to show nominally scaled differences. Orientation refers to the angle at which the symbol is placed on the map (see Figure 18.2). Different angles connote qualitative differences.

Shape is also useful in connoting qualitative differences in attributes. It is particularly effective when applied to line symbols. For example, shape variations let us define solid, dashed, and dotted lines so that they are easily distinguishable on the map (see Figure 25.12 below).

Some of the differentiating variables enhance map communication more than others. For example, orientation is more useful for symbolizing points or areas than for lines. Shape is best used for point or line symbolization and not for areas. Hue is useful for all three.

Secondary Visual Variables

As we learned in Chapter 18, there are also three secondary visual variables. These variables are created by repeating the basic graphic elements (marks), producing a graphic effect that we call pattern. These secondary visual variables—**arrangement (pattern)**, **texture (pattern)**, and **orientation (pattern)**, were illustrated in Figure 18.3.

Carefully applied, **texture** can be used as an ordering visual variable. The other two, **arrangement** and **orientation**, are differentiating visual variables. In all cases, cartographers should use these secondary visual variables to enhance a symbolization, not as the only visual variable to order or differentiate a feature attribute value.

SUMMARY OF THE SYMBOLIZATION PROBLEM

We can generalize by symbolization in two ways: (1) We can change the measurement level of feature attributes—that is, we can move from ratio in the direction of nominal (see Figure 18.4). Or (2) we can change the conception of the feature's dimensionality (see Figures 18.2 and 18.3).

Changing measurement level is the most common approach. Generalizing symbols by changing dimensions, as in Figure 25.1, is less common. The reason cartographers use this approach less often is probably related to their overriding concern with displaying "reality" on their maps. However, abstractions derived from a change of dimensions can provide many insights into a distribution's character or structure. Thus, we are seeing a slight trend toward cartographers using such abstractions more often.

After cartographers decide on measurement level and feature dimensionality, they select the visual variables they will use to encode the data. The visual variables themselves connote certain levels of measurement and vary in their usefulness for representing different feature dimensions and levels of measurement. Thus, choosing appropriate visual variables is the crux of cartographic symbolization. The success or failure of a map depends on this process.

Of secondary concern in this case is good design. Unquestionably, good design and execution of the chosen symbolization will enhance any map. Conversely, poor design can block the map's message from reaching the user. However, if the symbolization is incorrectly done, good design can't make the map useful.

We will assume that the cartographer has made the necessary decisions as to the amount of generalization that the symbolization will require. In the rest of this chapter, we will illustrate the use of the visual variables in symbolizing the results of those decisions.

SYMBOLIZING GEOGRAPHIC FEATURES

The wealth of mappable data is astounding. Spatial data tied to earth position is only a fraction of the total amount of mappable data. The medical profession, to name one, increasingly uses spatial images or three-dimensional models of organs and other body parts. The symbolization rules are similar. Likewise, architects, scientific modelers of all types, and engineers who design equipment face the same symbolization problem as does the cartographer.

Clearly, symbolization extends well beyond the boundaries of cartography. But we will limit ourselves here to discussing the symbolization of geographic features. Because of the nature of our environment, we must pay special attention to point, line, and area marks in symbolizing geographic data. Let's begin by seeing how cartographers map features conceived as points.

MAPPING FEATURES CONCEIVED AS POINTS

Many mapped phenomena exist at points, are referenced to points in reality, or are conceived of as points for mapping purposes. Examples abound. At large scales, telephone poles or geodetic control points may be shown as points. At small scales, villages, centers of areas such as counties, or nodes on a network may be mapped as points. Point feature attributes occur on all four scales of measurement (nominal, ordinal, interval, and ratio).

Each such feature has a locational attribute (x,y-coordinate) that specifies its position on the earth. Most features have additional attributes such as height, composition, a name, or a date. However, all mapped data have the location attribute, and we must portray that attribute by position on the map. We normally imply feature location by the positioning of a point symbol portraying some other (nonlocational) attribute of the same feature. Since most features that a cartographer may wish to map are multifaceted (have several attributes in addition to location), it is not unusual for these attributes to have values measured on different scales of measurement. For example, we can conceive of a telephone pole with both a nominally scaled attribute, composition (wood, cement or metal); and with a ratio-scaled attribute, height.

As noted earlier, we will want to symbolize point data differently depending on whether they represent qualitative or quantitative attributes. Let's look first at the problem of representing qualitative point data.

Box 25.A
Mimetic vs. Geometric Symbols

Many cartographic symbols can be constructed using the visual variables and the primary graphic elements. There is a continuum of symbols varying from purely **geometric symbols** (a dot, a line, or a solidly colored area) to highly pictorial symbols such as those shown in Figure 25.2. At some point on the continuum are symbols which resemble the feature attribute being symbolized (Figure 25.A.1). Pictorial symbols or those geometric symbols which we naturally associate with feature attributes are called **mimetic symbols**. Mimetic symbols are desirable on maps when they are unambiguous and easy to understand. But some mimetic symbols are so intricate that easy visual differentiation becomes difficult. Such symbols are not recommended. Figure 25.A.2 is an example.

Other symbols, like solid lines with cross ticks to designate railroads, or the color blue to represent water, are so standard that to violate the convention creates problems. Cartographers are advised to use mimetic symbols with

an awareness of their advantages and shortcomings. By the same token, purely geometric symbols require clear definition, usually in the legend, of the features they represent.

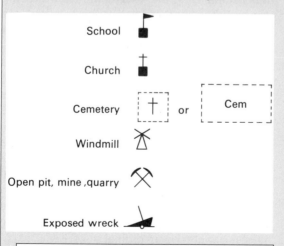

Figure 25.A.1 Mimetic symbols for nominally scaled point features used by the U.S. Geological Survey on topographic maps.

Figure 25.A.2 Detailed mimetic point symbols. (From the legend of a map, "Legend Resource Inventory." Department of Resource Development, State of Wisconsin, May 1964.)

Qualitative Point Symbolization

We saw earlier that when data are available at a nominal (qualitative) scale, we can symbolize them with the **differentiating visual variables**. These variables include **shape**, **hue (color)**, and **orientation**. All these visual variables are useful in portraying point data.

Notice the map in Figure 25.2. Here the visual variable of **shape** is used to form **mimetic** (pictorial) symbols. The illustrated differences are measured on a nominal scale. (See Box 25.A). If the map were printed in color, the visual variable of **hue** would also be a good choice for symbol differentiation.

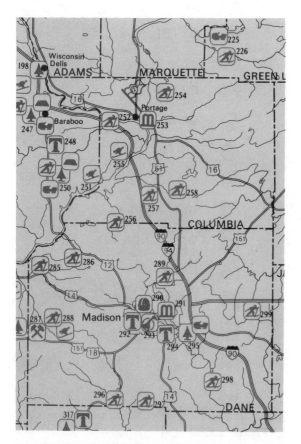

Figure 25.2 Nominally scaled pictorial symbols on a map promoting winter activities in a portion of the state of Wisconsin. The map legend lists 14 symbols. (From a class project, University of Wisconsin-Madison.)

The visual variable of **orientation** can also be used to differentiate nominal scale attributes of point features, as shown in Figure 25.3.

Figure 25.4 uses the visual variable of **shape** to symbolize the same data shown by orientation in Figure 25.3. If you compare Figures 25.3 and 25.4, it is clear that shape is more powerful than orientation for nominal scale point symbolization. Again, **hue** could help differentiate the shapes in Figure 25.4 if the map were in color.

Quantitative Point Symbolization

We saw earlier that when data are available at an ordinal or higher measurement scale, we can symbolize them with the **ordering visual variables**. These variables include **size**, **value (color),** and **chroma (color)**.

For our first example of symbolizing quantitative point data, we will focus on the primary visual variable **size**. We will use data from the Bureau of the Census, showing 1990 population (feature) of cities (feature instance) in northeastern Ohio. The columns in Table 25.1 represent attributes. The data in columns 1 and 2 display attribute values. By choosing a reduced scale for the map, we force ourselves to generalize. We do so by reducing the dimensionality of the geographic feature being mapped. The attribute values (population) are measured on a ratio scale. In reality we know that population covers an area of the earth's surface. However, at the reduced scale of the intended map, we are able to generalize by conceiving of the cities as **points** rather than areas. Point symbols are therefore selected to show the attribute and the attribute values at the location of the cities.

We can generalize still further by changing the measurement scale. Figure 25.5 illustrates mapping the ratio-scaled attribute values for 1990 populations of Ohio cities using point symbols. Figure 25.6 shows range-graded* mapping of the same attribute values. Figure 25.7 illustrates mapping the same values on an ordinal scale.

---

*The term range-graded measurement is used to denote data collected in categories such as age: 0-5 years, 6-10 years,..., 60–65 years, >65 years.

Figure 25.3 Nominally scaled symbols are used to indicate four classes of climatic stations with nonrecording gauges in Wyoming. The attributes of the classes exist on a higher level of measurement, but the depiction itself is nominal until the legend information is used.

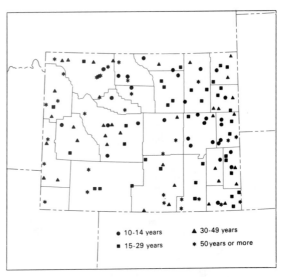

Figure 25.4 The use of the visual variable, shape, to portray the data given in Figure 25.3.

This series of three maps, Figures 25.5, 25.6, and 25.7, shows progressive generalization of attribute values by reducing the point symbols' measurement scales. It is therefore appropriate to state that Figure 25.7 is more generalized than Figure 25.5. Note that Figures 25.6 and 25.7 are visually identical. (See Box 25.B). In fact, all three maps—Figures 25.5, 25.6, and 25.7—could look identical and the appropriate legends would appear as shown in Figure 25.8. Only the legend text would differ and therefore so would the amount of generalization.

Figure 25.9 illustrates the use of value (color) as a primary visual variable to map the same data shown in Figures 25.5, 25.6, and 25.7. It is clear from Figure 25.9 that using size is more visually effective than using value (color).

Table 25.1
1990 Populations of Cities in Northeastern Ohio With More Than 50,000 Inhabitants

City	Population	Percent Black	Log n	Antilog of Log n x 0.5	Antilog of log n x 0.57	Scaled radii for Fig. 25.6	Fig. 25.B.2	Fig. 25.7
Akron	223019	24.51	5.34834	472.25	1118.29	1.05	1.16	1.00
Canton	84161	18.21	4.92511	290.10	641.47	0.64	0.67	0.75
Cleveland	505616	45.56	5.70382	711.07	1783.11	1.58	1.86	2.00
Cleveland Heights	54052	37.10	4.73281	232.49	498.54	0.52	0.52	0.50
Elyria	56746	13.68	4.75394	238.21	512.56	0.53	0.53	0.50
Euclid	54875	15.97	4.73937	234.25	502.85	0.52	0.52	0.50
Lakewood	59718	0.85	4.77610	244.37	527.69	0.54	0.55	0.50
Lorain	71245	13.77	4.85275	266.92	583.54	0.59	0.61	0.75
Mansfield	50627	18.08	4.70438	225.00	480.28	0.50	0.50	0.50
Parma	87876	0.73	4.94387	296.44	657.67	0.66	0.68	0.75
Warren	50793	21.33	4.70580	225.37	481.18	0.50	0.50	0.50
Youngstown	95732	38.11	4.98106	309.41	690.56	0.69	0.72	0.75

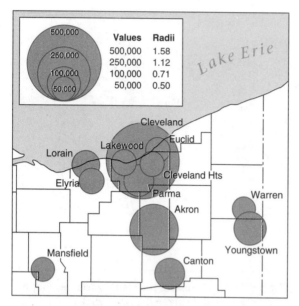

Figure 25.5 The population of some cities in northeastern Ohio. Symbols are proportionally scaled so that areas of the symbols are in the same ratio as the population numbers they represent (See Table 25.1).

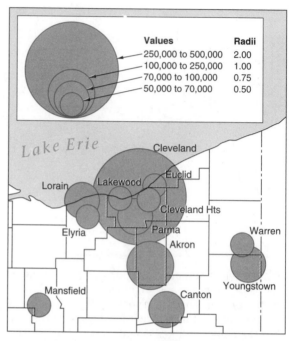

Figure 25.6 The population of some cities in northeastern Ohio. Symbols are range-graded to denote the population of the cities (See Table 25.1).

If we were to use an additional ratio-scaled attribute, percent of black inhabitants (see Table 25.1), we could combine size and value to symbolize the two attributes of the 1990 population. In this portrayal we could use size to show total number of inhabitants and value (color) to show percent of black inhabitants. (See Figure 25.10). Or we could use size to show percent of black inhabitants and value to show total number of inhabitants. (See Figure 25.11). The visualizations are startlingly different. Yet each map is legitimately symbolized. Which map cartographers make will depend on what story they wish to tell.

Figure 25.7 The population of some cities in northeastern Ohio. Symbols are ordinally scaled. Note the similarity with Figure 25.6. (See also Table 25.1). The legends are different due to the different levels of measurement.

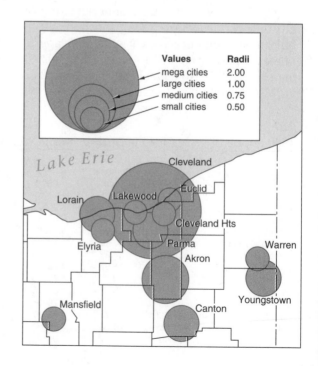

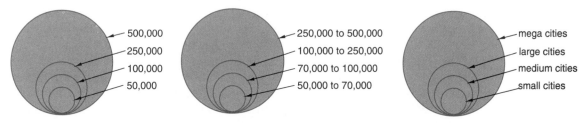

Figure 25.8 Three legends whose symbols are identical. The added information in the form of text puts one legend on an ordinal scale, one on a range-graded scale, and one on a ratio scale.

MAPPING FEATURES CONCEIVED AS LINES

Symbols portraying attributes of features conceived as lines are easy to find on maps. Such symbols include coastlines, rivers, administrative boundaries, railroads, and all kinds of flows or movements between locations. Examples of line symbols that can be used to portray these characteristics are shown in Figure 18.4.

Qualitative Line Symbolization

The primary visual variables of **hue (color)** and **shape** can be used to symbolize nominally scaled attributes of line symbols. **Shape** of line symbols is defined by the illustrations in Figure 25.12. Care must be taken to keep the size of these differing line shapes visually constant so as not to connote ordinal or higher scales of measurement. A combination of **size**, to denote impor-

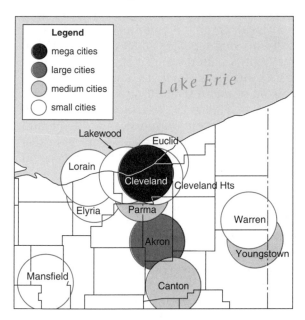

Figure 25.9 The population of some cities in northeastern Ohio. Symbols use the visual variable value (color) to order the data. Note: the map would not change appearance regardless of which level of measurement (see legends in Figure 25.8) is placed on the map.

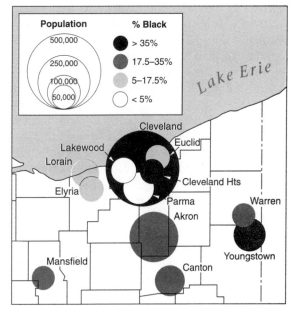

Figure 25.10 Characteristics of the population of some cities in northeastern Ohio. Total population is symbolized by size of the graduated circle symbols. Percentage of black inhabitants is symbolized by the value (color) of the graduated circle symbols.

tance (ordinal scale), and **shape**, to show difference in kind, can be extremely effective in portraying lines. See Figure 25.13.

The graphic element **orientation** presents a problem in symbolizing line features. Lines on the earth tend to have directions of their own, independent of their symbolization. Thus, it is difficult to vary the orientation of most line symbols. However, directions measured at places (for example, ocean currents or wind directions) can be symbolized by arrows, and the orientation of these arrows is usually significant. Such symbols may be considered point symbols rather than line symbols. The distinction between whether a symbol is a point or line symbol becomes rather subjective.

Quantitative Line Symbolization

The primary visual variables of **size** and **value (color)** are used to symbolize features conceived as lines much as we saw them used earlier to symbolize point data. As with point data, size is more effective than value in symbolizing line data. Figure 25.14 shows range-graded flow data, and Figure 25.15 is a ratio portrayal of the same data. In each illustration, the visual variable **size** is used as the ordering visual variable. (In both figures, the visual variable of **hue** is used to show immigrants' nation of origin. Note that this is a qualitative attribute).

In Figure 25.16, **value** is used to enhance size, and the qualitative attribute (nation of origin) is not symbolized. Comparing Figures 25.14, 25.15, and 25.16, it is evident that size is more visually prominent than value. If color were available, we could use **chroma** to enhance these flow maps. However, we would probably want to use chroma only to assist the primary use of size.

MAPPING FEATURES CONCEIVED AS AREAS

From climatology and soil science come examples of attribute values collected at points but conceived of as referring to areas and mapped by area symbols. For instance, the soils map shown in Figure 21.1 depicts areas of differing soil types based on an attribute value collected from a sample core. Likewise, the map in Figure 25.17, showing air masses over North America, is derived from measurements of attribute values at weather stations. These attribute values are measured on a ratio scale but generalized for visualization to a nominal scale.

The primary visual variable of size is far less effective when mapping features conceived as areas than when mapping features conceived as points or lines. However, the secondary visual variables associated with pattern (texture, arrangement, and orientation) are of considerably more importance in mapping features conceived as areas (as was illustrated in Chapter 21).

Qualitative Area Symbolization

Considering only nominally scaled attribute values, we can create effective area symbols using the visual variable of **shape** (which creates a pattern when repeated in an area). Some mimetic area patterns are so standard that they must be used only for that feature. For example, see the standard symbolizations for landforms shown in Figure 25.18. Likewise, we are all familiar with standardized swamp, desert, and forest symbols. (These are shown in Figure 18.4 in the cell representing symbols that have areal dimension and represent nominal data).

The secondary visual variables are also useful for nominal differentiation, especially if standardized patterns are available. (See Figure 21.2). Although **orientation (pattern)** is too subtle when used alone, it can be effectively combined with **arrangement (pattern)**, as shown in Figure 18.10.

When working with black and white in analog technology, be sure to check for standard symbolization schemes. Many disciplines have created such standardized ways of showing area patterns, and to violate these schemes would cause great confusion.

Fortunately, with digital technology and visual-

Box 25.B
Scaling Graduated Symbols

Point symbols have been scaled by symbol size for many years. When graduated circles were first used to symbolize attribute values measured on an interval or ratio scale, the sizes of the symbols were made proportional to the numbers they represented. Thus, if two features have values in a 1:2 ratio (that is, the magnitude of the second is twice that of the first), the second circle is constructed so that its area is twice that of the first. This has become known as the **square root method of circle symbol scaling**. The name comes from the fact that the area of a circle is πr^2. Therefore, scaling the radius for construction requires extracting the square root of r, in the equation $r = (area/\pi)^{1/2}$. Figure 25.5 is an example of using the square root method of circle symbol scaling.

Research into perception of cartographic symbols demonstrates that people's response to differences between graduated circle sym-

bol areas is not a linear function. Instead, the ordinary observer will *underestimate* sizes of larger symbols in relation to smaller ones. For example, if two quantities were in a 1:2 ratio, and the areas of two circle symbols representing them were in the same ratio, a map reader would think the larger circle was significantly *less* than twice the size of the first. When we make the areas of circle symbols strictly proportional to the numbers they represent, therefore, we reduce the apparent sizes of the larger circles in relation to the smaller. This is unfortunate, since the purpose of such maps is to give a realistic impression of the mapped distribution.

We may address this problem by scaling circle symbols to reflect the amount of expected underestimation by map users. This is called the **psychological scaling method**. Figure 25.B.1 shows the difference in sizes of two sets of circles constructed by the two

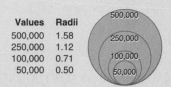

Values	Radii
500,000	1.58
250,000	1.12
100,000	0.71
50,000	0.50

Legend for 25.6.

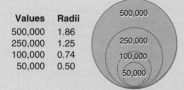

Values	Radii
500,000	1.86
250,000	1.25
100,000	0.74
50,000	0.50

Legend for 25.B.2.

Figure 25.B.1 Two nested sets of graduated circles prepared from the same data show the difference between (left) circles scaled in proportional relation to area (according to square roots), and (right) circles psychologically scaled to compensate for underestimation.

continued on next page

Box 25.B (continued)

methods. Figure 25.B.2 illustrates the psychological scaling method applied to the city data for northeastern Ohio. Table 25.1 gives the construction data for Figure 25.B.2.

Research indicates that map users do not have the same problem estimating linear symbol dimensions that they do with circles. Therefore, no compensation is needed for judging the width of linear symbols (for example, flow lines).

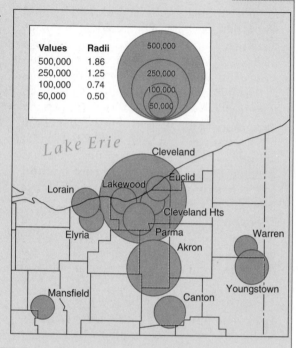

Figure 25.B.2 The population of some cities in northeastern Ohio. Symbols are psychologically scaled so that areas of symbols visually connote correct ratios between the population numbers they represent (see Table 25.1).

izations on color monitors, **hue** can be effectively used. When color is available, hue is the most effective visual variable to use to differentiate features conceived as areas. Attention should be paid to the effects of hue, as outlined in Chapter 21. The appropriate use of color on monitors is a topic that should be taught in any basic cartographic course.

Quantitative Area Symbolization

The visual variable of **value (color)** is effective for symbolizing quantities conceived of as relating to areas. When value is used in this way, a stepped surface or dasymetric map (see Chapter 26) is created. (See Figure 25.19). A more complete discussion of using value in symbolizing areas will be found in Chapter 26.

The visual variable of **chroma (color)** can also be used to show areal quantities, but it is effective only on color maps or color monitors, while value (color) is effective on black-and-white as well as color maps.

SUMMARY OF SYMBOLIZING GEOGRAPHIC FEATURES

We can summarize symbolization of geographic features in a two-dimensional matrix—in terms of (1) measurement levels and (2) feature dimensions. (See Table 25.2).

First, let's look at the use of visual variables with respect to the four **measurement levels** (nominal, ordinal, interval, and ratio). For nominal scale attributes, the visual variables of hue (color), shape,

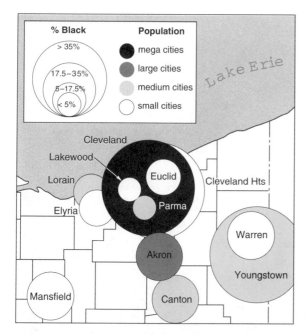

Figure 25.11 Characteristics of the population of some cities in northeastern Ohio. Percentage of black inhabitants is symbolized by size of the graduated circle symbols. Total population is symbolized by the value (color) of the graduated circle symbols. Compare to Figure 25.10.

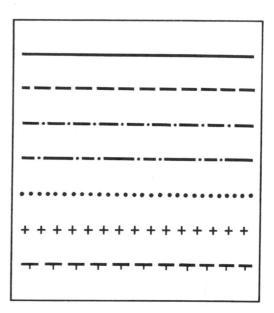

Figure 25.12 Examples of lines of differing character (the visual variable shape) which are useful for the symbolization of nominal linear data.

and orientation, along with the secondary visual variables of pattern (texture, arrangement, and orientation), may be used. Shape can be used for all feature dimensions, as can hue if color is available. In general, orientation is less useful to the cartographer than these other visual variables. For features conceived as areas, the secondary visual variables can be used effectively. For ordinal, interval, or ratio scale attributes, the visual variables of value (color), chroma (color), and size are used.

Second, let's look at the use of visual variables with respect to **symbol dimensions** (points, lines, areas, and volumes). For quantitative features conceived as points or lines, size is the most important visual variable. Value (color) can be used for quantitative features conceived

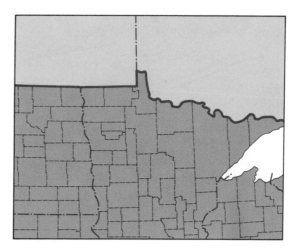

Figure 25.13 The use of line width (visual variable size) enhanced by the use of line character (visual variable shape) to denote the ordinal portrayal of civil administrative boundaries.

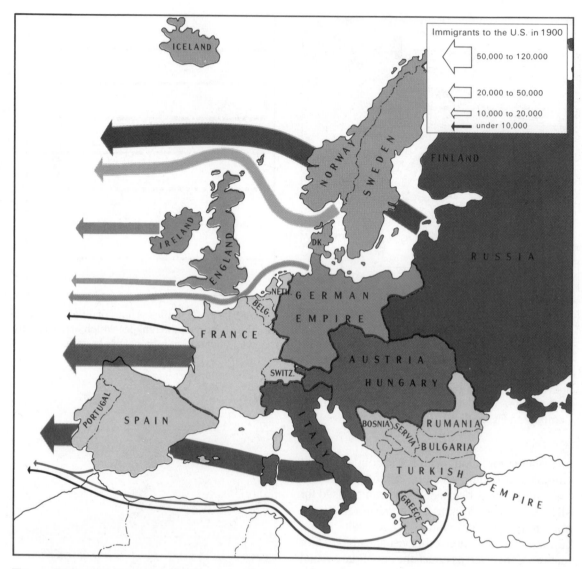

Figure 25.14 Range-graded line symbols. On this map of immigrants from Europe in 1900, lines of standardized width are used to represent a specified range of numbers of immigrants.

as points, lines, areas, or volumes, but is more effective in areas or volumes. Chroma (color) is often too subtle to be used except with areas or volumes.

As noted earlier, symbolizing volume data is more complex than symbolizing points, lines, or areas, and we need to examine it in more detail. We will do so in the next chapter.

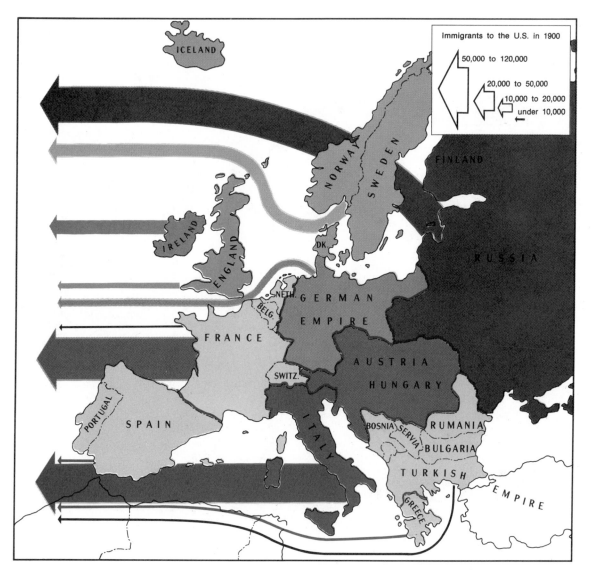

Figure 25.15 Ratio-scaled line symbols. On this map, which shows the same data as Figure 25.14, lines are scaled proportional in width to number of immigrants.

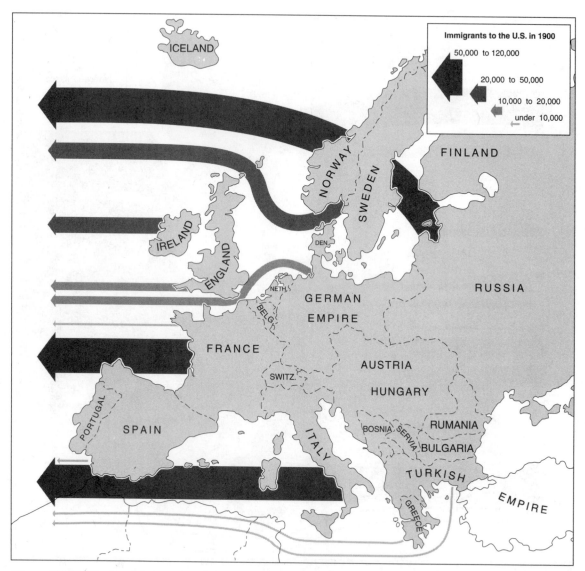

Figure 25.16 Ratio-scaled line symbols using size as the primary visual variable with enhancement provided by value (color). The lines are scaled proportional in width to number of immigrants.

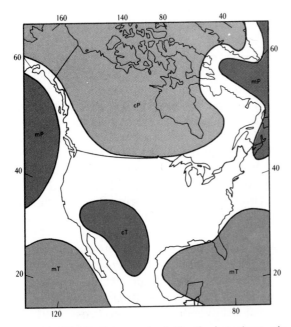

Figure 25.17 Portrayal of North American air masses and their source regions. Although data have quantitative characteristics, the intent of this illustration is simply to portray location of air masses. This can be accomplished by using nominal areal symbolization. (From *Fundamentals of Physical Geography* by Trewartha et al. Copyright 1977, McGraw-Hill Book Co. Used with permission of McGraw-Hill Book Co.)

STRUCTURE CONTROLLED LANDFORMS		LANDFORMS OF DEPOSITION	
	sandstone	blocks	
	quartzite	gravel, shingles	
	limestone	sand	
	dolomite	silt, mud	
	marl	clay	
	extrusive rocks	salt	
	crystalline rocks	gypsum	
	unconsolidated rocks	coral detritus	

Figure 25.18 Some standardized symbols for indicating lithologic data as suggested by the International Geographical Union Commission on Applied Geomorphology.

Table 25.2
Appropriate Uses of the Visual Variables for Symbolization

	Level of Measurement	
	Nominal	Ordinal/Interval/Ratio
Feature Dimension	Qualitative	Quantitative
Point	hue (color)	size
	shape	value (color)
	orientation	*chroma (color)*
Line	hue (color)	size
	shape	*value (color)*
	orientation	*chroma (color)*
Area	hue (color)	value (color)
	shape	chroma (color)
	pattern	*size*
	orientation	
Volume	hue (color)	value (color)
	shape	chroma (color)
	pattern	*size*
	orientation	

The visual variable in *italics* are of secondary importance.

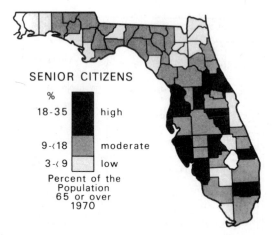

SENIOR CITIZENS

%		
18-35		high
9-‹18		moderate
3-‹ 9		low

Percent of the
Population
65 or over
1970

Figure 25.19   Map illustrating the range-graded classification of Florida counties. The use of the visual variable value (color) creates a stepped surface. (Courtesy, J.M. Olson.)

SELECTED REFERENCES

Arnheim, R., *Art and Visual Perception*, Berkeley: University of California Press, 1974.

Bertin, J., *Semiology of Graphics: Diagrams, Networks, Maps*, W.J. Berg, translator; Madison, WI: University of Wisconsin Press, 1983.

Dent, B. D., *Cartography: Thematic Map Design*, 2nd ed., Dubuque IA: William C. Brown Publishers, 1990.

DiBase, D., A. M. MacEachren, J.B. Krygier, and C. Reeves, "Animation and the Role of Map Design in Scientific Visualization," *Cartography and Geographic Information Systems*, 19, 1992: 201–214, 265–266.

Gill, G. A., "Experiments in the Ordered Perception of Coloured Cartographic Line Symbols," *Cartographica*, 25, 1988: 36–49.

Keates, J. S., *Cartographic Design and Map Production*, 2nd ed., Essex, U.K.: Longman Scientific and Technical, 1989.

Lloyd, R., and Steinke, T., "Comparison of Quantitative Point Symbols: The Cognitive Process," *Cartographica*, 22, 1985: 59–77.

McGranaghan, M., "Ordering Choropleth Map Symbols: The Effect of Background," *The American Cartographer*, 16, 1989: 279–285.

Monmonier, M. S., *Computer Assisted Cartography: Principles and Prospects*, Englewood Cliffs, NJ: Prentice-Hall, 1982.

Monmonier, M. S., *How to Lie with Maps*, Chicago: University of Chicago Press, 1991.

Muller, J. C., and Z. Wang, "A Knowledge Based System for Cartographic Symbol Design," *The Cartographic Journal*, 27, 1990: 24–30.

Robinson, A. H., and B. B. Petchenik, *The Nature of Maps: Essays Toward Understanding Maps and Mapping*, Chicago: University of Chicago Press, 1976.

Sorrell, P., "Optimal Mapping and the Determination of Cartographic Design Principles," *The Cartographic Journal*, 25, 1988: 128–138.

Wood D., and Fels, J., "Designs on Signs/Myth and Meaning in Maps," *Cartographica*, 23, 1986: 54–103.

SYMBOLIZATION: FEATURE ATTRIBUTE VOLUMES

*A*s we saw in Chapter 25, symbolizing point, line, and area data is rather straightforward. Mapping volume (or continuous surface) data is more complex, because volumes can occur at points (dollars in a bank), along lines (barrels of oil imported from Bahrain), or over areas (precipitation) and can be symbolized by point, line, or area marks. Whenever a volume occurs over an area, a statistical surface exists. In this chapter, we will detail the mapping of statistical surfaces.

This chapter is organized differently from Chapter 25. In this chapter, we are concentrating on symbolization methods that have been perfected over time to solve specific mapping problems. The situation addressed in this chapter is one in which the cartographer has already decided the feature dimensionality (volume) and level of measurement (ordinal, interval, or ratio). We therefore can use the primary visual variables of size, value (color), or chroma (color), and the secondary visual variable of texture (pattern). Additionally we have the ability to repeat symbols to connote volumes.

Geographic volumes are of recurring concern to those who use maps. A geographic volume, with its continuous surface, results from ordinal, interval, or ratio scaled attribute values referenced to a geographic area. Enumerations which refer to geographic areas are made, for example, by the Bureau of the Census. Cartographers aim to symbolize the areal distribution pattern of those enumerated attributes, or to give some impression of relative densities of the attribute in different parts of the region.

When symbolizing ordinal, interval, or ratio scaled attribute values referenced to areas, cartographers must first decide whether to use point, line, or area symbols. If they select point symbols, they may show the distribution by using the dot mapping technique. (See Figure 26.1). If they choose line symbols, either hachures or the isarithmic technique are possibilities. If they pick area symbols, two techniques are available: the choroplethic or the dasymetric. We will discuss all these techniques in detail in this chapter.

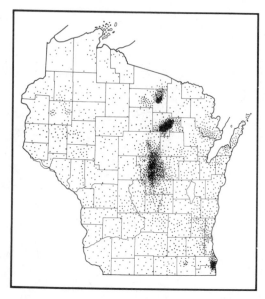

Figure 26.1 A well-rendered dot map in which each dot represents 16.2 hectares of land in potato production in Wisconsin in 1947.

CONCEPT OF A STATISTICAL SURFACE

The **statistical surface** is one of the most important concepts in cartography. A statistical surface exists for any distribution that is mathematically continuous over an area and is measured on an ordinal, interval, or ratio scale of measurement. These requirements form a geographic volume. But we are not as interested in mapping that volume as in portraying its surface shell. We refer to mapping undulations of the surface of a geographic volume as **mapping the statistical surface**.

It is possible to assume a statistical surface from attribute values recorded at points, along lines, or for areas, and to portray that surface with point, line, or area symbols. You can think of the magnitudes of the attribute values, referenced to unit areas, lines, or points, as having a relative vertical dimension. You can thus visualize these magnitudes as forming a three-dimensional surface. For example, Figure 26.2 shows an array of attribute

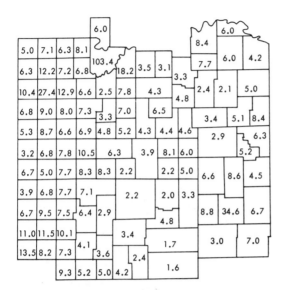

Figure 26.2 An array of attribute values for unit areas. The numbers are rural population densities for minor civil divisions in a part of Kansas. (Courtesy of G.F. Jenks and *Annals of the Association of American Geographers*).

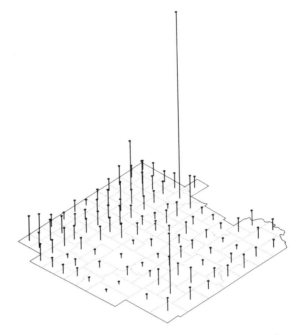

Figure 26.3 Elevated points whose relative height above the datum for each unit area is proportional to the attribute value in that unit area in Figure 26.2. (Courtesy of G.F. Jenks and *Annals of the Association of American Geographers*).

values that are a set of ratios (in this case population densities). We may arbitrarily assign each attribute value a location at the center of an enumeration unit area.

Figure 26.3 shows the erection of a column at the center of each unit area. The length of the column is proportional to the attribute value in that unit area in Figure 26.2. In Figure 26.4 the relative magnitudes of the attribute values in the unit areas are emphasized by erecting a prism over each area. This prism has a base shaped like the unit area and a height scaled the same as the length of the line in Figure 26.3. In a very real sense, Figure 26.4 is a three-dimensional form of the familiar two-dimensional histogram.

Figure 26.5 illustrates another way of showing the same statistical surface. Here the emphasis is on the magnitudes and directions of gradients on the statistical surface. Figure 26.5 consists of a series of equally spaced profiles across the interpolated shell of the statistical surface. As you can see, it gives a vivid picture of the statistical surface.

Figure 26.6 presents still another method, using line symbols to portray the statistical surface shell. Figure 26.6 is an isarithmic map of the data in Figure 26.2.

Figures 26.2 through 26.6 can all be thought of as representations of a statistical surface consisting of a set of points, each with an x, y, and z characteristic. The x and y values refer to the horizontal, or planimetric, locations. The z values refer to the assumed relative heights above some horizontal datum. Figures 26.3, 26.5, and 26.6 are symbolizations of the attribute values using line symbols. In contrast, Figure 26.4 is the symbolization of a statistical surface by varying only heights of area units.

The concept of a statistical surface is important because its visualization allows us to readily perform cartographic induction. For example, we can

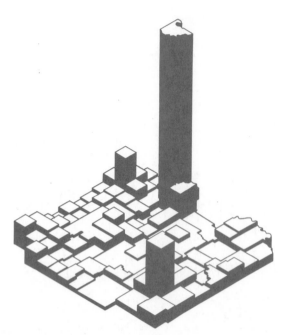

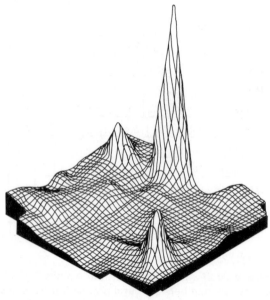

Figure 26.4 A perspective view of the statistical surface produced by erecting prisms over each unit area proportional in height to the attribute values shown in Figure 26.2. (Courtesy of G.F. Jenks and *Annals of the Association of American Geographers*).

Figure 26.5 A perspective view of a smoothed statistical surface produced by assuming gradients among all the attribute values shown in Figure 26.2. The elevated points in Figure 26.3 would all touch the surface shown here. (Courtesy of G.F. Jenks and *Annals of the Association of American Geographers*).

use it to derive many kinds of gradients, such as land surface slope and temperature variation. We can derive other complex data from concepts such as population or economic potentials. Such attributes may be thought of as **spatial vector quantities**. That is, at each point of the distribution on the map there is a direction, a magnitude, and a meaning.

Figure 26.7 shows the same statistical surface as Figures 26.2 through 26.6. This figure uses area symbols and the visual variable of value (color) to portray the same attribute values as Figure 26.2. Compare the visual impression of Figure 26.7 to the other portrayals of the same attribute value set. Note especially Figures 26.2 and 26.4. With practice, the portrayal in Figure 26.7 can become meaningful to a map reader, conveying patterns of areas of high or low attribute values. Figures 26.3, 26.4,

and 26.5 all give the visual impression of three dimensions. Figures 26.2, 26.6, and 26.7 do not. Planimetric position is more easily ascertained in Figures 26.2, 26.6, and 26.7, while the other three figures concentrate on giving an overall visual impression of the statistical surface. In the digital technological world of today, planimetric position is becoming less important than the visual impression presented. The precise planimetry is available in the digital database, if needed, not on the visualization.

In the remainder of this chapter, we will discuss the four main methods for portraying feature attribute value volumes. These four methods are the **dot mapping** method (Figure 26.1), the **isarithmic** method (Figure 26.6), the **choroplethic** method (Figure 26.7), and the **dasymetric** method (Figure 26.32).

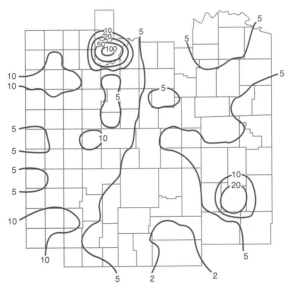

Figure 26.6 An isarithmic map of the statistical surface produced by assuming gradients among all the attribute values shown in Figure 26.2.

MAPPING THE STATISTICAL SURFACE WITH POINT SYMBOLS

For decades, cartographers have used the dot mapping method. With this technique, they repeat uniform point symbols to represent the value of an attribute referenced to an area. They equate each point symbol to a set number of feature instances in the distribution being mapped. Then, through the repetition and placement of these point symbols, they show the nature of the distribution in the mapped area. This repetition of uniform symbols contrasts with proportional or graduated symbols (see Box 25.B in Chapter 25) in which the size of point symbols is varied to show different magnitudes.

The dot map can show details of the geographic character of a distribution more clearly than any other type of map. Variations in pattern, such as linearity and clustering, become apparent. The dot map provides an easily understood visual impression of relative density, eas-

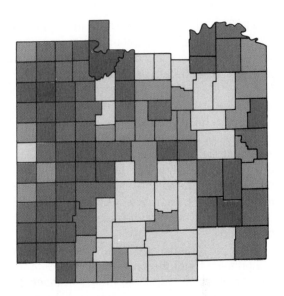

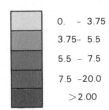

	0. – 3.75
	3.75– 5.5
	5.5 – 7.5
	7.5 –20.0
	>2.00

Figure 26.7 A choropleth representation of the data given in Figure 26.2 using five range-graded categories of the visual variable of value (color) to portray the data.

ily interpreted on an ordinal scale, but it does not provide any absolute figures.

Dot maps ordinarily show only one fact or attribute, such as population or hectares of cultivated land. But, by using different colored dots or different shaped point symbols, it is sometimes possible to include several different attribute distributions on the same map. (For examples of multivariate mapping, see Chapter 28).

DOT MAP

The data needed for a dot map consist of a count of feature instances for a set of unit areas. Civil divisions, such as states or counties, are commonly used as enumeration units. Rarely is it possible to show every single feature instance with a dot. Instead, we usually assign a number of feature instances to each dot. This number is called the **unit value** of the dot. To decide how many dots to place in each unit area on the map, we divide the total number of enumerated feature instances in each statistical division by the unit value. For example, if a county has a total of 6,000 hectares in corn, and we have chosen a unit value of 25 hectares per dot, then we would place 240 dots in the county to symbolize the area devoted to corn production.

Often the boundaries of enumeration areas are the primary, if not the only, additional data that cartographers have available. In such instances the cartographers must be sure that these boundary lines aren't left free of dots. If there aren't enough dots in these boundary line locations, they may show up markedly in the final map as white lines.

The dot map is best used for distributions that have distinct internal arrangements, such as linearity, or are tightly clustered and dense in one area and sparse in other areas. Attributes with uniform distribution characteristics are not well mapped by the dot technique.

Three things determine a dot map's success: (1) size of dots, (2) unit value assigned to a dot, and (3) location of the dots. Let's look at each of these factors.

CHOOSING DOT SIZE AND UNIT VALUE

If a dot map is to convey a realistic impression, dot size and value must be carefully chosen. (See Box 26.A). Figures 26.1, 26.8, 26.9, 26.10, and 26.11 have been prepared from the same data; only the size or number of dots used has been changed. The maps show areas of potato production in Wisconsin.

If the dots are too small, as in Figure 26.8, the distribution will appear sparse and insignificant, and patterns will not be visible. If the dots are too large, as in Figure 26.9, they will give an impression of excessive density that is equally erroneous. Note

how the dots coalesce too much in the darker areas of Figure 26.9. It's as if there is little room for anything but potatoes to grow in the region. Furthermore, when dots are so gross, they visually dominate the other information and result in an ugly map.

As important as choosing dot size is choosing a unit value for the dot. The total number of dots shouldn't be so large (small unit value per dot) that the map gives an unwarranted impression of accuracy. Nor should the total be so small (large unit value per dot) that the distribution lacks any pattern or character. These unfortunate possibilities are illustrated in Figures 26.10 and 26.11.

Now refer back to Figure 26.1. It is constructed from the same data as Figures 26.8, 26.9, 26.10, and 26.11. But its dot size and unit value were more wisely chosen.

CHOOSING DOT LOCATION

In addition to choosing a size and unit value for the dot, we need to choose the location of dots on the map. Theoretically, an ideal dot map would have a scale large enough so that each feature instance could be located. It is usually necessary, however, to make the unit value of the dot greater than one. The problem then is to locate the one point symbol that represents several differently located feature instances (see Figure 24.4).

When using manual methods to generalize, it is helpful to consider the grouping of several feature instances as having a kind of center of gravity. Then you can try to place the point symbol at that center. (See Figure 24.3). Usually, however, all you know is the total number of feature instances within each enumeration area and (once you select a unit value) the number of dots to be shown within that unit area. Consequently, you must use every available source of information to help you place the dots as reasonably as possible. Such sources may be topographic maps, remotely sensed images, and other distribution maps that correlate with the map you are preparing.

You may also produce dot maps by machine. Figure 26.12 was prepared by one of the first dot

Box 26.A
Determining Size and Unit Value of the Dot

Professor J. Ross Mackay developed a nomograph to help determine the desirable dot size and unit value. His graph (with metric additions) is shown in Figure 26.A.1.

Assume that a county contains 6,000 hectares of corn. We arbitrarily choose a unit value of 25 hectares per dot. As a result, we place 240 dots within the county boundaries.

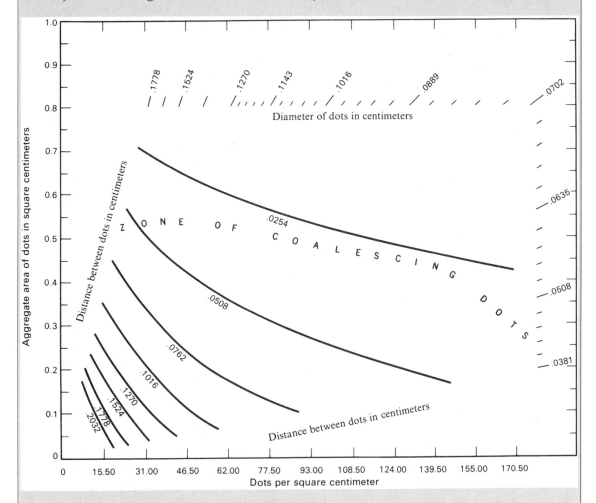

Figure 26.A.1 A nomograph showing the relationship between dot size and dot density. (Nomograph courtesy of J.R. Mackay and *Surveying and Mapping*, with metric notations added by author).

continued on next page

Box 26.A (continued)

Assume further that the county covers 4.0 cm² on the map. This means that we place the dots on the map with a density of 60 dots per square centimeter. An ordinate at 60 is erected from the X axis of the nomograph. A radial line from the origin of the nomograph to a given dot diameter on the upper-right circular scale will intersect the ordinate. The location of the intersection on the interior scale will show the average distance between the dots if they were evenly spaced. The height of the intersect on the Y axis indicates what proportion of the area will be black if that dot diameter and number of dots per square centimeter were used.

Also shown is the "zone of coalescing dots," at or beyond which dots will fall on one another. Ideally, the densest areas on the map should have dots that just begin to coalesce. If the initial trial seems unsatisfactory, we may change either the unit value or the dot size, or both, and plot a new ordinate or radial line.

mapping algorithms to be developed. The resulting map varies little from manually produced dot maps, but the savings in time and cost were large. Today dot maps are predominantly done by machine.

The advantages of producing dot maps by machine are speed and ease of execution. You can create several maps using different input parameters (such as one dot represents 10, 50,

Figure 26.8 A dot map in which the individual dots are so small that an unrevealing map is produced. Each dot represents 16.2 hectares in potatoes in 1947.

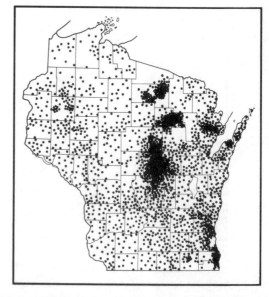

Figure 26.9 A dot map in which the individual dots are so large that an excessively "heavy" map is produced, giving an erroneous impression of abundant potato production. The same data and number of dots are used as in Figure 26.8.

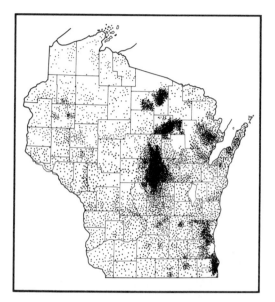

Figure 26.10 A dot map in which the unit value assigned to one dot is too small. Many dots must be drawn, causing an excessively detailed map. The dots are the same size as those in Figure 26.11. Each dot in this figure represents 6.07 hectares of potatoes.

Figure 26.11 A dot map in which the unit value assigned to one dot is too large. Few dots can be drawn, resulting in a barren map that reveals little pattern. Each dot in this figure represents 67 hectares of potatoes.

or 100 units). You can try several different dot sizes and choose the one that works best on the map you are making.

There are also useful algorithms available, which you should use when digitally preparing a dot map. These algorithms include statistical routines with varying probabilities for placing a dot at all possible map positions. You can use these algorithms to obtain initial placement of the dots. Then you can modify the placement, as shown on the computer screen, until you obtain the result you want.

Perceptual research shows that map readers have a slight tendency to underestimate the number of dots and differences between dot densities from one area to another. Thus, when preparing dot maps, you may want to add a legend note warning map users of potential problems associated with density estimation and comparison.

MAPPING THE STATISTICAL SURFACE WITH LINE SYMBOLS

The character of the statistical surface can be portrayed by four types of line symbols: hachures, profiles, oblique traces, and isarithms.

Profiles, oblique traces, and isarithms are conceptually similar. They all are obtained by intersecting an assumed surface shell with a series of parallel planes and mapping the lines of intersections. With all three of these techniques, we assume a series of parallel (usually but not always equally spaced) planes intersecting the statistical surface. The intersections define lines projected onto the datum of a map. The three results are as follows:

1. If the series of parallel planes intersecting the datum is at a right angle to the datum, the lines of intersection show **profiles** across the surface.

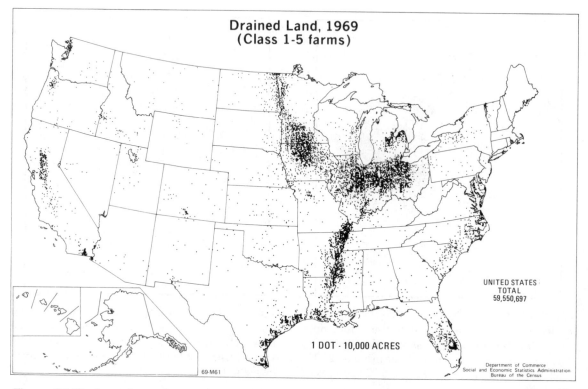

Figure 26.12 An early machine-produced dot map. (From Census of Agriculture, 1969, Vol. V, *Special Reports*, Part 15, Graphic Summary, p. 55. Washington, DC: U.S. Government Printing Office, 1973.)

2. If the series of parallel planes is situated so that the angle between the planes and the datum intersects between 0 and 90 degrees, the lines of intersection represent **oblique traces**.

3. If the series of parallel planes is parallel to the datum, and the intersection lines are orthogonally projected onto the map datum, a series of **isarithmic** lines results.

The fourth category of line symbols, **hachures**, are different from the other three. Hachures are short line symbols whose width or spacing depends upon the slope at a point on the statistical surface.

Let's look at all four types of line symbols in more detail, beginning with hachures.

HACHURES

When we consider the slopes on a statistical surface as attribute values existing at points, we may portray the surface with a line symbol called a **hachure**. Two options are available. In one, the short parallel hachures are equally spaced, and their thickness is varied to represent slope steepness (Figure 26.13). In the other variation, hachures are of even thickness, but their spacing is varied to represent slope steepness. The closer the spacing, the steeper the slope (Figure 26.14). These forms of line symbolization of a statistical surface are called **hachuring** (see Chapter 27), and have been traditionally used to depict the land surface.

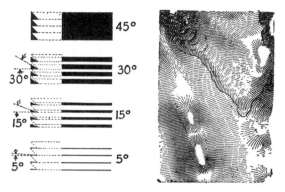

Figure 26.13 Line symbolization of slopes or gradients by using hachures. In this diagram, hachures are equally spaced but vary in thickness depending on the value of the gradient being portrayed (Lehman's system). (Used with permission of McGraw-Hill Book Co.).

PROFILES

Another way to portray a statistical surface is to use profiles. A **profile trace** results from the intersection of a plane perpendicular to the x, y datum and the statistical surface. A series of profile traces

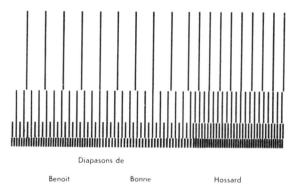

Diapasons de

Benoit   Bonne   Hossard

Figure 26.14 Line symbolization of slopes or gradients by using hachures. In this version, hachures are of equal thickness but vary in spacing depending on the value of the gradient being portrayed. Three systems are shown: Benoit, Bonne, and Hossard. The lines represent 5°, 15°, 25°, and 40° slopes. In these systems, hachure length is also significant. (From *Cartographie Generale*: Tome 1 by R. Cuenin, Copyright 1972, Editions Eyrolles, Paris, France).

placed parallel to one another allows the practiced observer to visualize a surface quite well. Figure 26.15 represents a series of profiles constructed along perpendicular directions. These surface profiles also form the basis for constructing an isometric block diagram. (See Chapter 27).

The steps to construct a profile are shown in Figure 26.16. You begin with an isarithmic map. On this map, mark the line along which you wish to construct a profile. On a blank sheet of paper, transfer the end points of the profile and the intersection of each isarithmic line with the profile line. Number a set of parallel ruled lines to include the isarithmic values of each line crossing the profile on the map. Mark the appropriate altitudes on the parallel lines (depending on the isarithmic line value). Then connect these altitudes by a smooth line on the ruled parallel line set.

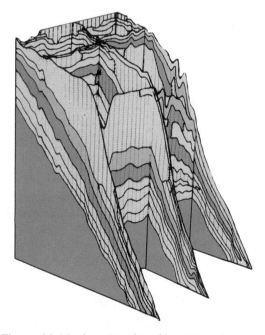

Figure 26.15 A series of profiles situated at right angles to one another gives a clear picture of the surface and its underlying strata. (Courtesy of John Wiley and Sons and J.W. Harbaugh and D.F. Merriam).

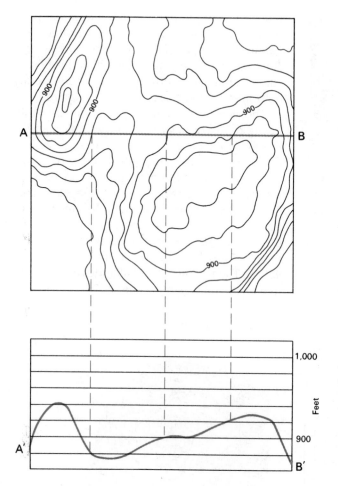

Figure 26.16 The construction of a profile from an isarithmic map. Line AB is drawn on the map. Intersections of the isarithms with⁰ line AB are perpendicularly projected to the appropriate line of a set of parallel ruled lines below. The resulting points are connected by a smooth line resulting in a profile A'B'.

OBLIQUE TRACES

Oblique traces result from the intersection of a series of planes with the base datum at some angle θ, where θ lies between 0° and 90°. These traces may be portrayed graphically in correct planimetric position or in perspective. Let's look at both types of oblique traces.

Planimetrically Correct Oblique Traces

A vertical profile across the statistical surface, when viewed from the side and placed on a map, uses planimetric position in two dimensions. Yet the profile's top and base are actually in the same planimetric position on the datum. It is impossible,

therefore, to use planimetric position to express vertical dimension without producing planimetric displacement. In order to overcome this problem, you may decide not to use the usual vertical or perspective profile across the surface. Instead, you may use a line that gives a similar appearance but is not out of place planimetrically. You may accomplish this cartographic legerdemain by mapping the traces of the intersections of the assumed surface with a series of parallel inclined planes (Figure 26.17). Constructing an inclined trace is not difficult. You can do so either manually or with computer assistance. (Robinson and Thrower, 1957; Yoeli, 1976).

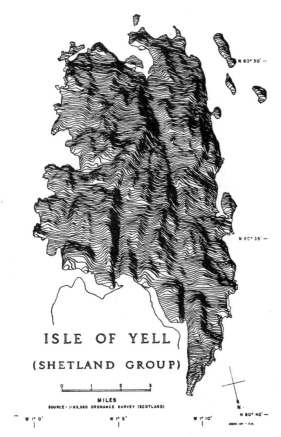

Figure 26.17 A reduced portion of a planimetrically correct terrain drawing of the Isle of Yell (Shetland Group). (Map drawn by P. Nicklin. Courtesy of Dr. N.J.W. Thrower).

Perspective Traces

Calculating and plotting **perspective traces** was one of the first cartographic methods to be computerized. Computer programs allow the cartographer to specify the following (see Figure 26.18):

- the rotation, θ, about the vertical axis of the statistical surface
- changes in the viewing distance, d (the distance of the viewer's eye from the corner of the map)
- changes in the viewing elevation, φ (the angle of the viewer above the horizon).

You can use the computer to draw perspective traces in either the x or y direction, or both. When the traces are drawn along two perpendicular directions, they form a surface grid, or "fishnet." This fishnet representation gives a realistic impression of surface form (see Figure 26.19A). Drawing the perspective traces along only the x or y direction yields a less dramatic but still effective surface representation (see Figure 26.19B and C). With any of these perspective trace methods, you can use either one-point or two-point perspective (see Figure 26.20).

For visualizations produced in color on a computer monitor, realistic portrayals using rasters of small pixel size are commonplace today. The ease of production and realistic impression produced by these methods insure their continued use in the future.

ISARITHMIC MAPPING

Symbolizing real or abstract three-dimensional surfaces is difficult. More time and effort have probably been devoted to this than to all other symbolization problems put together. The **isarithmic** method can be helpful in meeting this challenge.

There are two times when the isarithmic method is especially helpful. First, we use it when our interest in a geographic distribution is focused primarily on the form of the outside surface enclosing the geographic volume. (That is, we're interested in the arrangement of the magnitudes and the steepness and orientation of surface gradients). The second time we use the isarithmic method is when our focus is on the attribute values at points of a truly continuous distribution, such as land elevation.

The principles involved in delineating such a surface are best illustrated by the land surface. If the irregular land surface has been mapped in terms of planimetry, there will be an infinity of points, each of which has an x,y locational attribute. Associated with the attribute, location, can be an additional attribute, relative elevation. We can map relative elevation as a z coordinate

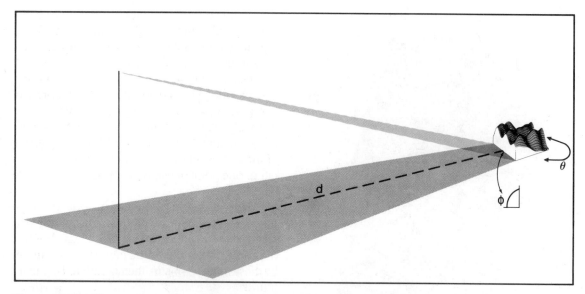

Figure 26.18 Three principal variables that the cartographer must specify when using a computer to plot a series of perspective traces. Theta (θ) is the angle through which the block is rotated in the *x,y* plane, φ is the viewing elevation, and *d* is the viewing distance.

position with respect to the datum surface to which the earth's surface has been orthogonally transferred (Figure 26.21). If an imaginary surface, parallel to the datum and a given *z* distance from it, is assumed to intersect the irregular land surface, it must do so at all points having that *z* value. The trace of the intersection of these two surfaces will be a closed line. When this closed line is perpendicularly projected onto the datum and orthogonally viewed, it shows the *x,y* locations of all points on the surface that have the particular *z* value. This line is an **isarithm** with a value of *z*.

Figure 26.22 shows a hypothetical island. In the top drawing, the island is shown in perspective, intersected by evenly spaced *z* levels. The lower drawing is an isarithmic map, showing distribution of *z* values on the island's surface. The configuration of the three-dimensional statistical surface is symbolized by the isarithms' shape and spacing patterns.

Smooth, steep, gentle, concave, convex, and similar gradients and surface forms may be readily visualized from isarithmic maps, as indicated in Figure 26.23. For example, the bends of contours always point upstream when they cross a valley; they always point down slope when crossing a spur. The angles of slope are shown by the relative spacings of contours provided by equally spaced *z* planes. Detailed topographic forms and often even genetic structural details are revealed by contour patterns (isarithms) on topographic maps.

Only since stereoviewing equipment was developed for air photographs (60 years ago) has it been possible to specify in its entirety the infinity of points on the land with their associated *z* (relative elevation) attribute values. If stereoviewing is not possible, we must infer the statistical surface from a limited number of attribute values (see *Spatial Sampling* in Chapter 9).

In theory, the infinity of possible *z* attribute values would constitute a statistical universe. Due to practical considerations, however, we usually use only a sample of these points to locate the isarithms. We have no way of knowing the actual

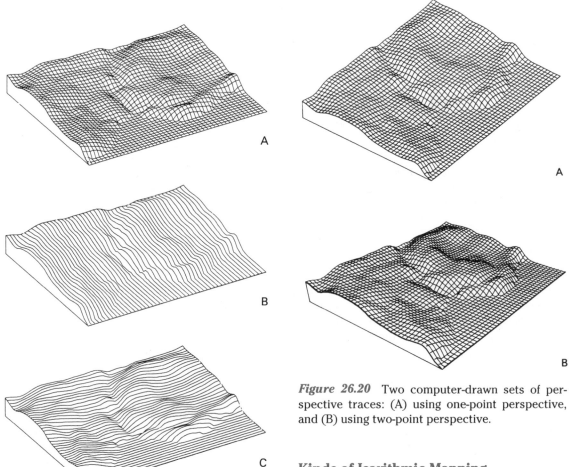

Figure 26.19 Three computer-drawn sets of perspective traces: (A) "fishnets" where traces are drawn in both the *x* and *y* directions, (B) traces drawn only parallel to the *x* direction, (C) traces drawn only parallel to the *y* direction.

Figure 26.20 Two computer-drawn sets of perspective traces: (A) using one-point perspective, and (B) using two-point perspective.

characteristics of the universe from which the sample was taken. Thus, extending the sample's characteristics to the universe by creating isarithms constitutes an inference. The accuracy and representativeness of the given sample values is critical to the success of the cartographer's inferences about the overall nature of the statistical surface.

Kinds of Isarithmic Mapping

There are two classes of *z* attribute values: (1) those that can occur at points and (2) those that cannot occur at points.

Attribute values that can occur at points may be either actual or derived values. *Actual values* are exemplified by elevations above or depth below sea level, temperature values, and thickness of a rock stratum. Only errors in observation or in specification of the *x,y* locational attribute of observation points can affect the validity of the sample values.

Derived values that can occur at points are of two kinds. One consists of averages or measures of dispersion, such as means, medians, standard deviations, and other statistics derived from observa-

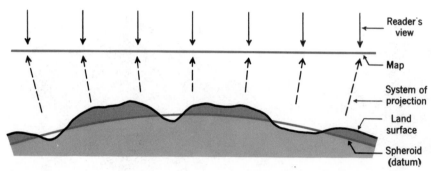

Figure 26.21 The structural basis of an isarithmic map. The positions of elevations are first projected orthogonally to a spheroid. The "map" on the spheroid is then reduced and transformed to a plane, which the reader in turn views orthogonally. Horizontal planes (see Figure 26.22) are then passed through the surface configuration. Curvature is exaggerated in this illustration.

tions made at a point. Examples are mean monthly temperature or average retail figures. A second kind of derived attribute value consists of ratios and percentages. Examples are the ratio of dry to rainy days at a weather station or percentage of precipitation that fell as snow. Such statistics or ratios are incapable of existing at any instant in time, but they do represent quantities that can be referenced to the points at which they are derived. They are subject to more error than measured actual attribute values such as elevations.

Quite different in concept are **attribute values that cannot occur at a point**. Again, these values may be either derived or actual. *Derived values* that cannot occur at points include percentages and other ratios that contain area in their definition directly or by implication. Examples are people per square kilometer, ratio of beef cattle to total cattle, or ratio of cropland to total farmland. With such quantities, we can derive an attribute value only for a unit area. Since each unit area represents an aggregate of x,y nodes, no single point possesses an attribute value. Nevertheless, in order to use isarithms to symbolize undulations of the implied statistical surface, it is necessary to assign derived attribute values to specific points.

Actual values that cannot occur at points include populations of minor civil divisions, number of cows in counties, or potato production in states. Just because their magnitude is affected by area, don't assume that they constitute a statistical surface. (For example, 1,000 persons in 20 km^2 is the same as 500 persons in 10 km^2). Only densities form a statistical surface.

As these examples make clear, there is a fundamental difference between values that can occur at points and those that can't. We will therefore want to make distinctions between mapping techniques to reflect these conceptual differences in the statistical data. Thus, the following terminology is important.

As we have seen, an **isarithm** (also called an **isoline** or **isogram**) is any trace of the intersection of a horizontal plane with a statistical surface projected orthogonally onto the map datum. It is thus the generic term for lines of equal value. But we want to make a distinction between lines of equal value representing data that can occur at points and data that can't. Isarithms showing distributions of values that *can* be referenced to points are called **isometric lines**. Isarithms showing distributions of values that *can't* exist at points are called **isopleths**.

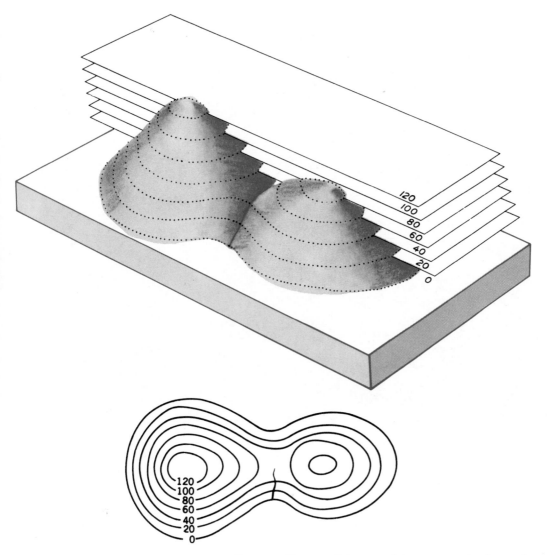

Figure 26.22   In the upper diagram, horizontal levels of given *z* values are seen passing partway through a hypothetical island. Traces of intersections of the planes with the island surface are indicated by dotted lines. In the lower drawing, the traces have been mapped orthogonally on the map datum and represent the island by means of isarithms.

Elements of Isarithmic Mapping

When we are called on to prepare an isarithmic map, whether we are working with an isometric or an isoplethic surface, we follow the same procedures. There are three elements involved in isarithmic mapping: (1) the location of control points, (2) the interpolation procedure used to predict values between control points, and (3) the number of control points. Let's look at each in turn.

Location of Control Points   The *x,y* location and each *z* attribute value of the assumed statistical surface is called a **control point**. The location of

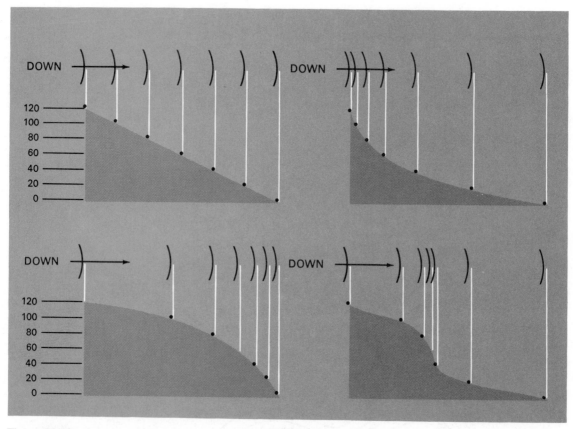

Figure 26.23 The isarithmic spacings above and profiles below illustrate in each diagram how the isarithmic spacings show the nature of the configurations from which they result. All the forms may be described as variations of the following rule: If isarithms are the traces of equally spaced z levels, then the closer the isarithms are, the greater the gradient will be.

control points is predetermined when the statistical surface being mapped is based on attribute values referenced to a point, as in *isometric* mapping. In this case, we merely use the location attribute of each observation as the control point location.

But locating control points is more difficult for *isoplethic* maps, since we are mapping distributions derived from ratios or percentages that involve area in their definitions. The attribute value resulting from these derived quantities refers to the whole areal unit. Therefore, we must arbitrarily assign x,y positions for control points so that the isarithmic lines may be positioned.

When the distribution of features is uniform over an area of regular shape, the control point may be placed at its center. If the distribution within the unit area is known to be uneven, the control point is normally shifted toward the concentration. We may consider the center of gravity as the balance point of an area having an uneven distribution of feature instances. Figure 26.24 illustrates this concept. The four diagrams of statistical divisions show possible locations of the center of gravity and the center of area for uniform and variable distributions. They also illustrate the problem of locating the control point in regularly and irregularly shaped unit areas.

Interpolation The second element involved in isarithmic mapping is the interpolation procedure

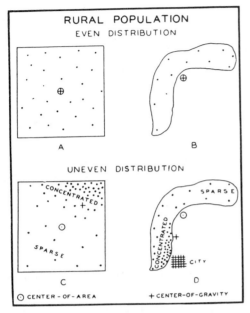

RURAL POPULATION
EVEN DISTRIBUTION

A

B

UNEVEN DISTRIBUTION

C

D

○ CENTER-OF-AREA   + CENTER-OF-GRAVITY

Figure 26.24 For isoplethic maps, problems in positioning the control points arise because either the center of area or the center of gravity may be used. These centers may not always coincide, or be representative of the distribution being mapped. (Courtesy of J.R. Mackay and *Economic Geography*).

used to predict values between control points. **Interpolation** is the process of estimating the magnitude of intermediate values in a series, such as the z values along the line in Figure 26.25.

As Figure 26.25 shows, if two attribute values at different x,y positions are depicted, the gradient between them may be represented by the straight line a or by some other gradient, such as the dashed lines b or c. Any of these lines may represent the true slope. Without further evidence, we must make a guess. Clearly, the curvilinear relations defined by b and c are more complex than the linear gradient shown by a. In science, whenever several hypotheses can fit a set of data, we choose the simplest. Consequently, in most manually drafted isarithmic maps, a linear gradient, a, is assumed in constructing the isarithms.

On the map in Figure 26.26, for example, z attribute values are located at x,y positions a, b, c,

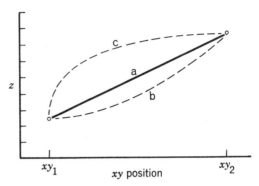

Figure 26.25 Three possible gradients which can be assumed between two control points.

and d. If the trace of the isarithm with a value of 20 is desired, it will lie 3/10 of the distance from a to b, 3/7 of the distance from a to c, and 3/4 of the distance from a to d. Lacking any other data, the dotted line representing the isarithm of 20 would be drawn as a smooth line through these three interpolated positions. In some cases there are known theories, and they should be used. For example, one theory holds that population density varies curvilinearly over space. (See b in Figure 26.25).

Figure 26.27 illustrates a common problem that arises when linearly interpolating among control points located in a rectangular pattern. It is called the **alternative choice problem**. Alternative choice arises when one pair of diago-

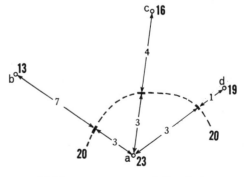

Figure 26.26 Linear interpolation between control points.

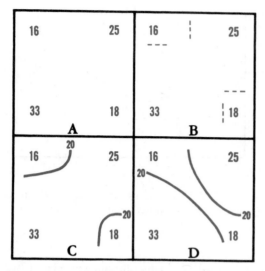

Figure 26.27   (A) The *z* values are arranged rectangularly. (B) Positions of isarithms of value 20 between adjacent pairs have been interpolated. (C,D) The two ways the isarithm of 20 could be drawn through these points. (Redrawn, courtesy of J.R. Mackay and *The Professional Geographer*).

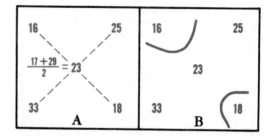

Figure 26.28   If the average of interpolated values at the center is assumed to be correct, the isarithm of 20 would be drawn as in (B). (Redrawn, courtesy of J.R. Mackay and *The Professional Geographer*).

nally opposite *z* attribute values forming the corners of the rectangle is above, and the other pair below, the value of the isarithm to be drawn.

Averaging the interpolated values at the intersection of diagonals will usually provide a value that will remove the element of choice (Figure 26.28). We are forced, in the absence of other data, to assume the validity of the average. This averaging method is often used in computer contouring routines.

Because of cost (time) constraints, manual interpolation is almost solely linear. With machine interpolation, on the other hand, both linear and nonlinear methods can be used with equal ease. The problem in machine interpolation, therefore, is how to choose among the many options. Since each interpolation model is a statement about the nature of reality, the choice (in theory) should be guided by the assumed surface variation of a distribution. In practice, however, the variations of only a few phenomena are actually known. This means that cartographers can face a major problem in selecting the best machine interpolation method.

Number of Control Points   The final element of isarithmic mapping is the number of control points. Initially, cartographers face one of two conditions: (1) they must collect the attribute values to be mapped, or (2) they are presented with a fixed set of attribute values, and there is no way to increase the number of observations or to improve their accuracy.

When cartographers gather feature attributes themselves, they can exercise some control over all three elements of the isarithmic mapping process. But when they are presented with a fixed set of attribute values, cartographers have more limited power. Fortunately, there are some guidelines for isarithmic mapping that permit assessment of a fixed data scatter and number of control points (Morrison, 1971).

In general, the larger the number of control points, the greater can be the detail of the isarithmic surface portrayal. Since the three elements of isarithmic mapping do interact, however, a large increase in control points is not always a blessing. There is a point of **diminishing returns** beyond which improvement in surface portrayal is not worth the effort to process additional control points. Sometimes adding control points only increases clustering, leading to uneven generalization. At other times, a denser clustering of control points occurs in an area of slight variation (smooth terrain) simply because the data are easy to collect in such an area.

Error in Isarithmic Maps

It is important for cartographers to know the error characteristics of isarithmic maps. Error can arise from:

1. the three elements of isarithmic mapping (discussed in the previous section)
2. the quality of the data
3. selection of the isarithmic interval.

Let's look briefly at each of these types of potential error in isarithmic maps.

Method-Produced Error   Errors arising from the three elements of isarithmic mapping (number of control points, their locations, and the selected interpolation model) are collectively called **method-produced errors**. These three elements interact to affect the accuracy of any isarithmic map.

Figure 26.29 illustrates the hypothetical relationship between accuracy and **number of points**. With few control points, the map isn't very accurate. A dramatic increase in accuracy occurs as control points increase. The diminishing returns mentioned earlier cause the curve to level off as more and more control points are added. The cartographer should strive to balance the accuracy level desired against the cost of obtaining or using more control points.

Figure 26.30 illustrates the hypothetical relationship between accuracy and **location of control points**. (The *x*-axis is scaled using the nearest neighbor statistic, where 1.0 represents a random distribution, less than 1.0 a highly clustered distribution, and 2.0 represents a square matrix of control points). As this figure shows, clustered control point scatters (near the left-hand edge of Figure 26.30) are to be avoided. Also, equispacing (2.0 on the abscissa) isn't as desirable as "not quite equispacing." The best results are usually obtained by a scatter of control point locations that is more evenly spaced than random (Morrison, 1974).

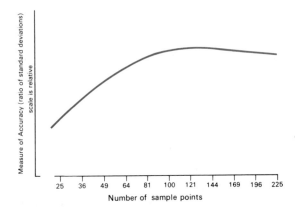

Figure 26.29   The characteristic curve of the hypothetical relationship between number of points and accuracy of the isarithmic map. The *y* axis scale is relative.

When control points are unevenly arranged so that they are denser in one area, inconsistent treatment will occur. We may end up with more detail than we want in areas of dense control points and less detail than we wish in areas of sparse control points. Hence, clustered control points tend to reduce overall surface portrayal accuracy.

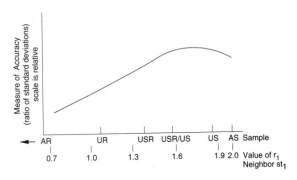

Figure 26.30   The characteristic curve of the hypothetical relationship between sample point scatter and accuracy of the isarithmic map. The *y* axis is relative. (The representative sample scatters are: AR, aligned random; UR, unaligned random; USR, unaligned stratified random; US, unaligned systematic; and AS, aligned systematic).

For *isometric* maps, the errors arising from number and location of control points are straightforward. But for *isoplethic* maps, the potential for error is greater. In isoplethic mapping, the number and location of control points are determined by the structure of the data collection areas. The sizes and shapes of these areas affect isopleth locations. Elongated unit areas can create strong gradients across the orientation of the elongation. These gradients will induce the isopleths to lie in the direction of elongation.

In addition to number and location of control points, the selected **interpolation model** can affect the accuracy of isarithmic maps. With computer-assisted cartography, it is practical to use nonlinear interpolation methods. But to use these methods effectively, we need to know something about the distribution at hand. Frequently, we have no theory about the nature of the surface variation. In this case, we must use the number and location of control points and the generalization level desired for the given map to help us determine the appropriate interpolation method. Under these circumstances, an assessment of method-produced error is difficult.

Quality of Attribute Value As Figure 26.31 shows, any error in the attribute value at a control point can have as much effect on the location of an isarithm as changing x,y attributes of the control point. If the z attribute value is correct but the x,y position is incorrect, the isarithm will be displaced. If the x,y position is correct but the z attribute value is in error, the same thing will happen. Clearly, both the positions of the control points and the validity of the value at the control point have a considerable effect on the accuracy of the statistical surface.

Four kinds of errors affect the reliability of the z attribute values from which an isarithmic map is made: (1) observational error, (2) sampling error, (3) bias or persistent error, and (4) conceptual error.

Observational error refers to the method used to obtain the attribute values. If the values are derived by means of instruments operated or read

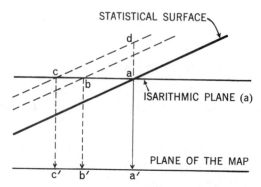

Figure 26.31 If a' is the x,y location of a control point with z value a, and if the isarithmic plane has the value a, then a' will be the orthogonal position of the isarithm on the map. If the z value a is incorrect and really should be d, then the a isarithmic plane would intersect the surface at c, and the isarithm on the map would be located at c'. If the a value were correct but the x,y position of a should be at b, then b' would be the map position of the isarithm.

by human beings, there is usually some inaccuracy, both in the instruments and in their reading by the observers.

Sampling error is of several kinds. The most obvious is the error associated with any map based on a few sample values that purports to represent the distribution of an entire set of features. In order to make a map of mean climatological values, for example, we must rely on a relatively small number of locational observations taken within a limited time span. Yet the finished map implies that we know much more about the distribution than we do.

Bias or **persistent error** may be of many kinds. Instruments may consistently record too low or too high. The majority of weather stations may be in valleys or on hilltops. People show preferences for certain numbers when estimating or counting. Such bias is hard to detect but may have great effect on measured attribute values.

Conceptual error differs from the previous three errors in the sense that it is related not so much to map error as to the validity of the concept presented by the map. Conceptual error may be illustrated by a map of mean monthly temperatures. The monthly temperatures will not be described very well by any mean value if the dispersion of attribute values around the mean is large. For example, the standard deviation of mean December temperatures in the central United States is around 4°F. About one-third of the time, therefore, an individual December mean will be more that 4° above or below the average December mean. Suppose that cartographers were making an isarithmic map to show distributions of mean December temperatures. They would be forced to locate each isarithm near the center of a map zone representing several degrees of temperature gradient. If individual year-to-year maps were made, approximately one-third of the time an isarithm would lie outside the zone delineated by an average of 8° of temperature gradient. On even a small-scale map, this zone of uncertainty would be noticeable. Clearly, if these temperature data were mapped with isolines exhibiting minute wiggles and sharp curves, the effect would imply more accuracy than is warranted.

It is important to take into account all four types of error in attribute values. Cartographers can overcome the effects of these errors and inconsistencies from one part of an isarithmic map to another by **smoothing the isarithms**. This smoothing can be done by removing small details or irregularities in the isarithm shape. (See Chapter 24).

Class Intervals in Isarithmic Mapping   So far we've considered error in terms of the locational accuracy of isolines. But we can also think of error in terms of problems people encounter when using isarithmic maps. For instance, relative gradients on isarithmic maps are easily judged only when there is a constant interval between isarithms. The rule that the closer the isarithms, the steeper the gradient, holds only under such circumstances. Therefore, if primary interest is in the configuration of the statistical surface, the interval should be constant.

However, irregular intervals are often used in isoplethic mapping. (See Figure 26.6). This requires much more mental effort and can lead to

error when map readers try to infer the form of the surface. Thus, cartographers should use irregular intervals sparingly and should let map readers know when they do so.

Summary of Isarithmic Method

Perhaps no cartographic method is as useful or as often requested for surface portrayal as the isarithmic method. There are many reasons for this usefulness. First, when equal-step intervals are used, the practiced reader can gain a good picture of the surface configuration. Second, map readers can use detailed cartometric methods to extract useful data from these maps. Third, many other cartographic methods (including profiles, oblique traces, hachuring, shading, and dasymetric maps) are derived from isarithmic portrayals.

The two forms of isarithmic map, the isometric and the isoplethic, look alike. The only difference is the form of the original data used for the map: Isometric maps result from point data, and isoplethic maps result from areal data values assigned to a center point in each data collection area.

MAPPING THE STATISTICAL SURFACE WITH AREA SYMBOLS

To map a statistical surface composed of quantitative attribute values referenced to areas, we have three choices. The **isoplethic** technique (which we discussed in the preceding section) uses line symbols. The **choroplethic** and **dasymetric** techniques use area symbols. These three techniques are compared in Figure 26.32. Note how differently the three maps represent the same information. In this section, we will compare the two methods that use area symbols. Let's begin with a discussion of choroplethic mapping.

CHOROPLETHIC MAPPING

The **choroplethic** method portrays a statistical surface with area symbols. These symbols coin-

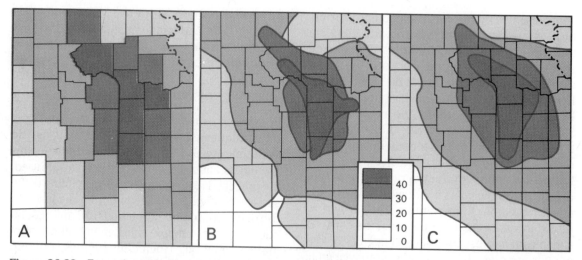

Figure 26.32 Examples of three ways we can map a set of *z* values that refer to enumeration districts or unit areas: (A) a simple choropleth map, (B) a dasymetric map, (C) an isoplethic map.

cide with the data collection regions. This gives the impression that there is uniformity within the data collection units and that breaks occur in the surface at the unit boundaries.

Kinds of Choroplethic Mapping

There are two choroplethic techniques. One is **simple choroplethic mapping**. The second is **classless choroplethic mapping**. Let's look first at the simple choropleth map.

Simple Choropleth Map Although tabular statistics are convenient for many purposes, we often want to visualize these statistics geographically and compare magnitudes in various places. Simple choropleth maps are useful for this purpose.

In **simple choropleth mapping**, the objective is to symbolize magnitudes of statistics as they occur within boundaries of unit areas (such as counties, states, or other kinds of enumeration districts). The simple choropleth map is a spatially arranged presentation of statistics that is tied to enumeration districts on the ground. (See Figure 26.7.) An area symbol is used to show the attribute value that applies to each enumeration

district. The symbolization scheme is usually range-graded.

The simple choropleth technique may be used for many kinds of densities and ratios. When statistical units are small, this technique may provide enough variability so that the map takes on the visual character of a continuous distribution. Ordinarily, however, this technique is used only when any other sort of map poses too many problems. In effect, the simple choropleth map presents only the spatial organization of the statistical data. No effort is made to insert any inferences, such as trends, into the presentation.

Absolute numbers alone should not be presented in a simple choropleth map. The quantity mapped is almost always some kind of average or density that is assumed to refer to the entire unit area. Where class limits call for a change of symbolization, it occurs only at the unit area boundaries. Since the boundaries are usually unrelated to variations in the phenomenon being mapped, this method exaggerates the true geographical variation of the distribution. Despite this exaggeration, the simple choropleth technique can give a useful first "glance" at a distribution.

Classless Choropleth Map The earliest choropleth maps were classless, since their makers tried to give each unit area attribute value a tonal value matched to its place on a continuous scale of values from lowest to highest. The first attempts were not successful because tonal gradations could not be controlled by the available production processes. The concept of a classless choropleth map has recently been revived, because computer-controlled output devices can produce the necessary tonal gradations.

A classless choropleth map is a ratio-scale portrayal similar in concept to proportionally scaled graduated point symbols (see Box 25.B in Chapter 25). The term choropleth map has traditionally referred to a range-graded portrayal of the attribute values; it is thus more generalized than a classless representation.

The effectiveness of a classless choropleth map compared to a simple choropleth map is not known. The classless map may appear more complex than its classed counterpart simply because it is less generalized. The classless choropleth map is sometimes the logical choice because it presents a clearer picture of the existing complexity.

In one sense, a raster image received from a remote sensing device is a classless choropleth map. Each pixel (unit area) in the image is assigned a numeric (attribute) value. Humans can't differentiate the large number of classes that a raster image can convey. The image may have geometric distortions that a conventional planimetric map does not. Also, the conventional map usually has generalization levels that the unprocessed image lacks.

Elements of Choroplethic Mapping

The basic elements in choroplethic mapping include (1) size and shape of unit areas, (2) number of classes, and (3) method of class limit determination. Let's look at each of these elements.

Size and Shape of Unit Areas The unit areas of a region form a pattern. This pattern acts as a spatial generalization filter. If unit areas are large,

the data's spatial variation tends to be reduced or averaged out. If unit areas are small, variation is preserved.

If unit areas vary greatly in size, variation is preserved in one part of the region and lost in another. The result is uneven generalization, an undesirable effect on a map.

Ideally, for the best use of the choropleth method, unit areas should be of relatively equal sizes (preferable small) and of similar shape. See Figure 26.33.

Number of Classes The most common choroplethic method results in a range-graded symbolization. The number of classes determines how detailed the mapped distribution will be. Ideally, we want to present the maximum number of classes that can easily be read. This maximum number depends on how complex the distribution is (see Figure 26.34). It also depends on whether the map is monochromatic or color. On a monochromatic map, readers can perceive relatively few distinct classes. Even when cartographers use both pattern and value to show different classes, readers can't perceive more than five to eight classes on a black-and-white map. When we use chroma progressions on a color map, however, we can show more classes (see Chapter 21).

The classless choropleth map, of course, uses as many categories as there are unique values in the mapped distribution. A classless choropleth map will probably not enable map readers to perceive an exact value for a unit area. On the other hand, the classless choropleth map can give map readers a good impression of the overall distribution. The effect may be similar to that created by hill shading (see Chapter 27), especially if there are many small unit areas.

Class Limit Determination Perhaps no aspect of choroplethic mapping has received more space in the cartographic literature than methods for determining class limits. After cartographers have decided on the number of classes for a choropleth map, they must set the limits of these

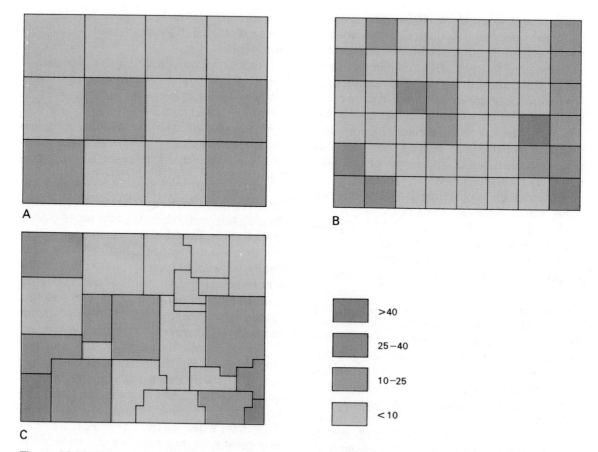

Figure 26.33 The same distribution mapped with three different sets of unit areas, using the same class interval scheme: (A) uses large units of equal area, (B) uses small units of equal area, and (C) uses irregularly sized unit areas.

classes. The methods they use to determine class limits are equally applicable to establishing isoplethic intervals or range-graded proportional point or line symbols.

Computer assistance has increased the number of methods available to determine class limits. (See Box 26.B). Because so many methods are available to the cartographer, it is relatively easy to select class-interval determination schemes that will range-grade the data in a way that presents a biased view. Of course, the potential to enhance a portrayal by wisely selecting class limits also exists.

When choosing class limits, cartographers should seek to highlight critical values in a distribution, if possible. For example, suppose a county qualifies for a federal program if 40 percent of families are below poverty level. In making a map of family income by county, then, it would make sense to choose this 40 percent value as one class limit.

If there are no critical values, however, cartographers should seek to maximize both the homogeneity within each class and the differences between classes. Fortunately, it is possible to state these criteria statistically and approximate them with computer assistance.

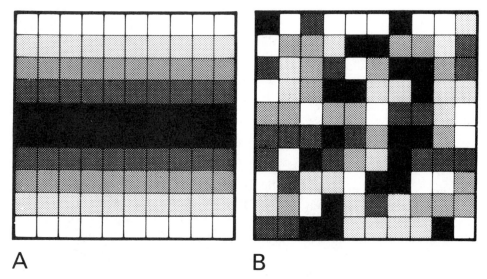

A   B

Figure 26.34 Both (A) and (B) show five classes. Because of the simple distribution pattern on (A), we could increase the number of classes used. The lack of pattern in the distribution of (B) would make it even more confusing to the map reader if more classes were used. (Maps courtesy of J.M. Olson).

Error in Choropleth Mapping

As we have seen, choropleth maps adhere to the structure of data collection units. Thus, the size, shape, and orientation of these units can influence the quality of the map. The logic of the choropleth method is that each unit area is homogeneous and that sharp breaks occur at unit boundaries. Error can occur if the structure of data collection units does not reflect the underlying geographic distribution.

Jenks and Caspall (1971) have studied the error characteristics that result from the number of classes and class limit selection in choropleth maps. They have proposed that error indexes be placed in the legends of all choropleth maps. Errors can be of three types, depending on whether the purpose of the map is to portray (1) an overview of the distribution, (2) the geographic positions of tabular values from the distribution, or (3) significant boundary lines for the distribution. According to Jenks and Caspall, the overview error, which some believe to be the most important, can be calculated as the volumes of prisms lying between the unclassed data surface (Figure 26.35) and a classed map model (Figure 26.36). Similarly, Jenks and Caspall present statistics for calculating tabular and boundary value indexes to address the other two types of error.

DASYMETRIC MAPPING

In the previous section, we saw that choropleth maps reflect the structure of data collection units. When these units are poorly matched to the character of a geographic distribution, the choroplethic technique can mask a great deal of information. In this situation, **dasymetric** mapping offers an effective alternative.

The dasymetric map is often made from the same data as the simple choropleth map. But in dasymetric maps, mapping unit boundaries are independent of enumeration boundaries. Dasymetric maps assume areas of relative homogeneity, separated by zones of abrupt change. To create these homogeneous areas, cartographers use other

Box 26.B
Class Interval Series

We can separate methods used to determine class intervals into three major groups:

1. Constant Series or Equal Steps

One way to determine class intervals is with **constant series** or **equal steps**. This method uses some kind of equal division of the attribute values or the geographic area. There are four types of constant series used for range-grading attribute values. These are (1) equal steps based on the range of attribute values, (2) parameters of a normal distribution (for example, the mean and standard deviation), (3) quantiles, and (4) equal-area steps or geographical quantiles. Three of these methods are illustrated along the bottom in Figure 26.B.1.

2. Systematically Unequal Stepped Class Limits

A second class is called **systematically unequal stepped class limits**. With this type of series, cartographers make the interval systematically smaller toward the upper or lower end of the scale. They choose the low or high end depending on which part of the distribution they wish to emphasize.

Two groups of series with unequal steps are (1) **arithmetic series**, in which each class is separated from the next by a stated numerical difference (not constant), and (2) **geometric series**, in which each class is separated by a stated numerical ratio. Both arithmetic and geometric series can take any of the following six forms:

1. increasing at a constant rate
2. increasing at an increasing rate
3. increasing at a decreasing rate
4. decreasing at a constant rate
5. decreasing at an increasing rate
6. decreasing at a decreasing rate.

The top three rows of maps in Figure 26.B.1 show selected class interval series for visual comparison. The three maps in the left column illustrate arithmetic progressions and the three maps in the right column illustrate geometric progressions.

3. Irregular Stepped Class Limits

A third class is the **irregular** or **variable series**. Cartographers use this kind of series when they wish (1) to call attention to internal characteristics of the distribution (such as values that may be significant in relation to other analyses), (2) to minimize certain error aspects, or (3) to highlight elements of the data range that would not be properly dealt with if they used a constant or a regular ascending or descending series.

We can divide irregular stepped interval techniques into two classes, based on the way they are derived: graphic techniques and iterative techniques. Three common **graphic** techniques are the **frequency graph**, the **clinographic curve**, and the **cumulative frequency curve**. (See Figure 26.B.2). In each of these methods, the attribute values are plotted and the cartographer selects breaks in the constructed curves as the class interval breaks. Maps resulting from these three graphic methods are shown in Figure 26.B.3.

Box continued on next page

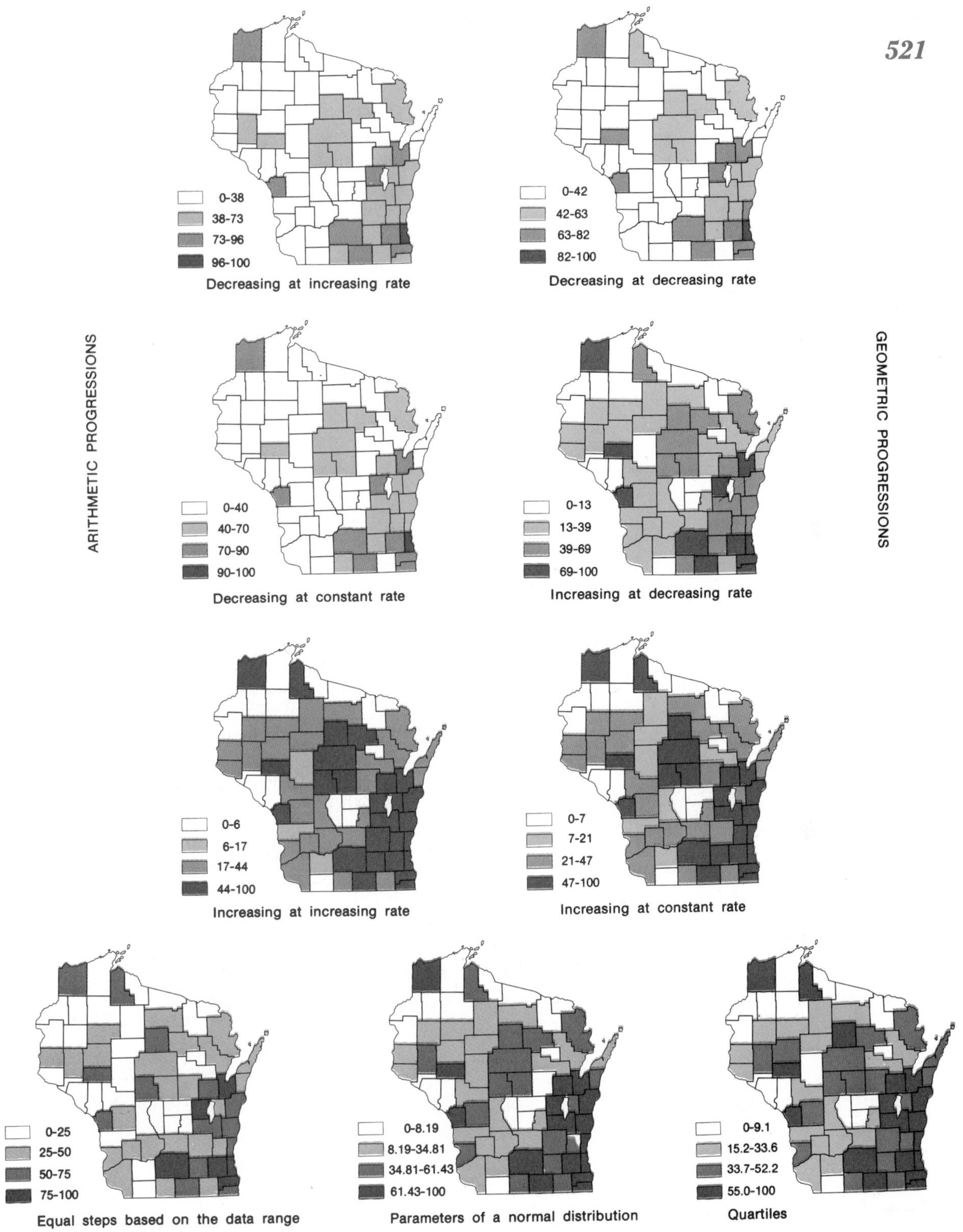

Figure 26.B.1 Collage of Wisconsin class interval maps illustrating the use of different class interval methods using the same data set. See Figure 26.B.3 for additional maps. Calculation methods are explained in text.

Box 26.B (continued)

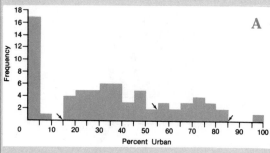

In contrast to graphic techniques, **iterative** techniques are based on repetitive calculations that progressively narrow the choice of class intervals. In analog cartography, iterative techniques were laborious to calculate and therefore were not very popular. With computer assistance, however, we can easily use iterative techniques to satisfy specified statistical criteria. Thus, the use of iterative techniques has increased at the expense of strictly graphic techniques.

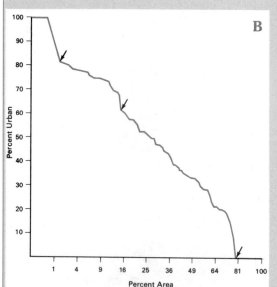

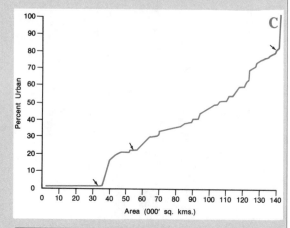

Figure 26.B.2 A frequency graph (A), a clinographic curve (B), and a cumulative frequency curve (C) of the same data mapped in Figure 26.B.1. The arrows denote the points on the graphs that the cartographer has selected to use for class limits.

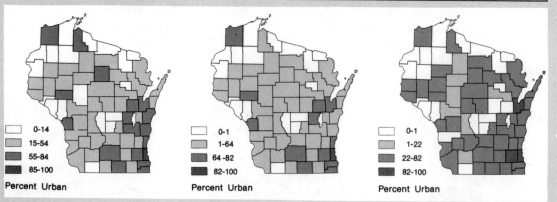

Figure 26.B.3 Maps of the same data graphed in Figure 26.B.2. The maps use class limits determined from the frequency curve (left), clinographic curve (middle), and cumulative frequency graph (right).

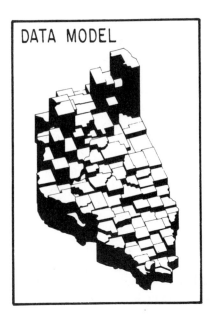

Figure 26.35 An unclassed perspective view of a statistical distribution. Each unit area is shown as a prism whose height is proportional to the value of the mapped distribution, and whose base is a scaled representation of the unit area. (Courtesy of G. Jenks and F. Caspall and the *Annals of the Association of American Geographers*, 1971).

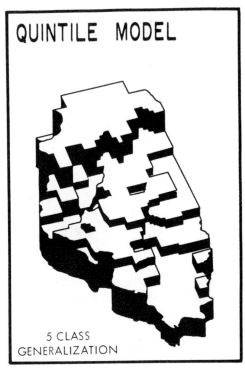

Figure 26.36 A five-classed, perspective view of the statistical distribution shown in Figure 26.35. (Courtesy of G. Jenks and F. Caspall and the *Annals of the Association of American Geographers*, 1971).

sources of information about the distribution, such as maps of related phenomena. Then, based on knowledge gleaned from these sources, the cartographers create new, more natural homogeneous units. Thus, they can approximate the true character of the surface variation better than is possible with simple choropleth mapping.

It should be emphasized that nothing in the original data will help cartographers decide where to create dasymetric map boundaries. Instead, this information must come from other knowledge about the spatial relationships involved. Geographic databases, which hold a variety of spatial data in a common format, are ideal for this purpose. Using geographic information systems (GISs) to access these databases has greatly simplified the dasymetric technique and is popularizing its use.

Supporting Information

The essential problem in dasymetric mapping using GISs is to peruse existing sources to select the best possible supporting information. This information should help us transform the arbitrary enumeration unit boundaries into boundaries that better reflect the nature of the distribution. Our task is to choose information that will provide insight into the character of the attribute values we are mapping. The information we select may be of two kinds: (1) limiting variables, and (2) related variables. Let's look first at limiting variables.

Limiting Variables **Limiting variables** set an upper limit on the quantity of mapped phenomena that can occur in an area. For example, suppose we are mapping the percentage of cropland.

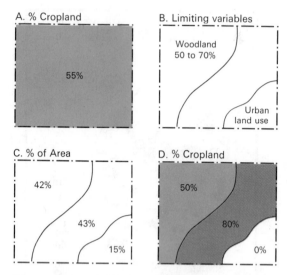

Figure 26.37 A knowledge of limiting variables can assist in dasymetric mapping. (See text for explanation.)

The data show that 55 percent of County A's total area is cropland. Our outside sources tell us that 15 percent of County A's area is devoted to urban land uses. Obviously, that area can't be cropland. Thus, we can exclude the urban area and distribute all the cropland in the remaining sections of the county.

Now suppose that a second limiting variable in County A is woodland. Our supporting information gives us the areal percentage of woodland for townships. Thus, we know that a portion of a township has 50 to 70 percent of its area in woodland. The maximum cropland it could have, therefore, is 30 to 50 percent. Consequently, other areas of the county need to have a much higher percentage of cropland to make possible the original statistic of 55 percent cropland for the whole county.

Figure 26.37 shows how cartographers can use limiting variables to add great geographic detail in dasymetric mapping. The choropleth map in Figure 26.37A depicts the county as a single class. But the dasymetric map in Figure 26.37D shows three classes for the same region. GIS functions let us perform these calculations almost instantaneously.

Related Variables The second type of supporting information in dasymetric mapping consists of related variables. Related variables are those geographic phenomena that show predictable variations in spatial association with the phenomenon being mapped.

Let's use our earlier example of percentage of cropland. Suppose one of our outside sources is a map showing that County A contains some level land and some steep hills. We can assume that there is a high positive correlation between level land and percentage of cropland. Thus, we can use this information to modify the cropland boundaries on the map.

The effects of related variables are less direct and more statistical than the effects of limiting variables. Thus, related variables are more difficult to use properly in dasymetric mapping. They are, however, very useful in helping the cartographer put some geographic "sense" into the bare statis-

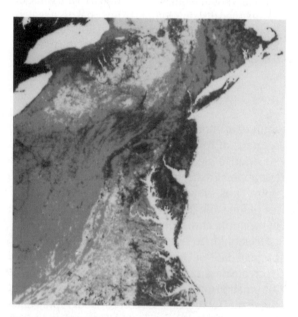

Figure 26.38 Part of a map of the U.S. detailing 159 categories of land charcteristics defined by clustering pixels of similar values. From darkest grey (agriculture, rangeland, forestland) to lightest grey (water, wetlands, barren land).

tics that are commonly gathered on the basis of political enumeration districts. Again, GIS functions can greatly speed the process of using related variables.

Dasymetric Maps from Point or Raster Data

The dasymetric technique can also be used to portray areas of uniform attribute values referenced to points. The cartographer creates these areas of uniformity by clustering or grouping contiguous similar attribute values into regions. Many different clustering techniques can be used. The pixels in raster data sets are often clustered in this manner. (See Chapter 12). This type of dasymetric map represents the results of a complex level of geographic analysis. (See Figure 26.38).

Error in Dasymetric Mapping

Each step of the dasymetric technique can add error to the map. Almost every step requires the cartographer to make some knowledgeable, but subjective, decision. With such subjectivity comes the possibility of human error. However, this error can be partially offset by the additional information offered by the related and limiting variables.

Unfortunately, no comprehensive study of the error characteristics of the dasymetric technique has been undertaken. However, errors in a GIS environment have been studied and may shed light on the nature of dasymetric error (Goodchild and Gopal, 1989).

MAPPING THE STATISTICAL SURFACE WITH LINE AND AREA SYMBOLS

For years cartographers have combined isarithmic maps with the visual variables that form color. The result is **layer tinting** (see Chapter 27). Layer tinting has traditionally been used to depict the terrain on small-scale atlas maps or wall maps. As cartographers become more skilled at using the capabilities of GISs, we can expect them to create more products in this category. Already they are routinely using layer tinting in GIS output of distributions other than terrain. A few examples follow.

SHADED ISARITHMS

When we shade the areas between isarithmic lines, we should use the visual variables of value or chroma (both elements of color). The isarithmic surface is quantitative, and (as we saw in Chapter 25) only value or chroma imply quantities when applied to areas.

SHADED CLASSLESS CHOROPLETHS

It is only a short move along the continuum from area data to point data when using area symbols to create a **shaded classless choropleth** map. As we noted earlier in this chapter, such a depiction is similar to a raster based image. As number of classes increases and unit area size decreases, a shaded classless choropleth map will come to look something like shaded relief. The major difference is that it will lack illumination highlights and shadows characteristic of terrain shading.

SELECTED REFERENCES

Bertin, J., *Semiology of Graphics: Diagrams, Networks, Maps*, W. J. Berg, translator; Madison, WI: University of Wisconsin Press, 1983.

Craig, W. J., and J. L. Adams, "User Control of Isarithmic Mapping," *Cartographica*, 28, 1991: 51–65.

Dent, B. D., *Cartography: Thematic Map Design*,

2nd ed., Dubuque IA: William C. Brown Publishers, 1990.

Gilmartin, P. P., "The Design of Choropleth Shadings for Maps on 2- and 4-Bit Color Graphics Monitors," *Cartographica*, 25, 1988: 1–10.

Gilmartin, P. P., and E. Shelton, "Choropleth Maps on High Resolution CRTs: The Effects of Number of Classes and Hue on Communication," *Cartographica*, 26, 1989: 40–52.

Goodchild, M. F., and S. Gopal (eds.), *Accuracy of Spatial Databases*, London: Taylor and Francis, 1989.

Jenks, G. F., and F. C. Caspall, "Error on Choropleth Maps: Definition, Measurement, Reduction," *Annals of the Association of American Geographers*, 61, 1971: 217–244.

Keates, J. S., *Cartographic Design and Map Production*, 2nd ed., Essex, U.K.: Longman Scientific and Technical, 1989.

Lavin, S., "Mapping Continuous Geographical Distributions Using Dot-Density Shading," *The American Cartographer*, 13, 1986: 140–150.

MacEachren, A. M., "Accuracy of Thematic Maps: Implications of Choropleth Symbolization," *Cartographica*, 22, 1985: 38–58.

Monmonier, M. S., "Flat Laxity, Optimization, and Rounding in the Selection of Class Intervals," *Cartographica*, 19, 1982: 16–20.

Morrison, J. L., *Method-Produced Error in Isarithmic Mapping*, Technical Monograph N. CA-5, Cartography Division, American Congress on Surveying and Mapping, 1971.

Morrison, J. L., "Observed Statistical Trends in Various Interpolation Algorithms Useful for First-Stage Interpolation," *The Canadian Cartographer*, 11 (1974): 149–159.

Slocum, T. A., S. H. Robeson, and S. L. Egbert, "Traditional Versus Sequenced Choropleth Maps: An Experimental Investigation," *Cartographica*, 27, 1990: 67–88.

Tobler, W. R., "Choropleth Maps without Class Intervals," *Geographical Analysis*, 5, 1973: 262–265.

Zoraster, S., "Honoring Discontinuities and Other Surface Features during Grid Processing on Vector Computers," *Cartographica*, 24, 1987: 37–48.

CHAPTER 27

PORTRAYING THE LAND-SURFACE FORM

*T*here is something about the three-dimensional (3-D) land surface that intrigues map users and sets it apart from other distributions portrayed on maps. The land surface is a visible "physical" continuous phenomenon which we experience every day. Thus, we are likely to be consciously or unconsciously more critical of its map portrayal than of maps representing other aspects of our surroundings.

During the past three decades, airline travel has become relatively commonplace in the industrialized world. This has given the map-reading public another perspective on the land surface. At the same time, technological advances in computing and remote sensing have given cartographers vast amounts of data about our planet's form. Digital terrain models are routinely extracted from imagery taken by airborne or satellite sensors and cameras. The same methods are being applied to the mapping of ocean depths and other planets. More precise data about the ocean bottom and the moon, Mars, and Venus are becoming available. Matrixes of precise elevation data and land images provide new possibilities in landform mapping, and we can anticipate a series of experimental portrayals during the coming decade.

Departures of the earth's surface up or down from the spheroid are actually very small, relative to horizontal distances. Take Everest, the highest mountain on earth, for example. It is 8,848 m. high. If it were represented accurately on a 3-D model of Asia at a scale of 1:10,000,000 (a map about 1.2 m. square), it would be less than 1 mm. high!

On the other hand, people are tiny in comparison to landforms. Thus, we think of even minor mountains and hills as being quite high. To match our subjective impressions, portrayals must greatly exaggerate the relative heights of terrain features. This presents a problem in large-scale mapping. If landforms are shown in enough detail to satisfy people's impression of their significance, they often obscure other features on the map. On the other hand, if cartographers include non-landform features which may be more important to the map's purpose, they can give only a suggestion of the land surface. Such an expedient is not likely to please either the mapmaker or the map reader.

It is equally difficult to portray landform on smaller-scale maps, atlas maps, wall maps, and other general reference maps. Showing the landform is especially hard on thematic maps for which regional terrain is an important element. The smaller scale requires considerable generalization as well as balancing the surface representation with other map features.

In this chapter, we will explore portrayal of the 3-D landform. We will discuss such attributes of landform as slope, aspect, and elevation above sea level. The principles apply equally to the solid surface beneath the earth's waters.

HISTORICAL BACKGROUND

Cartographers have always been keenly interested in portraying the landform. Throughout history, they depicted mountains as piles of crags. Occasionally, they showed ranges as a series of "fish scales" or simply as bands of color.

Until recently, however, they did not have accurate elevational data on which to base their portrayals. In the Western world, such measurement has come only in the last two centuries. As late as 1807, the heights of only about 60 mountains had actually been measured.

The story of how terrain portrayal developed is a search for suitable methods. The problem has been that the most effective visual techniques did not give precise terrain information. Likewise, methods that gave accurate terrain values were the least effective visually. Cartographers constantly sought ways to balance these opposing conditions.

Early terrain representations on smaller-scale maps showed only undulations of great magnitude. These depictions consisted of crude stylized drawings of hills and mountains, such as those in Figure 27.1.

From the 15th to 18th centuries, landform portrayal developed along with landscape painting of

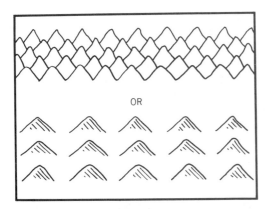

Figure 27.1 Examples of portraying hills and mountains in stylized form on early maps.

the period. Thus, perspective-like, oblique, or bird's-eye views for limited areas became popular. (See Figure 27.2).

Beginning in the 16th century, lines connecting points of equal elevation (**isarithms**) were used by Dutch, and later French, engineers and cartographers. They first used isarithms to portray forms of the underwater surface (Figure 27.3). Later, they used isarithms to show the dry-land configuration of the earth's surface (see Figure 27.4). These lines on land, *Hohenlinien* (German) or *courbes de niveau* (French), came to be known in English as **contour lines**.

Another metric method became known in 1799 when Johann Georg Lehmann, an Austrian army officer, systematized the use of short linear symbols. These symbols were called *Schraffen* in German and *hachures* in English and French.

During the 18th century, the great topographic surveys of Europe were initiated. Landform portrayal shifted from the **oblique** to the **plan** view (looking straight down). Symbols were developed to show the landform without planimetric displacement.

Another advance came when lithography was developed early in the 19th century. It thus became easy to produce continuous tonal variations or **shading**. Shading simulated the appearance of the irregular land surface. These developments,

Figure 27.2 A section of the 1712-1713 map of Switzerland, "Nova Helvetiae Tabula Geographica," by Johann Jacob Scheuchzer. Original scale ca. 1:230,000. (From facsimile published by De Clivo Press, 1971).

helped by perfection of the airbrush, led to a surge of interest by cartographers in **hill shading**.

At the same time, people began to understand the profound effects elevation has upon temperature, vegetation, and other aspects of the physical environment. Once they realized elevation's importance, they began to collect accurate altitude data. For the first time, cartographers could map precise elevation figures. Thus, **contouring**, which gives measurable information, emerged as a dominant landform portrayal method.

But strict contouring and hill shading could not satisfy all the needs for showing landforms on maps, particularly at small scales. Consequently,

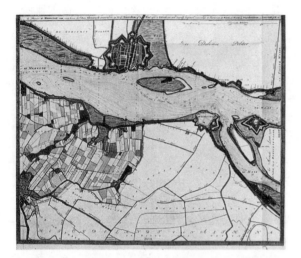

Figure 27.3 A portion of a 1729 map of the River Merwede by Samuel Cruquis. It is the earliest know map of isobaths, a form of isarithmic line symbol.

as early as the first half of the 19th century, cartographers began to experiment. Often these visualizations were adaptations of other methods. Thus, they are not easy to classify. However, in the discussion below, we will recognize three general

groups: perspective pictorial portrayals, morphometric maps, and terrain unit maps.

By the beginning of the 20th century, cartographers had agreed on three basic methods of presenting terrain on large-scale printed topographic maps. These methods were contouring, hachuring, and hill shading. The incompatibility between striving for visual realism on one hand and precise attribute values on the other was apparent. For the next several decades, the problem was how to combine techniques to achieve both ends.

The later third of the 20th century brought computers, space-borne cameras, and electronic sensors of high metric accuracy. These combined to provide cartographers with good data and new possibilities for landform portrayal. Digital terrain model (DTM) data (see Chapter 13) and images (see Chapters 8 and 12) are quickly surpassing conventional contouring and hill shading for maps produced by automation and appearing on a computer screen (not printed).

Frequently a DTM is used as a base to form a perspective surface upon which raster based image data are draped. (See Figure 27.5). Map readers can rotate or tilt the resulting view in whatever way

Figure 27.4 An early map using isarithms, in this case contours, to show the configuration of the land.

Figure 27.5 A photographic image of the ground near Harrisonburg, Virginia, has been "draped" over a DTM and plotted in perspective. (Courtesy of the USGS GIS Laboratory and R. DeAngelis).

they wish. They can also extract precise information from the database. This approach is revolutionizing landform portrayal. Hachures appear to be a method of the past, and cartographers now have only two distinct methodologies from which to choose—contouring and hill shading.

Recent printed topographic maps which combine hill shading with contours are the most effective ever produced by manual methods (Figure 27.6). The two methods combine to provide both a quantitative delineation and a realistic visual portrayal. The digital methods are not yet completely standardized, but perspective plots of the land surface which can be rotated in real time (see Chapter 29) and other recent developments promise even more effective portrayals. (See Figure 27.7).

As we stated at the beginning of Chapter 17, technology is once again changing the problem of balancing accurate terrain values with effective visual techniques. With the help of computer technology, we can obtain accurate terrain values from the digital database. By relieving the landform portrayal of this data storage function, we can concentrate on making the visual presentation more effective.

VISUALIZATION METHODS

Landform portrayal methods vary greatly in style and purpose. Some techniques focus on the landform's visual character. Other methods serve more analytical ends by giving precise elevation

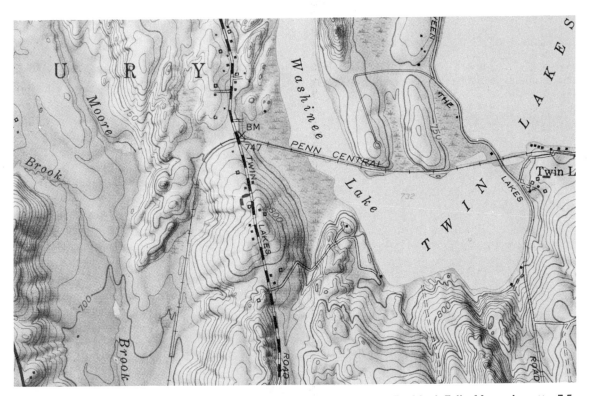

Figure 27.6 A section of a modern hill-shaded topographic map. From Bashbish Falls, Massachusetts, 7.5-minute quadrangle, U.S. Geological Survey, 1:24,000. The original color is much more effective than this monochrome reproduction.

Figure 27.7 A view of the Island of Hawaii using a new multidimensional technique of relief shading by computer (Courtesy of R. Mark of the U.S. Geological Survey).

and structural information. Still other techniques combine visual and analytical dimensions. We will consider some of these different approaches to landform representation in the remainder of this chapter.

PERSPECTIVE PICTORIAL MAPS

Even on ancient maps, major terrain features were represented pictorially. As artistic capabilities and knowledge of the earth's surface expanded, pictorial representations became increasingly effective. Within the past 50 years, this method of presenting land features has progressed rapidly. In the last 30 years, the computer's ability to manipulate landforms has added greatly to terrain visualization methods. In the next sections, we will look at three types of perspective pictorial maps, beginning with block diagrams.

Block Diagrams

During the 19th century, cartographers experimented with many ways of using perspective to show the landform. One such map, the block diagram, was used to portray geologic concepts. Today, the block diagram has become a standard form of graphic expression for geologists.

We can trace the origins of the block diagram to 18th-century geological studies and reports, which were illustrated with cross sections. The top edge of a cross section is a profile across the land surface. To add realism to the cross section, pictorial sketching was added. These sketches showed structural relations of subsurface formations.

On block diagrams, you view a "block" of the earth's crust as if you were looking at it from an oblique vantage point. From this perspective, the top as well as two sides of the block are visible.

Even the simplest block diagram is easily accepted as a reduced replica of reality and is remarkably graphic. (See Figure 27.8). Block diagrams can portray a great variety of information, ranging from structural information to successive stages in an area's geologic evolution.

Today, block diagrams are easily produced by computer-driven plotters. Consequently, almost any kind of surface distribution may be readily displayed on a perspective landform base. The viewing distance, elevation, and oblique angle of view are easily adjusted. Map users can take a tour through, over, and under the portion of the earth portrayed on the block diagram. (See Figure 27.9).

Block diagrams and similar perspective views are ordinarily restricted to small segments of the earth and do not take into account the curvature of the earth's surface. Extending the perspective view to larger areas leads to another perspective portrayal, the oblique regional view.

Oblique Regional Views

An oblique regional view is usually constructed on either an orthographic projection or an oblique

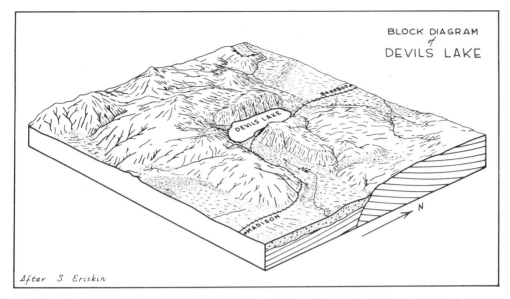

Figure 27.8   A simple block diagram prepared for student field trip use. The natural appearance of the surface forms on a perspective block makes the concepts easily understandable.

photograph of an actual globe. Such views are particularly useful in showing the relationship of a nation to its neighbors. The modeling is an artistic portrayal and thus requires artistic talent. It also requires understanding of the role of exaggeration

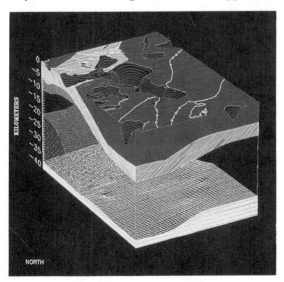

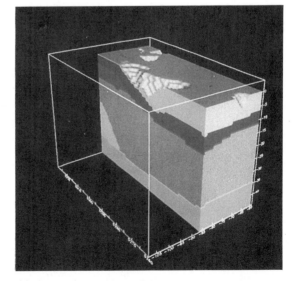

Figure 27.9   Examples of two block diagrams from a geologic transect starting in Quebec and extending eastward through Maine and into the Atlantic Ocean. (Courtesy of the USGS GIS Research Laboratory.)

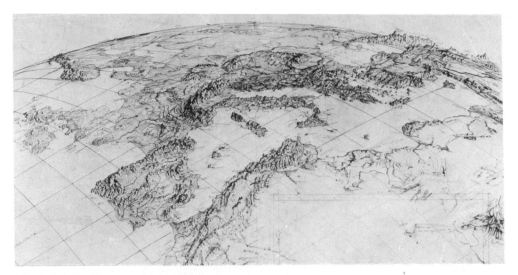

Figure 27.10 A reduced preliminary "worksheet" of an oblique regional view of Europe as seen from the southwest, prepared by R.E. Harrison to illustrate a strategic viewpoint during World War II. The final map appeared in *Fortune* in 1942.

in realistic landform portrayal. Remarkably graphic effects can be created by this method of pictorial representation. (See Figure 27.10). The computer has enhanced cartographers' capabilities to use this method. (See Figure 27.11).

Schematic Maps

Conventional planimetric maps are viewed orthogonally, and their scale relationships are systematically arranged on a plane projection. In contrast, block diagrams and oblique regional views are unconventional map forms. They are viewed obliquely and thus exhibit complex scale variation from one part of the map to another.

From block diagrams and oblique views, however, have come other map types, called **schematic** maps. These maps combine the perspective view of undulations of the land and the planimetric (two-dimensional) precision of conventional maps. On schematic maps, the pictorial treatment of the landform is highly symbolic.

One type of schematic map is called the **physiographic diagram**. On such maps, cartographers attempt to relate landforms to their geol-

ogy and geomorphology. They do so by varying value (darkness) and texture. Thus, they suggest the major structural and rock-type differences which have expression in the landforms. (See Figure 27.12).

A second version of schematic maps is called the **land-type map**. On these maps, more emphasis is placed on the character of the surface forms. This type of map often uses schematic symbols such as those developed by Erwin Raisz, to represent various classes of the varieties of landforms and land types. (See Figure 27.13).

There is no sharp distinction between physiographic diagrams and land-type maps. They share the trait that they are made on a conventional, planimetrically correct map base. The landform symbols themselves, however, are derived from their oblique appearance.

All physiographic diagrams and land-type maps have one major defect in common: Features are not in correct planimetric position. This is because the side view of a landform having a vertical dimension requires horizontal space. For example, if you draw a mountain as seen from the side or in perspective, the peak or

Figure 27.11 A visualization produced by draping a digital orthophoto over a digital elevation model and viewing the result in perspective. Area is near Harrisonburg, Virginia. (Courtesy of R. DeAngelis and the U.S. Geological Survey.)

base and its profile will be in the wrong place planimetrically.

Cartographers who draw physiographic diagrams and land-type maps recognize this fundamental defect. But they justify it, properly, on the ground that the realistic appearance outweighs the disadvantages of planimetric displacement. This is especially true on small-scale drawings in which planimetric displacement is not bothersome to the reader. At much larger scales, however, the relation between the perspective view and the consequent error of planimetric position may make it desirable to adopt other techniques.

Computers have helped immensely with this problem. On the computer screen, we can rotate the electronic map. If features are obscured from one perspective, we can change our viewpoint. Thus, we can select the best viewing angle before making a hardcopy map.

MORPHOMETRIC MAPS

Another type of map gives **morphometric**, or structural, information about the landform. This information includes average elevations, slope categories, relative relief, degrees of dissection, and so on. Some of these statistics are correlated with variations in human activity and may be used as "background data" in place of the simple elevations shown on general reference maps.

The ability to create contour, aspect, and slope maps from DTM data forms the modern basis for morphometric analyses. The portrayal of such data is straightforward: Area symbols may be used to reinforce either isarithms or dasymetric lines. Color variations on the screen allow cartographers to highlight any desired statistic relating to the land-surface form.

The major problem inherent in these methods is determining what to present, not how to present it. To use these methods, therefore, the cartographer must be a specialist about the distribution being mapped.

In addition to mapping precise elevations, cartographers often map **relative**, or local, relief on morphometric maps. Relative relief is the difference between the highest and lowest elevations in a defined area. Cartographers then portray these values with isarithms or by a simple choropleth or dasymetric map (see Chapter 26).

The relative relief method is valuable when applied to areas of considerable size, because it emphasizes the mapped distribution. It is unsuited for differentiating landform details that fall within the unit area chosen for statistical purposes. Thus, it is best adapted to relatively small-scale representation (Figure 27.14).

One useful type of morphometric map is the **slope zone** map. These maps show categories of slopes in several ways. For instance, they may show slopes in terms of average gradients, degrees of inclination, or percentage slopes. They may also

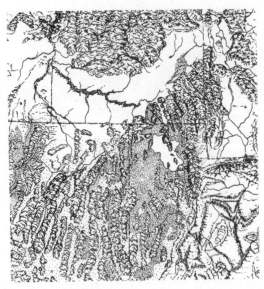

Figure 27.12 Physiographic diagrams are indeed diagrammatic. On the left is a relatively realistic portrayal of the region around Great Salt Lake (just to the right of center) and the Snake River valley, drawn by R.E. Harrison for the *National Atlas of the United States*. (Courtesy of U.S. Geological Survey). On the right is the same area from A.K. Lobeck's "Physiographic Diagram of the United States," which employs a schematic treatment to emphasize the geomorphic characteristics. (Courtesy Geographic Press, Hammond Company).

Figure 27.13 A portion of a small-scale landform map. Compare this with Figure 27.12. Note the inclusion of descriptive terms (from E. Raisz, "Landforms of Arabia," 1:3,600,000).

designate areas by such terms as flat, gently sloping, steep, and so on.

Slope zone maps are useful in evaluating relations between kinds of soil erosion, such as mud slides, gullying, or runoff rates. It is easier to make such evaluations with maps showing different categories of slopes than from maps showing only elevations.

Slope zone maps may be developed at relatively small scales. They are likely to be more useful at larger scales, however. Figure 27.15 is an example of an experimental slope zone map prepared by the U.S. Geological Survey.

Another type of morphometric map is the **slope-direction** or **aspect** map. **Aspect** is defined as the compass direction associated with the maximum rate of change of altitude or slope at a point. Aspect is computed in relation to facets created by fitting a plane to three or more neighboring elevational values from a DTM. A realistic impression of aspect variation can be rapidly produced by machine. (See Color Illustration 21.6 from Chapter 21). Im-

portant digital advances in terrain visualization using aspect have recently been reported by Moellering and Kimerling (see Chapter 21).

TERRAIN UNIT MAPS

The **terrain unit** method uses descriptive terms to give map users an idea of an area's landform. These terms may be as simple as "mountains," "hills," or "plains." Or they may be structural, topographic descriptions such as "maturely dissected hill land, developed on gently tilted sediments."

Terrain unit methods are essentially forms of dasymetric mapping (see Chapter 26). The lines bounding the area symbols have no meaning other than being zones of change.

You will find terrain unit methods used in textbooks, military terrain analysis, and regional planning. The method's main limitation is that it requires regional knowledge on the part of mapmakers and some geographic competence on the part of map readers.

HACHURES

As noted earlier, Johann Georg Lehmann proposed the hachure method in 1799. His idea was to position each individual line, or hachure, in the direction of greatest slope. That is, a hachure's orientation on the map is normal (at right angles) to the contours.

Lehmann proposed making the widths of these hachures proportional to the steepness of the slopes on which they lie. The steeper the slope, the thicker the line. The sense of down or up is not shown, however. Thus, on some maps, it is difficult to tell whether a blank area is a flat upland or lowland.

When many hachures are drawn close together, they collectively portray information about the surface configuration. The use of hachures turned out to be particularly useful on large-scale topographic military maps. For much of the 19th century, hachuring was widely employed. (See Figure 27.16 and Figure 27.17).

Variations of hachuring have developed. Instead of varying line width according to slope,

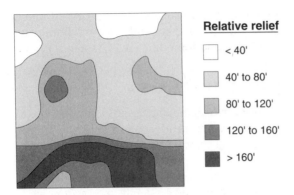

Relative relief

☐ < 40'

▨ 40' to 80'

▨ 80' to 120'

▨ 120' to 160'

▨ > 160'

Figure 27.14 An isarithmic map showing areas of similar relative relief for an area in southwestern Ohio. Values of relative relief calculated for square mile cells.

some cartographers use strokes of uniform width and make their spacing proportional to the slopes. Hachures have evolved so that they sometimes give the illusion of viewing an illuminated 3-D surface. On such maps, we can assume the light source to be orthogonal to the map (see Figure 27.16) or on one side (see Figure 27.17). On small-scale maps or maps of poorly known areas, hachuring degenerated into "hairy caterpillars." An example is shown in Figure 27.18. Fortunately, this map form is now nearly extinct.

CONTOURING

Contouring is the most precise way to give elevation information to map readers. Contours are isarithms—lines connecting points of equal elevation. They are the traces that result from passing parallel, equally spaced, horizontal surfaces through the three-dimensional land surface and projecting these traces orthogonally to the map surface. (See Chapter 26). The vertical distance between the parallel surfaces is called the **contour interval**.

Generally, contours are not visible on the land. Figure 27.19 shows the usual way we see three-dimensional form—by the interplay of light and dark—and, for comparison, a contour representation of the same feature.

Contours present a less dramatic visualization of the landform surface than hill shading (see *Hill Shading* later in this chapter). But they provide an immense amount of information to the skilled map analyst and interpreter. For analytical purposes, therefore, contouring is by far the most useful way of portraying the landform.

Contour Intervals

The usefulness of contours depends on their vertical spacing. Yet the choice of contour interval is not an easy task. Cartographers must play detail against space available to show contours on the map.

As contour interval is increased, more surface detail is lost between contours (Figure 27.20). Sometimes, because of small scale or lack of data, contour intervals must be excessive. In such cases, contours can only approximate the surface form. Rather than use contours in these cases, it may be better to use other methods of presentation, such as hill shading. If you do come across a small-scale contour map, you should be suspicious of its accuracy. Avoid using such maps for detailed interpretation and analyses.

Choice of contour interval is based on a number of factors. The first is the accuracy and completeness of landform data. (See the next section, *Contour Accuracy*). The better the data, the smaller the contour interval that can be justified.

A second major factor is the purpose of the map. A very large-scale topographic map, to be used for engineering and planning, requires a very small contour interval. A small-scale regional map, on which only major land-surface forms are needed, can be effective with a relatively large contour interval.

A third important factor is map scale. An interval that is too small for the scale will result in unwanted contour crowding.

Portrayal of slope (gradient) is one of the major objectives of contour maps. Consequently, contour intervals are almost always constant. (That is, they are equal-step progressions). When an uneven contour interval is used, slopes are difficult to visualize and calculate, and misleading impressions are likely.

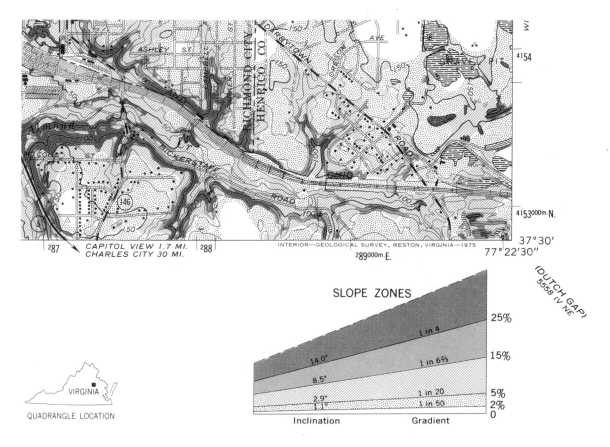

Figure 27.15 A portion of the experimental Richmond, Virginia, slope zone map, 1:24,000, prepared by the U.S. Geological Survey.

Differences in terrain in different parts of the world or in a large region make it impossible to have a constant contour interval for an entire map series. For example, a contour interval of 40 feet would be appropriate for the standard U.S. 1:24,000-scale topographic series in a high relative relief area, such as the western United States. But that same interval would allow only eight contour planes for the entire state of Florida (in which the maximum elevation is 345 feet). It is inconvenient to have different contour intervals on adjacent maps of a series. Thus, it is desirable to maintain regional consistency in contour intervals whenever possible.

One way to help attain such consistency is to use **supplementary contours**. These contours are usually established at a simple fraction of the basic interval. Supplementary contours are shown on U.S. Geological Survey maps as dotted or screened lines.

Supplementary contours are particularly useful on floodplains and other areas of low relief when the basic interval for the rest of the map is too large to delineate minor forms that are significant locally, such as natural levees and banks (Figure 27.21).

Choosing contour intervals and adding supplementary contours are problems associated with

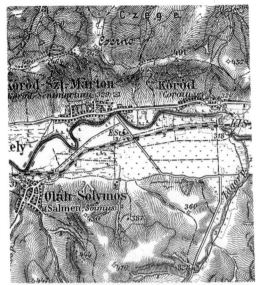

Figure 27.16 A section of a topographic map that portrays the land-surface form with vertically illuminated hachuring. (From Sheet 5473 (1894), Austria-Hungary, 1:75,000).

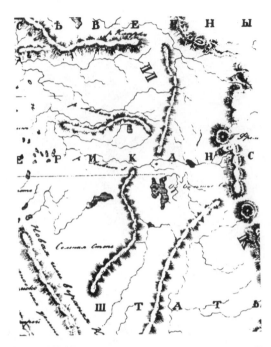

Figure 27.18 Genus hachure, species caterpillar. (From an old Russian atlas of western North America).

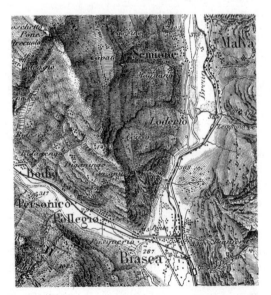

Figure 27.17 A section of a topographic map that portrays the land-surface form with obliquely illuminated hachuring. (From Sheet 19 (1858), Switzerland, 1:100,000, the "Dufour map").

the rigidity of traditional maps. Digital cartography can help us solve these problems. For example, digital methods give us the ability to easily change contour intervals. They also let us "zoom" in on features or "pan" around to gain a regional view.

Contour Accuracy

The quality of landform delineation by contours involves two kinds of accuracy—relative and absolute. **Relative accuracy** refers to the relationship of the contours with one another. Relative accuracy is important, because it gives us an idea of the shapes of terrain features.

Absolute accuracy gives us precise elevations. The absolute accuracy standard for the United States specifies that contour lines must be positioned within a band representing one-half the contour interval above or below the true elevation. This absolute accuracy standard is adequate for most engineering purposes but not for showing true slopes and terrain configurations.

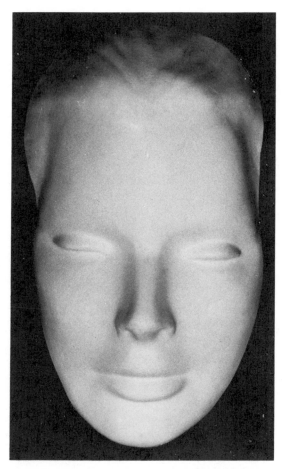 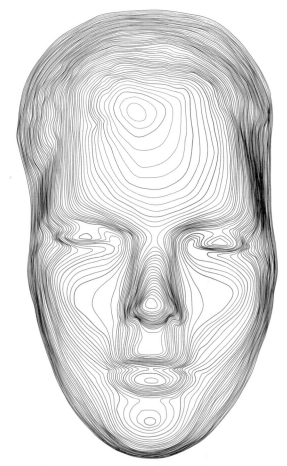

Figure 27.19 A vertically lighted plaster model and a precise contour map with a 1-mm. interval derived from it. The contours were obtained by photogrammetric methods. (Courtesy G. Fremlin).

For example, given a constant contour interval, a uniform slope should be delineated by evenly spaced contours. Suppose adjacent contours were spaced successively nearly a half interval above and below their correct elevations (which, remember, would meet absolute accuracy standards). The resulting contour pattern would imply terraces instead of a uniform slope! This deception could occur even though the map would comply with absolute map accuracy standards (Figure 27.22).

Both relative and absolute accuracy are important in giving us a feel for the landform. Fortunately, modern photogrammetric methods are precise enough that we can usually obtain both types of accuracy. Although absolute accuracy is not difficult to assess, relative accuracy is. Thus, current accuracy standards do not include any requirement for relative accuracy. The only way to check relative accuracy is to compare contours with a stereo model (see Chapter 12).

It's important to remember that not all contour maps are of the same order of accuracy. Before air photos were used to derive contours, the lines were drawn in the field with the aid of a few ground-survey "leveling" lines (see Chapter 7) and **spot heights** (a scattering of elevation figures). Consequently, they were often not accurately located.

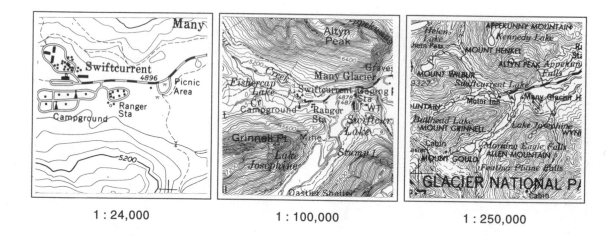

1 : 24,000   1 : 100,000   1 : 250,000

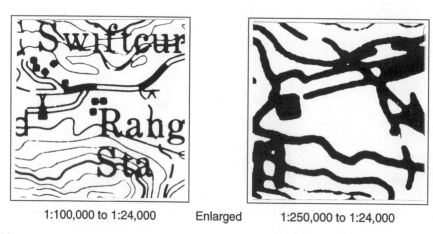

1:100,000 to 1:24,000   Enlarged   1:250,000 to 1:24,000

Figure 27.20 A comparison of contours for the same area at different scales of portrayal. (All sheets available from the U.S. Geological Survey).

Another accuracy problem occurs because of the arbitrary location of contour slices through the terrain. This may cause omission or distortion of small features. The horizontal planes that produce the contours are measured from an arbitrarily defined datum such as mean sea level. Where these planes intersect the three-dimensional land surface is strictly a matter of height above the datum. Whether a particular terrain feature is portrayed adequately is a matter of chance. Sometimes it is desirable to shift contours, within the limits of absolute accuracy, to more effectively portray a feature that is prominent in the landscape but not adequately revealed by the accident of geometry (Figure 27.23). The smaller the contour interval, the less likely it is that such omissions or distortions will occur.

Generalization of Contours

Contour lines have thickness, which limits their degree of curvature. Consequently, it is impossible for a contour line to reflect every variation in elevation. This means that all contours on maps are generalized, regardless of their scale. (See

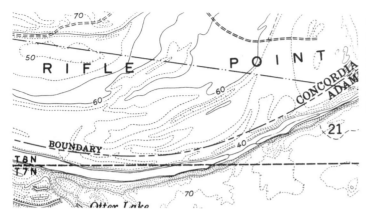

Figure 27.21 Supplementary contours at a 5-ft. interval, used to show detail on a section of the lower Mississippi floodplain. From Natchez, Mississippi-Louisiana, 7.5-minute quadrangle, U.S. Geological Survey, 1:24,000, for which the basic contour interval is 20 ft.

Chapter 24). The way we perform this generalization depends upon the terrain and the map scale.

When we generalize contours, we have three objectives: (1) to locate contours as accurately as possible, (2) to show as much detail as is consistent with the map scale, and (3) to effectively portray terrain features.

It is the set of contour lines, not the individual lines, which is generalized to depict the major masses, ridges, valleys, and depressions of a landscape. These landform features should retain their intrinsic character. The generalization should neither consist of lines smoothed into more rounded forms nor lines simply made more jagged. Attempts should be made to retain any sizable, sharply defined breaks in slope, such as crestlines and scarps.

We can produce generalized contour maps from a digital database, usually a DTM, in a variety of ways. Many studies are under way to assess the relative merits of these different methods. Increasingly, cartographers rely upon software to perform this generalization. However, such software does not absolve cartographers from responsibility. It is still up to them to carefully select which algorithm to use, based on map purpose.

Layer Tinting

Perhaps the most widely used method of presenting land-surface information on wall maps, in atlases, and on other printed "physical" maps is **layer tinting**. This method is also called **hypsometric coloring** or **altitude tinting**.

When cartographers use layer tinting, they apply area symbols to the zones between contours. On small-scale maps, the chosen contours are greatly simplified. Consequently, contours on small-scale maps are not particularly meaningful in presenting individual features. The process degenerates into a presentation of large areas of similar surface elevations.

We can make layer tints on small-scale maps more meaningful if we combine them with pictorial techniques or shading. Such maps present major structures as well as some details of form. Many modern atlases are excellent examples of the effectiveness of this combination.

Today's computer software algorithms enable cartographers to perform layer tinting easily. They are able to highlight areas of interest in a visualization by flooding these areas with color as they observe the map on the screen. They can easily perform this color-filling operation, for example, to the area lying above or below a given elevation or lying between two given elevation values. This is a quick and visually stimulating way to draw attention to a given elevation zone.

HILL SHADING

Contours accurately portray elevations, but a contour map doesn't look very realistic. A person well-trained in interpreting contour maps can visualize

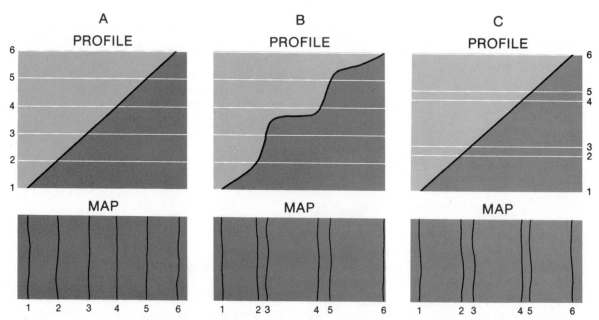

Figure 27.22   How displacement of contours can affect portrayal. The profiles in (A), smooth slope, and in (B), benches or terraces, are characteristically portrayed by contours on their respective maps when the contours are accurately located. If, as in (C), contours 2, 3, 4, and 5 were displaced as shown—less than half the contour interval, which meets map accuracy standards—then the resulting map would suggest benches, as in (B), instead of the true smooth slope, as in (A).

the landform symbolized by contour lines, but most casual map readers cannot. Instead, most people recognize shapes primarily by the interplay of light and dark. This interplay creates patterns which we recognize because they correspond with those in our experience.

There is nothing new about varying light and shade to artistically model shapes. *Chiaroscuro* is a standard art term derived from the Italian words *chiaro* (lightness) and *oscuro* (darkness). It describes any graphic method that derives its realistic effect from the gradation of light and dark.

As we have seen, several forms of hachuring were attempted during the 19th century to artistically portray the land-surface form. But it was not until the second half of the 19th century that cartographers could easily reproduce smooth tonal variations from light to dark. Not until then could they bring together photographic, engraving, lithographic, and halftone processes to gain realism. (See Figure 27.24).

The most realistic landform portrayal is thus a pictorial method that imitates the way we see shapes. Many elements, such as color, clarity, texture, and perspective are involved in creating this portrayal. But by far the most important element is the interrelation of dark and light. This method of portraying the land-surface form is called **hill shading**. It is also known as **relief shading**, **plastic shading**, **shaded relief**, or, simply, **shading**.

The highest-quality hill shading is produced by talented artists. These artists need to be skilled not only in handling the tools and materials but also in interpreting terrain characteristics from source information (usually contour maps and air photographs). A well-done manually shaded map is a portrait that captures the character of an area's landforms. (See Figure 27.25).

On the other hand, each artist will interpret an area differently. Thus, adjacent sheets of a series hill-shaded by different people are likely to look different.

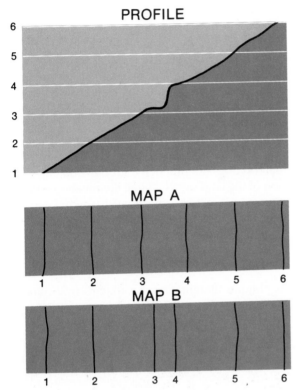

Figure 27.24 A portion of a shaded relief map titled "Rio Colorado of the West" by F.W. von Egloffstein, found in J.C. Ives, *Report Upon the Colorado River of the West*, Washington: Government Printing Office, 1861.

Figure 27.23 The prominent bench or cliff in the profile does not show in (A) because it occurs between two contour planes. It is desireable to shift contours 3 and 4 closer together on the map (each less than one-half the contour interval) as in (B), to improve the relative accuracy of the portrayal. (Based upon "Topographic Instructions of the U. S. Geological Survey").

There have been many attempts to make hill shading more mechanical and thus minimize the need for the talents and expense of artists. These attempts have increased as computer-generated graphics have become available. Today, realistic, photographic-quality hill shading is possible by computer. (See Figure 27.7).

Automated hill shading requires elevation data in the form of a digital elevation (terrain) model. Digital elevation models may be generated by interpolation from contour maps. Or, when high-quality photographic source materials are available, digital elevation data can be extracted auto-

matically. Other sensors (notably radar) can also help provide the necessary elevation data and structural information for the cartographer who is working manually.

Automated hill shading has several important advantages. It is fast and relatively inexpensive. Greater detail is possible than with manual hill shading. Also, hill shading of maps in a series will be consistent, and adjoining sheets will match.

Whatever the method of preparation, hill shading is essentially a map of brightness differences

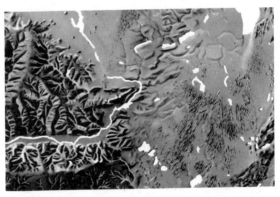

Figure 27.25 A portion of a shaded relief map of the state of Wisconsin. Rendered by D. Woodward (Courtesy of D. Woodward).

resulting from incoming (or **incident**) light being reflected to an observer. Producing effective
light-and-shade relationships depends upon certain assumptions. First, we assume that the light is
coming from a constant direction and elevation.
We also assume that the light is being reflected by
various orientations and slopes of an ideally reflective land surface. We assume that the observer is viewing the map orthogonally (from
directly overhead at all points). Thus, different
hill-shaded maps result from changing the direction and elevation of the light source (Figure
27.26).

In theory, the light source that produces the variations in reflectance can be located: (1) at the zenith that
is at the vertical viewing position, or (2) at any other
position, which will provide oblique illumination.
Let's look at both these illumination possibilities.

Vertical Illumination

Vertical illumination produces a pattern of reflectance in which horizontal surfaces are white, regardless of their elevation. The greater a surface
deviates from horizontal, the darker it appears.
Thus, shading darkness is directly related to only
one variable, slope steepness.

Although vertical illumination is analytically
useful if slope steepness is your main concern, the
method does not produce a strong visual impression of the landform. Oblique illumination—light
from any direction other than the zenith—produces a completely different and often more revealing pattern of reflectance.

Oblique Illumination

In oblique illumination, a surface's reflectance

Figure 27.26 The difference in appearance is due to the position of the light source relative to the viewer.
On the left, the light source is in front of the viewer; on the right, the light source is behind the viewer.

depends on its orientation relative to the direction of the incoming (incident) light. Ordinarily, oblique illumination is from the upper left (that is, northwest if the map is oriented with north at the top). If we assume a normally oriented map with illumination from the northwest, then a surface facing northwest with a slope perpendicular to the illumination will have the maximum reflectance (white). Any other northwest-facing surface having a greater or lesser slope will be proportionally darker. A horizontal surface will have a reflectance approximated by medium-gray shading.

Similarly, the more the aspect of a landform feature, such as a ridge, departs from a northwest orientation, the darker the feature's surface. Maximum darkness occurs when the steepest slope faces southeast. In effect, a southeast-facing slope is completely in shadow.

One of the problems facing the cartographer who plans to use shading is selecting the direction (or azimuth) from which the light is to come. When the light comes from a direction in front of the viewer, elevations generally appear "up" and depressions "down." A curious phenomenon is that with light from some directions (usually those behind the observer), depressions and rises will appear reversed.

In addition to providing the correct impression of relief, the direction of lighting is important in effectively portraying the surface being presented. Many areas have a "grain" or alignment of features that would not show up effectively if illuminated from a direction parallel to that grain. For example, a smooth ridge with a northwest-southeast trend would result in about the same illumination on both sides if the light were to come from the northwest.

The elevation of the light source also has a significant effect on the map. If we lower the elevation of the light source, we will emphasize low relief forms and increase apparent relief by producing greater contrasts. But at the same time we will decrease the visibility of detail in the darkened areas.

Computer software has made it possible to rotate the surface being portrayed and to raise or lower the elevation of the light source. Using manual techniques, these decisions had to be made prior to execution and were unchangeable once manual execution had begun.

Shading as Symbolism

Hill shading can be so realistic that there is a tendency to think of it as reality itself rather than a symbolization of it. In actuality, hill shading employs area symbols of differing values to produce a 3-D effect. (Sometimes the areas are so small that they may be considered points). Thus, it is as much a symbolization as any other map.

We must always keep in mind that a hill-shaded map is not an exact copy of the real world. It is a generalized and exaggerated portrayal of what we think our impressions of a denuded land surface might be.

When color is added to hill-shading, we can create an even more believable portrait of the land-surface form. Even just one color added to black-and-white shading lets us show more subtle distinctions in the land surface. For example, we can give glacial forms a blue tint or sand areas a buff tint.

Using full color, we can achieve extraordinary realism. Full-color renditions can incorporate vegetational, seasonal, and other variations in the surface coloring. The best examples of these maps, such as those created for Rocky Mountain ski resorts, almost achieve photo-realism. This type of mapping is an art form well beyond the scope of this book.

SELECTED REFERENCES

Chang, K., and B. Tsai, "The Effect of DEM Resolution on Slope and Aspect Mapping," *Cartography and Geographic Information Systems*, 18, 1991: 69–77.

Imhof, E., *Cartographic Relief Presentation*, H.J. Steward, (ed.), Berlin and New York: Walter de Gruyter, 1982.

Irwin, D., "The Historical Development of Terrain Representation in American Cartography," *International Yearbook of Cartography*, 16, (1976): 70–83.

Keates, J. S., "Techniques of Relief Representation," *Surveying and Mapping*, 21, (1961): 459–463.

Kraak, M. J., "Cartographic Terrain Modeling in a Three-Dimensional GIS Environment," *Cartography and Geographic Information Systems*, 20, 1993: 13–18.

Kumler, M. P., and R. E. Groop, "Continuous-Tone Mapping of Smooth Surfaces," *Cartography and Geographic Information Systems*, 17, 1990: 279–289.

Lobeck, A. K., *Block Diagrams*, 2nd ed., Amherst, MA: Emerson-Trussel Book Co., 1958.

Means, R., *Shaded Relief Technical Manual, Part 1*, ACIC Technical Manual RM-895. St. Louis, MO: Aeronautical Chart and Information Center, 1958; reprinted 1962.

Moellering, H., and A. J. Kimerling, "A New Digital Slope-Aspect Display Process," *Cartography and Geographic Information Systems*, 17, 1990: 151–159.

Robinson, A. H., and N. J. W. Thrower, "A New Method for Terrain Representation," *The Geographical Review*, 47, (1957): 507–520.

Tanaka, K., "The Relief Contour Method of Representing Topography on Maps," *The Geographical Review*, 40, (1950): 444–456; also in *Surveying and Mapping*, 11, (1951).

Thompson, M. M., *Maps for America*, 3rd ed., Washington, D.C.: U.S. Geological Survey, 1988.

Tuhkanen, T., "Computer-Assisted Production of Hill Shading and Hypsometric Tints," *International Yearbook of Cartography*, 27, 1987: 225–232.

Wallis, H. M., and A. H. Robinson, *Cartographical Innovations: An International Handbook of Mapping Terms to 1900*, Map Collector Publications, 1982.

Weibel, R., "Models and Experiments for Adaptive Computer-Assisted Terrain Generalization," *Cartography and Geographic Information Systems*, 19, 1992: 133–153.

Weibel, R., and M. Heller, "A Framework for Digital Terrain Modelling," *Proceedings of the 4th International Symposium on Spatial Data Handling*, Zurich, 1990: 219–229.

Yoeli, P., "Shadowed Contours with Computer and Plotter," *The American Cartographer*, 10, (1983): 101–110.

CHAPTER 28

MULTIVARIATE MAPPING AND MODELING

*O*n most thematic maps, a single variable is depicted over a framework of locational base information. This practice serves the purpose of studying the structure and spatial variation of an environmental phenomenon. But map users often want to consider several variables simultaneously. They may, for example, wish to estimate the degree or spatial pattern of cross-correlation between features.

In such cases, it is common for people to compare phenomena depicted on separate maps. This can be a difficult task, however, especially if the maps to be compared are not of the same scale, projection, orientation, or level of abstraction. The task grows rapidly more costly and difficult as the number of maps to be compared increases.

To overcome problems associated with comparing variables on separate maps, cartographers sometimes represent multiple variables on the same map. Although these multivariate maps vary greatly in conceptual sophistication and graphic complexity, they can be categorized into a few groups based on underlying mapping logic. In this chapter, we will discuss the basic approaches to multivariate mapping, along with issues of multivariate map fidelity and readability.

SUPERIMPOSITION OF FEATURES

The most common approach to multivariate mapping is to superimpose distributions, using different symbols or mapping techniques. The trick is to use symbol schemes that are relatively transparent yet visually distinct, especially as the number of phenomena to be superimposed increases. When only two variables are involved, there are many choices of compatible mapping methods. Dots and proportionate point symbols (Figure 28.1), or proportionate point symbols and isolines, can be effective combinations.

The advantage of the superimposition method is that it is conceptually simple. It can be quite effective when only two or three variables are involved. But map clarity suffers as the number of

variables increases. Also, it is difficult to convey the relative importance of the various phenomena. To some extent, these problems are overcome by adopting segmented symbols for multivariate mapping.

SEGMENTED SYMBOLS

Rather than employ different symbols or mapping methods for each phenomenon on a multivariate map, another approach is to map them separately using some form of segmented or divided symbols. The standard pie chart is often chosen for this purpose, probably because of its logical simplicity (Figure 28.2). Even more complex multivariate relations can be represented by a single map symbol through various visual elements, such as size, texture, and tone. Maps based on these more elaborate segmented symbols can be far less intuitive to use than those employing simple pie graphs, however (Figure 28.3).

One way to overcome problems associated with using abstract segmented symbols of many dimensions is to resort to familiar graphic forms. For example, we seem to be especially skilled in detecting and remembering complex facial expressions. If we could take advantage of these special skills by segmenting symbols using facial characteristics, map communication might be enhanced. Indeed, as can be seen in Figure 28.4, an immense amount of data is depicted with great clarity by using segmented faces.

The cartographer must exercise caution when designing segmented symbols. A symbol that takes little effort to construct may be difficult to use. Map users are known to have trouble estimating and comparing proportions, especially if the segments to be compared are of different shape and orientation. Neighboring symbols set up visual "field effects" that can alter perceptual judgments. It can also be difficult to compare parts of symbols that are widely separated on a map, particularly if there are many intervening and surrounding symbols to distract attention.

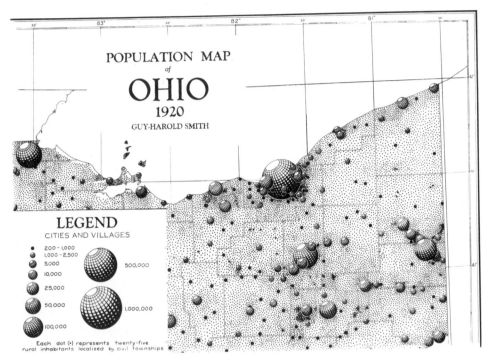

Figure 28.1 This multivariate thematic map achieves design clarity through superimposition of dots (rural population) and proportional point symbols (urban population). (Courtesy of G.H. Smith and *The Geographical Review*, published by the American Geographical Society of New York, 1920.)

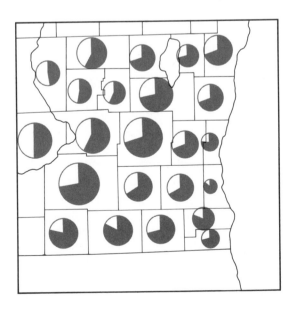

Given the potential difficulties that segmented symbols pose for map users, it is best that their design be kept as simple as possible. Elaborate multicomponent symbols may turn out to be more confusing than useful, resulting in a communication failure in spite of considerable map design effort.

Figure 28.2 The familiar pie chart is widely used in multivariate mapping. In this case land in farms (circle size) and the percentage of that land available for crops (shaded sector) by county is mapped for southeastern Wisconsin.

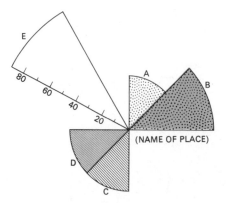

Figure 28.3 Examples of several ways to employ circle segments. (A) and (B) in the upper right quadrant indicate nominal (hue) and ratio (size) characteristics (for example, amount of passenger versus freight traffic). (C) and (D) in the lower left quadrant use texture/spacing to distinguish between nominal categories (for example, long grain versus short grain rice). (E), with ratio (size) scaling, can indicate yet another category, such as per capita income.

CROSS-VARIABLE MAPPING

The superimposition and segmented symbol approaches to multivariate mapping both represent variables separately, leaving the map user to judge their relationships. Another approach is actually to combine the separate variables into a single hybrid variable, which can then be represented by map symbols.

Cross-variable mapping, which is based on the choropleth mapping technique, is one method of creating and representing these hybrid measures. Two steps are involved. First, in the two-variable case, a rectangular legend is produced by plotting values from one variable against those of the other (See Figure 21.4 and Color Figure 21.3). Each axis of this scatter plot is then divided into classes, leading to a division of the legend into boxes. Each box represents a special relationship between the variables, such as high *x*/low *y*, high *x*/high *y*, and so forth.

The second step is to convert the numerical data

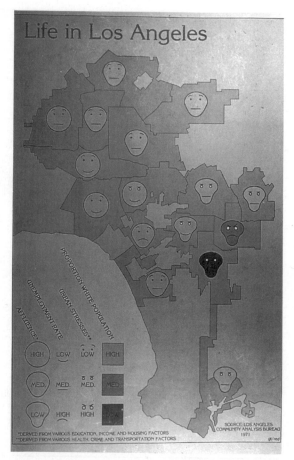

Figure 28.4 This dramatic multivariate map of Los Angeles effectively depicts four social status factors, each divided into three classes (low, medium, high). (Courtesy E. Turner.)

to graphic form. This is done by assigning graded symbols to each variable, so that each box has a distinctive visual character. Once this is accomplished, the map can be plotted. Each symbol on the map is drawn from the legend matrix so, in theory, the map user can read the multivariate information directly. If the legend is well-designed, the map should be relatively intuitive. Unfortunately, people seem to have difficulties using cross-variable maps. In fact, many map users say they prefer separate maps over the hybrid single display. At the least, map makers should keep the number of variable classes to two or three, since four classes for each variable would lead to an overwhelming 16 symbols. Also, they should choose symbols that provide a clear distinction between classes. Finally, they should emphasize off-diagonal and extreme (corner) classes in the legend, since they normally represent least expected relations and, therefore, hold the most information (see Color Figure 21.3).

Although a two-variable example is used here, cross-variable mapping is sometimes extended to three phenomena. The triangular graph and three-dimensional histograms have both been used to create a three-variable legend (Color Figure 28.1). Hybrid measures based on three variables are difficult for all but the most skilled and dedicated map users, however, so they are rarely attempted or encountered.

A word of caution about cross-variable maps is warranted. Since the map display itself looks no different from a choropleth map representing a single variable, it can be deceptive. For this reason, the legend should be a clear and prominent aspect of the map design. Every possible effort should be made to encourage the map user to use the legend as an initial step in the map reading process.

COMPOSITE INDEXES

A second and more complex approach to forming a single hybrid variable is to combine several measures into a single numerical index. This index can then be mapped as if it were a single variable. This type of mapping is called **cartographic modeling** or **composite variable mapping**. This type of multivariate mapping is growing rapidly in popularity, in part because complex environmental problems require more sophisticated maps, and in part because modern geographic information systems facilitate the demanding computational process needed to produce composite variables.

Note, for example, the groundwater contamination susceptibility map* in Color Figure 28.2. Since there is no instrument available to take into the

field to measure a concept such as susceptibility to groundwater contamination, a measure of susceptibility had to be modeled from phenomena that can be observed. Those deriving this model had to make three decisions: (1) which variables to include, (2) the importance to be assigned to each variable, and (3) how the weighted variable scores are to be combined into a single index. Each of these decisions warrants further consideration.

CHOICE OF VARIABLES

Ideally, the variables that contribute to the phenomenon we want to model are those that should be incorporated into a composite index. Unfortunately, we may be wrong in what we think is important. Furthermore, the high cost of primary data collection means that we often have to make do with available information. As a result, the choice of variables is often made more on practical than on theoretical grounds, and it would likely be different at some future date when we know more about the issue.

In the case of our map example, the index was created using data on surficial deposits, bedrock type, soils, depth to bedrock, and depth to groundwater. Conveniently, maps for each of these five variables were available for the state at the 1/500,000 scale at the time the project was undertaken.

In order to take advantage of the automated map overlay capability provided by geographic information system technology, all resource classes (regions) on the five maps had to be available as a digital coordinate record. Since the appropriate computer databases did not exist, they were laboriously created using manual table digitizers. The digital files were then edited and formatted for use with GIS software.

---

*This map was produced by the Wisconsin Department of Natural Resources and is based on work reported by Robin R. Schmidt in her master's thesis titled *Wisconsin Groundwater Pollution Susceptibiity Mapping: A Description and Analysis* (University of Wisconsin, 1986).

VARIABLE WEIGHTING

With the choice of variables made, the next step is to define the relative contribution of each variable to the overall index. This can be done by assigning numerical weights to attribute classes within each variable. It can also be accomplished by linking the weighted variables arithmetically in the final model.

The weighting method used with the example map is outlined in Table 28.1. Notice that a complex, two-stage weighting scheme is used. Each attribute for the bedrock type, surficial deposits, soil characteristics, and depth to water table variables is given a weighting value. Depth to bedrock, on the other hand, is considered to be so important in the model that it is used as a multiplier with the

Table 28.1
This is the weighting and linking model used to produce the final rating scheme for the groundwater contamination susceptibility map reproduced in Color Figure 28.2.

MULTIPLIER VALUES:	Type of Bedrock Multiplier	Surficial Deposits Multiplier	Soil Characteristics Multiplier	Water Table Multiplier
If Depth to Bedrock:				
0–5' (70% of Area)	13	0	1	1
0–5' (35%–70% of Area)	11	1	2	1
5'–50'	6	4	3	2
50'–100'	0	8	4	3
>100'	0	8	4	3

RATING SCHEME VALUE:

Resource Map	Attribute	Value
Type of Bedrock	Carbonate	1
	Sandstone	5
	Ign./metamorphic	6
	Shale	10
Depth to Water Table	0–20'	1
	20'–50'	5
	>50'	10
Soil Characteristics	Course texture/high permeability	1
	Medium course texture/high–medium permeability	3
	Medium texture/medium permeability	6
	Fine texture/low permeability	10
Surficial Deposits	No materials	0
	Sand and Gravel	1
	Sandy	2
	Peat	5
	Loamy	6
	Clayey	10

Formula for Composite Score:
[*Bedrock Type*(value) × *Bedrock Type*(multiplier)] + [*Surficial Deposits*(value) × *Surficial Deposits*(multiplier)] + [*Depth to Water Table*(value) × *Depth to Water Table*(multiplier)] + [*Soil Characteristics*(value) × *Soil Characteristics*(multiplier)] = *Composite Score*

other resource variables. Specifically, the depth to bedrock attributes are assigned unique multiplier values for each of the other resource variables. These multipliers are then used with the attribute values assigned to the other variables to derive a complex weight or score.

LINKS BETWEEN WEIGHTED VARIABLES

The third decision that must be made concerns the formula that will define the final composite score or index. In the example map, the individual weighted scores (attribute value × multiplier value) were merely added together. Whether weighted scores are added to, subtracted from, multiplied with, or divided by each other depends on the process underlying the concept we are modeling. Additive models are conceptually simplest and therefore most popular.

Once the formula for the composite index is defined, this information, along with the cartographic data files, can be input into GIS software. The task at this point is twofold. It is first necessary to perform polygon overlay between all variable attribute categories on each resource map to determine new hybrid map classes. Each hybrid class represents a certain combination of surficial deposits, soil characteristics, bedrock type, depth to bedrock, and depth to groundwater attributes.

The second task is to determine the composite score for each of the hybrid map attribute classes. These index values can then be displayed using an appropriate value-by-area symbol scheme. In the case of the example map, the legend ranks these symbols from low to high susceptibility.

FIDELITY OF CARTOGRAPHIC MODELING

The growing importance of cartographic modeling in environmental decision making raises the issue of the credibility of these multivariate maps. Concern with map accuracy traditionally focuses on positional errors. Thus, we have both horizontal and vertical map accuracy standards for government map series such as USGS topographic quadrangles. Factual errors, such as feature omission or mis-identification, fall beyond the realm of these accuracy standards. The effects of cartographic generalization are likewise ignored by conventional measures of map accuracy.

In the case of composite index maps, it generally is not practical to test for positional errors. Consider, for example, the problem of testing the accuracy of a groundwater contamination susceptibility map. Since boundaries between map classes cannot be found on the ground, field search is of little value. Similarly, the number, magnitude and dispersion of past toxic spills is not sufficient to provide a reliable means for evaluating map fidelity. Nor is it possible to create new toxic spills in order to generate the needed test data. Finally, the abstract, conceptual nature of the model makes it difficult to hold suitable test data out of the model development process.

Although we cannot easily test a cartographic model for positional accuracy, we can evaluate the sensitivity or stability of the map. The choice of variables, variable weights, and links between weighted variables each can be tested. The index can be recomputed after variables are added or dropped, weights are altered, and links (+, -, ×, /) are changed.

Small changes between original and recomputed index values would indicate low sensitivity and high stability. Map users could have high confidence that such a map reflected environmental conditions.

Conversely, if major changes occurred, the model would be judged to have low stability and a high sensitivity to data inputs. Such a map would likely be more an artifact of the modeling process than a true reflection of environmental conditions. An unstable map is at best suitable for abstract, regional decision making.

Given the importance of cartographic modeling, an indication of map sensitivity or stability should be included in the legend of these maps. Although this would add significantly to mapping cost, the inclusion of a map confidence statement should become an expected aspect of cartographic ethics. Maps lacking such an assessment should be treated with great caution.

READABILITY OF MULTIVARIATE MAPS

Multivariate mapping increases the amount of information a map carries. Sometimes such mapping is performed to cut expenses, since a single map costs less to reproduce than several separate maps. Multivariate mapping may also be done with the aim of facilitating variable comparison. Coding more information into a single map means fewer maps are needed when studying relations between environmental features. It is generally thought that comparisons between variables on a single map are more effective than comparisons between separate maps, although this is not well documented. A single map does, however, reduce problems associated with comparing maps with different scales, projections, and orientations.

Unfortunately, the readability of multivariate "supermaps" is often in doubt. Symbols can become so diverse in form and so densely packed that map clarity suffers. The complexity of the map symbols may not be appreciated, or much of the information contained by these symbols may be ignored. A map user who is able to crack the multivariate symbol code may be so overwhelmed by the volume of information unleashed that confusion rather than communication is the result.

At the least, the cartographer has an obligation to stop adding variables before the graphic quality of a multivariate map suffers. There is also a responsibility to provide a clear statement of the mapping process in the map legend. A common failure of multivariate maps is that an explanation of the complex data manipulation prior to mapping is not found on the map. Map users often lack access to the obscure documents that explain such matters, which greatly diminishes a map's value.

SELECTED REFERENCES

Benson, J. L., *Map Accuracy in Composite Variable Mapping: A Look at the Modelling Process*. Master's Thesis, Department of Geography, University of Wisconsin-Madison (1991).

Chang, K-T., "Multi-Component Quantitative Mapping," *The Cartographic Journal* 19 (1982): 95–103.

Eyton, J. R., "Complementary-Color, Two-Variable Maps," *Annals of the Association of American Geographers* 74 (1984): 477–490.

Fedra, K., "GIS and Environmental Modeling," Paper presented at the *First International Conference/Workshop on Integrating Geographic Information Systems and Environmental Modeling*.

Hopkins, L. D., "Methods of Generating Land Suitability Maps: A Comparative Evaluation," *Journal of American Institute of Planners* 43 (1977): 386–398.

Lodwick, W. A., Monson, W., and L. Svoboda, "Attribute Error and Sensitivity Analysis of Map Operations in Geographical Information Systems: Suitability Analysis," *International Journal of Geographical Information Systems* 4 (1990): 412–428.

McHarg, I., *Design With Nature*. New York: The Natural History Press, 1969.

Pelto, C. R., "Mapping of Multicomponent Systems," *Journal of Geology* 62 (1954): 501–511.

DYNAMIC/ INTERACTIVE MAPPING

Although the environment we map is always changing, maps by tradition have been passive, static representations. These freeze-frame snapshots of a dynamic environment are in part a useful abstraction, a way to look at space with time held constant. They are also a consequence of our limited mapping technology and resources.

Dissatisfied with these static images, mapmakers have explored various ways of adding a temporal dimension to map symbols. These efforts have focused primarily on the nature of mapped data, rather than on the map symbols themselves. The result is maps that depict environmental change, or the rate of change, over given time periods. The maps themselves are still static.

No matter how cleverly cartographers code time data into their map symbols, map users want more. They want maps set into motion. In the past, they could achieve this effect to some degree by walking around a physical model to gain the perspective of different vantage points. Or, they could manipulate the position of selected features on a static map base. Planners and military strategists, for example, often play out real-life and "What if?" scenarios by moving symbolic objects around on a base map.

But an even more useful solution is to put the entire map into motion. Mapmakers have produced such dynamic maps for some time using motion picture technology. Now, this task is greatly aided by applying video and electronic technologies that support dynamic mapping. Truly animated maps, as well as map-based navigation systems and simulations, are now a reality. These recent mapping developments are changing the relation between a map's maker and user, and opening new mapping opportunities.

ANIMATION

We can put maps into motion with the help of several animation techniques. Each involves first creating separate maps from different vantage points or time periods, or using different mapping techniques or parameters to create a series of maps. When we view these individual maps rapidly, in sequence, changes from one image to the next are transformed into a sense of motion.

The age of cinema introduced us to the potential of animated mapping. We can take a motion picture of the landscape by flying a camera over and around a region. Or we can rotate a movie camera around a physical model of the landscape rather than the environment itself. Both methods can produce dramatic results. Inexpensive video cameras (camcorders) have largely replaced movie cameras in recent years and brought new life to animated mapping.

But the most important advance in animated mapping came with the advent of high-speed, digital computers and their associated video display terminals. With the aid of computers and sufficient numerical data representing the terrain, we no longer need to photograph or videotape a physical model or the terrain surface itself. Instead, with appropriate software, we can create images from digital environmental databases through a series of calculations. These calculations take into account the vantage point of the observer, the orientation of the surface, and the source of illumination. If we wish to add natural textures or landcover, databases holding this information must be available. The images are then displayed on a video screen and may be recorded electronically in video mode for subsequent playback.

A powerful computer is needed to produce the 30 or so frames a second required for realistic, smooth motion. The realism of these animated maps depends primarily on the sophistication of available hardware, database, and software. Realism is also expensive in human and monetary terms. Less ambitious systems will show a rather skeletal terrain from a limited number of vantage points. More advanced systems will provide a realistic terrain view in full motion.

The least flexible animation methods involve playing through a fixed sequence of images from beginning to end. More sophisticated animation permits the viewer to interact with the system, by choosing the starting point as well as the sequence of images to follow. Let's look at each of these methods.

FIXED-SEQUENCE

The most inexpensive map animations display successive images in a predetermined or fixed sequence. We can, for example, focus on environmental change through time. Dramatic **temporal animations** have been made using GOES Weather Satellite images of the earth taken every half hour from 22,000 miles out in space (Figure 29.1). On the evening TV news, we can watch the cloud patterns build up, swirl across the land surface, and dissipate from day to day. Cloud pattern animations are enhanced by showing lightning-strike and Doppler-radar images.

If the differences in feature patterns from one image to the next are slight enough, and if enough frames are used per second, we get the impression of smooth motion. If too few images are used, the consequence is choppy or jerky motion. This often occurs in TV weather broadcasts, for instance, because only a few frames are available to track the rapid development of a storm system.

Fixed-sequence animated mapping need not be confined to temporal sequences, such as weather phenomena or landscape change. It is also possible to produce motion pictures of landscape change through space. If, for example, land-

form portrayals from a sequence of vantage points are animated, we get the impression that we are flying and the terrain is passing under and around us. The visualization of space and movement can be so realistic that we may actually feel pangs of airsickness! This **fly-through** effect occurs because our minds find it easier to accept our bodies moving than the terrain moving.

A good example of this second form of fixed-sequence animated mapping is the Los Angeles Basin* fly-through that has been widely broadcast in commercials and Cable News Network (CNN) "Science & Technology" segments. To produce this fly-through, thousands of scenes depicting the terrain surface were overlaid with a Landsat satellite image showing landcover. These scenes were computed from vantage points along a pre-chosen flight path (Figure 29.2). Since this fly-through is produced in video format, the sequence of images was set by its creator. With an appropriate VCR, you can achieve such special effects as freeze-frame, slow-motion, fast-forward, and reverse. But that is the extent of your control.

With a third form of fixed-sequence animated mapping, we use different mapping techniques or variations of a single mapping technique to show

---

LA—The Movie was created from a single Landsat TM scene (July 3, 1985) and a digital elevation model. It was produced by the Science Data Systems Group, Jet Propulsion Lab, 4800 Oak Grove Drive, Pasadena, California 91109 (Tel.: 818-354-4016).

Sunday, 7 a.m.   Sunday, 12 noon   Sunday, 6 p.m.

Figure 29.1   This sequence of still images of Hurricane Allen (August 1980) taken with NOAA's GOES Weather Satellite gives some idea of the power of animation to reveal change in environmental form through time.

Figure 29.2   Here we see three of over 3,000 frames used to make the Los Angeles Basin fly-through called *LA—The Movie*. (Photos Courtesy Jet Propulsion Laboratory.)

a set of data in its full light. For example, we can use animation to view an environmental setting through a range of map scales. Such animation through a hierarchy of scales is called **scale animation**. With scale animation, we can **zoom in** to a larger-scale view of the landscape in order to examine a smaller region in greater detail. Or we can **zoom out** to a smaller-scale view for a less detailed panorama of a broader region. The most effective scale animation involves maps that are appropriately generalized at each scale level. Simple enlargement or reduction of map artwork without a commensurate change in cartographic abstraction does not constitute a legitimate scale animation.

Alternately, we can view a series of maps of the same data created by manipulating the cartographic generalization parameters (see Chapter 24). We might, for example, animate the effects of altering the number of classes, or the specification of class limits (Figure 29.3). Similarly, we could animate the effects of giving different vertical dimensions to statistical surface renderings such as fishnets or three-dimensional histograms. Or, we might animate the effects of changing the dot size, unit value, and placement method when using the dot mapping technique.

We might even animate the landscape representation through a series of mapping techniques.

To do so, for example, we could draw on the transformational power of a **morphing** program (the word morphing is derived from the term metamorphose, which means to transform). This software would let us distort one image into another in a smooth sequence of steps. We might start with a photograph of the landscape, and then move to an oblique shaded rendering, a fishnet representation, and a contour depiction. Animations of this sort might be very helpful in learning the relationships between different mapping techniques, and understanding the relative strengths and weaknesses of each method.

A problem with many past attempts at fixed-sequence animated mapping is that individual static maps first had to be produced by hand, a la Walt Disney. Due to the immense effort required and in order to keep costs within reason, rather short sequences of simple line maps were generally used. Furthermore, mappers usually did not use enough maps to create the illusion of smooth motion; instead, the simulated motion on these maps tended to be jumpy.

The increasing use of video and computer methods is rapidly overcoming this problem. Of particular benefit is software that can gradually transform one image into another in a step-by-step sequence. Morphing programs are able to interpolate inter-

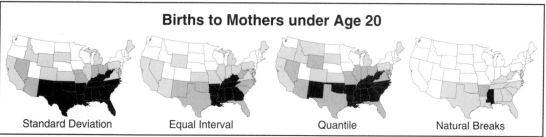

Figure 29.3 Individual frames in one map sequence (top panel) illustrate the effects of altering the the number of classes, while individual frames in the other sequence (bottom panel) illustrate the effects of choosing different class limit schemes. On all maps darker shades indicate higher percentages of births to mothers under 20 years of age (maps courtesy Michael P. Peterson).

mediate frames from a few key maps. Even with these modern computer methods, however, costs are still so high that the full potential of fixed-sequence animated mapping is far from being realized.

USER-SPECIFIED SEQUENCE

Although fixed-sequence animated mapping can provide dynamic environmental portrayals through space and time, their pre-programmed, linear character prevents the map user from directly interacting with the map beyond stopping, reversing, or changing the speed of the sequence. This inflexibility and lack of user control can be overcome by new technologies that support more user-directed animations.

Using the landform example, fixed-sequence animation would take you over the terrain on the path chosen by the animator. In contrast, a user-controlled system lets you call up any image onto the screen, in any sequence. You can simulate

movement realistically from one vantage point to any other by controlling the fly-over path. You might start up high to get the overall view, then fly in closer to get a better look at features of special interest. You can view the terrain from sequences that take you to as many heights and in as many directions as needed to get a feel for the nature of the landform.

Two approaches are being taken in creating user-controlled animations. One approach is based on stored images and the other is based on computed images. Let's take a closer look at each.

Stored Images

One approach involves computing views from a multitude of vantage points and storing them for subsequent retrieval by a system user. Such a system provides you with an interface, from which you select a three-dimensional path. Once you specify this path, the selected images are displayed in proper sequence.

The computational demands on such a system are enormous. Numerous scenes must be pre-computed, stored, and retrieved. This information must be displayed at the full-motion rate of 30 frames a second. The only way to perform the necessary computations and move the immense amount of data is to use a powerful graphics workstation or more powerful computer. Even then, you are limited by the availability of stored images.

Stored-image animation can be jerky and slow and may not let you see what you want. The reason is that system builders have to anticipate which scenes users might want to see, and then create and store these images. Searching a large image database takes time, and you can't see what is not there. Smooth motion, because it demands a large number of images, is an expensive proposition.

Computed Images

Problems with stored image systems are in part overcome with systems that provide the software and database needed to compute a sequence of views as you specify new vantage points. You may perform such a **real-time fly-through** by moving a joystick or by entering a series of vantage point coordinates from a keyboard.

The advantage of this computed-image approach is that you have complete flexibility in choosing a flight path. The drawback is that the computational demands far exceed those required by a stored-image system. An expensive supercomputer is usually needed to support the computations and data transfer rates required to compute images on demand at full-motion video rates. The costs involved limit this high-end approach to flight-simulation applications, special graphic effects in film-making, and other users who demand the best and have the resources to pay for it. Even then, in an attempt to save computational time, image quality is usually so degraded that it falls short of photo-realism.

NAVIGATION SYSTEMS

Maps and navigation have always been closely associated. Navigators use maps to assist in a variety of orientation, position-finding, and route-finding chores. Until recently, they did so manually.

But several decades ago, navigators of ships and airplanes began to use radio signals to fix positions and maintain a course. Revolutionary recent developments link maps to these radio signals and extend the technology to ground vehicles as well. Thus, you can now navigate your car by first observing your progress along a path on a map, and then making route decisions on the ground. Although several approaches are being used, in each case your changing ground position is plotted with a cursor on a map display.

With one type of system, called **inertial navigation**, the device monitors vehicle movements and uses this information to update the position of a "you-are-here" cursor on a map display. You must "initialize" (start) the system by positioning the cursor on the map at the beginning of your route. A drawback of this system is that each distance and direction computation involves some error. Since each subsequent computation is based on the previous one, errors accumulate. Thus, the estimated position tends to "drift" unless corrected periodically with some other source of locational information.

With the most primitive systems, the map backdrop may be a simple screen overlay. The user must initially position the map over the screen cursor. But more advanced technology uses digital cartographic data held in some form of disk storage. The maps may be stored in full image form as captured by an electronic scanner or video device from existing maps. Or map features may be held in digital files in coordinate form that can be plotted to the screen as needed.

The disadvantage of digital map display systems in inertial navigation is their relatively high cost compared to systems that require a map to be mounted over a cursor by hand. But this problem is offset by the advantage that digital display systems can automatically rotate the map on screen as the vehicle turns. This means the map is always oriented in the direction of travel, so a left turn on the map is also a left turn on the ground.

A second type of navigation aid, and one that seems to hold the most promise for the future, is based on **radionavigation** technology. Radio signals transmitted from known locations are used to compute distance and direction relations with respect to a receiver. This information can then be used to indicate your position on a map display.

Radionavigation has several advantages over inertial navigation. For one thing, radionavigation can provide an initial position fix, as well as subsequent fixes. If you are lost, the system will find your location. In addition, each new position determination along a route is computed independently of previous ones. Thus, errors do not accumulate as they do with inertial navigation.

Although ground-based radiobeacons are in widespread use, GPS-linked navigators are attracting the most attention (see Chapter 7). One reason is that they provide latitude-longitude coordinates which can be used to retrieve the appropriate map from disk storage. When the vehicle moves near the edge of one map frame, the next one in the direction of travel is automatically brought up on the display screen.

The major automobile companies are currently experimenting with GPS-based dashboard navigator units for their automobiles. General Motors Corporation, for example, has equipped 100 automobiles in Orlando, Florida, with its *TravTek* system (Figure 29.4). Many trucking, busing, delivery, and taxi companies are already using the technology. Ships and airplanes are switching to the technology as well.

SIMULATION

One of the great strengths of mapping is that it lets us engage in various "What if?" scenarios. Maps let us visualize how the environment might look in the future if we take certain actions now, or if current activities continue along some predictable course.

When it is not possible to manipulate the environment itself, the next best thing is often to manipulate a map representation. As we discussed earlier in this chapter, for example, planners and military strategists have long simulated future events through mapping. In recent years, scientists have used simulation to model environmental processes such as soil erosion, plate tectonics, global warming, ozone depletion, population dynamics, and so forth.

Maps have always helped us visualize how the environment might look from different vantage points in time and space. Electronic technology further enhances this ability by making it possible to change vantage points at will. We can do so with special machines called **simulators**, which are able to respond to a number of human commands. When animated maps are used in these machines, the effect is one of actually moving within or through a three-dimensional environment.

The term **virtual reality** is now commonly used for the modern "goggles and glove" approach to simulation. In this implementation, a person wears a glove equipped with motion detectors and goggles containing a miniature TV screen for each eye. Hand motions cause changes in the screen images, producing the sensation of body movement through a three-dimensional setting. This is the ultimate fly-through.

Although most simulation products today fall short of the promise of virtual reality, they have still proven useful in a variety of contexts. For example, simulators familiarize pilots with landing and take-off procedures before they climb into the cockpit of a real aircraft. In the same way, NASA astronauts were trained to land their spacecrafts on the moon by using stored video images. (These images were constructed using moon surface images taken during previous orbital flights).

Although the most sophisticated ground-based map simulations are still in the developmental, non-commercial phase, even the relatively primitive products in current application reveal exciting possibilities for this technology. These systems use an **optical disk** player controlled by a microprocessor. Initially, ground scenes are videotaped, and individual frames are transferred to

Figure 29.4 Photograph of a *TravTek* map screen mounted in a General Motors Corporation test car in Orlando, Florida (photo courtesy General Motors Corporation).

the disk. Map users then simulate movement through the video-mapped environment by giving commands with a joystick, mouse, or other system interface device. These commands call up the appropriate image sequence for display on a computer graphics terminal.

A number of state highway departments have now developed highway travel simulations based on interactive videodisk technology (Figure 29.5). The Wisconsin Department of Transportation's *Photolog* system, for example, lets you start at any point in the state highway system and then "travel" in either direction, viewing video frames at 100-foot intervals.* You can stop, change the speed, or reverse the sequence of images at any time. You can also call up a variety of highway information that is linked to the location of a video frame in associated digital databases.

It is easy to envision a future in which driving-test simulators will, in several minutes of "action," be

able to gauge your response to several lifetimes of traffic crises. This type of system will require the map database to be linked to a variety of other databases.

In technically advanced modern simulators, computer-controlled video frames (stored-image systems) are giving way to fully electronic computer graphics display systems. In this case, the image is created on command from digital data files which reside in the computer's memory and which represent the ground scene. The three-dimensional modeling procedures employed in these simulators incorporate knowledge of hidden surfaces, shading, shadow, and texture to create dramatically realistic maps.

---

*Photolog and similar systems used in other states is based on the Roadview® Video Image Capture System and Videologging Workstation technology developed and marketed by Mandli Communications, Inc., 2211D Parview Road, Middleton, WI 53562 (Tel.: 608-836-3344).

© 1994 Mandli Communications, Inc.

INTERACTIVE CARTOGRAPHY

Animation and simulation involving user-controlled change of vantage point entail interaction between the map and its user. Indeed, the map user becomes the mapmaker. The perspective of the map portrayal is determined by and changes with each user command. The map user can repeat processes or sequences until they are understood, and practice actions until problems are worked out and procedures are fine-tuned.

With simulation that permits change in vantage point, interaction is limited to perspective alone. Increasingly, however, it is possible to interact with other aspects of a map as well. In fact, the fully interactive map of our dreams will soon be a reality. It will permit a free question and answer exchange between map and map user. Through a series of prompts and queries, map users can ask the computer to determine such factors as the area of a lake, elevation at a point, distance along a route, direction between two points, least-effort path, and so forth (Figure 29.6).

Such interactive mapping will thrust cartography totally into the electronic age. Potential map users will be able to tailor map displays to their specific needs. They will also be able to gain access to the data used in mapping, or data that can be linked with map locations or features.

The shift from static to dynamic, interactive mapping is forcing cartographers to reconsider traditional design principles. Whereas maps have long been designed to serve data storage needs, this role can now be better served by an electronic information system in which maps are but one component. By unburdening the map of a primary data storage function, greater design emphasis

Figure 29.5 Three frames from a microcomputer-driven videodisk map of Interstate 5 and US Highway 14 in California are shown. Images were taken two weeks before the January 1994 earthquake. (Courtesy Mandli Communications, Inc.)

can be given to the role maps play as interface between people and environmental data.

MAP AS INTERFACE

Within the context of a geographic information system, an interactive map serves as a convenient visual interface to the numerical data stored in great detail in computer memory. The map display need only show enough information to assist its user in directing questions to the data files or records. The maps are used for visualization, while the map data are used for analysis. The computer provides the link between the two modes of representing geographic data.

In a fully developed interactive mapping environment, map analysis can be more natural and effective than is possible when working with static printed maps. In most current systems, the user can interact with the computer by merely pointing to map locations and features. In future systems, interaction with the computer map display will also be possible through voice commands.

Figure 29.6 With growing frequency, map users are able to analyze and interact with environmental data with the aid of electronic display screens. (Image courtesy U.S. Army Engineering Topographic Laboratories, Fort Belvoir, VA.)

SELECTED REFERENCES

Campbell, D. S. and E. L. Egbert, "Animated Cartography: 30 Years of Scratching the Surface," *Cartographica*, 27, 2 (1990): 24–46.

Cornwell, B., and A. Robinson, "Possibilities for Computer Animated Film in Cartography," *Cartographic Journal*, 3,2 (1966): 79–82.

Fox, D. *Computer Animation Primer*, New York: McGraw-Hill, 1984.

Gersmehl, P. J., "Choosing Tools: Nine Metaphors of Four-Dimensional Cartography," *Cartographic Perspectives*, 5 (1990): 3–17.

Langran, G., *Time in Geographic Information Systems*, New York: Taylor & Francis, 1992.

Magnenat-Thalmann, N., and D. Thalman, *Computer Animation: Theory and Practice*, New York: Springer-Verlag, 1985.

Moellering, H., "The Real-Time Animation of Three-Dimensional Maps," *The American Cartographer*, 7 (1980): 67–75.

Moellering, H., "Traffic Crashes in Washtenaw County Michigan, 1968–70," *Highway Safety Research Institute*, University of Michigan, Ann Arbor.

Monmonier, M. S., "Strategies for the Visualization of Geographic Time-Series Data," *Cartographica*, 27, 1 (1990): 30–45.

Peterson, M. P., "Interactive Cartographic Animation," *Cartography and Geographical Information Systems*, 20, 1 (1993): 40–44.

Peterson, M. P., *Interactive and Animated Cartography*, Englewood Cliffs, Prentice Hall (in press).

Tobler, W., "A Computer Movie Simulating Urban Growth in the Detroit Region," *Economic Geography*, 46 (1970): 234–240.

Weber, C. R., and B. P. Buttenfield, "A Cartographic Animation of Average Yearly Surface Temperatures for the 48 Contiguous United States: 1897–1986," *Cartography and Geographic Information Systems*, 20, 3 (1993): 141–150.

PART VII

MAP EXECUTION
AND DISSEMINATION

CHAPTER 30

MAP
REPRODUCTION

*F*ew maps are one-of-a-kind creations. Usually the high cost of mapping can only be justified if the product ultimately reaches a number of users. Thus, most maps are made to be duplicated. Indeed, much of the history of cartography is associated with a quest for faster, less costly, and higher quality ways to reproduce maps.

Processes used to duplicate maps have changed dramatically over the years. Long-standing manual processes were in large part replaced by mechanical processes with the development of printing in the 15th century. These mechanical processes were supplemented by photo-chemical processes in the late 1800s. In the mid-20th century, existing methods were further augmented by electronic processes, which now show signs of growing dominance.

These processes, individually and in various combinations, form the basis of a wide variety of map duplication methods. Although the procedures are given different names or commercial labels, they all involve transferring the map image from one form or material to another. Sometimes this transfer is completed physically, using inks, toners, or dyes. At other times it is accomplished through a photo-chemical reaction (as when silver salts are transformed into metallic silvers). It can even be performed through electro-chemical reactions, as when phosphor compounds are made to glow on a cathode ray tube screen through the stimulation of an electron beam.

REPRODUCTION GUIDES PRODUCTION

When duplicate copies of maps were made by hand, before the advent of mechanical, photomechanical, and electronic reproduction methods, the cartographer was free to prepare the map by any one of several procedures. Even today, if a map is not to be reproduced, the cartographer is under few constraints as to design or kinds of materials that may be used.

But the vast majority of maps are made to be duplicated. For these maps, factors such as cost, artwork requirements and the size limits of reproduction equipment place significant restrictions on the cartographer. Each reproduction method has special requirements. Unfortunately, maps are sometimes prepared by people who are unfamiliar with these restrictions and who give no thought to how the map is to be duplicated. In many cases it is discovered, too late, that there is no economical way to reproduce the map.

We cannot emphasize enough that the proper sequence in preparing artwork is first to determine the process by which it will be duplicated, and then to plan how the artwork can best be prepared to fit that method. For this reason, we will look at map reproduction in this chapter before we discuss map production in the next chapter.

When you select a duplication process, you must first consider the kinds of copies you need, how many copies are required, the quality desired, and the size. To make intelligent choices and to be able to plan for the proper artwork, a knowledge of common duplicating processes is a necessity.

The most practical way to classify reproduction methods is to group them on the basis of whether or not they involve a decreasing unit cost with increasing numbers of copies. In this chapter, therefore, we will group methods according to whether they are more appropriate for reproducing a limited number of copies or many copies. We will not discuss individual procedures exhaustively. Our aim is only to illustrate methods with a representative technique or two.

METHODS FOR A FEW COPIES

With some reproduction methods, the unit cost does not decrease significantly with increased numbers of copies. Since there is no price break in making many copies, these are called **limited-copy methods**. Even though each additional copy costs about as much as the first, the total cost of producing a few copies is still less than it would be if one of the many-copy methods was used (see *Methods for Many Copies* later in this chapter). The cost saving with limited-copy methods occurs because of their relatively low set-up expenses.

In recent years, a number of sensitized materials have been developed. These materials, which can be used under ordinary room lighting conditions, are invaluable for the cartographic process. These new materials and a trend toward having more duplicating apparatus in mapping companies permit the mapmaker to be much more involved in preparing the materials to produce a printing plate. It is not uncommon for the cartographer to furnish the printer with materials that have been proofed and are ready for platemaking (see *Lithography* and *Platemaking* later in this chapter).

As a first step in map reproduction, you need to decide if the few-copy method is the best approach. In making this assessment, you can start with the nature of the original that needs duplication. Is it available in the form of reflection (opaque) artwork, transmission (film) originals, or digital records? Let's look at each of these possibilities in detail.

REFLECTION ORIGINAL

To reproduce opaque artwork, you can use drawn, plotted, or printed materials. The only restriction is that the original map must be available on a single sheet of material that reflects light.

You can use reflection art to produce tangible copies, or you can project it onto a large screen. There are two methods for producing tangible copies: xerography and diffusion transfer. You can use both these techniques to enlarge or reduce the original and to duplicate color originals. Let's look at these two methods in more detail.

Xerography

You can reproduce reflection originals in a dry photo-mechanical process called **electrophotography** or **xerography** (for example, Xerox). Rather than using chemical reactions or solutions as is common among other photographic processes, xerography is an electrostatic process that involves photoconductivity and surface electrification.

Xerography begins by giving a special metal plate a positive charge which, upon exposure to the original through a lens, disappears in those nonimage areas subjected to light. The remaining positively charged surface constitutes a latent image, which is "developed" with a negatively charged black powder toner that adheres to the image area (recall that opposite electrical charges attract). Paper or a similar material, which has been given a positive charge, is placed against the plate. It receives the image by transfer of toner and is fixed onto the material with heat.

Because the xerographic process does not involve chemicals, you save the time of immersion in solutions and subsequent drying, and you can produce copies in just a few seconds. You can place the image on a variety of papers, plastics, or other materials (cloth, metal, wood), since there is a physical transfer of toner.

Standard black-and-white xerography can provide good copies of monochrome originals at a reasonable cost. But the quality of black-and-white duplicates of color originals is unpredictable, and usually disappointing.

Whereas only a single plate and toner are required for monochrome xerography, full-color xerographic reproduction of opaque artwork involves the use of separation plates. The standard procedure is to use a different plate for each of the subtractive primaries (cyan, magenta, yellow). Color separation is accomplished through a lens and filter (red, green, blue) arrangement. Line artwork and continuous tone imagery can both be copied.

Color xerography is more expensive and a little slower than monochrome xerography, but its price is very competitive when compared to other procedures for duplicating color originals. Reproduction color fidelity has been a serious problem in the past, but is steadily improving. The lack of a black plate (and toner) to give density and sharp definition to the image remains a problem. Yet, on balance, the advantages of xerographic reproduction for many cartographic purposes now seem to outweigh the disadvantages. Thus, you can expect this method of color reproduction in cartography to be used more frequently in the future.

Diffusion Transfer

You can duplicate reflection originals directly by using a photo-mechanical process called **diffusion transfer**.* First, you create a diffusion transfer print by exposing a sheet of light-sensitive carrier material (which serves as a negative) to the original in a process camera. You can use either line or continuous tone materials.

Next, place the carrier sheet on top of a sheet of positive print material (paper or film). Process it by running it through a machine containing a reusable activating solution (Figure 30.1).

A few seconds after the laminated materials emerge from the processor rollers, the transfer of the image from the carrier to the print is complete. You can then separate the two sheets of material, leaving a positive (film or paper) print. You can create halftones by using a special contact screen during exposure.

For a variety of reasons, cartographers find it convenient to use the diffusion transfer process. The results may not be quite as good as those obtained with photographic prints, but the quality is still high enough for it to be used as a basis for subsequent line and halftone printing operations. Diffusion transfer is relatively inexpensive, especially for reproducing continuous tone copy (for example, aerial photos). There is no need for a film negative (the carrier with its latent image is discarded), and the developer can be reused many times before it is exhausted.

The process is also fast, since in addition to eliminating a separate photo-mechanical step, there is no need for fixation or drying steps. The

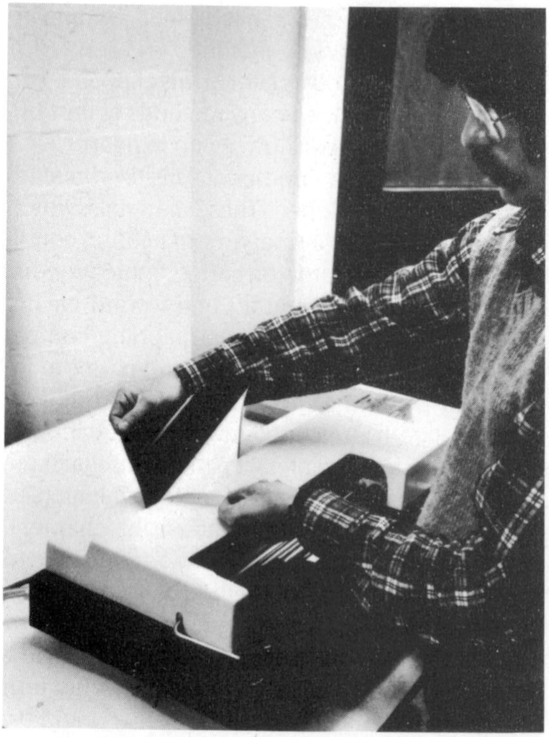

Figure 30.1 Diffusion transfer materials are processed through an activator bath and then pressed together between a set of rollers in a simple desk-top machine. (Courtesy AGFA-GEVAERT, Inc.)

possibility of making scale changes is another attractive feature, particularly for purposes of compilation or paste-up reformatting.

Still another useful trait of the diffusion transfer process is that if the original artwork lacks sharpness, you can actually improve it to some extent. In contrast, you can expect image degradation with most photographic processes.

You can also obtain high-quality color duplicates using diffusion transfer materials.** The process is identical to black-and-white diffusion transfer reproduction, except that you use a special

---

*Diffusion transfer materials are marketed under several trade names, including COPYPROOF (AGFA-GEVAERT, Inc., Graphic Systems Division, 150 Hopper Ave., Waldwick, NJ 07463) and PMT (Eastman Kodak Co., Rochester, NY 14650).

**Diffusion transfer color print materials from different manufacturers, such as COPYCOLOR (a product of AGFA-GEVAERT) and EKTAFLEX PCT (a product of Eastman Kodak Company), have similar handling characteristics.

color print material. You also use a camera, which makes scale changes possible. No screens or other exposure devices are needed.

Although these color materials are not widely used in map reproduction, their high quality and low cost make them an option worth considering. They could become particularly important in duplicating remote sensing imagery, since it is a great advantage for the cartographer to be able to control the color balance of such images.

TRANSMISSION ORIGINAL

Some cartographic artwork is produced directly in the form of transmission copy. An example is artwork constructed on a translucent or transparent drafting medium or created by negative scribing and electronic film plotting or recording. Reflection artwork can also be readily converted to film negatives or positives through photographic processing (see Appendix D).

Regardless of how the artwork came to be in transmission form, further processing of the image involves light transmission through the material rather than light reflection off the surface as is the case when processing opaque copy. The transmission artwork may be monochrome (black-and-white) or multiple color. Also, the artwork may consist solely of linework or may be continuous tone (at least in part).

There are several reproduction methods suitable for handling transmission artwork. For convenience, we will classify them into two types: methods that duplicate the original with a single sensitized coating, and methods that use multiple sensitized coatings to build up an image progressively in different colors. Let's look at each in turn.

Single Sensitized Layer

You can use printmaking paper, coated with a photographic emulsion, to reproduce maps if you have a transmission (positive or negative) original. You can make prints on any of a variety of special papers and films that provide good drafting surfaces and are dimensionally stable. Of the many materials available, we will discuss two that exhibit special properties.

Photographic Print Perhaps the most common method of reproducing maps from transmission originals is to make a standard photographic print. You make the copy by exposing the print paper in contact with the original in a vacuum frame. You then process the exposed paper in the same way as orthochromatic films (see Appendix D). Since photographic prints use a standard reversal emulsion, you need a negative original to produce a positive copy.

Dylux *Dylux 608 Registration Master Film** is available on paper or dimensionally stable plastic. It produces a blue line image immediately upon exposure to ultraviolet light through a negative or positive. The image requires no chemical developing and self-fixes under normal room lighting conditions. A pale yellow background disappears within an hour, but the process can be speeded by exposing the film to pulsed xenon or carbon arc lamps that are filtered with an ultraviolet blocking material.

The light source for the initial exposure must be exclusively ultraviolet to produce the maximum contrast. Best results are obtained with black light fluorescent lamps that have suitable phosphors yet screen out nearly all visible light. The density of the image is directly related to the length of exposure. You can give separation negatives different exposures to produce images in different tones. You can also use screens and masks as in normal film compositing.

Since ultraviolet light forms an image and visible light deactivates the emulsion, you can make positive images from positive film. To do so, make sure the film is in contact with the positive. Then make an exposure with a bright light, which removes the sensitivity in the nonimage areas. After removing the positive, expose the sheet to ultraviolet to produce the image.

---

*A product of E.I. Dupont de Nemours and Co., Wilmington, DE 19898.

Multiple Sensitized Layers

You can also reproduce film separations in positive, directly viewable form, using materials with separate emulsions for each color to be copied. These materials are called **proofing systems** in the graphic arts industry, because they are used to check (or "proof") the accuracy of artwork prior to many-copy reproduction.

With one such class of materials, called integral systems, the image is constructed by superimposing one layer of emulsion after another on an opaque base material. A second class of materials permits the overlaying of transparent sheets containing the separate emulsions. Let's look at both these proofing systems.

Integral Systems   Several integral systems are useful for reproducing a few copies of a colored map from film or scribe separations. The techniques vary in the way they handle their photosensitive emulsions. Some emulsions are available in liquid form, which must be wiped on a base material prior to exposure. Other emulsions are available in thin sheets that must be laminated onto a base material. To obtain multiple colors, you need to add repeated coatings or laminations.

Integral systems have several advantages that make them popular. They provide good color fidelity and high image resolution. They also are reasonable in cost.

Unfortunately, integral systems suffer from several drawbacks. First, some work only with special base materials. Second, when using several colors, problems with any one of the emulsion layers can easily ruin the quality of previous work and the overall result. Success in making one copy of a map through superimposition methods in no way guarantees that subsequent copies will be of comparable quality.

Several companies market **colored emulsion solutions** that you can apply to special plastic materials (including scribecoat).* The solutions come in a variety of colors, and you can mix the primary colors to obtain still other colors.

You will obtain the smoothest application of

color and therefore the best results with a mechanical whirler. Unfortunately, few small mapping establishments have access to such a device. With sufficient practice, however, you can apply the sensitizer smoothly by hand over a fairly wide area (Figure 30.2). The trick is to spread just the right amount of emulsion over a clean, dry surface with soft, even strokes until the emulsion is dry.

Next, with the coated material in contact with the negative artwork, make an exposure with light in the ultraviolet end of the spectrum. To develop the image, rinse the exposed material under a water tap, and dry it thoroughly.

Once the material is completely dry, you can add another color over the first and repeat the process. You can apply as many colors as you wish in this manner. If well done, the result is comparable to a map that has been printed in several colors by separate press runs.

You must take great care to ensure high-quality results, however. Not only must the coating be uniform, but you must rigidly control such environmental factors as dust, temperature, and humidity. For these reasons, wipe-on emulsions tend to be more suitable for generating a rough, quick preview copy of a map, or for transferring a guide image from one surface to another (as in preparation for scribing), than for reproducing finished maps from film separations in any numbers.

The problems of producing high-quality results using liquid emulsions are largely overcome by using **lamination processes**. You need special lamination equipment, however, and the cost of material is greater. But the quality of reproduction is excellent.

We will discuss two lamination processes here.** Each requires quite different procedures.

---

*Brite-Line, Camden Products Company, Box 2233 Gardner Station, St. Louis, MO; Kwik-Proof and Watercote, Direct Reproduction Corporation, 835 Union St., Brooklyn, NY 11215.

**Brand names of similar materials not discussed here include: 3M *Matchprint Transfer System*, Fuji *ColorArt*, Hoechst *Pressmatch*, Agfa *Agfaproof*, and Kodak *Signature*.

Figure 30.2 Wipe-on emulsions must be spread evenly over the surface if good results are to be achieved. To accomplish this by hand, use a block applicator covered with a soft pad in alternating series of horizontal and vertical strokes.

Figure 30.3 Lamination systems for color map reproduction normally require the use of a special mechanical laminator if consistent, high-quality results are to be achieved.

With the first process, the *3M Transfer Key System*,* you use a clear liner material that has been factory-coated with a color emulsion. You use pressure to laminate this coated layer to a special opaque base material (*Transfer-Base*). Although you can perform the lamination with a hand roller, the process is plagued with problems. For consistent, high-quality results, you should use a special mechanical laminator (Figure 30.3).

Once you complete the lamination, you can peel away the clear liner, leaving the color emulsion adhered to the Transfer-Base. You then make an exposure in a contact frame with a light source rich in ultraviolet.

With negative-acting Transfer Key, you can carry out development by hand, using 3M Brand Color Key/Transfer Key developer. The positive-acting material combines low-polluting aqueous development with rapid machine processing and yields a high resolution product. Once you have thoroughly dried the material, you can add more colors by repeating the process with other Transfer Key films of the chosen hues.

The second lamination process, the *Cromalin Color Display System*,** is a dry proofing process. The Cromalin material consists of a tacky photopolymer (light sensitive) layer sandwiched between a protective polypropylene sheet and a mylar cover sheet.

---

*3M Transfer Key is a product of the Printing Products Division, Minnesota Mining and Manufacturing Company, 3M Center, St. Paul, MN 55101.

**Cromalin is a product of E.L. Dupont de Nemours and Company, Wilmington, DE 19898.

Again, you use a mechanical laminator to apply heat and pressure. You can then discard the lower protective sheet, leaving the colorless sensitized layer adhered to the opaque base sheet.

After exposing it to a light source high in ultraviolet, you can strip away the top protective cover, leaving the clear emulsion on the base unprotected. Next, apply finely powdered color toner, which will cling to the sticky emulsion in the image area.

After you have cleaned all loose toner from the surface with a soft cloth, you can again run the base through the laminator to prepare it for another color. Repeat the process until all colors are in place. You can then add a protective covering by exposing a final lamination without any masking material in place.

Overlay Systems   A second type of proofing system makes use of transparent overlay materials containing colored emulsion coatings.* The clear sheets or foils of film are factory-coated with presensitized colored emulsions in standard hues, including the regular process colors (yellow, cyan, and magenta). You can use the materials under daylight conditions, and the only equipment you need is a pressure or vacuum frame. You also need a source of light that is rich in high-intensity ultraviolet.

Overlay methods share several features that are of special significance to cartographers. They are less expensive than integral systems. Another advantage is that you can create and view each color foil independently from the others. A third desirable feature is that the factory-applied emulsion coatings assure precise color fidelity and even color density, as well as permanent pigment colors.

To achieve results that most closely resemble the appearance of final printed maps, you should overlay the sandwich of foils in registry on a white base. Even then, however, the order with which you stack the foils will influence the overall color effects. If

you view the overlay sandwich over a light table or project it onto a screen, still different color effects will occur. Indeed, to some degree these variable colors under different viewing conditions may represent a disadvantage of overlay systems. Several products warrant special mention.

*Enco NAPS (Negative Acting Proofing System)*** uses a material that is exposed emulsion to emulsion in contact with a negative. When using this product, be sure the colored emulsion faces the source of illumination. Spread the developer gently over the exposed emulsion, using a piece of cotton and a light sweeping motion. After the background emulsion is removed (dissolved), rinse under a water tap and blow or blot dry. The result is a positive print in the chosen color.

You can prepare a series of positives in different colors, of separate flaps of a map, and then overlay them to form a full color proof of the map. To aid this process, the Enco NAPS film can easily be punched for pin registration. Screen tints and halftones are faithfully reproduced, and colors combine in the same manner as on a printing press.

*3M Color-Key Proofing Film**** is a negative-acting material that is similar to NAPS in handling and results. One difference, however, is that the 3M product is exposed with the color emulsion facing away from the source of illumination. Another difference is that Color-Key, unlike Enco NAPS, cannot easily be punched. Thus, to obtain pin registration, you need to tape prepunched strips or tabs of some other material to the 3M product. A special feature of Color-Key is that the orange emulsion is **actinically opaque** (opaque to the wavelengths of light that affect light-sensitive emulsions, but not visually opaque). Therefore, you can use it as you would use film negatives in subsequent processing of materials.

---

*Well-known examples of overlay proofs are 3M *Color-Key*, Dupont *CromaCheck*, and Enco *NAPS*.

**NAPS is a product of Azoplate, Division of American Hoechst Corporation, 558 Central Ave., Murray Hill, NJ 07974.

***3M Color-Key Proofing Film is a product of Minnesota Mining and Manufacturing Company, 3M Center, St. Paul, MN 55101.

DIGITAL ORIGINAL

Increasingly, the information needed to construct and reproduce a map image is stored in computer-readable form. Producing maps for direct viewing from these digital records can take several forms. One group of graphic output devices generates hard-copy maps on printers or plotters. Another class of machines produces soft-copy (ephemeral) images on graphic display terminals (screens). Each class of output device plays a special but different role in map reproduction. Each also involves special considerations for the cartographer.

Hard Copy

We have come to expect that graphic reproduction techniques will produce a tangible, physical product. Understandably, then, a variety of computer output devices fall in this category. These include raster devices that create a map by printing rows of small dots (dot matrix, electrostatic, laser, ink jet, dye sublimation, thermal wax, and solid-ink printers). Another class of hard-copy output devices is made up of vector plotters, which are capable of constructing map linework with pen and ink in a variety of colors.

These printing and plotting devices are discussed in detail in the next chapter. It's important to note here, however, that these devices are suitable not only to produce maps but also to reproduce maps. Each additional copy of a map produced on a computer graphics device will cost almost as much as the previous one, however. Thus, this approach is generally appropriate only when a few copies of a map are needed.

If very many duplicates are required, it may be more economical to use the computer output artwork as a master for some other form of reproduction. This is particularly true for artwork generated on vector plotters, because plotting time is usually lengthy and increases in direct proportion to the amount of detail in the map image.

Soft Copy

Despite the long tradition of reproducing maps in tangible form, the popularity of electronic map displays is growing dramatically (see Chapter 20). These soft-copy images add a new dimension to the way we think about map reproduction.

You can often obtain the information you need from a temporary map display. You can thus avoid some of the problems and expense associated with obtaining, storing, and retrieving individual map sheets. What this means is that a map need only be produced once. It can then be sent out on demand to dozens, even thousands, of viewers who have linked themselves together in a telecommunications network. In a sense this is the electronic version of reproducing maps in the form of slides or transparencies, except that the map image is sent out to the individuals rather than bringing them together for a group showing.

Map reproduction by means of electronic display has many benefits. You can center the map on any chosen geographic location and display artwork within a surrounding "window." Hence, you overcome the frustration of having your location of interest fall at the edge of a conventional map sheet. You can also zoom in on a portion of a map in order to get a blown-up picture of fine image detail. Possibly best of all, it is easier for the custodian agency to keep the master artwork current, since reproduction costs are minimized. This means that when you call up a map image on a display screen, it will probably be more up-to-date than conventional sheet maps usually are.

Growing interest in reproducing maps upon demand on a display screen is doing more than changing the way we duplicate cartographic artwork. It is also forcing us to rethink what we mean by the term "map reproduction." If, in the future, potential map users are provided with easy-to-use mapping software and ready access to digital records (cartographic and statistical databases), they can produce their own unique cartographic portrayals directly. As a result, the professional cartographer's role may shift, at least in part, from producing maps themselves. Instead, cartographers may concentrate on updating digital records, improving map design software, and raising cartographic literacy so that potential users can take full advantage of their newfound mapping capabilities.

Under these conditions, there may not be a need to produce the artwork for even a master map, except possibly for preview or data editing purposes. Consequently, there may also be far less need for reproducing maps in hard-copy form.

METHODS FOR MANY COPIES

When you need many copies of a map, you must use different reproduction methods than when you only want a few copies. Modern high-speed printing presses produce the first thousand copies of a map almost as cheaply as the first copy. In fact, as a rule, many-copy methods become more economical as the number of copies grows.

The most economical way to reproduce a map in large numbers is from a special plate using a mechanical printing process. Three approaches have been taken in creating these plates (Figure 30.4).

The earliest mechanical printing was done using a process known as **relief printing** or **letterpress**, in which paper receives ink directly from surfaces standing in relief. The initial method of forming raised features on the plate by mechanically cutting away the nonprinting areas of a smooth block of wood was in time replaced by photographic image transfer and chemical etching of plastic or metal plates.

A second development in mechanical reproduction, **intaglio printing**, also involved ink and an uneven surface. Grooves were cut in a flat metal plate, usually copper, and then filled with ink. The surface of the plate was then cleaned off and the plate, with its ink-filled grooves, was squeezed against a dampened sheet of paper. The paper "took hold" of the ink and, when removed from the metal plate, the pattern of grooves appeared as ink lines. In a sense, this process of printing from an intaglio surface is just the opposite of letterpress printing, since the inking area is "down" instead of "up."

A third important development in mechanical printing occurred in 1798 with the invention of **planographic printing** or **lithography**. This new

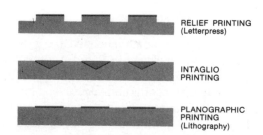

Figure 30.4 The basic processes of relief, intaglio, and planographic printing operate in different ways to produce a surface from which ink may be transferred to paper.

printing process was based on the incompatibility of grease and water. In contrast with printing from a relief or intaglio surface, the surface of the lithographic printing plate was a plane (hence the name planographic), with no significant difference in elevation between the inked and noninked areas. Although the early lithographic plates were made by drawing with a greasy ink or crayon directly on the smooth surface of a particular kind of limestone (which gave the process its name), this bulky medium has long since been replaced by thin metal, plastic, or paper plates.

All three of these plate-based printing techniques are built on an **all-or-nothing inking principle**. In other words, ink is either deposited in full amount (density) or not deposited at all. The image area is printed in solid black (or some other color), and the nonimage area receives no ink. This ink/no ink dichotomy causes no problem in reproducing simple line artwork. But the reproduction of continuous tone artwork, such as an aerial photograph or a shaded relief map, is another story.

Truly effective printing of continuous tone artwork requires special processing of the image prior to printing. We need some way to decompose the tonal variations in the image into fine, discrete marks that, when printed, give the illusion of continuous tone. During the 1880s, the mechanical halftone screen was devised for this purpose. Although these screens are still widely used, we can now achieve the same effect electronically by manipulating a digital record of the artwork (see Chapter 20). We can use either a laser imagesetter to produce a halftone negative or a laser platemaker to place the halftone image directly on the printing plate (see Chapter 31).

LITHOGRAPHY

In recent years, modern offset photo-lithography (hereafter referred to simply as lithography) has largely displaced other plate-based methods in the printing industry. With respect to cartography, lithography now accounts for over 99 percent of the maps that are printed from plates. Although other plate-based methods are of historical interest, they will not be discussed in the remainder of this chapter.

The overwhelming dominance of lithography can be attributed to the economical, high-quality results we can achieve with it. We can reproduce colors and fine lines with great fidelity. In addition, lithography is particularly well suited for printing large maps.

In the next sections, we will consider single-color and multiple-color lithography separately. If you can determine in advance the eventual printing strategy and the grade of paper to be used, you can design the best map for that particular combination.

Single Color

The lithographic reproduction of maps in a single color (usually black) takes several steps. First, you must process the original so that it is in transmission (film) or digital form.

The next step is platemaking. This involves contacting transmission separations in a platemaker, or writing the digital record to the plate with a laser or electron beam recorder. (For more details, see "Platemaking" later in this chapter).

Continuous-tone copy, such as relief shading or a photographic image, must be converted to a halftone negative prior to printing. With high-quality lithographic plates and presswork, halftone screens of up to 300 lines per inch can be reproduced on coated paper, although the use of 120- to 150-line screens is most common.

Once the printing plate is finished, it is mounted on the printing press. (For details on this step, see "Presswork" later in this chapter).

Multiple Color

The lithographic reproduction of maps in multiple colors does not differ fundamentally from black-and-white lithography except that a variety of colored inks is used. Each separate ink requires a separate printing plate, of course, and thus also requires a complete duplication of the steps in the whole lithographic process. Therefore, the cost of reproducing multiple-color maps by lithography is proportionately greater than that of single-color reproduction.

There are two different procedures for multiple-color lithography. One, called **process color printing**, involves creating a full range of colors on a continuous tone image using inks of the three subtractive primaries (cyan, magenta, yellow) and black. The other method, called **flat color printing**, involves printing each color symbol category on the map with a different ink, or combination of inks.

Process Color   When you want to reproduce colored photographs or artwork using different colors on a single drawing, you need to use process color printing. The process color method is based on the fact that almost all color combinations can be obtained by varying mixtures of the subtractive primaries (magenta, yellow, cyan) and black (see Color Figure 20.3). Inks in these colors are called **process inks**.

Colored artwork prepared directly by hand, a color slide or photographic print of a map displayed on a computer monitor, or a colored remote sensing imagery must be **color-separated** into the process colors before making the necessary printing plates. One way to color-separate the original colors is to use a camera equipped with green, red, and blue filters. You will need to make four exposures on separate pieces of film, using special filters and rotating a halftone screen to the proper angle for each exposure. The resulting separation negatives will record the three component additive primary colors (blue, green, red) of the original as black-and-white densities. You make the black separation by using partial exposures with each filter.

The modern approach to process color separation of colored originals involves the use of an electronic scanner. With a scanner, light reflected from or transmitted through the image is picked up by photocells covered with red, green, and blue separation filters. These photocells generate three electrical signals proportional to the intensity of the light they have received. These signals in turn are converted to digits and entered into a computer-compatible storage medium, from which the information can be retrieved and processed in ways that mimic the full range of manual photo-mechanical opera-

tions (including screening). The output signal from the computer is then used to control a laser beam focused on a sheet of photographic film (or a printing plate).

The three (positive) printing plates are halftones representing image components in each of the subtractive primaries. When these plates are printed together in **three-color process**, transparent inks perform their subtractive assignments, merge, and recreate the full color range of the original. Yellow ink mixed with magenta results in reds, yellow combined with cyan forms greens, magenta combined with cyan produces blues, and so forth (see Box 30.A).

Unfortunately, three-color process does not produce a true black. To gain the extra definition provided by black ink, a fourth printing plate is needed. The procedure is then called **four-color process printing**. The black plate includes the halftone that is used as a toner to increase the shadow densities and overall contrast.

Box 30.A
Understanding Process Color Effects

Understanding how process color reproduction works involves a quick review of color theory (also see Chapter 19). Before reproducing a colored remote sensing image, for example, we need to make halftones for each of the process colors and black. So, how does the procedure work?

Begin with the fact that a filter *transmits* light of its own color and *absorbs* the light rays of most other colors. Thus, when white light strikes the blue filter, the blue wavelengths are allowed to pass through to expose the film, and the other wavelengths are blocked. The resulting negative is least dense in the green and red (yellow light) areas. The positive printing plate made from this negative will print most ink in the

green and red areas. Therefore, it is inked with yellow, which is a mixture of green and red.

The green filter transmits green light and blocks blue and red. Since most ink will be deposited in the blue and red areas, magenta ink is used. Magenta, then, is a mixture of red and blue.

The red filter passes red light and absorbs green and blue. The resulting plate deposits cyan ink (a green-blue mixture) in the areas of green and blue.

The halftone negative for the black plate can be made by using three partial exposures with each of the three color filters. The resulting plate deposits black ink in shadow areas and improves contrast in the printed image.

Flat Color Flat color printing is the method most often used for reproducing line maps. Artwork for flat color reproduction is prepared in black and white and requires at least one separate flap for each colored printing ink.

Flat color reproduction takes two different forms. The most flexible approach is to separate the artwork so that a different color can be created for each map category or feature using the four process inks in combinations and tone values. The tone values are achieved by using appropriate screen tints (see Chapter 20).

This flat color printing technique is especially recommended if more than a half dozen or so areal tints are needed, as might be the case in reproducing a soils or land use map. The reason is that a total of four press runs suffices regardless of the number of distinct colors needed on the map.

But since all colors other than the three subtractive primaries and black are created by superimposition of some combination of the four process inks, extremely precise registration is required for high-quality color reproduction of fine linework or type. This degree of registration is usually not practical unless bi-angle screens and a modern four-color press is used. For this reason, the use of process inks to copy a map containing fine lines or type may not be the best choice.

The traditional approach to flat color reproduction and one that still is widely used today is to color-separate the artwork so that each category or feature can be printed with a special pre-blended ink of a conventional color. This is called the **spot color** method in the computer graphics industry. A wide selection of spot colors, each of which is uniquely identified by descriptive numbers, is available for this purpose.

In 1963 these conventional colors were standardized in a coordinated matching system for printing inks, called PANTONE MATCHING SYSTEM™.* (See Chapter 20). PANTONE colors are now a printing industry standard.**

The PANTONE MATCHING SYSTEM contains over 500 standard PANTONE colors, all of which were produced by blending eight basic colors, plus black and transparent white. For the convenience of the user, swatches of the PANTONE colors are printed with transparent inks on coated and uncoated paper and displayed in the handy PANTONE Color Formula Guide. The PANTONE color identification name or number and the complete blending formula for the color are provided with each color swatch. The inks and paper stock used in producing the PANTONE Color Formula Guide are representative of those found throughout the commercial printing industry. Thus, the PANTONE colors can be readily reproduced in map printing.

Since a spot color is printed directly (by a preblended ink of that color), the technique is especially well suited for color reproduction of fine lines and type (Table 30.1). Use of conventional colors is also convenient when a limited number of discrete areal tints are to be printed. But if the desired number of conventional colors grows much beyond four, it might be better to use process inks. Indeed, there is a practical limit on how many conventional colors can be printed economically. This is determined in large part by the fact that each additional ink requires another printing plate and another run of the map through the press. Extra time and materials are involved, and both cost money.

Combining Color Schemes There is no need to rely exclusively on either conventional or process

---

*Pantone, Inc.'s check-standard trademark for color reproduction and color reproduction materials.

**The Defense Mapping Agency (DMA), which is the largest map producer in the United States, uses an in-house system of Standard Printing Colors (SPC), which serves the same function for the government that the PANTONE colors do for the commercial printing industry. In the SPC system, each conventional color is uniquely identified by a five-digit number composed of the significant digits of dominant wavelength, purity, and luminous reflectance measures in the CIE system (see Chapter 20). Eight basic inks, plus black and transparent white, provide the colors from which all of the other inks are blended (Color Figure 30.1).

Table 30.1

Ink Colors for Features on U.S. Government Maps

USGS 7 ½-minute Quadrangles (Topographic)		U.S. DMA Aeronautical Charts	
Conventional Ink	Feature	Process Ink	Feature
Brown	Contour lines	Cyan	Heavy type
			Military grid
Green	Forest tint		Aeronautical annotations
			Rivers
Blue	Water bodies		Forest symbol
	Rivers	Magenta	Aeronautical information
	Grid ticks		Roads
Red	Roads		High elevation tints
	Urban areas		Heavy type
	USPLS grid		City tint
			Boundary tint
Magenta	Photo revisions	Yellow	Roads
Black	Type		Elevation tints
	Boundaries		Boundary tint
	Railroads		
	Building	Black	Relief shading
	Feature outlines		Railroads
	Point symbols		Roads
	Roads		Boundaries
	Grid ticks		Grid
			Fine type
			Contour lines
			Point symbols

inks, of course. To some degree the limitations of each approach can be overcome by combining the two methods. Thus, it is not uncommon to find the four process inks plus one or two conventional inks used on the same map.

Combining inks in this way lets you choose from the full range of colors made possible by the few process inks, while still having the advantage of using a preblended color or two for fine line features or type. The use of modern five-color presses encourages this sort of hybrid color use in mapping (see "Presswork" later in this chapter).

PLATEMAKING

A single plate can be successively exposed to several negatives. The process, called **double burning** or **double printing**, permits the printer to composite two or more separate flaps on one plate. For example, line and halftone negatives can easily be combined. Similarly, line negatives can be screened by interposing the mechanical screen between the negative and the sensitized plate. This accomplishes the same result on the printing plate as incorporating tints or patterns into the original artwork, or making composite negatives from open-window negatives and negative screens. However, better results are obtained with extra-fine patterns of lines or dots by applying them at this last stage, after photography, than by making them "stand up" through extra steps of the photographic process, particularly if precision graphic arts photography is beyond the cartographer's reach.

A difficulty with double burning of plates is that registry marks are not visible on the plate until it has been exposed separately to each negative and the composite image has been developed. Thus, to make certain that each negative is registered properly, a mechanical system, usually incorporating the use of registry pins, is used (see Chapter 23).

After proper exposures have been made, the plate is developed, placed on the press, and is ready to print. At this stage, most corrections on the plate are impractical if not impossible and are generally limited to deletions. If errors are found on the relatively inexpensive plate, it is usually discarded, the map artwork revised, a new negative made, and a new printing plate prepared.

PRESSWORK

Most lithographic presses incorporate an **offset** arrangement. With offset lithography, the impression from the plate is transferred to another cylinder covered with a rubber mat, called a blanket. The image is then transferred from the blanket to the paper (Figure 30.5).

This offset arrangement is so standard when using a lithographic plate that the terms "offset" and "lithography" have become synonymous. The soft rubber blanket that deposits the ink image on the paper in the offset process does not crush the paper fibers, and fine lines and dots can be faithfully reproduced, even on soft, absorbent paper. Of course, better results are usually achieved with finished or coated paper stock.

One way to classify lithographic presses is according to how many printing units they contain and, therefore, how many plates can be mounted simultaneously. Older presses contained one unit which restricted them to printing a single ink in a **press run**. A press of this type is quite appropriate for reproducing maps in a single color (for example, black). But the machine is handicapped if it is necessary to print another ink on the same map. To do this the press has to be cleaned, re-inked, a new plate mounted, and the process repeated.

In contrast, modern presses may contain up to half a dozen printing units arranged in tandem. These **gang presses** can print as many colors in

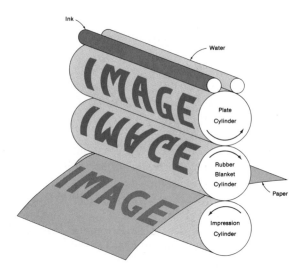

Figure 30.5 In offset lithography, the image is passed from the plate to a rubber blanket, and from the blanket to the paper.

sequence in a single press run as there are printing units. As might be expected, these multiple-unit presses are especially suited to reproducing colored maps. Two-color and four-color machines are the most common.

TRENDS IN MAP REPRODUCTION

There are several important trends in map reproduction:

1. Modern processes are eliminating intermediate steps in reproduction, thereby saving time and materials.

2. It is becoming more economical to obtain just a few copies of a map.

3. There is a trend from wet to dry processing of materials and a related movement away from materials that require darkroom processing.

4. There is growing interest in materials that do not contain forms of silver in their emulsions.

5. There is increasing use of processes that reduce the harmful effects of toxic by-products on the health of workers and the environment.

SELECTED REFERENCES

Bann, D., *The Print Production Handbook*, Cincinnati, OH: North Light Books, 1985.

Bann, D., and J. Gargan, *How to Check and Correct Color Proofs*, Cincinnati, OH: North Light Books, 1990.

Beach, M., S. Shepro, and K. Russon, *Getting It Printed: How to Work with Printers and Graphic Arts Services to Assure Quality, Stay on Schedule, and Control Costs*, Portland, OR: Coast to Coast Books, 1986.

Blair, R., (ed.), *The Lithographer's Manual*, Pittsburg, PA: Graphic Arts Technical Foundation, 1988.

Bruno, M. H., *Principles of Color Proofing*, Salem, NH: GAMA Communications, 1986.

Defense Mapping Agency, *Standard Printing Color Catalog for Mapping, Charting, Geodetic Data and Related Products*, March 1972.

Defense Mapping Agency, *Standard Printing Color Catalog (Process) for Mapping, Charting, Geodetic Data and Related Products*, June 1975.

Defense Mapping Agency, *Standard Printing Color Identification System*, ACIC Technical Report No. 72-2, February 1972.

Eckstein, H. W., *Color in the 21st Century: A Practical Guide for Graphic Designers, Photographers, Printers, Separators, and Anyone Involved in Color Printing*, New York: Watson-Guptill Publications, 1991.

Graves, F. W., and D. L. Des Rivieres, "Cartographic Applications of the Diffusion Transfer Process," *The American Cartographer*, 6 (1979), pp. 107–115.

Hannaford, S., "All the Proof You Need: Seven Ways to Check Color Before You Go To Press," *Publish*, 7 (1992), pp. 64–68.

International Paper Company, *Pocket Pal: A Graphic Arts Production Handbook*, 14th ed., New York: International Paper Co., 1989.

Keates, J. S., *Cartographic Design and Production*, 2nd ed., New York: John Wiley & Sons, 1989.

Lem, D. P., *Graphics Master 4*, Los Angeles: Dean Lem Associates, Inc., 1988.

Ovington, J. J., "An Outline of Map Reproduction," *Cartography*, 4 (1962), pp. 150–155.

Woodward, D., *Five Centuries of Map Printing*, Chicago, IL: University of Chicago Press, 1975.

MAP
PRODUCTION

*T*he quality of map copies is closely related to the nature of the original artwork. With the possible exception of image reduction by photographic or electronic means, duplicating processes cannot much improve the appearance of lines, type, or other map symbols. Flaws or irregularities in the artwork will usually be apparent on copies. To prepare proper artwork, we need to choose effective procedures and use the materials and instruments efficiently.

The graphic arts industry has made great strides during the past few years in perfecting new products and techniques. Some of these innovations have solved problems with which cartographers have struggled for years.

The most significant advances have been made in computer-assisted map construction (Figure 31.1). The bulk of map production is now done with the aid of computer screens. Automated printers and plotters can then be used to preview design alternatives quickly and cheaply prior to map reproduction.

When production must be done by hand, new sensitized materials are available (see Appendix F). These materials permit more efficient map construction. They also facilitate the preparation of a copy or two of the map as a design and production check.

In order to make use of new techniques and plan work properly, you must become acquainted with processes and materials in current use. No one procedure is the best for all maps, and the production plan should be tailored for each situation. The material presented in this chapter is built on the background that was developed in previous discussions of design, color, compilation, typography, and reproduction.* Your aim here should be to learn how to devise workable plans for the construction of well-designed maps that can be reproduced efficiently.

---

*For more technical information on materials, equipment, and procedures, you should consult trade manuals, material specification documents, and equipment catalogs.

FORM OF MAP ARTWORK

You can prepare maps in positive or negative form, or in combinations of both. When you prepare **positive artwork**, you construct the point, line, and area symbols, along with the lettering, on a relatively white or translucent surface. Since many map duplication processes call for negative materials, positive artwork is commonly converted to **photographic negatives** prior to the reproduction step.

Methods for constructing positive artwork vary considerably. Traditionally, manual drafting was used. Although hand drafting is fast disappearing from map production, automated plotted devices that use pens have become popular. Computer technology has introduced several additional methods for producing positive artwork as well. Printers, for example, that build up an image by printing large numbers of tiny dots, are now widely used in map production. In addition, maps are now routinely produced on video display screens.

Use of negative artwork generally improves cartographic quality and may reduce reproduction costs as well. The capability for the cartographer to produce very precise lines in negative form came with the scribing process, which was widely practiced in the past few decades. Recent decades also saw development of many materials that could be used in conjunction with the scribe coat and permit the cartographer to prepare artwork in negative form more easily. Computer-driven vector plotters can also be outfitted with scribing tools and materials, greatly reducing the need for manual scribing—assuming, of course, that the information to be mapped is available in the form of a digital record. Electronic technology has also made it possible to plot directly on film with a small spot of light or laser or electron beams. Some of these film plotters work on a vector principle, while others employ raster technology.

Negative artwork is usually made at reproduction size. But advances in microfilm technology have also made it possible for cartographers to

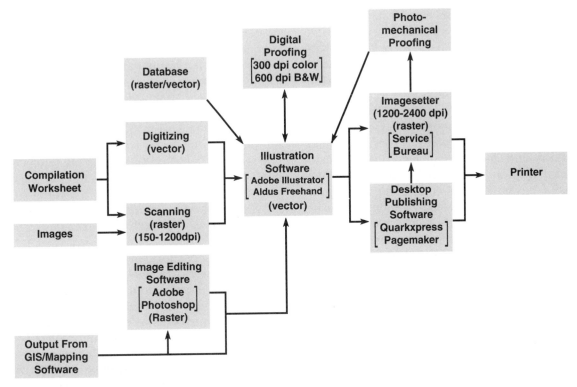

Figure 31.1 Diagram of the digital map production system currently used in the University of Wisconsin Cartographic Laboratory. Compare this diagram with Figure F.1, which outlines the previous analog mapping system now being phased out.

plot map images economically at a much reduced scale for later enlargement to reproduction size. The printer can then go directly from this negative artwork to the platemaking stage. By skipping the photographic step that is required to reverse positive artwork, the cartographer potentially has control of the materials much further into some (especially plate-based) reproduction processes than would otherwise be possible.

Negative artwork is prepared differently from positive artwork. You may produce the symbols automatically with a directed laser or electron beam, or you may produce them manually by cutting or scribing them into a coating that is actinically opaque.* You can use these **prepared negatives** in the same manner as a photographic negative in duplicating processes.

Before beginning the map, you need to decide which form of artwork—positive, negative, or a combination—will produce adequate copies at an economical cost. To make this decision, you must consider such factors as the type of reproduction to be used, size of the copies, amount of detail, kinds of symbols, amount of reduction, access to specialized equipment, and the desired quality of the finished product.

Whether artwork is prepared in negative or positive form or a combination of the two, it often consists of more than one sheet of material, called a flap (see Figure 23.3). If a map is to be printed in

---

*The negative need not be visually opaque. Instead, it can be actinically opaque—opaque only to the wavelengths of light that affect light-sensitive emulsions.

one color, only one sheet of material may be necessary. More than one color or certain special effects require separate flaps. These flaps are then combined at some stage in the production or reproduction process to produce the map on a single sheet.

The vast majority of mapping today is done by digital means. The map image is created on an electronic display screen using special mapping software installed on a computer. The actual mapping procedure varies with the cartographic intelligence coded into the software. Programs that are able to run existing databases through standard mapping algorithms are highly automated and require the least effort on the part of the map maker. Programs that require you to create your own database and build the map image on screen, symbol by symbol, are more demanding. We will explore these two categories of mapping software in the following sections.

ELECTRONIC DISPLAY SCREENS

With ever-increasing frequency, maps are designed and produced for direct viewing on display screens rather than in hardcopy form. The display device most commonly used today is a **cathode ray tube (CRT)**, although a variety of new display devices are attracting attention.*

CRTs work on the same raster scan principle used in familiar television technology.** The

---

*Display screens based on technologies other than cathode ray tubes are growing in popularity. Most of these displays are based on thin-screen technologies such as gas discharge, plasma panels, light emitting diodes (LED), and liquid crystal displays (LCD).

**Not all CRTs work in raster-scan fashion. Vector or random-scan displays create a map image line by line, much as might be done with drafting or scribing. An electron beam is moved across the phosphor-coated screen from endpoint to endpoint of successive line segments. These vector screens are relatively expensive, and their screen redraw speed decreases as image complexity increases. Thus, they are most popular for applications that require the plotting of tens of thousands of high-precision lines but that do not need dynamic picture manipulation.

map image is created when the phosphor-coated screen is excited by electrons. With monochrome screens, a single electron beam sweeps from left to right and top to bottom of the image, line by line. As the electron beam sweeps over each picture element (pixel), its intensity is modulated to create different tones.

Color screens work in the same way as monochrome screens, except that the single electron gun is replaced by three guns, and the screen is coated with dots of red, green, and blue phosphor (see Figures 20.25 and 20.27). By stimulating these dots with various electron gun intensities, we can make pixels appear to be different hues.

Because phosphor glows only briefly when excited by the moving electron gun, it must be restimulated or **refreshed** at a rate of about 72 times a second or the screen image may appear to flicker. Some less expensive screens achieve the desired refresh rate by scanning every other line in each pass of the electron gun(s). This procedure, called **interlacing**, can also lead to apparent screen flicker and is best avoided.

Since the refresh process is independent of image complexity, it takes the same time to refresh a simple map as a complex one. Data must be stored for each picture element (pixel) regardless of whether it is filled or empty. This means that data-processing requirements grow rapidly with greater screen resolutions.

On the negative side, the spatial and color resolutions of raster-scan displays currently in use fall far short of that achieved with printed maps. The reason is obvious. On a 20" (diagonal) monochrome screen of 1,024 x 768 resolution, 786,432 pixels must be altered in less than 1/72 second to prevent flicker. This yields a screen resolution of approximately 70 dots per inch (dpi), which is a little worse than newspaper-quality photo reproduction.

To achieve color imagery of photographic quality on the same screen, the monitor would have to be capable of 256 intensity levels (8 bits) with each of the red, green, and blue electron gun. In other words, 18,874,368 bits of data (786,432 pixels x 3 guns x 8 intensity bits) must be processed each 1/72 second. It takes a lot of computer power to process and move this large amount of information and, even then, the screen resolution is still about 70 dpi.

One of the greatest benefits of using a display screen is that it enables the cartographer to interact with the map that is being produced. Vast improvements in the technology make this interaction easier. Initially, keyboard commands were used, but they have been superseded to a large degree by a variety of pointing devices.

The mouse is by far the most common pointing device today. But for special applications, cartographers have turned to light pens that can be pointed at the screen, or a stylus that can be moved over a data tablet or menu. For some applications, a transparent, touch-sensitive panel is mounted on the screen. Possibly the ultimate development is audio communication, which permits hands-free interaction with the map.

Using these devices, the cartographer can in one way or another identify picture components to be operated upon, specify operations to be performed, or make additions to the image. Cartographers working in an interactive environment commonly are in control of several input devices, gaining tremendous flexibility and power in performing editing, data manipulation, design, and other mapping tasks.

CONSTRUCTION METHOD

There are several ways to prepare artwork on a computer monitor. The method used depends on the nature of available data and software, and on your mapping aim. The mapping package may contain a selection of base maps that you can display, edit, and send to a printing or plotting device. Or the software may consist of a graphic toolbox with a range of "electronic drafting" options that let you create artwork symbol by symbol. Still other software may be able to complete a map with only a few instructions and parameter speci-

fications on your part. Since these methods vary dramatically in the labor they require and the design flexibility they offer, you should be careful in your choice of approach.

We will now discuss the general classes of mapping software that have a significant role place in modern map production. You may find it helpful to refer back to Figure 31.1, which places individual software packages into a mapping system context. Clearly, a mapping project will commonly involve application of more than one class of software. And, further, cartographers rely on generic graphic arts software as well as specialized mapping products.

CLIP-ART MAPS

Outline and base maps are now widely marketed as **clip-art**.* After you select a clip-art map, it is "cut" from the database and "pasted" to the screen (Figure 31.2). You can send these maps to a printer "as is" for reproduction. Or you can tailor the map design on screen to suit your specific needs. Before sending the map on for reproduction, you can add or delete features, change color and other symbols, and modify the style and size of type.

Clip-art maps may fulfill the needs of those wanting to splice a reference map into a document. They may also serve the needs of amateur map makers. But collections of "cut-and-paste" maps are still too limited for most professional cartographers. For these individuals, a clip-art map best serves as a compilation worksheet, which provides a starting point for serious design work.

DATABASE/ALGORITHM MAPPING

To gain more flexibility than clip-art maps provide, professional map makers commonly use specialized mapping software that can import databases

---

*Limited collections of outline and general reference maps are commonly available as clip-art in wordprocessing, presentation graphics, and illustration software. In addition, specialized clip-art vendors sell vast collections of electronic maps (see Appendix C).

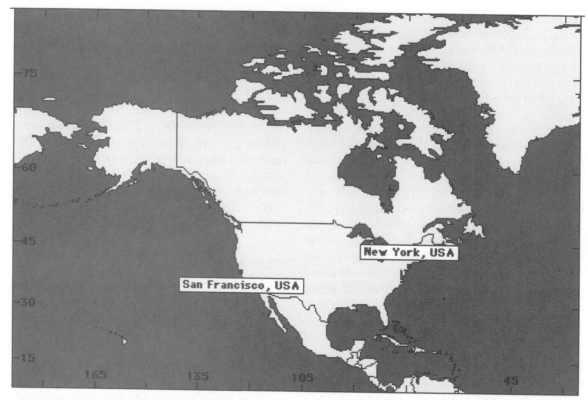

Figure 31.2   Here is a typical example of a clip-art outline map that is currently packaged with software targeted at graphic designers. Map taken from *MacGlobe* version 1.3.0 clip-art collection, Copyright ©1992 Broderbund Software, Inc.

representing geographic base data (coastlines, rivers, political boundaries, etc.) as well as thematic information. The type of data that can be used, and the way the data are handled, distinguish several categories of mapping systems.

Geographic Information Systems

Geographic information systems (GISs) provide the most integrated digital mapping environment. A GIS is a powerful tool for managing environmental information. High-end GIS software has provision for collecting (scanning and digitizing), editing, formatting, structuring, storing, analyzing, and representing environmental data (Figure 31.3).* The aim in GIS is to enhance compatibility

and, therefore, portability of data between software and hardware components of the system. To minimize problems, provision is made for import and export of files in various formats.

To appreciate the potential of mapping within an integrated GIS environment, you must realize that the bulk of mapping activity occurs before the graphic design and execution phase. Data gathering, formatting, structuring, and manipulation constitute the major portion of modern mapping activities. Mapping raw data no longer suffices for

---

*Examples of fully-functioned GIS software packages include *Arc/Info, Spans GIS, MicroStation GIS Enivronment (MGE)*, and *ERDAS IMAGINE*. (For vendor addresses see Appendix C.)

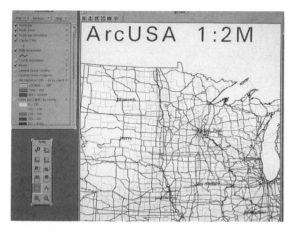

ArcUSA 1:2M

Figure 31.3 A geographic information system provides an integrated mapping environment designed to foster compatibility between hardware, software, and data. (Screen image of *ArcView* interface to *Arc/Info* system, courtesy ESRI.)

many purposes. Therefore, extensive pre-map data analysis has become particularly important to mapping success. GIS technology can serve these needs well.

GIS software provides for a variety of non-spatial statistical manipulations of environmental data (see Chapter 17). But GIS data are also purposely structured so as to preserve topological relations between data objects (points, lines, areas). This structuring makes the data amenable to spatial analysis, such as neighborhood searches, optimized siting and routing, intervisibility, buffering, and overlay. This spatial-analytic capability of GIS has popularized a variety of new types of maps, and greatly expanded the mapping audience in the process.

Not all systems use the same data model, exhibit the full range of desired functions, or are equally easy to use, however.* In choosing a GIS

for mapping purposes, you first should consider the relative strengths and limitations of raster and vector data models. Raster technology is generally less expensive, and is well suited to performing analytical functions that involve simple pixel counting. As a result, raster systems are widely used for resource inventory and management purposes. Since doubling the linear dimension of a pixel quadruples the amount of data that must be handled, raster applications tend to exhibit rather coarse resolutions.

But coarse pixels cause problems in representing non-areal phenomena, such as linear (roads, rivers, boundaries) and point (bridges, wells, towers) features. If you are concerned with all forms of environmental features, a vector system may be most appropriate. In particular, substitution of variable-sized polygons for equal-sized pixels gives vector systems added mapping flexibility. Since database and analysis requirements are greater for vector than raster information, a vector-based GIS is usually more complex and expensive than a raster system.

Once you decide which data model is most appropriate, you should consider the **ease of use** factor, since this is an important issue when dealing with complex GIS software. Some programs still use a **command-line interface** with its taunting c:\> DOS prompt. Although command-line software is difficult to learn, it is extremely powerful and flexible once you master the language and command structure.

At the other extreme is software that uses a **graphical user interface (GUI)**, with intuitive icons and menu prompts. The developmental trend is toward GUI because of its ease of use. However, the speed and functionality of current products is limited by the extra demands a GUI places on computing resources (programs, CPU, communication channel, monitor resolution).

The ease of using GIS software to develop and implement the logical aspects of map design does not necessarily carry over to ease of producing a high-quality graphic product. Indeed, crude map artwork is common with current GIS technology.

---

*You have a wide choice of GIS products beyond the ones mentioned in the previous footnote. Other software packages include *Atlas GIS*, *GisPlus*, *MapInfo*, *GRASS*, and *IDRISI*. (For vendor addresses see Appendix C.)

This need not be the case, however. The graphic limitations of a GIS can be circumvented by exporting the map files to specialized graphics software. Such software permits more sophisticated graphics work, such as an illustration or image editing program (see later sections).

Image Mapping Software

A GIS often can work with remote sensing images as well as graphic and statistical data, but it usually lacks a full complement of image-processing algorithms. Instead, the GIS is often able to import data files from a dedicated image mapping program, such as *ERDAS* (Figure 31.4).* Digital image processing serves two purposes.

One function of image processing is to enhance the overall quality of remote sensing imagery. Image processing is performed on a global, all-at-once basis, in contrast to the piecemeal pixel-by-pixel or subscene approaches taken with paint and image-editing programs (see "Paint Software" later in this chapter). Typical operations include correcting the geometry of the image (rectification), increasing image contrast, and sharpening the edge of features (see Chapter 12). Each of these digital enhancement procedures can increase the clarity and readability of a remote sensing image.

Image processing can be carried further, so that ground features are actually classified digitally (see Chapter 12). Once an image is classified into separate features, the features can be simplified so that the image is less detailed, or a feature or theme can be extracted from the image altogether.

Although a classified and simplified image is useful for many purposes, its utility if further enhanced if the image data can be related to geographic reference information, such as political boundaries. Some image mapping software is able to import and manipulate cartographic data files to perform the necessary data integration. But it is

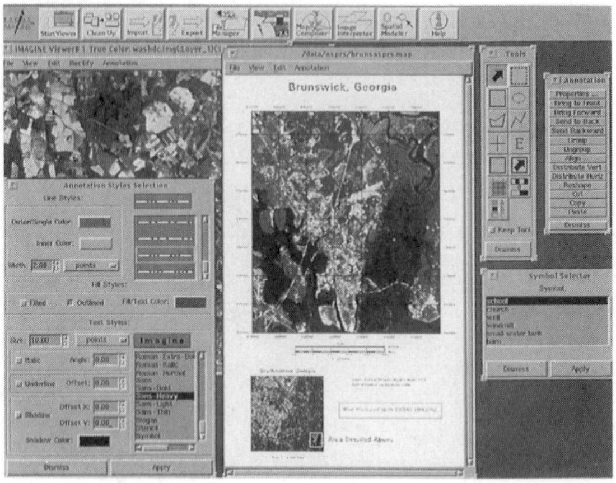

Figure 31.4   Image processing software facilitates a variety of image enhancement and manipulation operations. (Screen image courtesy ERDAS, Inc.)

also common to export remote sensing images to a GIS, which by nature is able to integrate the data layers (see previous section).

Statistical Mapping

Statistical mapping software shares with GIS an ability to manipulate quantitative data prior to graphic representation. But statistical mapping software is usually limited to use of non-spatial operations, such as computing measures of central tendency, correspondence, regression, and so forth (see Chapter 16).

In spite of these common characteristics, considerable variation exists among statistical mapping programs.** Most software is capable of creating dimensional portrayals, such as proportionate point and areal symbols. Other software adds representations that strongly imply three-dimensions, such as three-dimensional areal histograms, shadowed isolines, fishnet, or smoothly shaded surfaces (see Chapter 27).

---

*ERDAS is marketed by ERDAS, Inc. (See Appendix C for address.)

**Representative software packages include *MapInfo*, *Atlas*Graphics*, *MapViewer*, *Surfer*, and *ArcView*. (See Appendix C for vendor addresses.)

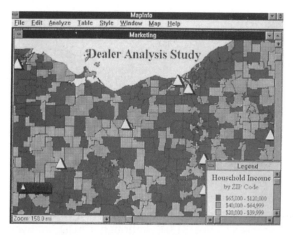

Figure 31.5 Statistical mapping software provides a means for manipulating spatial data and displaying the results using a variety of cartographic techniques. (Screen image courtesy MapInfo Corporation.)

Regardless of the functionality of statistical mapping software, the programs are able to manipulate and portray quantitative data associated with points, lines, or areas. These programs usually include at least some quantitative databases and geographic boundary files, such as country or state outlines. Some programs also require proprietary data files sold by the software vendor. But most statistical mapping programs make provision for importing boundary files and databases from other sources.

Statistical mapping software is highly automated, requiring the minimum of input from the map maker. Most programs use a convenient graphical user interface (Figure 31.5). Once you specify the mapping technique, the area to be mapped, and the data to be portrayed, the software often is able to give you a map by assigning **default values** to critical parameters.

Usually these maps well illustrate the saying, "It's easy to make a map with a computer; it's just difficult to make a good map." Thus, you will probably want to go beyond your "map by default" and explicitly tailor the map design to the data distribution and mapping purpose. You can do so by adjusting such variables as number of classes,

class intervals, and class symbols in two-dimensional representations. In three-dimensional representations, you must also consider the choice of vantage point and the apparent vertical component of the portrayal.

Single-Function Mapping

Mapping software evolved from single-function to multi-function programs. When cartographers began writing software in the early 1960s, the computers available were of relatively limited power. In order to keep mapping programs small enough to run on these machines, each mapping method had to be treated separately. Later, as the power of computers grew, related mapping methods were grouped together to take advantage of common interfaces, data, and algorithms.

Single-function mapping programs are still in use, however. Programs for plotting map projections are an example. Although a choice of projections is now included in most multi-function mapping programs, the selection is usually limited to a few common projections. Furthermore, you usually are not provided much flexibility in adjusting the projection parameters.

There will be times when you would like to use some other projection, or to manipulate more of the projection parameters than is possible with the multi-purpose software at hand. In these cases, a stand-alone projection program that exhibits in-depth scope and functionality may be ideal.* It is common for this projection software to provide dozens of projection choices with the full range of projection options (case, aspect, graticule spacing, etc.) available for each type.

Projection programs are similar in operation to statistical mapping software. In both cases, a prompting, menu-driven interface is standard. Furthermore, both types of programs require that you combine a database with the appropriate mapping algorithms.

---

*Programs that fall in this category include *World, Geocart, CAM,* and *MicroCAM.* (See Appendix C for vendor addresses.)

The difference is that the algorithms in a projection program are designed to solve the chosen projection equation, and the database represents geographic reference data (coastlines, rivers, political boundaries) rather than statistical data. Although it is possible with some projection programs to portray qualitative and quantitative information within the chosen projection framework, this function is more commonly carried out with a GIS or statistical mapping program (see earlier sections).

Your choice of database depends primarily on the desired projection scale. Commonly used data files include the Dahlgren World (8200 points) and United States (10,000 points) files, and the CIA's World Data Bank I (80,000 points) and II (6,000,000 points).* Many larger-scale databases are also available on a regional basis.

MAPS FROM SCRATCH

In contrast to working with mapping systems that use mapping algorithms to manipulate cartographic and thematic databases, you can also use a special class of graphics software that lets you create map artwork one symbol at a time. This category of software closely mimics traditional pen and ink procedures, albeit in a modern electronic mapping environment. Most programs permit you to import a digital cartographic database, which can provide a welcome boost to your "map from scratch."

CADD Software

Computer aided design and drafting (CADD) was one of the earliest graphic applications of electronic technology. This explains why *AutoCAD**,* the premier CADD package, is currently in Release 13 while most other graphic programs have yet to reach a half dozen versions. Due to the long history of CADD, you have a wide choice of software options. In fact, a variety of inexpensive and easy-to-use products provide much of the functionality of the high-end software***.

CADD software closely mimics vector-based mechanical drafting, and is widely used in the architecture, engineering, and construction industries. The software takes a **graphic toolbox** approach. Icons, menu keywords, and dialog boxes are used to manipulate a palette of drafting tools, editing functions, feature primitives, and graphic effects (Figure 31.6). Although CADD software initially was limited to two-dimensional representation, many programs now provide tools for three-dimensional representation and surface rendering as well.

With all CADD programs, maps can be made from scratch by adding one feature at a time. Since map construction usually follows a compilation worksheet, a pure "from scratch" approach is not very efficient. Fortunately, most CADD software also has provision for importing scanned or digitized files. By importing files, you can devote your CADD time to editing and enhance the artwork.

CADD systems are most widely used by engineering, planning, and utilities personnel, where a large collection of accurate base maps must be kept up-to-date. CADD is ideal for mapping qualitative features, such as roads, boundaries, and pipelines. Quantitative information, on the other hand, is not easily mapped with most CADD software. Since CADD data files are not topologically structured, the software also lacks the spatial-analytic capability of GIS. CADD files can be exported to both statistical and GIS software, however. Since CADD systems usually lack the graphic sophistication of more design-oriented graphic software, files are commonly exported to one of these programs if high-quality graphics are needed.

---

*The Dahlgren files were produced by A.V. Hershey at the U.S. Naval Weapons Laboratory in Dahlgren, VA; the World Data Bank files were created by the Cartography Division of the Central Intelligence Agency in Washington, DC.

**High-end CADD software includes packages such as *AutoCAD, MicroStation, Personal Designer,* and *CADvance.* (See Appendix C for vendor addresses.)

***Economical CADD software includes packages such as *FastCAD, Generic CADD, DesignCAD,* and *Drafix CAD.* (See Appendix C for vendor addresses.)

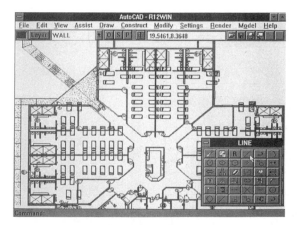

Figure 31.6 CADD programs provide a pallet of drafting and rendering tools that streamline the map design and construction process. (*AutoCAD* screen image courtesy Autodesk, Inc.)

Illustration Software

Of all the software available for digital mapping, vector-based **illustration programs** provide the greatest graphic design flexibility.* Although these high-end design programs were developed to fill the needs of graphic artists, they are widely used in cartographic laboratories as well. Once you master the range of software functions, you can do anything with current illustration programs that used to be done manually or photo-mechanically. Furthermore, a better-quality graphic product can be created faster and with less effort.

Illustration software, similar to CADD, takes a graphics toolbox approach. Basic drawing and type placement tools are represented by icons, and specifications for various graphic effects are provided for in extensive pull-down and pop-up menus (Figure 31.7). You are provided a selection of colors, screens, patterns, line types and widths, and fonts. You can also create special effects, such as curved type, graded or blended tints (vignettes),

reverses (drop-outs), outline symbols, and a fine overlap (trap) of adjacent colors. When the design is complete, you can separate the artwork by color, and export color composite files to a printer, imagesetter, or platemaker.

Although we have placed illustration software in the "from scratch" category, you have the flexibility of importing a wide variety of graphic file formats. You might, for example, import a vector file from an existing database, such as USGS DLG file (see Chapter 13). Or you may import a file from another mapping software package, such as a CADD drawing in .DXF format (see Appendix B). Since the imported graphic is in vector format, it can be directly enhanced using other data and the illlustration software's graphic tools and menu options.

You may also import data in raster format. This might be a scanned compilation worksheet as a .TIF file, a clip-art map or remote sensing image as a .PCX file. The imported raster file is not directly recognized by the vector-oriented illustration software. But it can serve as a template which you can trace by hand to extract feature or color layers.

Since hand tracing on the monitor is tedious, most illustration programs include an **autotrace** tool. This function is designed to let you quickly convert complex shapes to a vector outline. In practice, however, autotraced outlines of raster maps often require a great deal of hand editing to produce a clean vector file. The technology needs refinement before it will be a significant labor saver in mapping applications.

Paint Software

Paint programs represent low-end software for digital mapping.** These drawing packages produce bit-mapped images, which are difficult to enlarge or reduce without distorting the shape of graphic elements. The software "knows" only the location of pixels, not the nature of objects represented by the pixels. Furthermore, the quality of the output

---

*Illustration software popular with cartographers includes packages such as *Freehand, Illustrator, CorelDRAW*, and *Designer*. (See Appendix C for vendor addresses.)

**Examples of Paint software include *Publisher's Paintbrush, CA-Cricket Paint, Painter*, and *Halo Desktop Imager*. (See Appendix C for vendor addresses.)

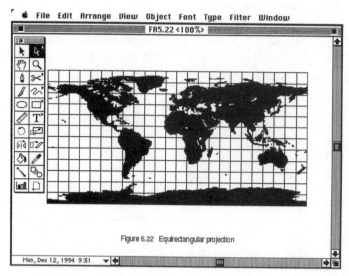

Figure 31.7 Illustration software provides a working screen filled with menu bars, drawing tools accessed through icons, and a selection of pop-up and pull-down menus. (Adobe *Illustrator* screen dump.)

depends on the resolution of devices such as graphics boards used to create the artwork, not the resolution of the output device.

Despite these disadvantages, paint programs can be very useful in "touching up" scanned or other bit-mapped images on a pixel-by-pixel basis. This is especially true if good-quality graphic devices are used to manipulate the high-resolution bit-maps. In fact, programs called **image editors** expand the functionality of common paint programs to include manipulation of pre-defined blocks of pixels representing image objects.* These image editors have largely replaced manual photo processing, and have ushered in a new era of image enhancement (Figure 31.8). Indeed, because of digital image editing, "seeing is believing" no longer holds true for photographs and other images. You can delete or add entire features, change the background, and perform a host of other creative procedures, with no visible evidence of tampering.

On high-end graphics devices, paint and image-editor programs provide a useful tool for improving the graphic quality of artwork created with less sophisticated graphic software. For example, you can import maps produced with CADD or GIS technology, enhance the image, and export the files to a high-quality printer or image setter.

DESKTOP PUBLISHING

Maps commonly appear in newspapers, magazines, brochures, and books. Good design in these media requires effective layout of text, tables, images and graphics. To perform these chores, layout specialists rely on a combination of "visualizing" and trial-and-error positioning.

Up until the past few years, the initial layout of positive artwork was done a page at a time through a physical "cut-and-paste" process called **pasteup**. The pasteup pages were then photographed using graphic arts film, leaving open windows where the illustrations were to appear (see Appendix D and Appendix F). Film negatives of the artwork were then **stripped** into these windows. Finally, printing plates were made from the final film negatives.

---

*Popular image editors include *PhotoStyler*, *PhotoFinish*, *Picture Publisher*, and *Photoshop*. (See Appendix C for vendor information.)

Today most layout is done electronically with the aid of a high-quality graphics monitor and specialized **desktop publishing (DTP)** software.* Operations such as cut-and-paste and stripping can be done digitally. The completed screen image is then sent to an imagesetter, where it is converted to film suitable for making a printing plate.

During the current transition from analog to digital technologies, hybrid systems are common. In this case the basic layout chores are usually done digitally. But rather than import digital records of map images into the screen display, the designer creates windows for subsequent stripping of conventional map filmwork.

OUTPUT OPTIONS

Display screen artwork may suffice for some mapping purposes. But a displayed image is not perma-nent. Because the map disappears when the screen is turned off, it is referred to as a **softcopy** map. At a later date or at some other location, another soft copy can be viewed by calling up the same digital record of data and mapping instructions.

Although ephemeral map display is becoming more popular due to advances in display screen, telecommunication, and image projection technology, people often prefer a more permanent map that they can handle. Physical, **hardcopy** maps generated on a variety of printers and plotters serve this need if you need only a few copies of a map.

You have a wide selection of printing and plotting devices suitable for map production. Although we call these devices **computer peripherals**, they usually include on-board computing and

*Popular desktop publishing software includes *PageMaker*, *Corel VENTURA*, and *QuarkXPress*.

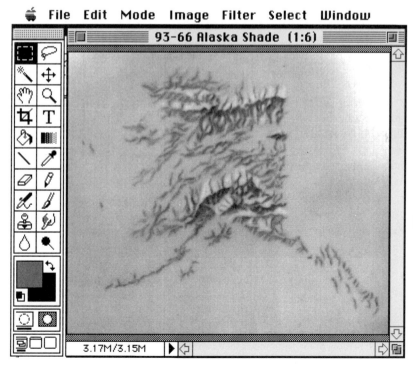

Figure 31.8 Image editors are used by cartographers to perform a variety photo processing tasks previously done photomechanically in a darkroom. This class of software is especially useful for cleaning up problems with scanner output. (Adobe *Photoshop* screen dump.)

memory capacity capable of processing the large digital records associated with maps. Their suitability for mapping is also determined by the type and size* of media they can handle, and their spatial resolution, color capability, and speed. For the most part, there is a direct relation between cost and functionality.

RASTER DEVICES

Printers operate on a scan-line or raster principle, building up images by printing evenly spaced rows and columns of small dots. As a result, the detail or complexity of the artwork being produced has no effect on the speed of production. The printer must receive instructions for each cell or printing element regardless of whether it is to be printed or left empty.

Printers used for mapping are usually electronic in design, producing their dots using electrostatic, thermographic, electromagnetic, and ink-jet technologies. These printers vary substantially in their resolution. Lower-resolution printers produce artwork that lacks density and has rather fuzzy or ragged edges, but the finest-resolution printers produce artwork of extremely high quality. As a rule, the higher-cost machines operate with the best resolution and greatest speed.

A computer normally sends digital cartographic data to a printer with the help of a special **page description language**. The industry standard is *Adobe PostScript*, which handles both type and graphics in vector format.** (See Appendix B). The printer then uses **raster image processing (RIP)** technology to convert the graphic objects defined by the page description language into pixel-by-pixel instructions the printer can understand. RIP requires an on-board microprocessor. The more powerful this computer, the greater the possible speed, resolution, and media size of the

printing device. More powerful microprocessors are also more expensive, of course.

Electrostatic Printers

Electrostatic printers have one or more linear arrays of conducting nibs (pin-like electrodes). These nibs place static charges (dots) a row at a time on a specially coated (non-conductive) paper or film. The paper or film moves at a constant rate beneath the nibs, producing row after row of static dots (Figure 31.9). Resolution up to 400 dots per inch and rates up to six inches per second can be achieved.

Each nib is independently controlled by the digital record to give the paper or film a small round negative charge. We can "develop" this latent electrical image by passing a positively charged toner over the paper or film. The toner adheres to the paper or film where the charges were deposited to form the positive image dots. Limited tints and patterns can be achieved by defining small texture components with different numbers of dots, a procedure known as dithering (see Chapter 20).

Electrostatic printing is usually done in black and white, but full color is also possible. Full-color printing is more expensive because it requires color-separated data files. It also requires that the medium (paper or film) pass under a separate nib array for each of the magenta, cyan, yellow, and black toners.

A disadvantage of electrostatic printing is that the machines and the special image media are relatively expensive. Humidity control and close toner monitoring are necessary to avoid streaks, smudges, and random dots. Images produced at

---

*The graphic arts industry uses the following standard letter codes to define a range of drawing widths: A-size (8.5"), B-size (11"), C-size (17"), D-size (22"), and E-size (34"). These standard codes are used with both sheet and roll media.

**PostScript* is a product of Adobe Systems, Mountain View, CA (415-961-4400). PostScript Level 1 (released in 1985) brought electronic publishing to the desk top and revolutionized the printing industry. PostScript Level 2 (released in 1991) unifies and extends the functions of the original PostScript. Related products include Hewlett-Packard Co.'s Printer Control Language (PCL).

Figure 31.9 Electrostatic plotters are non-impact devices that use one or more linear arrays of printing nibs to build up an image a row of dots at a time. High-quality linework is achieved by offsetting the printing heads to produce overlapping dots. (Image of CalComp 6800 Series Color Electrostatic Plotter courtesy CalComp).

lower resolutions, such as 200–300 dpi, may also lack density, contrast, and sharpness.

Yet for many purposes, the great speed of producing positive artwork on an electrostatic printer far outweighs these disadvantages. For large-format map output on a daily basis, electrostatic printing is often the best choice.

Ink-Jet Printers

Ink-jet printers literally spray-paint the artwork onto the image medium (uncoated paper or clear film). A stream of ink is broken up into microscopic droplets, and shot from a nozzle at a rate of tens of thousands of drops per second. Under the control of the digital record, the droplets acquire an electrical charge. Then, passing between charged plates, they either are allowed to strike the paper, or they are deflected away from the paper to an ink reservoir. By using multiple nozzles, ink-jet printers can simultaneously print in several colors of ink. If cyan, magenta, yellow, and black inks are used, the result looks very much like a process color halftone printing from lithographic plates.

The droplets of ink used in ink-jet printers play the same role as electrons in a cathode ray tube. Unfortunately, ink droplets are more troublesome to work with than electrons. The mechanical system of ink reservoirs, pumps, tubes, and nozzles is subject to clogging and other failures. Furthermore, the discrete ink droplets can be placed imprecisely, potentially compromising the quality of the print.

Ink-jet printers are slower than electrostatic printers. But they are also less expensive. Since they use the same technology found in desktop ink-jet printers, true resolution is limited to 300 dpi or less. By using special resolution enhancement technologies, however, machines with a true resolution of 300 dpi can be made to produce virtual resolutions of 600 dpi on media up to 36 inches wide (Figure 31.10). Printing can be done on off-the-shelf media, such as paper, polyester film, and vellum. When water-based ink is used, it can soak through paper and puddle up on the film surface, leading to smearing or smudging unless carefully handled. Ink-jet output tends to be fuzzy and lacking in ink density.

But if colored artwork (including transparencies) needs to be done quickly and relatively inexpensively, ink-jet printing may be the best choice. As with electrostatic methods, however, the image is not as high in quality as might be produced by photo-mechanical means. Yet for many mapping purposes it is adequate.

Laser Printers

Laser printers work much like a standard xerographic copier. Thus, some refer to this printing process as **electronic xerography**.* Higher-qual-

---

*The term "electro-photography" is probably more accurate than laser printing, since some units use light emitting diode (LED) arrays or some other non-laser energy source.

Figure 31.10 Printers based on ink jet technology can produce large-format maps in color at relatively low cost. (Image of NOVAJET II courtesy ENCAD, Inc.)

expensive than electrostatic and ink-jet printers. What they lack in speed and economy, however, they more than make up for in quality. Machines with a resolution of 300 rasters per inch are the current standard for general applications. Higher-end machines with true resolutions of 600 dpi (360,000 dots per square inch) are found in many map production shops (Figure 31.11 and Figure 31.12). With the help of various resolution enhancement technologies, vendors have pushed these 300–600 dpi laser printers to virtual resolutions up to 1,200 dpi. The quality of graphic output from these top-end laser printers approaches that of offset presses.

Thermal Printers

Thermal printers use a page-sized inking ribbon, a coated receptor material, and a heat-producing printhead (Figure 31.13). In **thermal transfer printing**, the ribbon is coated with soft wax or plastic impregnated with color pigments. The ribbon is divided sequentially into pages of cyan, magenta, yellow, and black.

Most thermal transfer printers use the PostScript page description language. The printhead moves

ity printers use the PostScript page description language. To produce an image, a laser beam is routed through a modulator that turns it on or off for each dot position according to available digital information for the artwork. A series of mirrors then leads the modulated beam to a scanning prism that sweeps it across the photo-sensitive drum line by line in raster fashion. The image on the drum, now a pattern of positive charges, is next dusted with negatively charged toner particles. A positively charged sheet of paper rolled over the drum attracts the toner particles, which then are fused onto the surface under heat and pressure.

Although laser printers typically produce black-and-white output, machines that create color output are growing in popularity. As a rule, these laser devices tend to be slower and more

Figure 31.11 Desktop laser printers are convenient proofing devices and are now common fixtures in cartographic production units. (Image of 600 dpi LaserJet 4P courtesy Hewlett-Packard Company.)

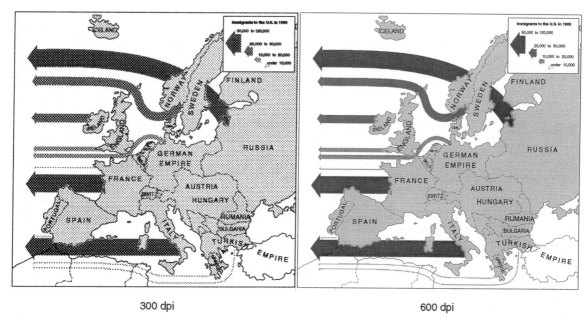

300 dpi   600 dpi

Figure 31.12   Laser printer output at 300 dpi (A) is visibly inferior in graphic quality compared with output at 600 dpi (B). Compare these examples with Figure 25.16, which was output on an imagesetter at 2400 dpi.

across the colored ribbon in raster fashion. At each dot, the coating can be heated to one of 256 temperature levels, melting a proportionate amount of pigment. The colored liquid is then transferred with a roller to a coated receptor material, where it cools into a glossy spot of color. Since the pigment coating is not melted in background areas, ink transfer occurs only at image locations.

Thermal dye-sublimation (DS) printers use a technology similar to thermal transfer printing, except that the four-page ribbon is coated with a special dye instead of with color-impregnated wax or plastic. A DS printer's thermal head converts (sublimes) the solid ink to a gas, which liquifies on the surface of the receptor material. The more heat applied at a dot, the greater the amount of ink transferred at that point.

The output quality of thermal printers surpasses that of ink-jet, but equipment and material costs are also substantially higher. Thermal printers can achieve tonal effects by both dot-dithering and

Figure 31.13   Desktop thermal printers are commonly used in cartographic production units for color proofing maps of relatively small size. (Image of Shinko CHC-746i ColorStream/Plus courtesy Mitsubishi International Corp.)

halftoning. With resolutions as high as 300 dpi, and 256 intensity levels per color, it is possible to create continuous tones and over 16 million colors. Prints from high-end thermal devices equal the quality of photographs.

Imagesetters

The highest-quality graphic output is obtained when a map is plotted in raster fashion directly on photographic film or paper with devices called **imagesetters**. Imagesetters may be stand-alone machines, or they may be high-quality electronic scanners equipped to "write out" as well as to "read in." Imagesetters generally operate with a laser or electron-beam writing head capable of extremely fine spatial resolution. As a result, imagesetters can produce output of unmatched graphic quality. The product may be finished composite separations directly suitable for platemaking. Alternatively, output may consist of open windows and masks that require compositing and screening at the platemaking stage (see Appendix F).

Because imagesetters are relatively expensive, they are not usually found in small mapping establishments. Small firms usually have their film recording done by commercial **service bureaus**. These private firms are widely distributed and work on a contract basis. By operating 24 hours a day and serving a number of clients, service bureaus are able to provide products at competitive prices despite a large investment in equipment.

Prior to film recording, software has to convert the coordinate descriptions of the map artwork to instructions the device can understand. This is done using PostScript or some other page description language. An electron or laser beam then sweeps across the film line by line, turning on (to expose) or off as directed by the raster scan file. Output from imagesetters is usually proofed (see later section) and then sent on to a printer for preparation of printing plates (see *Composite to Plate* later in this chapter).

Some imagesetters work in large format at map reproduction scale, while others operate in microfilm format. Large-format imagesetters generally operate at resolutions of 1,200 lpi for paper output,

or 2,400 lpi for film output. Local service bureaus commonly have machines that can handle film widths up to 18 inches. If a larger film format is needed in map production, data usually have to be sent to a large, regional service bureau that has a high-end imagesetter.

By taking advantage of microfilm technology, imagesetters can also produce film separations much smaller than the desired final map size (see Figure F.6). This requires a **computer output on microfilm (COM)** unit, which can plot extremely fine (sub-micron) lines on high-resolution film. The microfilm output is subsequently enlarged to reproduction size for making printing plates. When good equipment and materials are used, microfilm separations can be enlarged many times with no evidence of image degradation.

VECTOR PLOTTERS

Anything that used to be drafted by hand can now be plotted mechanically with a **vector plotter**. Indeed, in many ways these relatively inexpensive machines mimic hand drafting. The plotting bed may be a flat surface or a rotating drum (Figure 31.14). Although it is possible to adapt the plotting head for mechanical scribing and film plotting, usually it holds a selection of drafting pens (Figure 31.15).

For the most part, these are "incremental" plotters, which means that as the plotting tool moves to draw a diagonal line, it must be "stepped" in the horizontal and vertical directions. This **stepping** is done at the resolution of the plotter.

Stepping produces zigzag increments along diagonal lines, an effect known as **aliasing**. On low-resolution plotters, aliasing can be quite noticeable. It is usually of some help to plot at a larger scale and then reduce the plot photographically to final size. But if high-quality linework is needed, a better approach is to use a higher-resolution plotter or a more expensive non-incremental machine that can plot diagonal lines without stepping.

Early vector plotters were slow because each line segment was plotted as a separate entity. This meant that the plotter head had to accelerate from (and decelerate to) stops at the endpoints of each

Figure 31.14 Vector plotters are commonly used in proofing or for producing a few copies of a large-format map in color. (Image courtesy Calcomp.)

Figure 31.15 The drawing head on vector plotters commonly consists of a turret of four to eight drafting pens. Available pen and media options may include roller ball, liquid ink, and fiber tip for plotting on bond, polyester film, or vellum.

line segment. Today's plotters are able to determine the direction of upcoming line segments and adjust their velocity accordingly. Since the plotter only slows down if it has to, plotting speed is greatly increased. Even with these high-speed plotters, however, plotting time is directly proportional to the complexity of the map artwork.

Vector plotters produce a map with the help of a digital database, a few user-supplied plotting specifications, and software capable of converting artwork details to plotting instructions. For example, you may have to specify such things as the width and type (plastic or paper) of plotting medium, scale of the plot, choice of line weights and form (solid, dashed, etc.), type of pen (ball point, fiber tip, fluid ink), ink color, desired artwork (feature or color) separations, and location of registry marks.

Since plotting instructions must be understood by the plotting device, they require a special **plotter language**. For this purpose, the second generation of the *Hewlett-Packard graphic language (HP-GL/2)* is a de-facto standard (see Appendix B). Even such well-known competing plotter manufacturers as Calcomp and Houston Instruments support HP-GL/2.

Once the plotter is given proper instructions, the biggest concern is to ensure that the pen is working properly, a problem that has long plagued manual drafting. If a pen clogs or runs out of ink during a plot, the artwork is ruined and must be redone. To minimize problems, plots are commonly supervised, but this is inconvenient and expensive. Several plotter manufacturers now offer options on their more sophisticated machines that greatly reduce the need for plot supervision. Truly unattended pen plotting appears to be close at hand.

DIGITAL PRODUCTION PROBLEMS

Digital map production circumvents many problems associated with making maps using other technologies. But digital production introduces some new problems as well. Some problems arise because the map is constructed in one medium and reproduced in another, and the two media are not fully compatible. Other problems occur when there is significant difference between the scales of data gathering and map construction, or between map construction and reproduction scales. Let's look at both these problems.

MIXED-MEDIA DEGRADATION

When maps designed and constructed in one medium are reproduced in another medium, fidelity problems are common. For example, maps designed for printed media that are displayed electronically commonly suffer severe degradation. One reason is that the spatial resolution of even the most expensive computer monitors falls far short of the resolution of print media, causing fine details to be lost. In order to compensate for this loss of detail, coarser symbols must be used on maps designed for viewing on a monitor.

Color fidelity problems are also common. A map designed on a computer monitor will often look different when it is printed on paper because there is poor color matching between the displayed and printed versions (see Chapter 20). Equipment vendors offer various **color matching schemes** to help rectify problems of color coordination between devices, but a satisfactory solution is not yet at hand. In the meantime, problems can be minimized by using color conservatively when the media of map construction and reproduction differ. It's also a good idea to provide a color check by proofing the color prior to reproduction (see "Proofing" later in this chapter).

SCALE ISSUES

Two scale issues confront the map maker. One concerns the relation between the scales of data collection and map construction. Since data collection entails abstraction, there is no problem as long as the map is equally or more abstract than the original data. It is legitimate to map at scales equal to or smaller than the data resolution.

But one of the most basic principles of cartography is that maps should not be made at scales much larger than the resolution of the data. Maps made to appear less abstract than the data used in mapping are potentially deceptive, because environmental details that would normally be included in the map are not available in the data. For example, a digital database derived from 1:100,000 maps will exhibit greater environmental detail than a database derived from 1:250,000 scale maps. Thus, if the database taken from maps generalized for reproduction at 1:250,000 is used to produce maps at scales larger than 1:100,000, the result may imply a lack of geographic detail and lead to a false impression of the character of the region mapped.

You have a responsibility to determine the scale of data collection prior to mapping, and to match the map scale to the resolution of the data. Avoiding "blowing up" some layers of information in a database to fit the scale of others. It's fine to "zoom in" on database features for construction and editing functions, but always drop back to an appropriate reproduction scale.

A second scale issue concerns the construction of artwork at one scale for reproduction at another scale. Artwork should not be enlarged for the same reasons databases should not be mapped at larger scales than they were prepared. Enlargement of artwork also makes graphic irregularities more apparent and accentuates registry problems.

Reduction of artwork for reproduction is another matter. Although reducing artwork may sharpen an image and reduce registry problems to some extent, it presents a design challenge.* When preparing artwork that is to be reduced, people have a ten-

---

*Reduction of artwork is far less common in digital production and scribing than it was when map making was a pen-and-ink affair (see Appendix F). However, reduction can sometimes improve the graphic quality of maps produced from low-resolution databases, or output on low-end printers or plotters.

dency to add so much detail or such fine detail that it is either lost or difficult to read at the reduced size. As a rule, you should avoid reducing artwork over 50 percent. At these reductions, it is difficult to anticipate how the finished artwork will look.

Reduction of artwork is most troublesome when it is not planned for by the map designer. Unfortunately, it is rather common for publishers to reduce map artwork to fit available space in a newspaper or magazine column despite the fact that map legibility is significantly degraded. Although last-minute publishing decisions must sometimes be made, you should make every effort to ensure that your maps do not end up published at unreadable scales. Since small, fine marks associated with symbols, type, and screens are the first details to be lost upon reduction, avoid using these small features if there is a possibility your artwork will be reduced for reproduction.

COMPOSITING SEPARATIONS

You need a good knowledge of the possibilities and limitations of the printing process in order to visualize how the various map separations are to be used. You must keep in mind, for example, that a print or printing plate cannot be exposed to two negatives at the same time, unless one is a negative contact screen. Negatives usually require successive exposures.* On the other hand, more than one film positive, or film positives and a negative, can be overlaid for a single exposure.

On one print or printing plate, then, there can be an image that resulted from a series of exposures, any of which can incorporate more than one piece of film. The result of multiple exposures is a **composite**. These composites are either made on film prior to platemaking, or they are made directly on the printing plate.

*Using contact tint screens to screen contact pattern screens is an exception to this rule, since both contact screens are usually in negative form.

COMPOSITE TO FILM

When map artwork is produced on a film recorder, it is commonly in the form of a film composite. You give pixel-by-pixel instructions to the imagesetter using a page description language such as Adobe PostScript. This special language provides for the definition of all windows, screens, traps, reverses, and other special effects.

The most efficient approach in digital production is to produce a composite for each of the spot colors to be printed, for each of the four process inks (cyan, magenta, yellow, and black), or for some combination of the two color systems. Since a composite is made in one scan of the film recorder's intense laser or e-beam, image quality is outstanding and needs little of the "touch-up" associated with photo-mechanical compositing (see Appendix F).

Although film compositing is efficient and growing in popularity, it does have a few potential drawbacks. For example, if you are not satisfied with the resulting map, or decide to modify the artwork for other maps, any design change will usually entail redoing one or more composites.

Another potential problem is that compositing to film prior to platemaking leaves the printer little flexibility to make adjustments to compensate for difficult paper, ink, and press considerations. As we advised in Chapter 30, it is best to tell the printer the effect you want to achieve, and let the printer decide how this is best done. Fortunately, most service bureaus are well tuned to the needs of printers, and the composites produced from imagesetters are of such high quality that problems can be minimized by getting these professionals talking to each other.

COMPOSITE TO PLATE

The alternative to film compositing is to composite to the printing plate. This was common practice when photo-mechanical production dominated mapmaking (see Appendix F). It provided the cartographer and the printer with a great deal of flexibility in reusing separations and in achieving the highest quality map reproduction possible.

To minimize the potential for problems when compositing to plate, you need to clearly label the artwork flaps. In addition, you should provide the printer with a complete set of instructions on how the separations should be composited. A standard instruction sheet is handy for this purpose (Figure 31.16).

But compositing to plate also has drawbacks. Cartographers tend to use far more separations than printers are used to handling. If compositing instructions are unclear or read improperly, the result will be incorrect. Furthermore, each time a flap or screen is handled under darkroom conditions, there is potential for dust to degrade the image and for compositing mistakes to occur.

The trend is toward **direct-to-plate** compositing. In this case, a plate replaces the film in an imagesetter, thus avoiding the film step entirely. The direct-to-plate approach is still in development, and is presently used primarily for reproducing small-format maps.

Direct-to-plate has its own problems. Either a costly press run must be made to get an idea of how the printed map will look, or a full composite must be made by some non-press printing or plotting device. This need for a check or proof copy deserves further discussion.

UNIVERSITY OF WISCONSIN CARTOGRAPHY LAB
PRINTING INSTRUCTIONS

MAP NUMBER ________ PRINTING SIZE ________

NUMBER OF FLAPS ________ COLORS ________

Flap (a) - Color ____, Solid or ____%, Reverse from ____, Halftone ____
Flap (b) - Color ____, Solid or ____%, Reverse from ____, Halftone ____
Flap (c) - Color ____, Solid or ____%, Reverse from ____, Halftone ____
Flap (d) - Color ____, Solid or ____%, Reverse from ____, Halftone ____
Flap (e) - Color ____, Solid or ____%, Reverse from ____, Halftone ____
Flap (f) - Color ____, Solid or ____%, Reverse from ____, Halftone ____
Flap (g) - Color ____, Solid or ____%, Reverse from ____, Halftone ____
Flap (h) - Color ____, Solid or ____%, Reverse from ____, Halftone ____
Flap (i) - Color ____, Solid or ____%, Reverse from ____, Halftone ____
Flap (j) - Color ____, Solid or ____%, Reverse from ____, Halftone ____
Flap (k) - Color ____, Solid or ____%, Reverse from ____, Halftone ____
Flap (l) - Color ____, Solid or ____%, Reverse from ____, Halftone ____

Additional Comments:

Figure 31.16 **A standard instruction sheet is a handy way to minimize potential compositing problems when sending complex artwork involving multiple separations to a printer for compositing to plate.**

PROOFING

When artwork is completed, it must be carefully and thoroughly checked before being sent for duplication. If the map is constructed with more than one flap, a composite of the separations is used as the proofing vehicle. This composite is called a **pre-press proof** if the map is to be reproduced by plate-based printing. As the name implies, a pre-press composite is made prior to the actual printing stage.

As a rule, it is wise to proof a map if it consists of more than one flap. Even one-color maps can consist of 10 or more flaps, greatly increasing the chances for error and making editing difficult. A properly prepared proof will closely resemble the copies that will be printed from the artwork and can therefore evaluate the success of the design as well as reveal errors.

DIGITAL CHECK-PLOTS

If map artwork is available in the form of digital records, you can use the full range of computer-driven screens, printers, and plotters to make pre-press proofs. Color screens are particularly well suited for this task. With appropriate hardware, software, and data files, you can alter the digital records interactively and view the corrected image almost instantly. Thus, you not only save time and effort, but you avoid the high materials cost associated with photo-chemical methods (see next section). There is a tradeoff in programming, computer charges, and capital costs, of course.

Since fine details and color distinctions can be difficult to assess on CRT screens, many people prefer a hardcopy proof. Computer-driven printers and plotters provide output that is suitable for this purpose. The terms check-prints or check-plots are commonly used to describe proofs made by these devices.

PHOTO-MECHANICAL PROOFS

Artwork prepared on translucent material, such as output from a film recorder, is also suitable for pre-press proofing. But in this case the proofing is done photo-chemically in a laboratory setting. Most materials require a platemaker (see Appendix D), but darkroom facilities are not necessary (see Chapter 30).

In photo-chemical proofing, exposures are usually made through negative separations. During each step of the operation, pin registration must be used. For those materials that are not easily punched, strip registration is required. Strip registration is achieved by using gummed prepunched tabs or by punching strips of discarded film and taping them to the material sheets.

INVOLVING THE PRINTER

If a map is to be reproduced by plate-based methods and facilities are not available for proofing, you can ask the printer to furnish pre-press proofs. Have the printer provide these proofs after making the negatives but before making the printing plates. Errors found at this stage can often be corrected on the negative. If the artwork must be revised, you have lost only the cost of the negatives and not the cost of the plates, paper, and press time.*

For complicated colored maps, you may wish to request **press proofs**. In this case, the actual plates that will be used to print the final copies are used to print a few copies of the map. Errors found at this point save the cost of paper and press time. But there is the added expense of making and mounting new plates. It is particularly expensive if the press is reserved but idle while corrections are made.

---

*The cost of the negative is not much different from the cost of the plates these days. What is expensive is the cost of labor to remake the plate.

SELECTED REFERENCES

Aaland, M., and R. Burger, *Digital Photography*, New York: Random House, 1992.

Anson, R. W., (ed.), *Basic Cartography for Students and Technicians*, Volume 2, New York: Elsevier Applied Science Publishers, 1988.

Beach, M., S. Shapiro, and K. Russon, *Getting It Printed*, Portland, OR: Coast To Coast Books, 1986.

Beach, M., "The Burden of Proofs: To Save Money and Meet Deadlines, Choose the Right Proofs for Your Printing Jobs," *Aldus Magazine*, 4, 2 (1993): 57–60.

Byrne, K., "Desktop Map Design: Some Odysseys of Form and Flow, *Cartographic Perspectives*, 15 (Spring 1993): 13–20.

Curran, J.P., (ed.), *Compendium of Cartographic Techniques*, New York: Elsevier Applied Science Publishers, 1988.

Edwards, K., and R. M. Batson, "Preparation and Presentation of Digital Maps in Raster Format," *The American Cartographer*, 7 (1980): 39–49.

Fink, P., *PostScript Screening: Adobe Accurate Screens*, Mountain View, CA: Adobe Press, 1992.

Fink, P., *How to Make Sure What You See is What You Get: Expert Tips for Success in PostScript Output*, Washington, D.C.: Peter Fink Communications, Inc., 1992.

Green, D. R., "Map Output from Geographic Information and Digital Image Processing Systems: A Cartographic Problem," *The Cartographic Journal*, 30, 2 (1993): 91–96.

Holzgang, D. A., *Understanding PostScript*, 3rd Ed., Alameda, CA: SYBEX Inc., 1992

Leonard, J. J., and B. P. Buttenfield, "An Equal Value Gray Scale for Laser Printer Mapping," *The American Cartographer*, 16, 2 (1989): 97–107.

Muehrcke, P. C., "An Integrated Approach to Map Design and Production," *The American Cartographer*, 9 (1982): 109–122.

Peterson, M. P., "Chapter 5: Graphic Processing" in *Interactive and Animated Cartography*, Englewood Cliffs, NJ: Prentice Hall (in press).

Schokker, P. W. M., "Desktop Publishing in a GIS Environment," *ITC Journal*, 3/4 (1989): 221–224.

Wallace, P., "Persuasive Printouts," *PC/Computing*, 6, 9 (1993): 170–192.

Whitehead, D. C., and R. Hershey, "Desktop Mapping on the Apple Macintosh," *The Cartographic Journal*, 27, 2 (1990): 113–118.

For up-to-date information see graphic arts and printing industry trade journals and newsletters, such as *Aldus Magazine, American Printer, IN-PLANT Reproductions, Desktop To Press, Graphic Arts Monthly* and *Publish*.

FEDERAL GEOGRAPHIC DIGITAL DATA PRODUCTS

SOURCES OF DIGITAL DATA

  U.S. DEPARTMENT OF AGRICULTURE (USDA)
  U.S. DEPARTMENT OF COMMERCE (USDC)
  U.S. DEPARTMENT OF THE INTERIOR

ELECTRONIC ACCESS

*T*his appendix describes federal geographic digital data products that are national in scope and commonly distributed to the public. Geographic data products include digital map data, aerial photography and multispectral imagery, earth science products, and other geographically referenced data sets.

SOURCES OF DIGITAL DATA

The data products listed here provide a sample of commonly used digital data sets. They cover only part of the information available either from the governmental or private sector. In addition, rapidly changing geographic data technologies, especially digital technologies, are constantly resulting in new products and new forms of existing products. Contact the offices listed here to ask about new products.

U.S. DEPARTMENT OF AGRICULTURE (USDA)

Soil Conservation Service (SCS)

The Soil Conservation Service (SCS) collects digital map databases representing soil-unit boundaries. Field mapping methods, based on national standards, are used to construct the soil maps in the **Soil Survey Geographic Database (SSURGO)**.

SSURGO is the most detailed soil mapping done by SCS. Mapping scales range from 1:12,000 to 1:31,680. SSURGO digitizing duplicates the original soil survey maps. This level of mapping is designed for use by landowners, townships, and county natural resource planning and management.

Soil maps for the **State Soil Geographic Database (STATSGO)** are made by generalizing the detailed soil survey or SSURGO geographic database. The mapping scale for STATSGO is 1:250,000. The level of mapping is designed for broad planning and management uses covering state, regional, or multi-county areas.

Soil boundaries of the **National Soil Geographic Database (NATSGO)** are formed from the **major land resource area (MLRA)** and **land resource region (LRR)** boundaries. The MLRA and LRR boundaries are developed from state general soil maps. The NATSGO map is digitized at a scale of 1:7,500,000.

Technical, Ordering, and Availability Information

National Cartographic and Geographic Information Systems Center
USDA—Soil Conservation Service
P.O. Box 6567
Fort Worth, TX 76115
Telephone: (817) 334-5559
FAX: (817) 334-5290

U.S. DEPARTMENT OF COMMERCE (USDC)

Bureau of the Census

The Bureau of the Census collects population and housing statistics, organized hierarchically from national to census block level. Data reports are further organized by **complete-count subjects** (asked of all residents) and **sample subjects** (asked of about 17.7 million randomly selected housing units).

During the 1990 Census of Population and Housing, the Census Bureau developed the **TIGER (Topologically Integrated Geographic Encoding and Referencing)** system. TIGER has spawned a new line of public-use extract products. These products contain digital data for map features, boundaries, names, codes, coordinates, and (in urban cores of the most populous counties) address ranges and ZIP codes.

Technical, Ordering, and Availability Information

Customer Services
Bureau of the Census
Washington Plaza, Room 326
Washington, D.C. 20233
Telephone: (301) 763-4100
FAX: (301) 763-4794

National Oceanic and Atmospheric Administration (NOAA)

National Environmental Satellite, Data, and Information Service (NESDIS)

National Climatic Data Center (NCDC) The National Climatic Data Center (NCDC) collects and maintains all United States weather records. NCDC is the largest climatic center in the world. It is a unique central source of historical weather information and related products. It manages and operates the **World Data Center A (WDC-A)** for Meteorology, which provides international exchange of meteorological and climatological data.

NCDC holds all weather records routinely collected by the U.S. federal government. It also maintains data acquired from foreign sources, from cooperative exchanges with state or local agencies, and from research activities. These data include surface observations from land stations, ocean weather stations, and moving ships. Daily climatological observations from cooperative observing stations, upper air observations (including balloon and aircraft reports), and radar observations are also archived at NCDC.

Technical, Ordering, and Availability Information

NOAA
National Climatic Data Center
Climate Services Division
NOAA/NESDIS E/CC3
Federal Building
Asheville, NC 28801-2696
Telephone: (704) 259-0682
FAX: (704) 259-0876

National Geophysical Data Center (NGDC) The National Geophysical Data Center (NGDC) combines, in a single data center, the fields of seismology, geomagnetism, marine geology and geophysics, solar phenomena, the ionosphere, and snow-and-ice research. NGDC collects, organizes, publishes, and distributes solid-earth, solar-terrestrial physics, and snow-and-ice data from worldwide sources. NGDC also operates the World Data Center A (WDC-A) for Solid Earth Geophysics, the WDC-A for Solar Terrestrial Physics, and the WDC-A for Marine Geology and Geophysics. NGDC contracts with the University of Colorado to operate the WDC-A for Glaciology (Snow and Ice) and the **National Snow and Ice Data Center (NSIDC)**. Information is portrayed in a variety of classification schemes, standards, data formats, reference systems, accuracy, resolutions, and time frames. Data are derived from sources applicable to geophysical, solar-terrestrial, and snow-and-ice research.

Technical, Ordering, and Availability Information

National Geophysical Data Center
Information Services Division
NOAA/NESDIS E/GC4
325 Broadway
Boulder, CO 80303-3328
Telephone: (303) 497-6958
FAX: (303) 497-6513

National Oceanographic Data Center (NODC) The National Oceanographic Data Center (NODC) maintains a large, multidisciplinary marine database. NODC's activities include acquisition, processing, storage, and retrieval of oceanographic data. It manages and operates the World Data Center A (WDC-A) for Oceanography. Major types of data exchanged by the WDC-A are serial oceanographic station data, bathythermograph, current, biological, and sea surface observations. NODC also archives and distributes **U.S. Navy Geodetic Satellite (GEOSAT)** data. GEOSAT provided the first long-term global observations of sea-level, windspeed, wave-height, and ice topography.

Technical, Ordering, and Availability Information

National Oceanographic Data Center
User Services Branch
NOAA/NESDIS E/OC21
1825 Connecticut Avenue, NW
Washington, DC 20235
Telephone: (202) 606-4549
FAX: (202) 606-4586

National Ocean Service (NOS)   The National Ocean Service (NOS) collects coastal marine data from the continental United States, including the Great Lakes, Alaska, Hawaii, Puerto Rico, U.S. Virgin Islands, United States trust territories, and other islands in the Atlantic and Pacific Oceans. Digital bathymetric maps (topographic maps of the sea floor) use contours to show the size, shape, and distribution of underwater features. Nautical charts are produced at various scales for navigation of the earth's water surface. There are four classifications of nautical charts:

Sailing Charts, published at a scale smaller than 1:600,000, are used by mariners to plan and fix their position as they approach the coast from the ocean or as they sail along the coast between distant ports. On these charts, the shoreline and topography are generalized. Only offshore soundings, major navigational aids, outer buoys, and landmarks visible at considerable distances are shown.

General Charts of the coast, published at scales from 1:150,000 to 1:600,000, are used for coastal navigation when a course is far offshore but can be fixed by landmarks, lights, buoys, and characteristic soundings.

Coast Charts, published at scales from 1:50,000 to 1:150,000, are used for close navigation inside outlying reefs and shoals, in entering or leaving bays and harbors of considerable size, and in navigating the larger inland waterways.

Harbor Charts, published at scales of 1:50,000 and larger, are used for navigating in harbors and smaller waterways and for anchorage.

Technical, Ordering, and Availability Information

For orders:
Distribution Branch
National Ocean Service
NOAA
Riverdale, MD 20737
Telephone: (301) 436-6990

For information:
Coast and Geodetic Survey
National Ocean Service, NOAA
1315 East-West Highway, Station 8620
Silver Spring, MD 20910-3282
Telephone: (301) 713-2780

U.S. DEPARTMENT OF THE INTERIOR

U.S. Geological Survey (USGS)

The United States Geological Survey (USGS) provides a wide array of digital data products. To obtain these products, USGS conducts earth science investigations, including regional geologic studies, geophysical surveys, geochemical surveys, offshore geologic investigations, mineral resource investigations, topographic mapping, and water resource investigations. Major product lines include **Digital Line Graph (DLG)** data and **Digital Elevation Models (DEM)**.

USGS provides many sources of digital imagery through its **EROS (Earth Resources Observation System) Data Center.** These images include:

- **Side-Looking Airborne Radar (SLAR)** data. SLAR penetrates cloud cover and enhances shadows of subtle terrain features.

- **Advanced Very High Resolution Radiometer (AVHRR)** data. AVHRR is useful for tracking vegetation growth on a regional basis.

- **Digital Orthophoto Quad** data, with a ground resolution of approximately one meter.

- **Land Satellite (LANDSAT)** data, which include **Thematic Mapper (TM)** images, **Multispectral Scanner (MSS)** images, and **Return Beam Vidicon (RBV)** images.

- **Systeme Probatoire d'Observation de la Terre (SPOT)** data. SPOT provides both panchromatic and MSS images.

USGS also distributes products created by other agencies. Digital data produced by the **Defense Mapping Agency (DMA)** are distributed by USGS. These data include:

- digital elevation data derived from 1:250,000-scale maps.
- **Digital Chart of the World** data, derived from 1:1,000,000-scale **Operational Navigational Charts (ONC).**

USGS also distributes **National Wetlands Inventory (NWI)** data produced by the **U.S. Fish and Wildlife Service**. NWI data provide planners and resource managers with information on wetland location and type.

Technical, Ordering, and Availability Information

Earth Science Information Center
U.S. Geological Survey
507 National Center
Reston, VA 22092
Telephone: (703) 648-5920
FAX: (703) 648 5548
Toll Free Number: 1-800-USA-MAPS

Sioux Falls ESIC
U.S. Geological Survey

EROS Data Center
Sioux Falls, SD 57189
Telephone: (605) 594-6151
FAX: (605) 594-6589

ELECTRONIC ACCESS

A growing number of data sets are available electronically via the **Internet**. Browsing software, such as Mosaic, make it easy to search electronically for public-domain information. The number of sites providing electronic access is growing rapidly. You should, therefore, check first with the USGS site. It contains up-to-date references to other sites with digital cartographic data.

For those on using Mosaic on World Wide Web, the address for the USGS home page is:

http://info.er.usgs.gov

The Internet address for FTP access is 130.11.54.41 (oemg.er.usgs.gov).

APPENDIX B

DATA AND GRAPHIC FILE FORMATS

DATA FORMATS

- ASCII
- BINARY

GRAPHIC FILE FORMATS

- VECTOR FILE FORMATS
- RASTER FILE FORMATS

COMPUTER GRAPHICS STANDARDS

*C*artographers often must store maps as digital data in a graphic file format that will allow the map to be viewed and modified at a later date on the same or another computer mapping system. This involves storing positional and attribute data in either **ASCII** or **binary** format, depending on the data input options in the mapping system.

DATA FORMATS

Data describing vector mode maps are often stored in ASCII text files that can be read into and edited in word processing programs as well as computer mapping systems. In contrast, standard raster image map formats are often binary representations of the digital map, sometimes in a compressed form that greatly reduces data redundancy. We will first describe these two data formats, and then turn to commonly used graphic file formats.

ASCII

ASCII (American Standard Code for Information Interchange) has become an international standard for the coding of alphanumeric and graphic characters used as computer input-output data. Characters typed on the keyboard or listed on a monitor screen are transmitted and stored as ASCII integer number codes.

Decimal integer codes for two ASCII character sets are listed in Figure B.1. Codes 0-32 do not appear since they represent characters, such as the ENTER key, used to control printers, monitors, and other input-output devices. The standard ASCII alphanumeric character set contains characters for integer codes 0-127. The top half of the full 256 character set is not standardized and is software dependent. Additional character sets have been devised for other languages and for symbols commonly used in mathematics, music, and cartography. The third column shows the cartographic symbol character set developed for the *CorelDRAW* illustration package.

BINARY

All numbers and characters manipulated by computers must be in a **binary** (base 2) code that the central processing unit of the computer can understand. All ASCII characters entered from the keyboard, digitizing tablet, or an ASCII text file are internally converted into binary integers. Data such as map coordinates are often stored directly as binary integers or real numbers.

Binary Integers

Binary integers are composed of **bits** (binary digits) that can be either 0 or 1. Several binary integers are listed in Table B.1, along with the corresponding decimal value. Notice that binary integers composed of two bits can only range in decimal value from 0 to 3, eight-bit binary integers can only range from 0 to 255, and so on. We also commonly represent binary integers in shorthand notation by their **hexadecimal** (base 16) values, such as the corresponding hexadecimal values in the third column.

Binary integers are stored and manipulated in groups called **bytes.** Each byte is made up of eight bits. ASCII characters, for example, are converted in the computer to single-byte binary integers representing decimal numbers ranging from 0 to 255.

Table B.1

Binary Integers with Corresponding Decimal and Hexadecimal Values

Binary Integer	Decimal	Hexadecimal
0	0	0
1	1	1
10	2	2
11	3	3
100	4	4
1000	8	8
10000	16	10
100000	32	20
1000000	64	40
10000000	128	80
11111111	255	FF
100000000	256	100

Graphic Character Chart

Figure B.1 shows columns of: Decimal code, Alphanumeric character, Graphic character.

Decimal code	Alphanumeric	Decimal code	Alphanumeric	Decimal code	Alphanumeric	Decimal code	Alphanumeric	Decimal code	Alphanumeric	Decimal code	Alphanumeric
		059	;	092	\	0125	}	0192	À	0225	á
		060	<	093	]	0126	~	0193	Á	0226	â
		061	=	094	^	0161	¡	0194	Â	0227	ã
		062	>	095	_	0162	¢	0195	Ã	0228	ä
		063	?	096	`	0163	£	0196	Ä	0229	å
		064	@	097	a	0164	¤	0197	Å	0230	æ
		065	A	098	b	0165	¥	0198	Æ	0231	ç
033	!	066	B	099	c	0166	¦	0199	Ç	0232	è
034	"	067	C	0100	d	0167	§	0200	È	0233	é
035	#	068	D	0101	e	0168	¨	0201	É	0234	ê
036	$	069	E	0102	f	0169	©	0202	Ê	0235	ë
037	%	070	F	0103	g	0170	ª	0203	Ë	0236	ì
038	&	071	G	0104	h	0171	«	0204	Ì	0237	í
039	'	072	H	0105	i	0172	¬	0205	Í	0238	î
040	(	073	I	0106	j	0173		0206	Î	0239	ï
041	)	074	J	0107	k	0174	®	0207	Ï	0240	ð
042	*	075	K	0108	l	0175	¯	0208	Ð	0241	ñ
043	+	076	L	0109	m	0176	°	0209	Ñ	0242	ò
044	,	077	M	0110	n	0177	±	0210	Ò	0243	ó
045	-	078	N	0111	o	0178	²	0211	Ó	0244	ô
046	.	079	O	0112	p	0179	³	0212	Ô	0245	õ
047	/	080	P	0113	q	0180	´	0213	Õ	0246	ö
048	0	081	Q	0114	r	0181	µ	0214	Ö	0247	÷
049	1	082	R	0115	s	0182	¶	0215	×	0248	ø
050	2	083	S	0116	t	0183	·	0216	Ø	0249	ù
051	3	084	T	0117	u	0184	¸	0217	Ù	0250	ú
052	4	085	U	0118	v	0185	¹	0218	Ú	0251	û
053	5	086	V	0119	w	0186	º	0219	Û	0252	ü
054	6	087	W	0120	x	0187	»	0220	Ü	0253	ý
055	7	088	X	0121	y	0188	¼	0221	Ý	0254	þ
056	8	089	Y	0122	z	0189	½	0222	Þ	0255	ÿ
057	9	090	Z	0123	{	0190	¾	0223	ß		none
058	:	091	[	0124	|	0191	¿	0224	à		

Figure B.1 ASCII decimal codes for alphanumeric and graphic character sets. Codes 0-32 are used for special control characters. The graphic character set is that developed for the *CorelDRAW* PC illustration package.

Data may be initially stored as one-byte (short), two byte, or four-byte (long) binary integers.

The greater the number of bytes used for each number, the wider the range of possible numbers, as shown in Table B.2. In this table, we also see that binary integers normally are stored in what is called 2s complement form, allowing for an equally large range of negative and positive integers. One byte binary integers may range from -128 to 127 in value, for example.

Binary Real Numbers

Real numbers cannot be stored and manipulated as binary integers, since the fractional part of each number cannot be represented. Real numbers entered as a series of ASCII characters must be converted to binary real numbers prior to computer processing. Binary real numbers are commonly encoded in six-byte blocks in scientific

Table B.2

Decimal Number Ranges for 1, 2, and 4 Byte Binary Integers

2s Complement Decimal Range	
1 byte integers	-128 to 127
2 byte integers	-32768 to 32767
4 byte integers	-2147483648 to 2147483647

notation form. One byte is used for the exponent, one bit of the remaining 40 for the sign of the number (0 positive, 1 negative), and the remaining 39 bits for the number itself. Eleven significant digits can be encoded in this manner over a numerical range from approximately 10^{-38} to 10^{38}.

GRAPHIC FILE FORMATS

The internal binary data format unique to each piece of mapping software is not appropriate for graphical information transfer between dissimilar systems. Numerous graphic exchange file formats have been developed by software companies and others (see Box B.1). These can be grouped into formats for the transfer of raster format and of vector format graphic data (although at least one format allows for both raster and vector data transfer).

Several of these data transfer formats have come to be used as "standards" for graphical data exchange. A file format becomes a "standard" when it is adopted as an input-output option in several mapping systems. Formats devised for widely used CAD software packages and plotting devices are

---

Box B.1
Graphic File Formats

Vector Formats

Acronym	Full Name	Developer/Major User
CGM (ISO)	Computer Graphics Metafile	*Int. Standards Org.*
DXF	Drawing Exchange File	*Autodesk*
HPGL	Hewlett-Packard Graphics Language	*Hewlett-Packard*
IGES	Initial Graphics Exchange Specification	*U.S. Nat. Bur. Stds.*
MAC	MacPaint	*Apple*
PICT	Picture Format	*Apple*
PS &EPS	PostScript	*Adobe*
SCODL	Scan Conversion Object Description Language	**Imagesetting Devices**
WPG	Wordperfect Graphic Format	*Wordperfect*

Raster Formats

Acronym	Full Name	Developer/Major User
BMP	Bitmap File Format	
GIF	Graphics Interchange Format	*Compuserve*
IFF	Interchange File Format	*Amiga*
MSP	Microsoft Paint	*Microsoft*
PCX	Picture Exchange Format	PC Paintbrush
PIC	Picture Format	*Lotus*
PS & EPS	Postscript	*Adobe*
TGA	Targa Image Format	*Targa*
TIFF	Tag Image File Format	**Desktop Scanners**

often incorporated into programs used in computer cartography, as are standard formats accepted by the *American National Standards Institute (ANSI)*, *International Standards Organization (ISO)*, and similar organizations.

There are seven graphic file formats that you are likely to encounter as exchange standards in computer mapping systems. This diversity of formats illustrates the saying that "the nice thing about standards is that there are so many from which to choose!" In the following discussion, we have grouped formats by vector and raster data applications. You will find that some formats are unique to vector or raster data, while other formats are applicable to both types of data.

VECTOR FILE FORMATS

Let's look first at the five formats that are commonly used when working with vector data. We will describe the structure of each format and then illustrate it with a common example so that the formats can be compared.

Initial Graphics Exchange Specification (IGES)

IGES was released in 1981 by the *U.S. National Bureau of Standards* as the first standardized graphics exchange format for CAD/CAM systems. This standardization was deemed essential to the Department of Defense, where contractors were faced with transferring thousands of graphics files among the large number of CAD systems used in weapons design. IGES **graphic elements** reflect this mechanical engineering heritage, although later versions (2 and 3) have added elements important in other fields, including cartography.

The IGES file is composed of five **sections** of ASCII records:

- a **START** section containing a file label indicating which IGES translator was used.
- a **GLOBAL** section listing the output file name, IGES translator version, the company or agency producing the file, and general data characteristics such as the number of bits per integer and the precision of real numbers.
- a **DIRECTORY ENTRY** section containing general information about each of the more than 70 graphic entities that have been defined. Entities typically found in digital map files are listed in Table B3. Each record contains an entity type code and points to where more detailed data reside in the Parameter Data section. Each record also includes codes indicating the line pattern, layer index value, transformation matrix, line weight, and color.
- a **PARAMETER DATA** section providing a detailed description of each entity, including coordinate and attribute data.
- a **TERMINATE** section consisting of a single record giving the number of records in each of the preceding four sections.

IGES records may be in a fixed field or compressed ASCII format. The fixed field length format is based on 80-character records divided into 10 fields of eight characters. The records required to define the bar scale in Figure B.2 are listed in Table B.4, along with annotations explaining important fields.

Table B.3

IGES Entities Commonly Found in Digital Map Files

Number	Description
100	circle or circular arc
102	composite curve
108	polygon
110	line
112	parametric spline curve
114	parametric spline surface
116	point
124	transformation matrix
158	sphere
168	ellipsoid
212	text string
228	general symbol
304	line font
310	text font
314	color definition

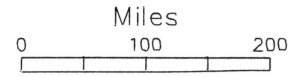

Figure B.2 Simple bar scale used to illustrate graphic file formats.

Table B.4
IGES (.IGS) File Required to Define the Bar Scale in Figure B.2

START Section
IGES file generated from an AutoCAD drawing by the IGES   S0000001
translator from Autodesk, Inc., translator version IGESOUT-3.04.   S0000002

GLOBAL Section
drawing name   file name   creator   IGES version
,,8HBARSCALE,18HA:BARSCALE.IGS.IGS,14HAutoCAD-R11 c2,12HIGESOUT-3.04,32,G0000001
    ID   **units**   **date/time**
38,6,99,15,8HBARSCALE,1.0,1,4HINCH,32767,3.2767D1,13H920629.090243,   G0000002
        author   **organization**
5.5D-9,5.5,21HCARTOGRAPHIC SERVICES,24H OREGON STATE UNIVERSITY,6,0;   G0000003

DIRECTORY ENTRY Section

type	PD row		line font		Seq. #
110	1	1	1		00000000D0000001
type		**line count**			**Seq. #**
110		1			D0000002
110	2	1			00000000D0000003
110		1			D0000004
110	3	1			00000000D0000005
110		1			D0000006
110	4	1	1		00000000D0000007
110		1			D0000008
110	5	1	1		00000000D0000009
110		1			D0000010
110	6	1	1		00000000D0000011
110		1			D0000012
110	7	1	1		00000000D0000013
110		1			D0000014
212	8	1	1	0	00000100D0000015
212		2			D0000016
212	10	1	1	0	00000100D0000017
212		2			D0000018
212	12	1	1	0	00000100D0000019
212		2			D0000020
212	14	1		0	00000100D0000021
212		2			D0000022

Table   B.4 (continued)

PARAMETER DATA Section

type x y z x y z		seq. #	
110,1.0,5.0,0.0,3.0,5.0,0.0;		1P0000001	
110,3.0,5.0,0.0,3.0,5.1,0.0;		3P0000002	
110,3.0,5.1,0.0,1.0,5.1,0.0;		5P0000003	
110,1.0,5.1,0.0,1.0,5.0,0.0;		7P0000004	
110,1.5,5.1,0.0,1.5,5.0,0.0;		9P0000005	
110,2.0,5.1,0.0,2.0,5.0,0.0;		11P0000006	
110,2.5,5.1,0.0,2.5,5.0,0.0;		13P0000007	

type bounding box x	seq. #	
212,1,1,6.6929133858268D-2,0.1,1,,0.0,0,0,9.6259842519685D-1,	15P0000008	

y z text	seq. #	
5.15,0.0,1H0;	15P0000009	
212,1,3,2.1417322834646D-1,0.1,1,,0.0,0,0,1.8834645669291D0,	17P0000010	
5.15,0.0,3H100;	17P0000011	
212,1,3,2.1968503937008D-1,0.1,1,,0.0,0,0,2.8862204724409D0,	19P0000012	
5.15,0.0,3H200;	19P0000013	
212,1,5,4.9606299212598D-1,0.15,1,,0.0,0,0,1.7377952755906D0,	21P0000014	
5.35,0.0,5HMiles;	23P0000015	

TERMINATE Section

	seq. #	
S0000002G0000003D0000022P0000015		T0000001

Computer Graphics Metafile (CGM)

A **computer graphics metafile (CGM)** contains all the digitally defined elements in a "picture," including positional and attribute data for lines, polygons, text, and point symbols. The CGM, first released in 1987, is envisioned as an international standard (ISO 8632) in three file formats: ASCII characters, compressed ASCII code, and binary. The standard consists of graphic element names and acronyms (called opcodes), and rules as to where elements may be placed in the file.

A CGM file is structured hierarchically. The metafile is composed of an introductory section giving characteristics of the entire file, followed by the graphic elements constituting one or more maps and other "**pictures.**" Each picture, in turn, has an introductory section giving general picture characteristics, followed by the graphic elements making up the picture body (points, lines, text).

Positional and attribute data are listed for each element, along with control elements such as coordinate precision.

CGM elements and corresponding ASCII text acronyms commonly encountered in metafiles describing digital maps are listed in Table B.5. These are divided into delimiter, general metafile descriptor, general picture descriptor, and picture body elements. A sample CGM file defining the bar scale in Figure B.2 is listed in Table B.6.

Table B.5

CGM Elements Commonly Encountered In Digital Map Metafiles

Element Name	Opcode Acronym
BEGIN METAFILE	BEGMF
END METAFILE	ENDMF
BEGIN PICTURE	BEGPIC
BEGIN PICTURE BODY	BEGPICBODY

Table B.5 (continued

END PICTURE	ENDPIC
METAFILE VERSION	MFVERSION
METAFILE DESCRIPTION	MFDESC
INTEGER PRECISION	INTEGERPREC
REAL PRECISION	REALPREC
INDEX PRECISION	INDEXPREC
COLOR PRECISION	COLRPREC
COLOR INDEX PRECISION	COLRINDEXPREC
COLOR VALUE EXTENT	COLRVALUEEXT
FONT LIST	FONTLIST
CHARACTER SET LIST	CHARSETLIST
SCALING MODE	SCALEMODE
COLOR SELECTION MODE	COLRMODE
LINE WIDTH SPECIFICATION MODE	
LINEWIDTHMODE	
EDGE WIDTH SPECIFICATION MODE	
EDGEWIDTHMODE	
BACKGROUND COLOR	BACKCOLR
POLYLINE	LINE
TEXT	TEXT
POLYGON	POLYGON
CELL ARRAY	CELLARRAY
RECTANGLE	RECT
CIRCLE	CIRCLE
CIRCULAR ARC 3 POINT	ARC3PT
ELLIPSE	ELLIPSE
ELLIPTICAL ARC	ELLIPARC
LINE TYPE	LINETYPE
LINE WIDTH	LINEWIDTH
LINE COLOR	LINECOLR
TEXT FONT INDEX	TEXTFONTINDEX
TEXT COLOR	TEXTCOLR
CHARACTER HEIGHT	CHARHEIGHT
CHARACTER ORIENTATION	CHARORI
TEXT ALIGNMENT	TEXTALIGN
CHARACTER SET INDEX	CHARSETINDEX
INTERIOR STYLE	INTSTYLE
FILL COLOR	FILLCOLR
HATCH INDEX	HATCHINDEX
PATTERN INDEX	PATINDEX
EDGE TYPE	EDGETYPE
EDGE WIDTH	EDGEWIDTH
EDGE COLOR	EDGECOLR
PATTERN TABLE	PATTABLE
PATTERN SIZE	PATSIZE
COLOR TABLE	COLRTABLE

Table B.6
Computer Graphics Metafile to Create the Bar Scale in Figure B.2

```
BEGMF;
MFVERSION 1;
MFDESC AJK 1992;
REALPREC 0,10000,4;
FONTLIST helvetica bold;
BEGPIC;
SCALEMODE abstract,25.4;
LINEWIDTHMODE scaled;
BEGPICBODY;
LINETYPE 1;
LINEWIDTH 1.0;
LINE (1.0,5.0)(3.0,5.0)(3.0,5.1)(1.0,5.1)(1.0,5.0);
LINE (1.5,5.0)(1.5,5.1);
LINE (2.0,5.0)(2.0,5.1);
LINE (2.5,5.0)(2.5,5.1);
CHARHEIGHT 0.15;
TEXT 0.975,5.15,final,"0";
TEXT 1.9,5.15,final,"100";
TEXT 2.9,5.15,final,"200";
CHARHEIGHT 0.3;
TEXT 1.75,5.35,final,"Miles";
ENDPIC;
ENDMF;
```

Drawing Exchange File (.DXF™)

The Drawing Exchange File (DXF) format was devised for the exchange of drawing files between AutoCAD© program users.* A DXF file is an ASCII text file that contains a complete description of the AutoCAD drawing. This complete description, coupled with the widespread use of AutoCAD and similar design and drafting programs with DXF output options, has made this file format a standard way of exchanging cartographic drawing data.

A full DXF file consists of the following four **sections**:

*AutoCAD is a product of *Autodesk, Inc.*, Sausalito, California.

- a **HEADER** section containing general information about drawing parameters, which are identified by their **variable name**. One hundred thirty-seven header variables are found in the standard AutoCAD (release 12) DXF output file.
- a **TABLES** section containing definitions of graphic items in the drawing, such as linetypes, layers, text styles, the user coordinate system, and the viewpoint configuration.
- a **BLOCKS** section describing all Block entities used in the drawing file.
- an **ENTITIES** section defining all drawing entities, including two- and three-dimensional lines and polylines, text, circles and ellipses, a large variety of point symbols, area patterns, and any blocks used in the drawing.

The annotated DXF file segment in Table B.7 shows how the sections are defined in a typical cartographic drawing file, the DXF file for the pen plotted bar scale in Figure B.2. The BLOCKS section does not appear because no blocks were required in this very simple drawing. Only small parts of the HEADER, TABLES, and ENTITIES sections are listed, but the listing still shows the basic form of all sections. Note the general approach of using a numerical code specifying the type of entity to follow, and then the entity description in text string or numerical form.

Table B.7
Annotated DXF File Segment

DXF Content	Meaning
0	entity follows
SECTION	entity is a section
2	section name follows
HEADER	section name is HEADER
9	variable name follows
$EXTMIN	minimum drawing extent
10	x coordinate follows
0.966535	x coordinate
20	y coordinate follows
5.0	y coordinate
30	z coordinate follows
0.0	z coordinate
9	variable name follows
$EXTMAX	maximum drawing extent
10	x coordinate follows
3.109843	x coordinate
20	y coordinate follows
5.5	y coordinate
30	z coordinate follows
0.0	z coordinate
9	variable name follows
$TEXTSTYLE	text style
7	text style name follows
HB	HB (helvetica bold)
0	entity follows
ENDSEC	end of section
0	entity follows
SECTION	entity is a section
2	section name follows
TABLES	section name is TABLES
0	entity follows
TABLE	entity is a table
2	table name follows
LTYPE	name is LTYPE (linetype)
70	number of table entries follows
1	one entry in table
0	entity follows
LTYPE	entity is a linetype
2	linetype name follows
CONTINUOUS	continuous line
70	linetype flag follows
64	table entry is used in drawing
3	linetype description follows
Solid line	a solid line
72	line alignment code follows
65	code
73	number of dashes in line follows
0	zero dashes
40	total line pattern length follows
0.0	zero pattern length
0	entity follows
ENDTAB	end of table
0	entity follows
TABLE	entity is a table
2	table name follows
LAYER	table name is LAYER
70	number of table entries follows
1	1 entry in table
0	entity follows

Table B.7

LAYER	entity is a layer	10	beginning point x coordinate follows
2	layer name follows	1.0	x coordinate
0	layer name is "0"	20	beginning point y coordinate follows
70	layer flag follows	5.0	y coordinate
64	layer was used at least once in drawing	30	beginning point z coordinate follows
62	layer color number follows	0.0	z coordinate
7	color number 7	11	ending point x coordinate follows
6	layer linetype follows	3.0	x coordinate
CONTINUOUS	continuous line	21	ending point y coordinate follows
0	entity follows	5.0	y coordinate
ENDTAB	end of table	31	ending point z coordinate follows
0	entity follows	0.0	z coordinate
TABLE	entity is a table	•	
2	table name follows	•	rest of lines
STYLE	table name is STYLE	•	
70	number of text styles follows	0	entity follows
1	one style used in drawing	TEXT	entity type is text
0	entity follows	8	name of layer for text follows
STYLE	entity is a text style	0	layer name is "0"
2	style name follows	10	insertion point x coordinate follows
HB	helvetica bold	1.737795	x coordinate
70	style flag follows	20	insertion point y coordinate follows
64	text style used at least once in drawing	5.35	y coordinate
		30	insertion point z coordinate follows
40	text height follows	0.0	z coordinate
0.15	0.15 inches	40	text height follows
41	text width factor follows	0.15	0.15 inches
1.0	1.0 factor	1	text string follows
50	text obliquing angle follows	Miles	string is "Miles"
0.0	0 degrees	7	text style follows
71	number of text flags follows	HB	style is HB
0	zero flags	72	horizontal justification code follows
42	last text height used follows	1	text is centered
0.15	0.15 inches	11	text alignment point x coordinate follows
3	font file name follows		
HB	helvetica bold	2.0	x coordinate
4	"big-font" file name follows	21	text alignment point y coordinate follows
	none used		
0	entity follows	5.35	y coordinate
ENDTAB	end of table	31	text alignment point z coordinate follows
0	entity follows		
ENDSEC	end of section	0.0	z coordinate
0	entity follows	•	
SECTION	entity is a section	•	rest of text
2	name of section follows	•	
ENTITIES	section name is ENTITIES	0	entity follows
0	entity type follows	ENDSEC	end of section
LINE	entity type is line	0	entity follows
8	name of layer for line follows	EOF	end of DXF file
0	layer name is "0"		

PostScript©(.PS) and Encapsulated PostScript (.EPS) Files

PostScript* is a graphics programming language developed to define text, graphic objects, and raster format images on pages displayed or printed by raster devices. Regular PostScript files may be used to send maps to PostScript output devices such as raster-scan monitors, dot-matrix and ink jet printers, and high resolution laser imagesetters. Encapsulated PostScript files are specially formatted for insertion into graphics illustration programs and often include a TIFF format "icon" image. A computer mapping or image processing program with a PostScript output file option contains built-in software routines to convert the digital map drawing or image to an ASCII text file PostScript language description of the map or image. The PostScript page description is, in turn, interpreted by software in the hardcopy device and converted into a raster binary format specific to the device. This allows text, linework, graytones, digitized remote sensor images, and digital halftones to be combined on one page.

PostScript page files (with a .ps or .eps extension) are normally structured into **prolog** and **script** sections. The prolog contains file lineage (see Chapter 14), graphic image dimension, and text font information, followed by definitions of variables and procedures that are stored by keyword in "dictionaries." The script section contains the sequence of PostScript operations (commands) and associated data that define all graphic elements on the page. Several hundred operators exist to carry a variety of functions, including:

1. arithmetic, trigonometric, boolean, matrix, coordinate system, and text string data manipulations.
2. text definition and placement.
3. definition and placement of graphic characteristics including linework, graytones and colors.
4. data file input/output.

---

*Commercially implemented and distributed by *Adobe Systems Inc.*

Figure B.3   Bar scale used to illustrate the power of the PostScript page description language.

The following annotated PostScript page descriptions produce the bar scale plotted in Figure B.3. We can begin to appreciate the power and complexity of the PostScript language by studying and prolog and the basic operators in the script.

```
!PS-Adobe-2.0
%%BoundingBox: 69.59 360.00 223.91 396.00
%%Creator: AJK
%%Title: TESTMAPS.PS
%%CreationDate: 6/24/1992 10:35:45
/m /moveto load def              %move procedure
/s { show newpath } bind def          %text display
/l /lineto load def              %line draw
/c /closepath load def           %close polygon
/ss { setlinewidth stroke } bind def   %set linewidth
/scr { currentscreen 3 1 roll pop pop setscreen } bind def
/sp { gsave dup stringwidth pop grestore} bind def
/sr { exch sub 0 rmoveto } bind def
%%EndProlog
newpath
% Laser printer page size: 7.9 10.4
156.37 15.12 translate          %move to page center
72 72 scale                     %scale to inches,not pts
/cmtx matrix currentmatrix def        %define graphic
                                          elements
.5 setgray 150 45 scr           %graytone specification
1.5000 5.1000 m                 %define first gray area
1.5000 5.0000 l                 %boundary
1.0000 5.0000 l
1.0000 5.1000 l
c fill                          %fill with screen
2.5000 5.1000 m                 %define second gray area
2.5000 5.0000 l                 %boundary
```

2.0000 5.0000 l

2.0000 5.1000 l

c fill   %fill with same screen

0 setgray   %make new symbols black

1.0000 5.0000 m 3.0000 5.0000 l 0.01 ss   %draw black
lines

3.0000 5.0000 m 3.0000 5.1000 l 0.01 ss

3.0000 5.1000 m 1.0000 5.1000 l 0.01 ss

1.0000 5.1000 m 1.0000 5.0000 l 0.01 ss

1.5000 5.1000 m 1.5000 5.0000 l 0.01 ss

2.0000 5.1000 m 2.0000 5.0000 l 0.01 ss

2.5000 5.1000 m 2.5000 5.0000 l 0.01 ss

[0.1000 0.000 0.1000 0] (_Helvetica-Bold) ts %plot text

0.9626 5.1500 m

(0) sp 2 div 0.0374 sr s

1.8835 5.1500 m

(100) sp 2 div 0.1165 sr s

2.8862 5.1500 m

(200) sp 2 div 0.1138 sr s

[0.150 0 0.000 0.150 0 0] (_Helvetica-Bold) ts

1.7378 5.3500 m

(Miles) sp 2 div 0.2622 sr s

showpage   %send page to output device

restore

Hewlett-Packard Graphics Language (HPGL)

Hewlett-Packard Graphics Language (HPGL) is the ASCII graphics language used to control *Hewlett-Packard* and other pen plotters that can be configured to read HPGL files. Over 50 two-letter commands with required and optional parameters exist to initialize the plotter and plotting conditions, as well as to control the plotting of lines, text, point symbols, circles, rectangles, and other basic graphic elements. Many programs employed in computer mapping allow HPGL files to be saved on disk, and many word-processing and graphic illustration programs can read HPGL files as a part of the page composition process.

An annotated subset of HPGL commands often used in cartography is presented in Table B.8.

Several of these are used in the following HPGL command list that will produce the plotted bar scale in Figure B.2.

IN;

SP1;

LT;

PU;PA1000,5000;

PD;PA3000,5000,3000,5100,1000,5100,1000,5000;

PU;PA1500,5000;

PD;PA1500,5100;

PU;PA2000,5000;

PD;PA2000,5100;

PU;PA2500,5000;

PD;PA2500,5100;

SI 0.15,0.2;

PU;PA970,5150;PD;LB0e$_x$

PU;PA1900,5150;PD;LB100e$_x$

PU;PA2900,5150;PD;LB200e$_x$

PU;SI 0.25,0.3;PD;LBmilese$_x$

PU;PA0,0;SP;

Table B.8
HPGL Commands Commonly Used in Computer Cartography

Command	Meaning
IN	initialize plotter
SP x	select pen # x
SC xmin,xmax,ymin,ymax	establish user coordinate system
PU	pen up
PD	pen down
PA x,y	move pen to coordinate x,y
CI rad	draw circle of radius rad
AA x,y,ang	draw circular arc centered at x,y from current pen position for an angle of ang°
FT x	fill an area with fill pattern x
XT	draw a vertical X-tick
YT	draw a horizontal Y-tick
SM x	draw symbol x
LT x	use line type x
LB...e$_x$	plot text string until e_x
SI w,h	set character width and height
CP s,l	move pen s text spaces and l text lines
DI x,y	plot text at angle defined by $\tan^{-1}(x/y)$

RASTER FILE FORMATS

Now let's look at the four graphic formats that are commonly used with raster data. You will notice that two of these formats are also used with vector data (see previous section). Again, each format is described and illustrated with a common example for purposes of comparison.

PostScript (.PS) and Encapsulated PostScript (.EPS).

As mentioned in the description of PostScript as a vector graphic file format, remote sensor images and digital halftones in raster format may also be defined in the PostScript language. The following annotated PostScript page description produces the digital image test segment plotted in Figure B.4. Notice that the image is stored as one long string of hexadecimal numbers.

Tag Image File Format (.TIF)

The Tag Image File Format (TIFF) has become a widely used standard for the binary storage of scanned images, particularly in desktop publishing programs. A TIFF file has three parts: a header, an image file directory, and the image data. The header gives the byte order for integer data, the TIFF version, and a numerical pointer to the location of the first image file directory. Each image file directory begins with the number of entries, followed by each entry. Each entry is 12 bytes in length and consists of a numerical **tag** code defining what the entry is, followed by a code giving the data type (ASCII, integer, real, etc.) of any tag attribute, and finally the attribute code. Selected tag codes and definitions are listed in Table B.9.

The tag system makes TIFF probably the most complete raster image data format in terms of the kinds of raster images that can be encoded and the number of encoding and data compression schemes supported. The header, image file directory, and image data needed to encode the sample raster image in Figure B.4

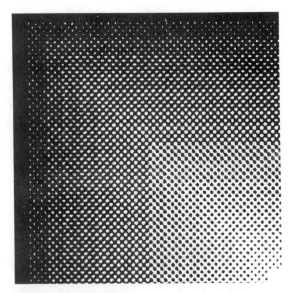

Figure B.4 Test image used to illustrate the PostScript page description language.

```
newpath              %begin a new sequence of operations
/picstr 16 string def %define image string called picstr
300 300 translate              %move to center of page
72 72 scale        %scale image to 1 in. (72pt) square
16 16 8            %image dimensions #cols,#rows,#bits
[16 0 0 -16 0 16       %define image coordinate system
{currentfile           %read image data from program file
picstr readhexstring pop}       %read each hexadecimal pair
image                           %form image on page
                                %hexadecimal image data
00000000000000000000000000000000
00111111111111111111111111111111
00112222222222222222222222222222
00112233333333333333333333333333
00112233444444444444444444444444
00112233445555555555555555555555
00112233445566666666666666666666
00112233445566777777777777777777
00112233445566778888888888888888
00112233445566778899999999999999
00112233445566778899AAAAAAAAAAAA
00112233445566778899AABBBBBBBBBB
00112233445566778899AABBCCCCCCCC
00112233445566778899AABBCCDDDDDD
00112233445566778899AABBCCDDEEEE
00112233445566778899AABBCCDDEEFF
showpage                %send page to output device
restore
```

Table B.9
Selected TIFF Tag Codes and Definitions

Tag Code	Tag Definition	Tag Description
254	SUBFILETYPE	Type of File
256	IMAGEWIDTH	Width in Pixels
257	IMAGELENGTH	Height in Pixels
258	BITSPERSAMPLE	Number of Bits Per Pixel
259	COMPRESSION	Data Compression Method
262	PHOTOMETRIC	Gray Tone/Color System
269	DOCUMENTNAME	Name of Image Document
270	IMAGEDESCRIPTION	Information about Image
271	MAKE	Scanner Make
272	MODEL	Scanner Model
281	MAXSAMPLEVALUE	Maximum Value in Image
282	XRESOLUTION	Resolution in X Direction
283	YRESOLUTION	Resolution in Y Direction
284	PLANARCONFIG	Image Plane Type
285	PAGENAME	Name of Document Page
286	XPOSITION	X Direction Page Offset
287	YPOSITION	Y Direction Page Offset
290	GRAYRESPONSEUNIT	Gray Scale Curve Accuracy
296	RESOLUTIONUNIT	Units of Measurement
297	PAGENUMBER	Page Number of Multipage Image
305	SOFTWARE	Type of Software Used
306	DATETIME	Date and Time of Image Creation
315	ARTIST	Image Author
316	HOSTCOMPUTER	Type of Host Computer

are listed below. Decimal equivalents of the hexa-decimal representation of the binary TIFF file are given in parentheses.

File Header

Hex Code	Meaning
49 49	integer byte order code (most to least significant)
2A 00 (42)	TIFF version 42
08 00	Pointer to first image file directory

Image File Directory

| 08 00 (8) | Eight entries in directory |

FE 00 04 00 01 00 00 00 01 00 00 00
  Tag 254, type 4=byte
(254)   (4)   (1)   (1)   Attr. 1=full resolution
00 01 03 00 01 00 00 00 10 00 00 00
  Tag 256, type 3=long int.
(256)   (3)   (1)   (16)   width=16 columns

01 01 03 00 01 00 00 00 10 00 00 00   Tag 257, type 3
(257)   (3)   (1)   (16)   height=16 rows
02 01 03 00 01 00 00 00 08 00 00 00   Tag 258, type 3
(258)   (3)   (1)   (8)   8 bits per sample
03 01 03 00 01 00 00 00 01 00 00 00   Tag 259, type 3
(259)   (3)   (1)   (1)   Attr. 1=no compression
06 01 03 00 01 00 00 00 01 00 00 00   Tag 262, type 3
(262)   (3)   (1)   (1)   Attr. 1=min. value black
19 01 03 00 01 00 00 00 0F 00 00 00   Tag 281, type 3
(281)   (3)   (1)   (15)   max. image value=15
1C 01 03 00 01 00 00 00 01 00 00 00   Tag 284, type 3
(284)   (3)   (1)   (1)   Attr. 1= single plane
00 00   Offset to next image
(0)   file directory
Image Data
(Same hexadecimal image data as in PostScript example above)

Picture Exchange (.PCX) Format

The Picture Exchange (PCX)* format, originally developed for the PC Paintbrush© program, is a popular means of exchanging raster image data. Like the widely used Tag Image File Format (TIFF), PCX files consist of a header record followed by compressed binary image data.

The PCX header record is 128 bytes in length, with the first 72 bytes presently used. These bytes contain numerical codes and integer values for such image information as the number of color planes, the number of bits representing each pixel, the image dimensions and resolution, and the number of bytes required for each scan line. For eight-bit byte pixels, the following run-length compression method is used, beginning with the first pixel in the first scan line:

1. If the highest two bits in the byte are set to 1, the lower six bits specify the number of consecutive pixels having the same numerical value. The next byte gives the numerical value of the pixels. For example, the binary bit pattern 11010000 00000000, or D000 in hexadecimal notation, means that 16 consecutive pixels have a value of 0. The greatest number of consecutive pixels that can be encoded in this manner is 111111, or 63.

---

*Developed by *ZSoft Corporation.*

2. If the highest two bits are not set to 1, there are no consecutive pixels of the same value. In this case, the byte gives the value for the pixel.

The following hexadecimal representation of the binary image compression for the image segment illustrated in Figure B.4 shows the PCX method, which achieves an average compression of 25 percent. Note that in no case is the compressed file longer in bytes than the original.

D000
00CF01
0001CE02
000102CD03
00010203CC04
0001020304CB05
000102030405CA06
00010203040506C907
0001020304050607C808
000102030405060708C709
00010203040506070809C60A
00010203040506070809A C50B
00010203040506070809A0BC40C
00010203040506070809A0B0CC30D
00010203040506070809A0B0C0DC20E
00010203040506070809A0B0C0D0E0F

COMPUTER GRAPHICS STANDARDS

The community of researchers involved with automated cartography and geographic information systems are not the only group interested in establishing exchange standards for graphic data. The scientists and engineers involved in computer graphics, particularly computer aided design (CAD), have been preparing exchange standards for over two decades.

The CAD community is far larger than the cartography and GIS community, both in terms of people involved and software sales. It is imperative, therefore, that we understand the work being done in the CAD community and be able to judge if that work is applicable to GIS. Information on computer graphics standards is available in a number of references (Enderle, et al., 1986; Hopgood, et al., 1986; National Computer Graphics Association, 1989).

Computer graphics standards can be categorized into several major types. First there are **device-level standards**, such as the Computer Graphics Interface (CGI). CGI provides a device-level interface that links applications with particular devices. CGI implements the interface between a virtual device defined in a CGI-compatible application program and a particular hardware device, such as a graphics display or plotter. CGI is a draft ISO standard (ISO 2nd DP 9636, 1988).

Another device-level standard we have just seen is the Computer Graphics Metafile (CGM). CGM is an ISO standard (ISO 8632, 1987). It defines a standard file content and format for describing pictures (images) in a device-independent fashion. Thus, images can be exchanged electronically between different types of hardware and applications software.

Two **programming interface standards** are widely used in computer graphics. One is the Graphical Kernel System (GKS). It has been approved as an ISO standard in both two-dimensional (ISO 7942, 1985) and three-dimensional (ISO 8805, 1988) implementations. The second is the Programmer's Hierarchical Interactive Graphics Standard (PHIGS). It is also an ISO standard (ISO 9582, 1989).

Both GKS and PHIGS provide a uniform programmer's interface for graphics input and output. They consist of subroutine libraries providing graphics subroutines, graphics workstation management, and interactive input in a device-independent fashion. The subroutines are callable from high-level languages. These standards provide routines that simplify applications development.

Although GKS and PHIGS overlap in their functions, there are some unique differences. GKS initially was restricted to two-dimensional applications. (Extensions have been added so that GKS can now handle three-dimensional data). PHIGS was designed from the start for three-dimensional applications. PHIGS also can support hierarchial representation of graphical objects.

The standards most relevant to exchange of geographic data files are **graphic data exchange standards**. We have already looked at the first of these was the Initial Graphics Exchange Specification, an ANSI standard (ANSI Y14.26M, 1987) first published in 1981. Now in its fifth version, the main application of IGES is the exchange of two-dimensional engineering drawings.

In 1984, the French developed an alternative to IGES known as the Systeme d'Echange et de Transfer (SET). Compared to IGES, SET offers a simplified internal structure (Aerospatiale Direction Technique, 1984).

In the early 1980s, work on a successor to IGES began. This proposed standard is called the Product Data Exchange Specification (PDES). PDES will include capabilities not only for the interchange of 2-D drawing files that IGES handles, but also for the exchange of databases and knowledge bases describing complete 3-D projects.

PDES will support dynamic (or active) data exchange, distributed database definition, and knowledge-based system definition. In the definition of knowledge-based systems, the PDES scheme will contain both the required data and the rules for using those data. Of particular interest to GIS specialists is the fact that PDES has capabilities to handle topology and complex features (Warthen, 1988).

PDES is also the United States' contribution to an international project to create a single, internationally accepted data exchange standard. This ISO project is known as the Standard for the Exchange of Product Data (STEP). STEP is planned to be an aggregation or merger of related activities, such as IGES, PDES, and SET.

The goals of STEP mirror the goals of PDES. Both require exchange completeness, archiving completeness, extensibility, efficiency, separation of data content from physical format, and logical classification of data elements. Both PDES and STEP are described in an ISO draft standard (ISO DP 10303, 1988). Three layers are defined in STEP:

- a physical layer that deals with file format and data structure

- a logical layer that contains generic entities and application specific entities
- an application layer that deals with various applications, such as electrical systems, piping systems, and, potentially, geographic information systems.

As a result of joint development, PDES and STEP will probably be identical, at least where they share common functionality. However, there will be separate approval processes for PDES and STEP, so the two standards could diverge (Warthen, 1988).

SELECTED REFERENCES

AutoCAD© Release 11 IGES Interface Specifications, Sausalito, CA: Autodesk, Inc., 1990.

AutoCAD© Release 11 Reference Manual, Sausalito, CA: Autodesk, Inc., 1990.

Enderle, G., Grave, M., and Lillehagen, F., eds., *Advances in Computer Graphics I*, New York: Springer-Verlag, 512 p.

Henderson, L.R. and A.M. Mumford, *The Computer Graphics Metafile,* London: Butterworth & Co. Ltd., 1990.

Hopgood, F. R. A., and Hubbold, R. J., eds., *Advances in Computer Graphics II*, New York: Springer-Verlag, 1986, 186 p.

Interfacing and Programming Manual: HP7475A Graphics Plotter, San Diego: Hewlett-Packard Co., 1983.

National Computer Graphics Association*, Standards in the Computer Graphics Industry,* Fairfax, VA. 1989.

PostScript Language Reference Manual, Adobe Systems Inc., Reading, MA: Addison-Wesley Publishing Co., 1990.

Technical Reference Manual (.PCX format). Marietta, GA: ZSOFT Corporation, 1991.

Warthen, Barbara, "Move Over IGES: Here Comes PDES/STEP," *Computer Graphics Review*, Nov-Dec, 1988, pp 34–40.

SELECTED LIST OF MAPPING SOFTWARE VENDORS

CLIP-ART

MacGlobe
Broderbund Software, Inc.
500 Redwood Boulevard
P.O. Box 6121
Novato, CA 94948

MapArt
Cartesia Software
Lambertsville, NJ

GEOGRAPHIC INFORMATION SYSTEMS

Arc/Info
Environmental Systems Research Institute
380 New York Street
Redlands, CA 92372
(714-793-2853).

Spans GIS
Intera TYDAC Technologies Inc.
1600 Carling Avel, Ottawa, Ontario K1Z 8R7, Canada
(613-722-7508)

MicroStation GIS Enivronment (MGE)
Intergraph Corp.
Huntsville, AL 35894
(205-730-2000).

ERDAS IMAGINE
ERDAS, Inc.
2801 Buford Highway
Atlanta, GA 30329
(404-248-9000).

GRASS
GRASS Information Center
USACERL
ATTN: CECER-ECA
P.O. Box 9005
Champaign, IL 61826-9005

Atlas GIS
Strategic Mapping Inc.
4030 Moorpark Ave.
San Jose, CA 95117-4103
(408-985-7400)

GisPlus
Caliper Corp.
1172 Beacon Street
Newton, MA 02161
(617-527-4700)

MapInfo
MapInfo Corp.
200 Broadway
Troy, NY 12180-3289
(518-274-6000)

IDRISI
The Clark Labs for Cartographic Technology and Geographic Analysis
Clark University
950 Main St.
Worcester, MA 01610
(508-793-7526).

MapGrafix
ComGrafix, Inc.
620 E Street
Clearwater, FL 34616
(813-443-6807)

IMAGE PROCESSING

ERDAS
ERDAS, Inc.
2801 Buford Highway
Atlanta, GA 30329
(404-248-9000).

EASI/PACE
PCI Remote Sensing Corp.
1925 N. Lynn Street
Arlington, VA 22209
(703-243-3700)

STATISTICAL MAPPING

MapInfo
MapInfo Corp.
200 Broadway
Troy, NY 12180
(518-274-8673)

Atlas *Graphics
Strategic Mapping Systems
4030 Moorpark Avenue
San Jose, CA 95117
(408-985-7400)

MapViewer and Surfer
Golden Software, Inc.
809 14th Street
Golden, CO 80402-0281
(303-279-1021)

ArcView
ESRI
380 New York Street
Redlands, CA 92373
(714-793-2853).

PROJECTIONS

World (World Mapping and Projection Program)
P.M. Voxland
Social Science Research Facilities Center
269 19th Avenue South
University of Minnesota
Minneapolis, MN 55455
(612-625-8556)

Geocart
Terra Data Inc.
Bramblebush
Croton-on-Hudson, NY 10520

CAM (Cartographic Automated Mapping)
Central Intelligence Agency
Washington, DC

MicroCAM (Microcomputer Automated Mapping)
Microcomputer Specialty Group
Association of American Geographers
1710 16th Street, NW
Washington, DC 20009-3198.

COMPUTER AIDED DESIGN (CAD)

AutoCAD
Autodesk, Inc.
2320 Marinship Way
Sausalito, CA 94965
(800-445-5415)

MicroStation
Intergraph Corp.
Huntsville, AL 35894-0001
(800-826-3515)

Personal Designer

Computervision Corp.
100 Crosby Drive
Bedford, MA 01730
(800-248-7728)

CADvance

IsiCAD
1920 W. Corporate Way
Anaheim, CA 92803-6122
(714-533-8642)

FastCAD

Evolution Computing
437 S 48th Street
Tempe, AZ 85281-9936
(800-874-4028)

Generic CADD

Autodesk Retail Products
(206-487-2233)

DesignCAD

American Small Business Computers
327 South Mill St.
Pryor, OK 74361
(918-825-4844)

Drafix CAD

Foresight Resources
10725 Ambassador Dr.
Kansas City, MO 64153
(816-891-1040)

ILLUSTRATION

Freehand

Aldus Corp.
411 First Ave. S.
Seattle, WA 98104
(206-628-2320)

Illustrator

Adobe Systems Inc.
1585 Charleston Rd.
Mountain View, CA
(800-833-6687)

CorelDRAW

Corel Systems Corp.
1600 Carling Ave.
Ottawa, Ontario
Canada K1Z 8R7
(613-728-8200)

Designer

Micrografx
1303 Arapaho Rd.
Richardson, TX 75081
(214-234-1769)

PAINT

Publisher's Paintbrush

Z-Soft Corp.
450 Franklin Rd.
Suite 100
Marietta, GA 30067
(404-428-0008)

CA-Cricket Paint

Computer Associates
One Computer Associates Plaza
Islandia, NY 11788
(516-342-5224)

Painter

Fractal Design Corp.
335 Spreckels Drive
Aptos, CA 95003
(408-688-8800)

Halo Desktop Imager

Media Cybernetics, Inc.
8484 Georgia Ave.
Silver Springs, MD 20910
(301-495-3305)

IMAGE EDITORS

PhotoStyler

Aldus Corporation
411 First Ave. S.
Seattle, WA 98104-2871
(206-628-2320)

PhotoFinish

Z-Soft Corporation
450 Franklin Rd
Marietta, GA 30067
(404-428-0008)

Picture Publisher

Micrografx
Richardson, TX
(800-272-3729)

Photoshop

Adobe Corporation
Mountain View, CA
(415-961-4400)

APPENDIX D

GRAPHIC ARTS PHOTOGRAPHY

RECORDING MEDIUM
- COLOR SENSITIVITY
- CONTRAST

EXPOSURE CONTROL
- PROJECTION COPYING
- CONTACT COPYING
- ELECTRONIC IMAGING SYSTEM

FILM REGISTRATION
- CAMERA WORK
- CONTACT WORK

LIGHT SOURCES
- FLOOD SOURCES
- LASER SOURCES

PROCESSING PHOTOGRAPHIC MATERIALS

RETOUCHING AND ALTERATIONS

We have discussed the photographic process in cartography in relation to environmental remote sensing (Chapter 8) and image processing (Chapter 12). The photographic process is also an integral part of map production (Chapters 20 and 31) and reproduction (Chapter 30), in which case it is called **graphic arts photography**.

Field and laboratory applications of the photographic process have many similarities but also some major differences. We will organize further discussion of cartographic applications under the headings of recording medium, exposure control, and material processing.

RECORDING MEDIUM

Photography includes films and papers that are sensitized to the visual part of the electromagnetic spectrum. The recording medium consists of several layers (Figure D.1).

In the case of film, the base material can be acetate or some other plastic material such as polyester or polystyrene. Since acetate is not dimensionally stable, it is usually unsuitable for cartographic work.

The film base is usually sandwiched between two coatings. On the front is one (for black-and-white rendition) or more (for color rendition) layers of light-sensitive emulsion. Traditionally, this emulsion has been composed of minute crystals or grains of silver halide (a salt) suspended in a solidified gelatin matrix.* The back of the film base may be coated with an **antihalation material**. This material absorbs any light rays that penetrate the base during exposure, thereby preventing reflection back to the emulsion.** Also, this coating tends to mini-

---

*Numerous emulsions in current use or under development are sensitive to radiant energy but do not contain silver halides. Although, strictly speaking, these are nonphotographic materials, they may perform the same function and have similar handling characteristics. In fact, clear distinctions are becoming more difficult each year.

**Not all films used by cartographers have antihalation backing, however.

Figure D.1 This exaggerated cross section of black-and-white film shows the position of the sensitized emulsion and the antihalation layer relative to the film base.

mize curling and compensates for distortion or dimensional changes caused by changes in the emulsion when it absorbs moisture.

COLOR SENSITIVITY

The color sensitivity of films is built in during manufacture. It is indicated in film data specifications by a diagram called a wedge spectrogram (**Figure D.2**).

The height and extent of the colored area in each spectrogram indicates the sensitivity of a particular film to the various colors. Although only the bands of blue, green, and red wavelengths are labeled in Figure D.2, the film sensitivity to other colors can be determined. For instance yellow, an additive mixture of red and green, falls where those two colors merge. Since films of different sensitivity serve different purposes, a variety is used in map reproduction.

Blue-Sensitive Emulsions

A blue-sensitive emulsion will record high negative densities for blue areas of an original, whereas greens, yellows, and reds will appear as low negative densities. On a contact positive, the tones are reversed, and the blue areas will be transparent while the greens, yellows, and reds will be opaque. Since the film is blind to most of the spectrum, a bright yellow or red safelight can be used in the darkroom without exposing (or fogging) the film.

Orthochromatic Emulsions

An **orthochromatic emulsion** is sensitive only to blue, green, and yellow, and is blind to red. This permits use of a red safelight and also causes red to

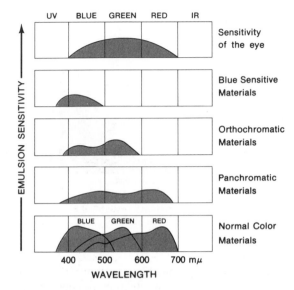

UV BLUE GREEN RED IR

Sensitivity
of the eye

Blue Sensitive
Materials

Orthochromatic
Materials

Panchromatic
Materials

BLUE GREEN RED

Normal Color
Materials

400 500 600 700 mμ

WAVELENGTH

EMULSION SENSITIVITY

Figure D.2 Wedge spectrograms in which the sensitivity of each film to the various colors is indicated by the height and extent of the shaded area. (Yellow falls where red and green merge.)

register the same as black on the film. Both positive- and negative-acting emulsions are available. On standard reversal film, both red and black produce transparent areas (low densities) on a negative, while blue, green, and yellow cause the film to be opaque.

In contrast, duplicating film is positive acting. You can use it to make duplicate copies of positive artwork or film. That is, you can make a negative directly from a negative, or you can make a positive from a positive.

If a sheet of this film is developed with no exposure to light, it will be completely opaque. Exposure to white light, however, removes density, so that light passing through transparent areas of a negative will cause transparent areas on the duplicating film.

To produce duplicate negatives or positives that have the same orientation (emulsion-image-reading relationship) as the original, you should place the base (nonemulsion side) of the original in contact with the emulsion of the film. With a somewhat longer exposure, you can achieve the

same result by placing the emulsion of the original in contact with the base of the duplicating film.

Orthochromatic emulsions are by far the most important sensitized materials used in map reproduction. Their high contrast makes them ideally suited for processing line artwork (inked or scribed) and for photo-mechanical manipulation of separations, masks (positive and negative), and screens in preparation for plate-based printing (such as offset lithography).

Several types of orthochromatic emulsions are available. Regular **lithographic (litho) film** has long been the industry standard, but **stabilization film** (also referred to as **direct or rapid-access film**) is becoming extremely popular. In practical terms, the primary difference between these two materials lies in the way they are processed after exposure.

Regular litho film is highly sensitive to development conditions, including the temperature and strength of the developer, and the time in the chemical bath. Thus, litho film gives the cartographer control of product quality during both exposure and development. Although such control can be an advantage in the hands of a skilled cartographer, it can produce problems of overdevelopment or underdevelopment in unskilled hands.

Stabilization film, on the other hand, is not very sensitive to development conditions and, therefore, must be controlled through exposure alone. Although this may reduce the chance of making mistakes, it also restricts the special effects that can be achieved. Thus, the choice of one material over the other may have a direct effect on the cost of production and the quality of the final product. To complicate matters even further, materials from competing manufacturers exhibit variations in speed, thickness, dimensional stability, quality, exposure and development range, cost, and so forth.

Panchromatic Emulsions

Emulsions that are sensitive to all visible colors and render colored artwork in tones of gray are called **panchromatic emulsions**. A dark area on the artwork, which reflects little light, will appear more translucent than a lighter area that reflects more light. Making a print from the negative reverses the relationship and produces a replica of the original.

Because of the wide range of sensitivity, this film must be handled and processed in complete darkness. When compared with the widespread usefulness of orthochromatic emulsions, the cartographic applications of panchromatic emulsions seem rather limited. Black-and-white aerial photography is the major application. Laboratory use of panchromatic emulsions is appropriate when duplicating continuous tone imagery or artwork (such as shaded relief portrayals).

CONTRAST

Sometimes it is useful to categorize films on the basis of their **characteristic curves**, which are defined by the relation between the exposure time and resulting film density. A characteristic curve for a particular film is obtained by plotting the exposure time on a horizontal logarithmic scale and the corresponding densities produced on a vertical arithmetic scale (Figure D.3).

The central section of the graph is of special interest. In this straight-line portion of the curve, there is a constant relationship between the resulting densities and the exposure lengths. The tangent of the angle between this part of the curve and the horizontal is a measure of the steepness of the curve and is referred to by the Greek letter **gamma (γ).**

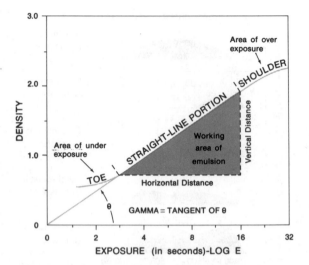

Figure D.3   The straight-line portion of the characteristic curve shows a constant relationship between the resulting densities and the length of exposure. The tangent of the angle between the straight-line portion and the horizontal is a measure of the steepness of the curve and is referred to by the Greek letter gamma (γ)

duction techniques depend upon exposure to negatives, on which the image area in transparent and the background is opaque, or to film positives, on which the image area is opaque and the background is transparent. And, finally, high-contrast films are the only ones suitable for making plates for lithographic printing (see Chapter 30).

High-Contrast Film

The characteristic curve for high-contrast film has a steep straight line and is said to have **high gamma**. Small changes in exposure cause great variations in density. Indeed, there are virtually no intermediate tones. The result is a clean break between the image and nonimage areas. Thus, high-contrast films are excellent for copying line art on which there is a strong contrast between image and background.

High-contrast film is very popular in cartographic work. One reason is that it can sharpen up messy originals. Another reason is that many map repro-

Low-Contrast Film

The characteristic curve for low-contrast film has a gently sloped straight line and is said to have a **low gamma**. Small changes in exposure cause small variations in density. Both panchromatic and color film fall into this category. For this reason, they can faithfully copy small variations in tones of gray or color.

Low-contrast film is needed to reproduce standard photographs, electronic images, and shaded drawings. Because all tones from light to dark are possible, these are called **continuous tone** media. Low-contrast films are not designed for copying sharp line originals, however, because they produce soft (transitional) breaks between image and nonimage areas.

The continuous tone negative produced with panchromatic film or color film cannot be used directly with many methods of map reproduction for two reasons. First, many sensitized materials used in proofing and map reproduction act in the same way as high-contrast films. This means that they will not record tonal differences faithfully.

Second, the plates used in conventional printing (lithography, letterpress) must be divided into two kinds of surfaces, one that takes ink and one that does not (see Chapter 30). If the printing plates are to be made photographically, the negative must be composed of either opaque areas or transparent areas, with nothing in between.

The way cartographers get around this problem of reproducing continuous tone artwork is to change the color shading or gray tones into varying sized dots on the negative. The result is called a **halftone** (see Chapter 20). The goal is to produce a halftone negative on which the size of the clear dots relative to the opaque areas between them is directly related to variations in the darkness and lightness of tones on the original copy (see Figures 20.10 and 20.11, and Color Figure 20.5).

Traditionally, this conversion of a continuous tone image to a reproducible dot structure has been done by using a special mechanical screen to produce what is called a **halftone negative**. Orthochromatic (high-contrast) film is required for this purpose. A **contact halftone screen** is used to produce the desired dot structure.

Contact halftone screens are made on flexible film and are used in contact with the exposed emulsion of the photographic material. The screens produce varying sized dots* through modulation of light by the optical action of a **vignetted dot pattern** of the screen acting on the film emulsion (see Figure 20.12).

The vignetted dots are produced by a dye that is thicker near the centers of the dots. The size of the halftone dot depends on the amount of light that is able to pass through the dye. Magenta screens are generally preferred for reproducing black-and-white originals, while gray screens are used for color reproduction.

The latest development in halftone production is **electronic screening**. Conventional mechanical screens are replaced by a laser scanner, controlled by a computer algorithm (see Chapter 3). This algorithm arranges clusters of microscopic points to form the individual screen dots. As the electronic record of the map image is exposed to film or sensitized plates, each digitally specified intensity is translated into an appropriate dot size and shape, according to the screen ruling specified.

The number of opaque lines per inch on halftone screens varies from about 75 to 300. All other things being equal, the closer the lines, the smaller and closer together the dots will be. The closer together they are, the more difficult it is for the eye to see them individually, and the smoother and more natural the result will appear. Finer screens demand higher quality reproduction media, procedures, and equipment, of course.

You can create different visual effects by printing halftones with their lines of dots oriented at different angles.** For this reason, it's important to be careful when orienting screens for halftone work. With monochrome reproduction of continuous tone copy, for example, the least visual disturbance occurs when the halftone negative is made with the screen oriented at 45° to the horizontal.

It's even more important to be careful in positioning the screens when making halftones for process color reproduction. The reason is that the lines of dots for the different printed colors must have an angular separation of 30° in order to prevent an undesirable **moiré** effect (see Figure 20.6).

Electronic screening simplifies the problem of producing halftones with different screen angles.

---

*The "dots" on these halftone screens may be round, square, or elliptical. Each of these dot structures produces slightly different effects in the final product.

**Screen angle is defined by the angle at which the rows of dots or lines hit the edge of the screen.

Desired screen angles need merely be specified prior to activating the laser plotter.

The process is a little more involved if you are using mechanical screens. You can take one of two approaches. First, you can rotate a single screen to the desired angle for each exposure. When you are using a single contact screen, a special **screen angle guide** (template) greatly simplifies the screen angling process (see Figure 20.7).

Alternatively, you can use a set of pre-angled screens. Commercial sets of pre-angled contact halftone screens of rectangular shape are customarily used. Since these screens reduce the chance of making errors, they greatly ease the chore of producing halftones in a darkroom.

EXPOSURE CONTROL

For many cartographic purposes, the goal of photographhy is to produce an image (either negative or positive) at the desired reproduction scale. Since the originals and their resulting copies on photographic media may be large, the equipment required for controlling exposure to the emulsion must also be large. Devices must be available for handling both reflection and transmission copy, and for making precise enlargements or reductions if that is necessary. Equipment must also be available for handling digital records. These needs are served by process (graphic arts) cameras, vacuum frames, platemakers, and electronic imaging systems. In order to accommodate emulsions ranging widely in sensitivity, a variety of light sources is used in conjunction with this specialized equipment.

PROJECTION COPYING

The most versatile piece of photographic equipment for the cartographer is the large-format **process camera**.* It is designed to copy large pieces of artwork onto equally large sheets of film and can make precision enlargements or reductions if needed.

The parts of the process camera are fairly standard, although in some machines they are arranged vertically and in others horizontally. The components include a copy frame (copyboard) that holds the artwork in place and flat under pressure or vacuum, a lens (or lens system),** a vacuum-activated film or plate holder, a light source for frontlighting reflection copy or backlighting transmission copy, and a set of controls for changing scale and adjusting exposure (Figure D.4).

Both the copyboard and the lens are movable. Thus, you can vary the distance between lens and film, and between lens and copyholder, to produce enlargements or reductions while maintaining sharp focus.

CONTACT COPYING

If you want to change the scale of the artwork, you must use a camera. But if no scale change is necessary, contact copying is the best method. You can make contact copies on either film or paper from negatives or positives or from translucent artwork. During exposure, you must be sure that the unexposed film is in tight contact with the piece being reproduced.

Vacuum Frame

Contact photography is usually accomplished in a vacuum frame equipped with a rubber pad, a glass lid, and a pump to remove air from inside the frame. As air is exhausted, atmospheric pressure holds the pieces of film snugly together and against

---

*Devices known as enlargers are sometimes used in cartographic laboratories to make enlargements from small format negative artwork (such as 35 mm. slides). Although use of these machines has been limited by a lack of appropriate map artwork in the past, increasing involvement with remote sensing imagery and photographs of maps displayed on electronic screens may generate more interest in enlargers in the future.

**The lens normally used in a process camera is color corrected and specially designed for copying flat originals. In general, the longer the focal length, the less geometric distortion in the result.

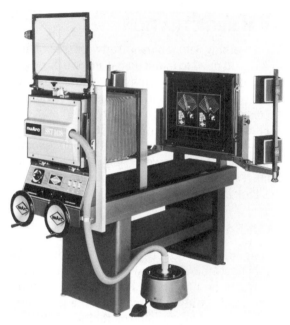

Figure D.4 The basic components of a camera are a film holder, which keeps material flat; a lens; and a copyholder, which can be moved perpendicular to the plane of the film. The positions of the lens and copyholder control the amount of enlargement or reduction. (Courtesy nuArc company, Inc.)

Figure D.5 A vacuum frame is commonly used for contact photography when a relatively low intensity exposure to a point light source is sufficient. (Courtesy nuArc Company, Inc.)

the glass. Length of exposure is controlled by an accurate timing device. Light sources vary, depending on exposure requirements for different emulsions. For low-intensity exposures, a point light source is usually suspended over a horizontal contact frame (Figure D.5). Filters and a rheostat can be used to regulate wavelength and brightness characteristics.

Platemaker

When a high-intensity light source is needed, a platemaker is commonly used. With this device, the light source and a contact frame are enclosed in a box. The light source is usually positioned in the bottom of the box. The photographic material is held in a downward-facing vacuum frame at the top (Figure D.6).

Because of the short distance between the light source and the photographic material, an array of lights rather than a single bulb is customarily used to give even illumination of the full copy area. The lights themselves are high-intensity sources rich in ultraviolet rays. Among other things, the enclosing box helps prevent potentially harmful exposure to the operator.

With flip-top platemakers, you can rotate the contact frame around a central axis so that it faces upward or downward. This arrangement makes it easier to load and unload the vacuum frame. You can gain added exposure flexibility by suspending a point light source over the platemaker. When you do so, you can use the flip-top platemaker as a standard vacuum frame when in one position (facing up), and as a platemaker when in the other position (facing down).

ELECTRONIC IMAGING SYSTEM

Laser or electronic cameras have been around since the mid-1970s and are now coming of age. The more complete laser cameras are more appropriately characterized as electronic imaging systems, since they consist of such components as a

Figure D.6 A platemaker consists of a box that contains a vacuum frame and a high-intensity source of illumination. It is widely used with graphic arts materials and emulsions that can be safely handled under normal daylight conditions. (Courtesy nuArc Company, Inc.)

job-planning station, data entry station, laser scanner, output laser film recorder, and a film processor (see also *Imagesetting* in Chapter 31 and *Matrix Arrays* and *Videography* in Chapter 8). These devices convert images to digits and then store them on discs or magnetic tape. At that point, you can use these machines to perform all manner of electronic manipulations of the image, including screening of up to 300 lines.

In the past, a small film format in laser cameras limited cartographic applications. But technical advances in equipment design have solved this problem except for the largest maps. In contrast to standard process cameras, for which exposure is given in seconds, input and output exposures with an electronic imaging system are stated in terms of inches or square feet per minute. Output can be in the form of negatives or positives, and multicolor images can be processed electronically as black-and-white separations.

FILM REGISTRATION

When working with separation artwork (see Chapter 23), you need to maintain registration throughout graphic arts photography. The principles and procedures are similar to those used in handling the registration of original artwork.

Filmwork, particularly when it is carried out under darkroom conditions, does introduce some special problems, however. Since camera and contact registration require different procedures, let's look at each separately.

CAMERA WORK

You need a high-quality graphic arts camera to assure that separation artwork is still in accurate registry at the film stage. Poorly maintained or inexpensive equipment may lack the precision necessary to make repeated matching exposures. Thus, unless you have access to a precision camera that can return to the same exact setting time after time, it is best to do all the photography of multiple-flap artwork in a single session. The actual procedure for maintaining registry through the camera varies with the type of registration system used on the original artwork.

When graphic registration marks are the only means of registration used with multiple-flap artwork, you have two choices. You can photograph the flaps one after another, with no concern for registry, and later register the processed film sheets visually. This registration chore grows more difficult and inaccurate as the size of the material and the number of flaps increase. Furthermore, the subsequent map reproduction process may require pin registration (see Chapter 23), which means that you must attach pre-punched strips or tabs of stable material to the film sheets. You must make this conversion from graphic mark to pin registration while the film sheets are properly aligned, using the graphic marks. Since this strip registration, or stripping, process is tedious, it is better avoided, if possible.

The preferred alternative to graphic mark and strip registration of film is to maintain pin registry through the camera. Since this requires pin-registered separation artwork, materials that are registered solely by graphic marks should first be converted to this form.

The next step is to punch the sheets of film that you will be using so that the holes match those of the original artwork.* Then, tape registry pins to the copyboard so that they will accept the artwork separations, as well as to the filmboard, so that they will accept the pre-punched film sheets. It may take some juggling of the pins to get the image area to fall properly on the film, especially if you're not using a piece of film of generous size.

Finally, photograph each flap with both film and artwork positioned on its respective pins. After processing and remounting on pins, the film sheets should be in proper registry.

CONTACT WORK

Film registration in contact photography is far less complicated than it is in camera work. Punch registration is again the preferred system.

The first step is to see to it that the material containing the unexposed emulsion and the materials being used to control the exposure are identically punched and properly registered.** Then, tape appropriately spaced pins to the bed or mat of the vacuum frame or platemaker.

Finally, superimpose pre-punched materials on the pins in the desired order, and make the exposure. The result will be a sheet of exposed material that is fully registered with the other materials.

---

*As an alternative, you can achieve punch registry by stripping prepunched tabs or strips of stable material to the film sheets. But take care that the stripped materials do not fall off during the wet film processing that is to follow. Tape containing a special adhesive is available for this purpose.

**Sheets of material that are used to control exposure in contact photography are called masks. They are discussed in detail in Appendix F.

LIGHT SOURCES

The terms **light source** and **light-sensitive emulsion** are used rather loosely in cartography. Radiation generated by so-called light sources often extends beyond the visible spectrum into the longer infrared and shorter ultraviolet wavelengths. Similarly, the emulsions of many materials used routinely in the graphic arts industry are affected by infrared and ultraviolet energy as well as by visible light wavelengths.

Don't let this nonliteral use of terms confuse you. What is important is the relationship between the light sources and the cartographic materials used in production and reproduction. You must be able to match the wavelength composition of different light sources to the photographic characteristics of the original maps. You also must be able to match the light characsitics to the sensitivity of the materials used in map reproduction.

Two types of artificial light sources are used in cartography: flood lamps and laser beams. Let's take a look at each.

FLOOD SOURCES

Two types of flood light sources are used in cartographic applications of graphic arts photography. Their wavelength characteristics are illustrated in Figure D.7.

Incandescent lamps are one common type. They exhibit a continuous spectrum of emission. They range from relatively low-intensity tungsten filament lamps (including the "photoflood" type) to the high-intensity tungsten-halogen (or quartz-iodine) lamp.

The second, more specialized, category of light sources includes electric or vapor discharge lamps. Carbon arc lamps (enclosed or open), mercury-vapor lamps (including the fluorescent lamp modification), and pulsed xenon arc lamps all fall in this second category. Although some of these artificial sources of illumination (carbon arc, fluorescent) exhibit a continuous spectrum of emission, others

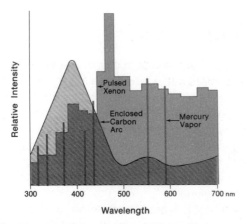

Figure D.7 The sources of illumination used in graphic arts photography exhibit very different wavelength characteristics. This makes it possible to take advantage of specialized emulsions possessing a wide range of sensitivities.

(mercury vapor and pulsed xenon arc) generate energy only at certain wavelengths or within limited wavebands of the spectrum. To some degree, you can tailor the energy discharge characteristics of the various light sources through use of appropriate filters.

LASER SOURCES

Ordinary light sources emit light in many different colors (wavelengths). In contrast, lasers produce an intense beam of light of a pure single frequency or color.

Many types of lasers have been built since the first was constructed in 1960. Cartographers have taken advantage of the fact that light produced by lasers is far more monochromatic, directional, and powerful than that generated with other light sources.

One current application is the laser scanner used to transform graphic artwork to digital form. Another is the laser recorder used to change digital records to graphic form. Since many new laser-based products are being introduced to the mar-

ketplace each year, there is little doubt that lasers will be used extensively throughout the high-quality graphic arts industry (including cartography) during the coming decade.

PROCESSING PHOTOGRAPHIC MATERIALS

When photographic materials are exposed to light, a photo-chemical reaction takes place within the emulsion, creating an invisible latent image. Careful chemical processing is then required to bring out (develop) and give some degree of permanency (stability) to this image. With emulsions containing silver halide crystals, the visible image is formed by reducing the exposed silver salts to grains of metallic silver that appear black. Unexposed silver salts are removed, leaving the film transparent in nonimage areas.

You can use either wet or dry processes in handling photographic materials. In **dry processing**, the necessary chemicals are built into the photographic material or a material "sandwich" that is subsequently separated. The process is convenient but expensive. Since you have little control over the processing, high-quality results are sometimes difficult if not impossible to obtain.

In contrast, **wet processing** entails a five-step procedure of developing, stopping, fixing, washing, and drying. Traditionally these steps were carried out manually with the aid of a special developing sink and chemical trays. Automated processors now perform the same chores with greater speed and consistency (Figure D.8).

Two types of films—lithographic and stabilization films—are available for wet processing. They need to be handled differently.

With lithographic films, development time is critical and depends on the strength and temperature of the lithographic developing solution. Material that is taken out of the developer too soon will be underdeveloped (details will be fuzzy and the background will be gray rather than black). Material that is left in the developer too

Figure D.8 Automated wet film processors perform the sequence of developing, stopping, fixing, washing, and drying operations with great speed (about 60 seconds) and consistency. (Photo courtesy Eastman Kodak Company).

long will be overdeveloped (fine details will be lost). Careful development can produce an extremely high-quality image, however. Moreover, you can alter development time to compensate for exposure errors.

With stabilization or rapid-access films, time in the developer and strength and temperature of the chemical bath are not so critical. Material placed in the chemical bath develops up to a point and then stops. Furthermore, you can reuse the chemical bath a number of times before you need to replace it. To gain this convenience, however, you must give up something: the ability to create special effects through careful control of development time and procedure.

RETOUCHING AND ALTERATIONS

Touching up or manipulating (making additions, deletions, or corrections to) film negatives or positives to improve the quality of the image is an important step in the cartographic technique. Some

results can be obtained better at this stage than in the drafting stage. No matter what modifications you plan to make, however, you must first bring the negative or positive to perfection so that the image is sharp and clear (Figure D.9).

To make your modifications, first place the film on a light table. Remove all pinholes or blemishes in the dark emulsion area by covering them with a water-soluble paint called **opaque** or with a special **retouching fluid**.* Remove dark spots and blemishes by scraping them with a sharp blade. If the film is to be composited with a screen, do the opaquing and scraping before you do the screening, because the film will be more difficult to touch up after a screen (a fine pattern of lines) is added.

Deletions are easy to make. On positives, you can simply scrape off unwanted marks. On film negatives, you can make deletions by covering transparent areas with opaque or retouching fluid or by covering the areas with red lithographer's tape. Be sure to do this work on the base side of the film to prevent accidental damage to the emulsion.

Additions of names or symbols are more difficult to make than deletions. One possibility is to remove the emulsion from the areas you wish to change by cutting or scraping it away from the film base. You can then "strip" a piece of emulsion containing the new artwork into place on the negative.

Another way to make additions is to "engrave" them into the negative emulsion. The emulsion is quite brittle, however, and tends to chip off, often resulting in ragged lines.

A final way to add artwork is to lay it out on a piece of transparent base material. Then you can add this material as a unit in a clear area on the film.

---

*Disposable, fiber-tipped retouching pens are available in several widths and fluid colors (opaque black and light safe red). The availability of fast-drying, high-density retouching fluid in pens is a great convenience. A representative brand name is *Formopaquer* (marketed by Graphics Products Corporation, Rolling Meadows, IL 60008).

Figure D.9 Before being stripped into position for platemaking, pinholes or unwanted words or symbols are removed by applying a water-soluble opaque or retouching fluid to the film.

SELECTED REFERENCES

Bann, D. *The Print Production Handbook*, Cincinnati, OH: North Light Books, 1985.

Cogoli, J. *Graphic Arts Photography Black-and-White*, Pittsburgh, GATF, 1981.

E.I. Du Pont de Nemours and Co., Inc., *The Contact Screen Story*. Wilmington, Del.: E.I. Du Pont de Nemours & Co., 1972.

Eastman Kodak Company, *Halftone Methods for the Graphic Arts*, Rochester, N.Y.: Eastman Kodak Co., 1982.

Eastman Kodak Company, Basic *Photography for the Graphic Arts*, Rochester, N.Y.: Eastman Kodak Co., 1982.

Eaton, G.T., *Photographic Chemistry*, 4th ed., Dobbs Ferry, NY: Morgan & Morgan, 1986.

International Paper Company, *Pocket Pal: A Graphic Arts Production Handbook*, 12th ed., New York: International Paper Co., 1980.

Keates, J.S., *Cartographic Design and Production*, 2nd ed., New York: John Wiley & Sons, Inc., 1989.

Kidwell, R., et al., "Experiments in Lithography from Remote Sensor Imagery," *Technical Papers*, ACSM Annual Meeting, Washington, D.C., 1983, pp. 384–393.

Mertle, J.S., and Gordon L. Monsen, *Photomechanics and Printing*, Chicago: Mertle Publishing Co., 1957.

Rosenthal, R.L., "Digital Screening and Halftone Techniques for Raster Processing," *Technical Papers*, ACSM Annual Meeting, St. Louis, 1980, pp. 126–135.

Schlemmer, R.M., *Handbook of Advertising Art Production*, Englewood Cliffs, N.J.: Prentice-Hall, Inc., 1966.

Stefanovic, P., "Digital Screening Techniques," *ITC Journal*, Special Cartography Issue, 1982, pp. 139–144.

Stevenson, G.A., *Graphic Arts Encyclopedia*, 2nd ed., New York: McGraw-Hill Book Company, 1979.

Urbach, J.C., T.S. Fisli, and G.K. Starkweather, "Laser Scanning for Electronic Printing," *Proceedings of the Institute of Electrical and Electronic Engineering* 70, 1982, pp. 597–618.

USEFUL ADDRESSES FOR REMOTE SENSING PRODUCTS

GOVERNMENTAL AERIAL PHOTOGRAPHY, ORTHOPHOTOQUADS, HUMAN SPACECRAFT PHOTOGRAPHY, LANDSAT IMAGERY AND DIGITAL DATA

National Cartographic Information Center Offices

Eastern Mapping Center-NCIC
U.S. Geological Survey
536 National Center
Reston, VA 22092

Mid-Continent Mapping Center-NCIC
U.S. Geological Survey
1400 Independence Road
Rolla, MO 65401

Rocky Mountain Mapping Center-NCIC
U.S. Geological Survey
Box 25046, Stop 504 Federal Center
Denver, CO 80225

Western Mapping Center-NCIC
U.S. Geological Survey
345 Middlefield Road
Menlo Park, CA 94025

NCIC
U.S. Geological Survey
National Space Technology Laboratories
NSTL Station MS 39529

HIGH-ALTITUDE AERIAL PHOTOGRAPHY, SLAR IMAGERY, HUMAN SPACECRAFT PHOTOGRAPHY, LANDSAT IMAGERY AND DIGITAL DATA

U.S. Geological Survey
EROS Data Center
User Services Center
Sioux Falls, SD 57198

ACSC, USFS, AND SCS PHOTOGRAPHY

Aerial Photography Field Office
Agricultural Stabilization and Conservation Service
U.S. Department of Agriculture
P.O. Box 300100
Salt Lake City, UT 84125

SPACE SHUTTLE LARGE FORMAT CAMERA PHOTOGRAPHY

Chicago Aerial Survey, Inc.
LFC Department
2140 Wolf Road
Des Plaines, IL 60018

HISTORICAL AERIAL PHOTOGRAPHY

National Archives and Records Library
Room 2W
8th and Pennsylvania Ave. N.W.
Washington, D.C. 20408

CANADIAN AERIAL PHOTOGRAPHY

National Air Photo Library
Surveys and Mapping Branch
Dept. of Energy, Mines, and Resources
615 Booth St.
Ottawa, Ontario K1A OE9

SIDESCAN SONAR IMAGERY

NOAA National Geophysical Data Center
325 Broadway / Code E-GC3
Boulder, CO 80303

LANDSAT IMAGERY AND DIGITAL DATA

Customer Service Department
Earth Observation Satellite Company
4300 Forbes Blvd.
Lanham, MD 20706

SPOT IMAGERY AND DIGITAL DATA, SOYUZKARTA IMAGERY

SPOT Image Corporation
1897 Preston White Drive
Reston, VA 22091

WEATHER SATELLITE IMAGERY AND DATA, PASSIVE MICROWAVE IMAGERY

NOAA Satellite Data Services Division
Room 100
Princeton Executive Center
Washington, D.C. 20233

PHOTO-MECHANICAL MAP PRODUCTION

DRAFTING

SCRIBING

    GUIDE IMAGE
    IMAGE ORIENTATION
    SCRIBING INSTRUMENTS
    SCRIBING TECHNIQUES

COMPLEX ARTWORK

    OPEN WINDOW NEGATIVES
    NEGATIVE SCREENS
    POSITIVE MASKS

*T*he dominance of strictly hand map construction was lost with the development of photo-sensitive materials and related chemical technologies. The combination of hand and photo-chemical procedures that replaced hand methods is called **photo-mechanical production**. This technological hybrid dominated map construction in the past half century (Figure F.1), but is now rapidly losing influence to digital procedures (see Figure 31.1).

Photo-mechanical procedures have not disappeared yet, however. In fact, they are still widely practiced. This is due in part to the competitive advantage inherent in this mature technology. It can also be attributed to inertia associated with an immense investment by schools, government, and industry in materials, equipment, and skilled personnel.

In this appendix, we will look at some of the photo-mechanical procedures that either are still used or serve as a useful model for understanding their electronic replacements. Graphic arts photography (see Appendix D) is closely related to the procedures discussed in this appendix.

DRAFTING

Hand drafting enjoyed many advances in tools, media, and techniques that helped prolong its dominance of map construction for centuries. But it is now fast fading from the mapping scene and, therefore, is not emphasized in this book. Still, drafting concepts are relevant in cartography because they are mimicked by electronic plotting devices and manual scribing. Indeed, a great deal of current mapping practice has its roots in drafting.

The basic tool of drafting is the technical pen (Figure F.2). This marvel of modern engineering is still a useful tool when you need to quickly alter existing artwork or create a special symbol. But drafting is relatively frustrating, tedious, and inflexible when compared with modern electronic production methods. Skilled draftspeople are an aging group, with few young recruits. And what is killing drafting more than anything else is the "dead end" nature of the product. Except for simple scale changes and recompositing of separations, the in-

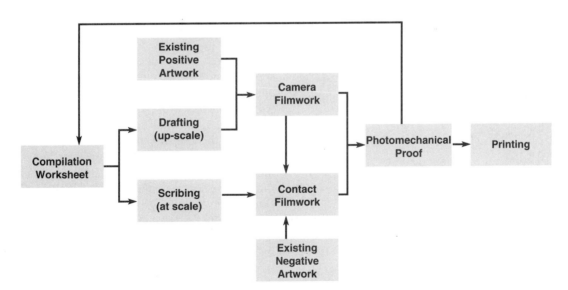

Figure F.1 Chart outlines photomechanical map production system currently being phased out at the University of Wisconsin Cartographic Laboratory.

vestment in drafting cannot be recycled into other applications. In drafting, the production intelligence is held in the mind of the draftsperson and is lost immediately upon completion of each task.

In digital map construction, on the other hand, much of the production intelligence is codified externally in the data format and structure and in the software. This makes it relatively easy to recycle the production effort and parlay it into various other applications. The digital approach is more versatile and amenable to a sharing of resources.

For these reasons, hand drafting tends to be competitive today only in low-volume situations in which labor is more affordable than technology, and an isolated end product is the aim. These situations are growing rarer every day. Each drop in the price of computer technology moves drafting a step closer to the fate of earlier procedures that are no longer practiced, such as copper engraving (see Chapter 30).

SCRIBING

Hand scribing has much in common with drafting and, ultimately, faces the same fate. Its decline is merely taking longer. For at least the next few years, scribing still has a place, especially in large-format mapping and in updating existing separations for map reprinting. When both map dimen-

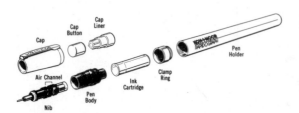

Figure F.2 Technical reservoir pens are precision drafting tools, both for manual drafting and automated machine plotting. (Courtesy Koh-l-Noor Rapidograph, Inc.)

sions exceed 24 inches, scribing is often more economical than electronic film recording. Indeed, it is an irony of modern mapping that large-format computer plots are still commonly used as worksheets for scribing. In other words, in special situations a mix of manual and electronic technologies is less costly than a pure digital approach.

In contrast to drafting, in which you add ink to the drawing surface, you perform scribing by removing material. You begin with a sheet of hard plastic film to which a soft, actinically opaque coating has been applied. Then, working over a light table, you remove the coating by cutting and scraping with special engraving instruments to produce the lines and symbols. When it is finished, the sheet has the same general appearance as a negative made photographically.

Because of its many advantages, scribing is still widely practiced by commercial and government mapping establishments. The tools are sharp enough to produce extremely fine, clean lines. In fact, scribed artwork is of such high quality that it is normally done at reproduction scale. Since the product is a negative, it can be used directly to make proofs and printing plates.

GUIDE IMAGE

Several steps are required to create scribed map separations. First, you need to create a guide image and transfer it onto the scribing material. As with positive artwork, you compile a worksheet from which the final artwork is to be scribed. Although the translucent quality of scribe coat permits tracing the worksheet on a light table, this method is seldom used, because tracing an image through the scribe coat is hindered by light diffusion. Instead, scribing is more easily and accurately accomplished if an image of the worksheet is first placed directly on the scribe coat. Even then, however, most scribing is still done on a light table to enhance the contrast between scribed lines and the guide image.

The worksheet is usually produced at the proposed scribing scale by hand or machine plotting in positive form on translucent material. This permits transfer of the image to the scribe coat base without the use of a camera or darkroom facilities. If, for some reason, it is necessary to prepare the worksheet at a larger scale, it can be reduced to the desired size by photography and then contacted to the scribe base.

You can place a guide image on the scribe coat in two ways. The simplest procedure is to obtain one of the commercially available scribe sheets with a surface that is precoated with a sensitized emulsion. Scribe sheets sensitized with a photographic emulsion for use with line image worksheets are available in both negative- and positive-acting form. You expose the worksheet in a platemaker in contact with the scribe coating to transfer the guide image. Since some scribers find a negative guide image easiest to work with over a light table, you may want to experiment to find the procedure that best suits your needs.

If you wish to scribe from a continuous tone guide image, such as an aerial photograph, you will need a special emulsion. The Keuffel and Esser Company produces one such product with the brand name *Stabilene Contone Film*.* This material is made up of a photo-mechanical emulsion on the surface of a double-layer (white and rust) scribe coat. When it is necessary to produce a line negative in exact registry with a photographic image map, this product is superior.

Although presensitized scribe sheets are convenient, they have several drawbacks. For one thing, they are expensive. Also, their shelf life is limited, making it difficult for the low-volume user to obtain consistent results. In addition, their emulsion can only be developed once, limiting the scriber to a monochrome guide image.

You can overcome these drawbacks to some degree by purchasing unsensitized scribe coat and applying the sensitizer yourself. The best way to perform this do-it-yourself sensitizing is to use bichromate emulsions that you can wipe onto the scribe

coat (see Chapter 30). Since these emulsions are negative acting, they require a negative worksheet if a positive guide image is desired. Bichromate emulsions are available in several colors. Thus, you can build up a color-separated guide image through successive applications and developments of the colored sensitizers. To accomplish this end, of course, you would need to separate the worksheet artwork on the appropriate flaps.

IMAGE ORIENTATION

Scribing procedures initially may seem strange, because the scriber must learn to "think in reverse." The guide image should be scribed wrong-reading (Figure F.3). This is because a negative image is created directly, rather than indirectly through a camera step as in positive ink drafting. Since the camera reverses positive map copy, the scriber must perform this same function.

The key point is that lithographic plates are right-reading. To achieve the best printing quality, the printing plate must be made with the scribe coat in contact with the printing plate emulsion. Since the plate emulsion is facing up, the scribe coat must face down. Reverse scribing will be right-reading when the scribe coat is placed down against the printing plate.

Figure F.3 For best results, guide image orientation on the scribe coat should be wrong-reading for reproduction by offset lithography.

---

*Kratos/Keuffel and Esser Company, 20 Whippany Road, Morristown, NJ 07960.

SCRIBING INSTRUMENTS

Although scribing tools are available in many design variations, there are only a few basic kinds of instruments. Regardless of design, each of these instruments has two components: a cutting point and a point holder called a graver or scriber. The type of point and the way it is held (or guided) are crucial in scribing. To become a skilled scriber, you must learn to make the proper match of cutting point and point holder to the different scribing tasks.

Freehand Scribing

Simple hand-guided gravers are sometimes used for scribing fine lines and for doing final "touch-up" work. These fine-line gravers may be nothing more than a chuck to hold the scribing point and a pen-type handle (see top illustration in Figure F.4). For this reason, this type of instrument is sometimes referred to as a **penholder graver**. More elaborate freehand models incorporate such items as a magnifying glass or an offset point chuck, which holds the point in a near-vertical position and keeps the handle out of the scriber's line of sight when the tool is held in normal writing position.

Fine-line gravers are relatively difficult to use. Considerable practice may be required to develop the proper "touch." Too much pressure makes the fine point gouge the scribe base. Too little pressure means that the desired line weight is not obtained. Variable pressure leads to inconsistent line weight. If the scribe point is not held in a near-vertical position, all three problems may occur. Because of the difficulties of freehand scribing, it's best to use fine-line gravers sparingly, at least until you have attained sufficient skill in their use.

Tripod Point Holders

The problems associated with freehand scribing are largely overcome by using a tripod-type point holder. The **rigid graver** is the simplest instrument of this type. The standard version consists of a horizontal plate with a vertical center rib for a handgrip, supported by two legs and the cutting

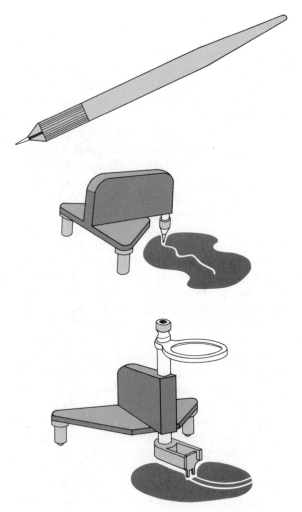

Figure F.4 A variety of tools are used to hold scribing points in proper orientation. Each design serves a special purpose.

tip, which functions as a third leg (see middle illustration in Figure F.4). The scribing point is held in an adjustable chuck, which permits the use of different cutter sizes. The chuck can be raised or lowered to level the instrument (to keep the cutting point or blade vertical).

On some rigid gravers, a third supporting leg is used and the chuck is spring-loaded. This arrangement makes it possible to preset the desired scribing pressure by adjusting the spring tension. On still other models, a turret head holding a selection of point sizes is incorporated into the instrument.

You can mount both conical and chisel-edged scribing points in a rigid graver (Figure F.5). If you use a conical point, you can move the graver in any direction. Unfortunately, scribing difficulty increases in direct relation to the size of a conical scribing point. In practice, conical points larger than .006 inch are rarely used. You can get around this problem by mounting a chisel-edged point in the rigid graver. However, when you do so, the instrument is good only for straight-line scribing along a straight edge, because you must keep the cutting edge at a right angle to the direction the point is traveling.

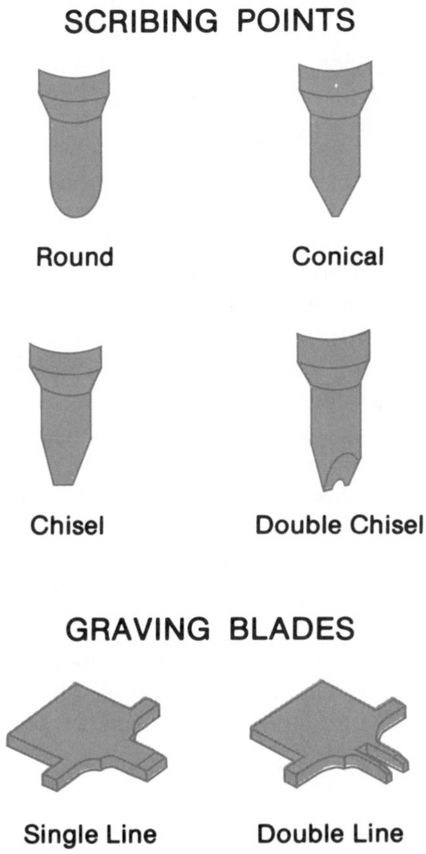

Some of the problems associated with rigid gravers are overcome by using a tripod graver that incorporates a swivel chuck (see bottom illustration in Figure F.4). This **swivel graver** arrangement permits both conical points (needles) and chisel-edged points (both single and multiple tipped points or blades) to be used in the instrument. Since the chuck "swivels" into proper position as the instrument is moved, the direction of movement is not a concern once the point is correctly aligned. You do have to take care, of course, that the point is properly aligned before you apply cutting pressure to the graver. For the expert scriber, the greater cost of a swivel head and the extra effort required to master a swivel cutter are more than compensated by the higher quality of the resulting scribe work. Swivel gravers are particularly suited to scribing double irregular lines, such as roads.

SCRIBING POINTS

Round

Conical

Chisel

Double Chisel

GRAVING BLADES

Single Line

Double Line

Figure F.5 Scribing points are designed for two types of cutting action. To be used effectively, they require point holders of different design.

SCRIBING TECHNIQUES

To take full advantage of scribing, you need to know how to handle the special instruments and how to alter scribed artwork. We will discuss these techniques in the following sections.

Handling Instruments

If the hard, outer layer of the base material is penetrated, the scribing point will gouge the material. It will then become difficult to control the path of the scribing tool, resulting in a ragged, irregular line. Thus the **scribing touch** is critical. Once you acquire a "feel" for the material and instrument, however, gouging presents no real problem. With the proper touch and correctly sharpened points, you will be able to remove the scribe coating smoothly and cleanly.

Each scribing instrument and each point style and size requires its own special handling technique. Most people find it easiest to start with a rigid graver in the .004 to .006 inch point size range.

As you gain skill, you can then work to smaller and larger sizes. Each point size will take a different amount of pressure, which may be adjusted by bearing down on the instrument with different force, or by moving the fingers closer or farther from the point. It is difficult to start cutting lines over .02 inch with a round point unless the graver is tilted slightly in the direction of the intended cut when the line is started.

Once you have mastered the rigid graver, go on to the harder-to-use swivel gravers, and repeat the process. Finally, go to the pen-type gravers, which are still harder to learn to use.

The direction a scribing instrument is moved is largely a matter of personal preference. Some people prefer to pull the instrument toward them, while others find that pushing the instrument away works best. You should experiment with both approaches before adopting one or the other. Most scribers like to have the scribe sheet loose (not fixed) on the light table, so that it can be freely rotated and moved about during scribing. An exception might be when scribing long straight lines. In that case, a fixed scribe sheet tends to be more convenient.

Making Alterations

Sometimes you may want to alter a scribed image. You can make changes by filling in unwanted scribed lines with an opaque material that can itself be scribed. Specially prepared liquid and crayon-type opaques are available for this purpose. Unfortunately, scribing lines through hand-applied opaque is generally less satisfactory than working with the original scribe coat, so the practice is best avoided whenever possible.

COMPLEX ARTWORK

Once you have a good knowledge of duplicating and printing methods, you can take advantage of their capabilities to produce a variety of special effects that improve map design and legibility.

Various photo-mechanical procedures can assist you in this process. The following sections describe some of these procedures and provide examples of their use.

OPEN-WINDOW NEGATIVES

You can use several photo-mechanical processes to produce tints, patterns, or colors within regions on a map. Although you could use positive artwork, photo-mechanical methods usually produce higher quality results.

First you need a negative that is actinically opaque except in the area in which you want to add the areal symbol. These are called **open-window negatives**, or **negative masks**. The idea is to make the border of the open area fall exactly on the bounding line of the region to be filled. You can construct the required negatives manually, mechanically, photo-mechanically, or electronically.

Stick-up

Preparing an open-window negative completely by hand can be a tedious job. Thus, you should probably avoid it in most situations. But when the artwork consists of only a few areas with simple shapes, then it may be most practical to prepare the open-window negative by hand. One way to do so is with the stick-up approach. With this method, you use an adhesive-backed masking film.* Place the material over the map surface, cut along the border of area features, and remove the unwanted material. The product can be a negative, but more commonly is done in positive. If you perform the procedure at scale on a translucent base (usually drafting film), you can use the artwork directly in a contact frame. Artwork done at other scales requires camera work to produce the proper sized open window.

---

*Several products are available for this purpose. *Parapaque* is marketed by Para-Tone, Inc., 512 Burlington Ave., LaGrange, IL 60525. *Kimoto Mask* is marketed by Kimoto USA, Inc. (East), 1116 Tower Lane, Bensenville, IL 60106.

Cut and Peel

A somewhat more convenient way to prepare open-window negatives by hand is to use special materials consisting of a thin masking film laminated to a clear polyester base.* You perform this procedure with the masking material registered over the map worksheet on a light table. Create the window by cutting around the region border with a sharp blade and then peeling the masking layer away from the base. The cut-and-peel method is fast and accurate if regions have simple geometric shapes or rather smooth curved boundaries (or thick-lined borders). But it becomes progressively more tedious and less satisfactory as a region becomes more detailed or complex in shape.

Etch and Peel

The easiest way to prepare open-window negatives is to do it photo-chemically. Several brands of sensitized materials, called **peel coat**, are available for this purpose.** They are alike in appearance, each having a thin ruby film adhered to a transparent polyester base sheet. No mechanical cutting is involved. Instead, you expose the guide image (containing boundary lines) to the sensitized material in a vacuum frame with a source rich in ultraviolet light. After exposure, you develop and chemically etch the sheet.

After you have etched the sheet, but before you peel it, you must block out unwanted lines. You can do so with a special orange, water-soluble opaquing material.*** You can apply the opaque by hand with a squeegee or by mechanical whirler.

---

*Rubilith and Amberlith are manufactured by Ulano Graphic Arts Supplies, Inc., 610 Dean St., Brooklyn, NY 11238. Kimoto Strip Coat is marketed by Kimoto USA, Inc. (East), 1116 Tower Lane, Bensenville, IL 60106.

**Striprite is produced by Direct Reproduction Corporation, 811-13 Union Street, Brooklyn, NY 11215. Peelcoat Film is available from the Keuffel and Esser Company. Kimoto Peel Coat is produced by Kimoto USA, (East), 1116 Tower Lane, Bensenville, IL 670106.

***A Keuffel and Esser Company product called Mask Kote is widely used for this purpose.

Once the opaque has dried, you can carry out the peeling step as with other peelable masking films. For easy peeling, use a knife to lift a corner of the material to be removed, and strip it off with a piece of adhesive tape. As a final step, use a moist cotton swab (a Q-tip is ideal) to remove the opaque **trap line** that remains along the edges of an open window after peeling. Thus, the area symbol will overlap the boundary line, and you will avoid white gaps if there is some misregistry of separations in subsequent compositing and printing.

Peel coat materials may be processed under normal room lighting conditions and can, therefore, be used in cartographic drafting rooms that have limited equipment. Most materials are processed using negatives. All materials produce good-quality negatives that can be registered almost perfectly. A pin registry system is used throughout the process, of course, to insure registry of each open-window negative with the lines on the scribe coat.

Film Plotting

A fourth way to produce open-window negatives is to create them directly from a digital record using a film plotter. In this case, a spot of light or laser beam is scanned back and forth across the film in raster fashion, while it is being modulated by the electronic record of the map data. The film is exposed line by line until the image is completed (Figure F.6).

You may perform the film plotting at reproduction scale, or you may plot it on microfilm and later enlarge it for use as an open window in print or platemaking (see Chapter 30). Once the film has been plotted, you handle the negative in the same way as those produced by other means.

NEGATIVE SCREENS

Making an open-window negative is only the first step in placing area symbols on maps photo-mechanically. Next, you add the actual pattern or tint to the map, using a negative contact screen in conjunction with the open-window negative. You

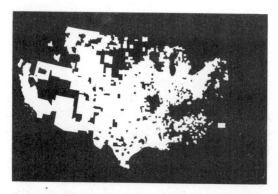

Figure F.6 One of a number of open-window negatives that were produced as computer output on microfilm in conjunction with the reproduction of a complex colored map. (Courtesy, U.S. Census Bureau).

can do so on paper or film in a vacuum frame or at the later platemaking stage. You need only insert the appropriate screen between the carefully registered negative and the sensitized material.

By this contact-screen stage, any reduction of the artwork has already been accomplished. Thus, there is no danger of tint and pattern details disappearing, as there is when preprinted areal symbols are placed on original artwork (Figure F.7 and Figure 20.1). As a result, contact tint and pattern screens generally produce higher quality results than are achieved with equivalent positive areal symbols.

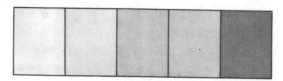

Figure F.7 Screen tints are printed, plotted, or stuck-up on positive artwork to give an illusion of gray. The effect can be destroyed, however, if the artwork is subsequently reduced for reproduction. When screen tints are introduced at the film compositing or platemaking stage, the amount of reduction of the original artwork need not be a consideration. The tints shown above were labeled as 10, 20, 30, 40, and 50 percent.

Area Patterns and Tints

Pattern contact screens are the same as their printed or plotted counterparts, except that they are in negative rather than positive form (see Chapter 20). Otherwise, our previous discussion of positive patterns applies to contact pattern screens. Also, handling procedures for pattern screens are much the same as those outlined for screen tints in the following paragraphs.

In contrast to pattern screens, tint contact screens are composed of fine dots or closely spaced lines that produce the impression of gray on the printed copies. As with printed or plotted tints (see Chapter 20), contact screen tints are specified by two measures: texture in lines per inch (lpi) and density as percent blackness. The texture of contact screen tints is defined in the same way as with positive tints. Note, however, that you can use finer contact screen tints than positive tints because image degradation associated with an extra photographic step has been eliminated.

The density rating of the tone produced by screen tints is designated in percentages of ink coverage. Therefore, a 10 percent screen will produce a very light tone, and an 80 percent screen will give a very dark tone. Screen tints are considered to be in negative form. Hence, an 80 percent screen actually has very small opaque dots, with most of the film transparent, whereas a 10 percent screen is mostly opaque with small transparent openings.

Commercial sets of screen tints are usually graded in 10 percent intervals from 1O to 90, although some series also include 5, 7.5, 15, and 25 percent. Less common are screens calibrated throughout at a 5 percent interval.* Since this limited selection of screen values may not match those you need, you may want to combine screens

---

*You can obtain brochures detailing product lines from the manufacturers, including Borrowdale (250 West 83rd St, Chicago, IL 60620); LogEtronics, Inc. (7001 Loisdale Rd, Springfield, VA 22150); or ByChrome Co. (P.O. Box 1077, Columbus, OH 43216).

to form composite tones. When you do so, however, you must contend with dot overlap and possible moiré effects (see next section).

Multiple Screens

In monochrome printing, up to three screen tints can be printed on one area. In flat color printing, as in full-color halftone printing, a fourth screen is often added (see Chapter 30). Whenever you use more than one screen, you must position the screens very carefully to prevent undesirable moire effects (see Figure 20.6 and Color Figure 20.3). Superimposing screens within a single color is a satisfactory method only if you rotate the screens so that the lines are separated by angles of 30 degrees. If you use more than one color to print a map, you should assign each color a screen angle 30 degrees from the others (see Chapters 20 and 30). You can then superimpose screens of different colors.

You can use a mechanical **screen angle gauge** or a **graphic template** to format standard screens. You can then readily position the screens at the proper angle (see Figure 20.7). The alternative to using guide devices is to use **preangled screens**. You can purchase them from companies that manufacture standard screen tints, or you can make them yourself from standard screen tints.*

The ability to superimpose screens means that you may not need to prepare a separate flap for each tone. If, for example, you wanted values of 20, 40, and 60 percent, you would need only two flaps. On flap number one, you would opaque the areas of 20 and 60 percent and specify a 20 percent screen. On the second flap, you would include the areas of 40 and 60 percent and specify a 40 percent screen. Since the two flaps overlap, the 60 percent area would receive 20 plus 40 percent ink (Figure F.8). You would thus reduce the cost by one less negative and one less plate exposure.

A possible drawback of this method of creating

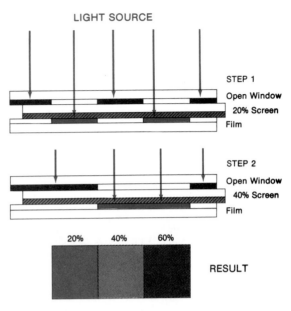

Figure F.8 The ability to superimpose screens means that a separate flap may not have to be made for each tone desired on the map. In this example, only two flaps are required to produce three values.

hybrid tints by superimposing screens is that the resulting blackness (percent area inked) will be somewhat less than the sum of the screen percentages involved. Thus, in the previous example, the combined 20 and 40 percent screens do not actually produce 60 percent inked area. The reason for this discrepancy is that some of the dots from one screen will overlap with dots from the other screen. Kimerling has analyzed the cross-screening relationship, and his values are given in Table F.1. Fortunately, the deviation will probably be small when compared with your design thresholds and therefore may not be a serious problem.

When you place two gray tints next to each other, you will probably not be able to join them exactly. Therefore, you will need to put a black line between them. If the tones are a result of the superimposition of screens, however, the black line may not be necessary (Figure F.9). When a screened area is adjacent to a solid area, the artwork should always carry the tint into the solid.

---

*See J. M. Olson, "A Simple Technique for Pre-Angling Screens," *The American Cartographer,* 9 (1982), pp. 81–83.

Table F.1
Percent Area Inked for Various Combinations of Dot Screen Tints

Screen 1 (%)	10	20	30	40	50	60	70	80	90
10	19								
20	28	36							
30	37	44	51						
40	46	52	58	64					
50	55	60	65	70	75				
60	64	68	72	76	80	84			
70	73	76	79	82	85	88	91		
80	82	84	86	88	90	92	94	96	
90	91	92	93	94	95	96	97	98	99

Screen 2 (%)

Source: A. Jon Kimerling, "Visual Value as a Function of Percent Area Inked for the Cross-Screening Technique," *The American Cartographer 6*, no. 2, (1979):141–48.

If you attempt to make them join, white spaces are almost sure to occur.

Screened Symbols

Solid patterns that have been applied to artwork tend to look harsh (Figure F.10). A useful technique is to screen the areas of these patterns so that they appear more subdued. Also, superimposed lettering or symbols that might otherwise be lost will show clearly through the screened patterns.

Again, you will need to prepare a separate flap for screening. Then place an open-window negative of this flap in contact with both the pattern screen (above) and the screen tint (below) in a vacuum frame (Figure F.11). The result will be patterned artwork that lacks strong contrast and looks rather gray in appearance.

When you screen linework, type, or other symbols, the screen tint dots can reduce the sharpness of feature edges. You can improve this situation somewhat by using finer textured screens or **bi-angled screens**,* or by choosing coarser patterns. It is also possible for a moiré to develop when screening patterns are made up of fine dots or certain other marks. As a precaution, test the orientation of patterns and screen tints by superimposing them on a light table.

Lines, lettering, and symbols can also be screened, but very narrow lines and thin serifs on letters may be lost. Lettering is probably most successfully screened with line screens, which do not produce the ragged edges caused by dot screens. Bold lines and sans-serif lettering, when

---

*Bi-angled screens are created by superimposing two standard dot screens. Thus, they exhibit an irregular, random, or dot rosette pattern. Whereas a single (standard) screen will pick up parts of lines oriented in some directions better than in others, bi-angled screens have a better chance of preserving sharp edges on features oriented in all directions.

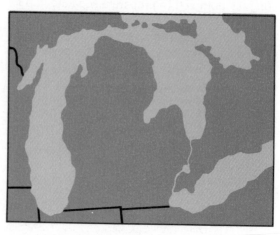

Figure F.9 It is extremely difficult to make two screen tints match precisely (*left*). A tint superimposed on another (*right*), however, causes no difficulty.

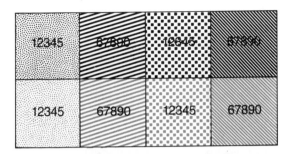

Figure F.10 The patterns in the lower row have been screened to 30 percent of black. The patterns appear less harsh, and lettering or symbols remain legible.

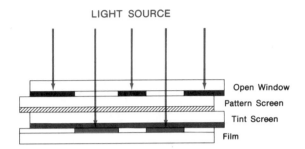

Figure F.11 Patterns can be screened in a vacuum frame by using this arrangement of open-window negative, pattern screen, screen tint, and film. Note the emulsion orientations.

screened, can add much to the utility and overall appearance of a map.

POSITIVE MASKS

The opposite of an open-window negative (or negative mask) is a **positive mask**. Open windows are used to create special effects by permitting light to expose particular parts of a sensitized material. In contrast, positive masks block light from reaching certain parts of a sensitized material. The result is called a **reverse**, or a **knock-out**, because the effect is one of looking through a gap in the surrounding areal symbol.

You can construct a mask in positive on a translucent surface for direct contact work. Alternatively, you can create a positive mask by exposing positive artwork to duplicating film or by exposing an open-window negative to standard reversal film.

You can use positive masks to create a variety of special effects, such as vignettes, outline symbols, and reverse artwork. Only the reverse artwork application is discussed here because producing vignettes and outline symbols involves a long, complicated process that is now done very simply by digital means.

Reverses in Tints and Patterns

It is useful for the map designer to have the option of producing white lines or type on a tint or pattern background. Unfortunately, white lines and type are difficult to produce directly on the artwork. The most satisfactory method is to reverse artwork at the print (film) or platemaking stage. Reversing lines and type produces sharp images from conventionally prepared artwork.

Use a separate flap to prepare the artwork that is to be reversed. One of the other flaps will, of course, be an open-window negative to produce either a tone or solid color for the background. Instruct the printer that a given flap is to be reversed from another. When the plate is exposed, a film positive of the flap to be reversed is placed, in registry, between the open-window negative and the plate (Figure F.12). The opaque positive artwork blocks the light and produces a nonprinting area on the plate. The white paper then shows through the tone or solid color, and white lines or type is the result.

Reverses in Images

You can also use masks to block out unwanted portions of continuous tone images. Suppose, for example, that you are making a shaded relief map for an area in which numerous lakes are to have a smooth gray tone. You can shade the terrain by hand or machine on the first flap without regard for the lake outlines. The shading actually should overlap into the lake area.

Then prepare a second flap, on which all the lakes are opaque. This second flap will provide an

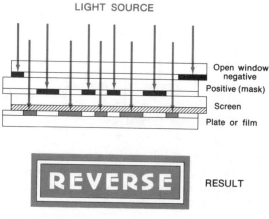

Figure F.12 White (or reverse) lettering and symbols on a dark background can be produced by using a positive mask during the film compositing or plate-making stage.

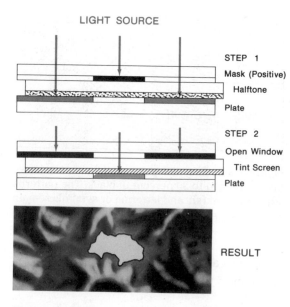

Figure F.13 The shaded relief was done without regard for the lake. A film positive (mask) was made of the flap to be used to deposit the tint in the lake. This mask was used to reverse the halftone dots out of the shaded relief (step 1). An open-window negative (mask) of the lake flap then was screened and deposited an even tint in the lake (step 2).

open-window negative to be screened to produce an even tint (Figure F.13).

When making the plate, place a film positive of the second (lake) flap over the halftone negative made from the terrain drawing. The plate will be unexposed in the lake areas. Then use the open-window negative of the lake flap with an attached screen to lay the even tint in the masked areas.

The area surrounding open artwork has a critical effect on the success of the technique. The same is true of solid artwork, of course. There must be adequate contrast between a symbol and its surroundings, or you will not be able to recognize the symbol. Any dark background tint down to about 30 percent is adequate for reversing, while any tint up to about

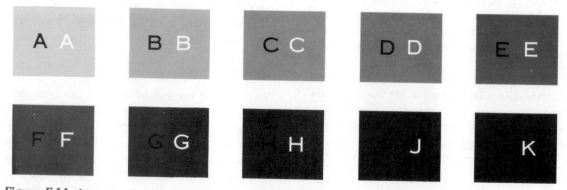

Figure F.14 Lettering has been superimposed and reversed from screen tints ranging in equal value increments from 10 percent to solid black. We can determine from this illustration the point at which reversing or overprinting will be unsatisfactory. When the background is variable in tone, special problems can arise.

60 percent is satisfactory for overprinting artwork in solid black (Figure F.14).

These percentages change somewhat when the background or artwork is printed in colors other than black. A background that is variable in tone presents special problems. For instance, reversed type or linework on continuous tone aerial photographs or shaded relief maps may have enough visual contrast in some areas but not in others. The result is that type or lines may fade in and out from one part of the image to another, especially if they are rather thin in the first place.

A similar problem may arise when printing symbols in solid colors over existing artwork. This difficulty is sometimes overcome by using a positive mask to open up rectangles. You can then place solid lettering in these rectangles. Unfortunately, this not only destroys background detail but is not very attractive. A better method is to outline the reverse or solid artwork.

SELECTED REFERENCES

Keates, J.S., *Cartographic Design and Production*, 2nd ed., New York: John Wiley & Sons, Inc., 1989.

Kers, A.J., "Flow Diagrams in Map Production," *ITC Journal, Special Cartography Issue*, June, 1982, pp. 37–48.

Kimerling, A.J., "Visual Value as a Function of Percent Area Inked for the Cross-Screening Technique," *The American Cartographer*, 6 (1979), pp. 141–48.

Shearer, J.W., "Cartographic Production Diagrams: A Proposal for Standard Notation System," *The Cartographic Journal*, 19 (1982), 5–15.

INDEX